Gravitation in Astrophysics

Cargèse 1986

NATO ASI Series

Advanced Science Institutes Series

A series presenting the results of activities sponsored by the NATO Science Committee, which aims at the dissemination of advanced scientific and technological knowledge, with a view to strengthening links between scientific communities.

The series is published by an international board of publishers in conjunction with the NATO Scientific Affairs Division

A	**Life Sciences**	Plenum Publishing Corporation
B	**Physics**	New York and London
C	**Mathematical and Physical Sciences**	D. Reidel Publishing Company Dordrecht, Boston, and Lancaster
D	**Behavioral and Social Sciences**	Martinus Nijhoff Publishers
E	**Engineering and Materials Sciences**	The Hague, Boston, Dordrecht, and Lancaster
F	**Computer and Systems Sciences**	Springer-Verlag
G	**Ecological Sciences**	Berlin, Heidelberg, New York, London,
H	**Cell Biology**	Paris, and Tokyo

Recent Volumes in this Series

Volume 150—Particle Physics: *Cargèse 1985*
edited by Maurice Lévy, Jean-Louis Basdevant, Maurice Jacob, David Speiser, Jacques Weyers, and Raymond Gastmans

Volume 151—Giant Resonances in Atoms, Molecules, and Solids
edited by J. P. Connerade, J. M. Esteva, and R. C. Karnatak

Volume 152—Optical Properties of Narrow-Gap Low-Dimensional Structures
edited by C. M. Sotomayor Torres, J. C. Portal, J. C. Mann, and R. A. Stradling

Volume 153—Physics of Strong Fields
edited by W. Greiner

Volume 154—Strongly Coupled Plasma Physics
edited by Forrest J. Rogers and Hugh E. Dewitt

Volume 155—Low-Dimensional Conductors and Superconductors
edited by D. Jérome and L. G. Caron

Volume 156—Gravitation in Astrophysics: *Cargèse 1986*
edited by B. Carter and J. B. Hartle

Series B: Physics

Gravitation in Astrophysics
Cargèse 1986

Edited by

B. Carter

CNRS—Observatoire de Paris
Paris, France

and

J. B. Hartle

University of California
Santa Barbara, California

Plenum Press
New York and London
Published in cooperation with NATO Scientific Affairs Division

Proceedings of a NATO Advanced Study Institute on
Gravitation in Astrophysics,
held July 15-31, 1986,
in Cargèse, France

Library of Congress Cataloging in Publication Data

NATO Advanced Study Institute on Gravitation in Astrophysics (1986: Cargèse, Corsica)
Gravitation in astrophysics, *Cargèse* 1986.

(NATO ASI series. Series B, Physics; v. 156)
"Proceedings of a NATO Advanced Study Institute on Gravitation in Astrophysics held July 15-31, 1986 in Cargèse, France"—T.p. verso.
"Published in cooperation with NATO Scientific Affairs Division."
Includes bibliographies and index.
1. Astrophysics—Congresses. 2. Gravitation—Congresses. 3. Cosmology—Congresses. 4. Gravitational radiation—Congresses. I. Carter, B. (Brandon) II. Hartle, J. B. (James B.) III. North Atlantic Treaty Organization. Scientific Affairs Division. IV. Title. V. Series.
QB460.N38 1986 521.'1 87-7810
ISBN-13: 978-1-4612-9056-8 e-ISBN-13: 978-1-4613-1897-2
DOI: 10.1007/ 978-1-4613-1897-2

Softcover reprint of the hardcover 1st edition 1987
A Division of Plenum Publishing Corporation
233 Spring Street, New York, N.Y. 10013

REVIOUS CARGESE SYMPOSIA PUBLISHED IN THE ATO ASI SERIES B: PHYSICS

olume 150 PARTICLE PHYSICS: *Cargèse 1985*
edited by Maurice Lévy, Jean-Louis Basdevant, Maurice Jacob, David Speiser, Jacques Weyers, and Raymond Gastmans

olume 130 HEAVY ION COLLISIONS: *Cargèse 1984*
edited by P. Bonche, Maurice Lévy, Phillippe Quentin, and Dominique Vautherin

olume 126 PERSPECTIVES IN PARTICLES AND FIELDS: *Cargèse 1983*
edited by Maurice Lévy, Jean-Louis Basdevant, David Speiser, Jacques Weyers, Maurice Jacob, and Raymond Gastmans

olume 85 FUNDAMENTAL INTERACTIONS: *Cargèse 1981*
edited by Maurice Lévy, Jean-Louis Basdevant, David Speiser, Jacques Weyers, Maurice Jacob, and Raymond Gastmans

olume 72 PHASE TRANSITIONS: *Cargèse 1980*
edited by Maurice Lévy, Jean-Claude Le Guillou, and Jean Zinn-Justin

olume 61 QUARKS AND LEPTONS: *Cargèse 1979*
edited by Maurice Lévy, Jean-Louis Basdevant, David Speiser, Jacques Weyers, Raymond Gastmans, and Maurice Jacob

olume 44 RECENT DEVELOPMENTS IN GRAVITATIONS: *Cargèse 1978*
edited by Maurice Lévy, and S. Deser

olume 39 HADRON STRUCTURE AND LEPTON–HADRON INTERACTIONS: *Cargèse 1977*
edited by Maurice Lévy, Jean-Louis Basdevant, David Speiser, Jacques Weyers, Raymond Gastmans, and Jean Zinn-Justin

olume 26 NEW DEVELOPMENTS IN QUANTUM FIELD THEORY AND STATISTICAL MECAHNICS: *Cargèse 1976*
edited by Maurice Lévy and Pronob Mitter

olume 13 WEAK AND ELECTROMANGETIC INTERACTIONS AT HIGH ENERGIES: *Cargèse 1975* (Parts A and B)
edited by Maurice Lévy, Jean-Louis Basdevant, David Speiser, and Raymond Gastmans

PREFACE

With the discovery of pulsars, quasars, and galactic X-ray sources in the late 60's and early 70's, and the coincident expansion in the search for gravitational waves, relativistic gravity assumed an important place in the astrophysics of localized objects. Only by pushing Einstein's solar-system-tested general theory of relativity to the study of the extremes of gravitational collapse and its outcomes did it seem that one could explain these frontier astronomical phenomena. This conclusion continues to be true today.

Relativistic gravity had always played the central role in cosmology. The discovery of the cosmic background radiation in 1965, the increasing understanding of matter physics at high energies in the decades following, and the growing wealth of observations on the large scale structure meant that it was possible to make increasingly detailed models of the universe, both today and far in the past. This development, not accidentally, was contemporary to that for localized objects described above.

From these twin developments in compact objects and cosmology there emerged the subject of relativistic astrophysics—the application of relativistic gravity to realistic astronomical phenomena. That area is the subject of this volume. In it we have attempted to document and review the recent progress and consolidation in the subject so that it might be more accessible to a new generation of research students and to scientists with a general astrophysical background. Although *manifest astrophysical relevance* has been one of our principal criteria in the choice of subject matter, we have limited our attention essentially to *basic theory*, leaving to other more astronomically oriented schools the detailed coverage of actual observational results and the experimental programs for the detection of gravitational radiation. We have stressed those applications in which gravity is involved clearly and fundamentally, only briefly treating, for example, the subject of particle physics in cosmology which has received extensive attention elsewhere.

In order to save space for coverage of more advanced topics, and for inclusion of the most recent developments, the written versions of the lecture notes differ in some cases from the corresponding lectures as originally presented at Cargèse, by the omission of a certain amount of introductory material that is easily available elsewhere in standard textbooks. The presentation of this volume is based on the same principle as the organization of the school in being divided into two parts. The first is concerned with *gravitation in localized systems* (including topics such as black hole theory, gravitational radiation theory, and the Newtonian theory of many-body systems). The second is concerned with *gravitation in cosmology* (including aspects such as the theory of inflation, the origin of inhomogeneities, and the quantum process of creation of the universe itself).

The editors wish to conclude by thanking all who have contributed to the volume in various ways, starting of course with the NATO Scientific Affairs Division, which provided the basic funding for the Advanced Study Institute, and not forgetting the universities and other public institutions in many NATO countries that have provided the necessary long term support for the other authors and themselves, as well as sustaining much of the relevant underlying research over the years. More particularly they wish to thank the University of Nice, which made available the material facilities of the Institute des Etudes Scientifiques at Cargèse, and also specially to thank the French C.N.R.S., which (in addition to supporting several of the authors) has continued a longstanding tradition that has in practice been indispensible for the smooth running of the Institute in Cargèse, by providing the excellent administrative services of Miss Marie-France Hanseler. At a personal level we should also like to express our strong appreciation to the permanent Director of the Institute at Cargèse, Professor Maurice Lévy, and to the principal Scientific Directors of this particular Advanced Study Institute, Jean Audouze and Hubert Reeves, who not only took the initiative in the choice of the subject matter (and of the editors) at the outset of the planning, but also provided continous help and supervision thoughout all the subsequent stages of realization. Finally we wish to thank the other authors for their enthusiastic cooperation, and (last but not least) to thank the students and other participants for providing a stimulating and interactive environment.

B. Carter
J.B. Hartle

CONTENTS

SPECIAL TOPICS I

SPECIAL TOPICS II

PART I :

GRAVITATION IN LOCALIZED SYSTEMS

AN INTRODUCTION TO THE THEORY OF GRAVITATIONAL RADIATION

Thibault Damour

Groupe d'Astrophysique Relativiste
CNRS – Observatoire de Paris
92195 Meudon Principal Cedex, France

§1. INTRODUCTION

1.1. Scope of these lectures

The aim of these lectures is to provide an introduction to the aspects of gravitational radiation theory which are relevant to astrophysics. This way of thinking about gravitational radiation theory is relatively recent. For a long time, gravitational radiation has been thought of as a phenomenon of great theoretical interest, but of no relevance to the real world. Many articles, or textbooks, estimated the numerical order of magnitude of the gravitational-wave "energy flux" emitted by some real sources (usually rotating rods and/or binary stars) and, explicitly or implicitly, concluded from the extreme smallness of their result that both the emitted gravitational waves, and their back reaction on the source, would always be completely negligible. This state of mind seems to have been fairly common up to the early sixties. For instance, Pirani concludes his classic 1964 Brandeis lectures on gravitational radiation theory[1] by the remarks that "the primary motivation for the study of this theory is to prepare for quantization of the gravitational field", and that "unless [the classical theory] is eventually brought into the contemporary domain by quantization, the theory of gravitational radiation cannot have much to do with physics."

As far as I know, Dyson[2] was the first to point out that the usual formula giving the gravitational-wave "energy flux", F_{GW}, from a binary star,

$$F_{GW} \sim \frac{G}{c^5} m^2 a^4 \omega^6, \tag{1.1}$$

leads, in the *extreme relativistic case,* to the prediction of a stupendous amount of radiation of the order of

$$F_{GW} \sim \frac{c^5}{G} = 3.629 \times 10^{59} \text{ erg/sec.} \tag{1.2}$$

This remark gave an extra stimulus to the pioneering experimental work of Weber[3] (which had started as early as 1959) and prompted the physics, and astrophysics, communities to consider gravitational radiation as a phenomenon of great potential importance in the real world. This view is now commonly held and gravitational radiation is considered as one of the important elements of contemporary astrophysics. In particular, two aspects of the role of gravitational radiation are of immediate interest: on the one hand, the dynamical effects on the matter, correlated with the emission of gravitational waves, and, on the other hand, the information about the motion and structure of the matter, contained in the emitted gravitational waves. The first aspect ("gravitational radiation reaction") has become especially important after the discovery by Hulse and Taylor[4] in 1974 of the binary pulsar PSR 1913+16. We shall discuss, in §6 below, the dominant gravitational radiation reaction effects in binary systems, with particular attention to their observation in the binary pulsar. Another radiation reaction effect which has potentially significant astrophysical implications is gravitational radiation recoil. However this is a subdominant effect which is not yet fully understood (see references in §3 and §5 below). The second aspect of gravitational radiation ("emission of gravitational waves") has the potential of becoming an essential tool of tomorrow's astrophysics. Indeed, the detection of gravitational wave signals, if successful, will open a new window on our Universe, and this opening of a new information channel will, very probably, increase considerably or even revolutionize our comprehension of the Universe, by giving access to completely new information, unobtainable via the electromagnetic or neutrinic windows.

The exciting prospect of starting this new "Gravitational Wave Astronomy" motivates a very active world-wide effort, which comprises theoretical and experimental research. The theoretical research concerns mainly the problem of the generation of gravitational waves, and it uses either analytical or numerical methods. The experimental research concerns the problem of the detection of gravitational waves, and it uses either Weber-type cylinders or optical interferometers as gravitational wave receivers. These lectures do not aim to review all aspects of this ongoing research. To get an overall picture the reader may consult the reviews of Thorne[5] and Brillet *et al.*[6] A more detailed picture of the field is contained in the books edited by De Sabbata and Weber,[7] Smarr,[8] Hawking and Israel,[9] Edwards,[10] Held,[11] Ruffini,[12] Meystre and Scully,[13] Hu Ning,[14] Deruelle and Piran,[15] Le Laboratoire "Gravitation et Cosmologie Relativistes",[16] Ruffini,[17] and Sato and Nakamura.[18]

In these lectures we shall restrict our attention to the analytical investigations of the theory of gravitational radiation. There exist already several reviews concerning this topic (e.g. by Pirani,[1] Sachs,[19] Thorne,[20] Newman and Tod,[21] Walker,[22] Ashtekar,[23] or Schutz[24]), and, in particular, a recent detailed review, by Thorne,[25] where gravitational radiation theory is put in a form suitable for astrophysical studies. This is why we shall limit the scope of these lectures to supplement the existing reviews (and textbook treatments, like e.g. Misner *et al.*[26]) in two ways. First, we would like to present, as clearly as possible, both the basic concepts of gravitational radiation theory, and the precise conditions, as well as the limitations, of validity of some of the well-known results in this theory. Indeed, as these results have been, or will be, applied in astrophysics, it is important to have clearly in mind both what they mean, and when they can be legitimately applied. Second, we would like to present a progress report on some of the ongoing analytical research in gravitational radiation theory.

1.2. Conventions and Notation

We shall follow the sign conventions of Misner *et al.*,[26] in particular our signature is $-+++$. We also follow essentially their notation (in particular greek indices = 0,1,2,3; latin indices = 1,2,3). However we shall denote the *flat* metric by $f_{\mu\nu}$, instead of the usual $\eta_{\mu\nu}$:

$$f_{\mu\nu} := \mathrm{diag}(-1,+1,+1,+1). \tag{1.3}$$

Also we denote,

$$g := -\det(g_{\mu\nu}), \tag{1.4}$$

so that $g > 0$, and, for the Einstein tensor we use

$$E_{\mu\nu} := R_{\mu\nu} - \tfrac{1}{2}Rg_{\mu\nu}. \tag{1.5}$$

§2. WHAT IS A GRAVITATIONAL WAVE?

2.1. First viewpoint: propagation of discontinuities

From the physical point of view, a very general way, valid in many different theories, of characterizing the phenomenon of "wave propagation" is to require the possibility of the transmission of some kind of information between a source and several detectors. Moreover, any kind of information broadcasting implies the possibility of some discontinuity in the quantity measured by each detector, corresponding, say, to some on/off transition in the source (which is the minimal behaviour expected in an emitter of information). From the mathematical point of view, this means that the theory considered should admit solutions containing discontinuities. When this is the case, the surfaces across which the solutions can be discontinuous, which are called *"characteristic surfaces"* by mathematicians, can be thought of as what physicists would call *"wavefronts."* Therefore a mathematical criterium for the presence of waves in the theory considered is the existence of (real) characteristic surfaces.

Applying this viewpoint to the theory of General Relativity, we are led to study the purely mathematical question of the existence of solutions of the Einstein equations containing some discontinuities. In order to put this study on a clear mathematical basis, we need first to state precisely what are the usual regularity conditions assumed for the metric tensor $g_{\mu\nu}(x^\lambda)$.

Let us first recall that a real function on R^n is said to be of class $C^p(\mathrm{R}^n)$ if it is p-times continuously differentiable in R^n, i.e. if all its partial derivatives of order $\leq p$ exist and are continuous in R^n. Now let us consider the *closed* half-spaces $\mathrm{R}^n_+ := \{(x^0, x^1, \ldots, x^{n-1}) \mid x^0 \geq 0\}$, and $\mathrm{R}^n_- := \{(x^0, x^1, \ldots, x^{n-1}) \mid x^0 \leq 0\}$ such that $\mathrm{R}^n_+ \cup \mathrm{R}^n_- = \mathrm{R}^n$, and $\mathrm{R}^n_+ \cap \mathrm{R}^n_- = S$, S being the hyperplane $x^0 = 0$. The notion of partial derivative can be made meaningful in R^n_+ by taking as neighbourhoods of the independent variable the restriction of neighbourhoods of R^n to R^n_+ (like in the definition of a *right derivative*). Now a real function on R^n, $f(x)$, is said to be piecewise of class C^p if its restrictions to R^n_+, and R^n_-, are, respectively, p-times continuously differentiable within R^n_+, or R^n_- (in the sense just defined). This definition is equivalent to saying that each restriction of $f(x)$, say to R^n_+, can be considered as the restriction of some function, say $f_+(x)$, which is of class C^p all over R^n; it is also equivalent to saying that $f(x)$ is of class C^p in each *open* half-space $\mathrm{R}^n_+ - S$, $\mathrm{R}^n_- - S$, *and* that all the partial derivatives of $f(x)$ admit locally *uniform* limits when $x^0 \to 0^+$, and $x^0 \to 0^-$. Evidently the values (or limits) of f, and its partial derivatives, on each side of S are in general different, so that f is, in general, *not* $C^p(\mathrm{R}^n)$, and in fact, not even $C^0(\mathrm{R}^n)$. The previous definition is easily extended to the case of an arbitrary (but sufficiently regular) "surface of discontinuity" S, and also to the case where there are several surfaces of discontinuity: this defines the general class of piecewise C^p functions.

Now the usual regularity conditions, due to Lichnerowicz,[27] for the metric coefficients, $g_{\mu\nu}(x^\lambda)$, are motivated by the fact that one should admit discontinuous jumps of the matter density (keeping, however, the latter everywhere finite). They state that the $g_{\mu\nu}(x^\lambda)$ are everywhere C^1 but piecewise C^3 (and that the manifold V_4 is correspondingly C^2 and piecewise C^4). We are now in position to discuss the existence of solutions of the vacuum Einstein equations,

$$R_{\mu\nu}(g, \partial g, \partial^2 g) = 0, \tag{2.1}$$

which contain some discontinuities (wavefronts of pure gravitational waves).

Let S be a looked for hypersurface of discontinuity, and let the equation defining S be

$$S : \; u(x^\lambda) = 0. \tag{2.2}$$

We can always introduce (locally) a new coordinate system, x'^{λ}, *adapted* to S, in the sense that

$$x'^{0} = u(x^{\lambda}). \tag{2.3}$$

It is easily deduced from the Lichnerowicz regularity conditions that not only $g'_{\mu\nu}$ and $g'_{\mu\nu,\lambda}$ but also their "tangential" derivatives $g'_{\mu\nu,ij}, g'_{\mu\nu,0i}, \ldots$ will be continuous and that the first discontinuity can only appear in the second "normal" derivatives: $g'_{\mu\nu,00}$. Now an easy calculation shows that the components of the Ricci tensor have the structure

$$R'_{ij} = -\tfrac{1}{2}g'^{00}g'_{ij,00} + \ldots, \tag{2.4a}$$

$$R'_{i0} = +\tfrac{1}{2}g'^{0j}g'_{ij,00} + \ldots, \tag{2.4b}$$

$$R'_{00} = -\tfrac{1}{2}g'^{ij}g'_{ij,00} + \ldots, \tag{2.4c}$$

where the three dots denote some functions of $g'_{\mu\nu}$, $g'_{\mu\nu,\lambda}$, $g'_{\mu\nu,ij}$, $g'_{\mu\nu,0i}$, which are therefore continuous across S. The four quantities $g'_{00,00}, g'_{0i,00}$ do not appear in the Ricci tensor, and can in fact be changed at will by a suitable coordinate transformation. In particular they can be made continuous across S. Then a real discontinuity will be present if $g'_{ij,00}$ has different values on the two sides of S. As seen immediately from equation (2.4*a*), a necessary condition for such a discontinuity to exist is (in adapted coordinates):

$$g'^{00} = 0. \tag{2.5}$$

As $g'^{00} \equiv g'^{\mu\nu}(\partial x'^{0}/\partial x'^{\mu})(\partial x'^{0}/\partial x'^{\nu})$, the necessary condition (2.5) can be rewritten, using equation (2.3), in an arbitrary coordinate system as:

$$g^{\mu\nu}\frac{\partial u}{\partial x^{\mu}}\frac{\partial u}{\partial x^{\nu}} = 0. \tag{2.6}$$

Equation (2.6) means that the hypersurface S must be *null*, i.e. tangent in each point to the local light cone. Therefore the preceding argument shows that the vacuum Einstein equations admit real "characteristic surfaces," which are the null hypersurfaces of the solution $g_{\mu\nu}$. A more detailed study of the full system (2.4) shows that the jumps of the $g'_{ij,00}$ must have a special algebraic form (of the type of equation (2.31) below), and that these jumps satisfy some differential equations (or "propagation" equations) along the trajectories of the 4-vector

$$\ell^{\mu} := g^{\mu\nu}u,_{\nu} \tag{2.7}$$

which is both normal and tangent to S. As is well-known, and easily checked from (2.6), the trajectories of ℓ^{μ} are some null geodesics contained in S which can be called the "rays" of S (they are called "bicharacteristics" by mathematicians).

For full details about pure gravitational wavefronts (or "shock waves"), and also about combined gravitational-electromagnetic wavefronts, the reader may consult Pirani,[1] Papapetrou[28] Lichnerowicz[29] and references therein.

In conclusion, the first viewpoint allows one to say that Einstein's theory can admit some pure gravitational "wavefronts" and that the "information" (in the sense of discontinuity jumps) propagates along null geodesics. This is one of the very few exact results of General Relativity which can be interpreted by saying that General Relativity admits, in a rigourous sense, some gravitational wave propagation with the (local) "velocity of light."

Intuitively, one would like to define a gravitational wave as a "ripple" in the curvature of spacetime. The problem with this definition is that there is no well-defined way of decomposing the (real) metric in a background plus a superimposed "ripple." However there is one way in which one can meaningfully formalize this idea, namely by a generalization, to nonlinear partial differential equations, of the WKB approximation method. Such a generalization has been devised, and applied to General Relativity, by Choquet-Bruhat.[30] It is of the "two-timing" type, i.e. it consists of looking for solutions of the Einstein equations that are given as formal series of the type

$$g_{\mu\nu}(x^\lambda) = \hat{g}_{\mu\nu}(x^\lambda) + \sum_{n \geq n_o} \omega^{-n} h^{(n)}_{\mu\nu}(x^\lambda, \omega u(x^\lambda)), \tag{2.8}$$

where $u(x^\lambda)$ is a scalar function and ω is a parameter which tends to infinity. In the usual WKB method one puts

$$h^{(n_0)}_{\mu\nu}(x^\lambda, \omega u) = \text{Real}(a_{\mu\nu}(x) e^{i\omega u(x)}), \tag{2.9}$$

so that it appears that $\omega u(x)$ plays the role of the *phase* of the "high-frequency waves" $h^{(n)}_{\mu\nu}$. However the assumption (2.9) is unnecessarily restrictive and the whole method can be developed from (2.8) without having to assume a sinusoidal dependence on ωu (the latter can, however, still be called the "phase").

Replacing (2.8) in the vacuum Einstein equations (2.1) leads to a set of conditions to be satisfied by the "phase" $u(x)$, by the "ripples" $h^{(n)}$, and also by the "background" $\hat{g}(x)$. It is found that the "phase" u must satisfy the equation

$$\hat{g}^{\mu\nu} \frac{\partial u}{\partial x^\mu} \frac{\partial u}{\partial x^\nu} = 0. \tag{2.10}$$

In other words, the "surfaces of constant phase" must be null with respect to the background. The "ripples" $h^{(n)}$ must satisfy both algebraic relations, which restrict their structure (to the type (2.31)), and differential relations, which say essentially that the $h^{(n)}$'s "propagate" along the trajectories of the 4-vector

$$\hat{\ell}^\mu := \hat{g}^{\mu\nu} u_{,\nu}. \tag{2.11}$$

In other words, the "ripples" propagate along some null geodesics of the "background" $\hat{g}$.

As concerns the conditions that the "background" must satisfy, they depend crucially on the lowest value n_0, taken by the integer n in the formal expansion (2.8). If $n_0 \geq 2$, which was the case treated originally by Choquet-Bruhat,[30] $\hat{g}$ must satisfy the *vacuum* Einstein equations. On the other hand if $n_0 = 1$, a case first treated by Isaacson,[31] one finds that the high-frequency gravitational waves give rise to an "effective stress-energy tensor" which curves the "background" (this result generalizes previous ideas and results due to Wheeler[32] and Brill and Hartle[33]). More explicitly, the Einstein tensor of the background (see eq. (1.5)) must satisfy:

$$\hat{E}_{\mu\nu} = \frac{8\pi G}{c^4} T^{\text{effective}}_{\mu\nu}, \tag{2.12}$$

where the "effective stress-energy tensor" of the high-frequency waves is

$$T^{\text{effective}}_{\mu\nu} = \frac{c^4}{32\pi G} \sigma(x) u_{,\mu} u_{,\nu} , \tag{2.13}$$

with

$$\sigma(x) := \lim_{Y\to\infty} \left[\frac{1}{Y} \int_0^Y dy\ \left(\hat{g}^{\alpha\gamma}(x)\hat{g}^{\beta\delta}(x) - \tfrac{1}{2}\hat{g}^{\alpha\beta}(x)\hat{g}^{\gamma\delta}(x)\right) \frac{\partial h^{(1)}_{\alpha\beta}(x,y)}{\partial y} \frac{\partial h^{(1)}_{\gamma\delta}(x,y)}{\partial y} \right]. \tag{2.14}$$

For full details about the high-frequency gravitational waves the reader is referred to the works of Isaacson,[31] Choquet-Bruhat,[34] Misner *et al.*,[26] Madore,[35] Taub,[36] Gerlach,[37] and references therein. For further references, and discussion of some of the nonlinear wave interactions contained in the higher order terms of (2.8) see ref. 38.

A word of caution may be necessary: the expansion (2.8) has, as far as I know, only the sense of a formal series. There remains the difficult problem of controlling the difference between (2.8) and some exact solution. For results in the case of linear partial differential wave equations see e.g. some works of Leray.[39,40] The general case might be appreciably different, and might lead to secular instabilities (due, for instance, to wavetails and nonlinearities). Therefore the direct applicability of the expansion (2.8) over "astronomical" distances is somewhat dubious.

In conclusion, the second viewpoint allows one to say, at least *formally* (i.e. in the sense of formal series), that Einstein's theory admits some high-frequency gravitational waves which propagate along null geodesics of a (low-frequency) background. Moreover, within this approach, it is possible to attribute to the high-frequency waves an effective (average) stress-energy tensor with contributes to curving up the background. If we compare this conclusion with the one of the last sub-section, we see that we went down in rigour, but that we gained more precision in the knowledge of what are the effects of a gravitational wave. The next sub-section will amplify this tendency.

2.3. Third viewpoint: weak gravitational waves on a flat background

In order to study in more detail the structure and the propagation of general "ripples" of curvature, let us make the usual linearized post-Minkowskian approximation, i.e. let us assume that in some finite region there exist approximately Minkowskian coordinates such that

$$g_{\mu\nu}(x^\lambda) = f_{\mu\nu} + h_{\mu\nu}(x^\lambda), \tag{2.15}$$

where $f_{\mu\nu}$ is the flat metric (1.3) and where

$$|h_{\mu\nu}(x^\lambda)| \ll 1 \tag{2.16}$$

is so small that it is sufficient to keep only first order terms in h ("linearized approximation," or "first post-Minkowskian approximation").

Then the connection coefficients $\Gamma^\mu{}_{\alpha\beta}$ are small of the first order in h, as well as the curvature tensor $R^\mu{}_{\nu\alpha\beta}$. Therefore the Bianchi identities

$$R^\mu{}_{\nu\alpha\beta;\gamma} + R^\mu{}_{\nu\beta\gamma;\alpha} + R^\mu{}_{\nu\gamma\alpha;\beta} \simeq 0, \tag{2.17}$$

can be approximately written as (neglecting terms quadratic in h):

$$B^\mu{}_{\nu\alpha\beta\gamma} := R^\mu{}_{\nu\alpha\beta,\gamma} + R^\mu{}_{\nu\beta\gamma,\alpha} + R^\mu{}_{\nu\gamma\alpha,\beta} \simeq 0. \tag{2.18}$$

The contraction $\mu = \beta$, i.e. $B^\mu{}_{\nu\alpha\mu\gamma}$, gives

$$-R_{\nu\alpha,\gamma} + R_{\nu\gamma,\alpha} + R^\mu{}_{\nu\gamma\alpha,\mu} \simeq 0 \tag{2.19}$$

so that, *in vacuum* (i.e. $R_{\mu\nu}=0$), one has

$$R^{\mu}{}_{\nu\gamma\alpha,\mu}\simeq 0. \tag{2.20}$$

Now by computing the (flat) divergence over γ of the Bianchi identities, i.e. $f^{\gamma\delta}B^{\mu}{}_{\nu\alpha\beta\gamma,\delta}$, one finds, after using the vanishing of the divergences of $R_{\mu\nu\alpha\beta}$ (equation (2.20)):

$$\Box_f R_{\mu\nu\alpha\beta}\simeq 0, \tag{2.21}$$

where

$$\Box_f := f^{\alpha\beta}\partial_{\alpha\beta}=\Delta-\frac{1}{c^2}\frac{\partial^2}{\partial t^2} \tag{2.22}$$

is the usual flat space wave operator.

Now, it is well known that, in linearized approximation, the components of the curvature tensor are invariant under small coordinate transformations (i.e. transformations $x'^{\mu}=x^{\mu}+\xi^{\mu}(x)$ compatible with (2.15)–(2.16)). Therefore the approximate wave equation (2.21) shows directly that there exist, in linearized approximation, "curvature waves" which have an invariant meaning and which propagate in vacuum "with the velocity of light."

Once the existence and invariant meaning, in linearized approximation, of such gravitational waves has been established by concentrating on the propagation properties of the curvature tensor (an approach first suggested by Eddington[41]) it is useful to construct special coordinate systems which are well adapted to studying these waves in detail. Unfortunately there are no coordinate systems (also called "gauges") which are adapted to the description of all aspects of gravitational waves. The propagation aspects and, generally speaking, the behaviour of gravitational waves on arbitrarily large spatial scales is conveniently described in the "transverse traceless gauge" or "TT gauge" (which can also be called: Gaussian coordinates, synchronous coordinates or "temporal gauge"). On the other hand the physical effects of a gravitational wave on small spatial scales (i.e. on distances much smaller than a typical wavelength) are more conveniently described in other coordinate systems: some "proper" or "Fermi-like" coordinate systems. This multiplicity of coordinate systems to describe one given gravitational wave can be confusing. In order to help the reader to distinguish them clearly we shall use lower case x's to denote "TT coordinates" and capital X's to denote Fermi coordinates.

Let us first state more precisely what we shall call a "gravitational wave" in the linearized approximation. Let us consider some spatially finite region of spacetime (for instance the solar system) and let us assume that the (linearized) metric deviation (2.15), within this region, can be decomposed in a sum,

$$h_{\mu\nu}=h^{\text{local}}_{\mu\nu}+h^{\text{wave}}_{\mu\nu}\ , \tag{2.23}$$

where h^{local} is a functional of the distribution of energy within the considered region [that satisfies the (linearized) inhomogeneous Einstein equations], and where h^{wave} satisfies the vacuum Einstein equations, *and* was *zero* before some time in the past. The last condition means, on the one hand, that we neglect the "non-radiative" tidal effects due to the matter outside the considered region, and on the other hand, that we assume that the externally generated "radiative" tidal effects turn off when going back in the infinite past.

Now, the effect of a small coordinate transformation $x'^{\mu}=x^{\mu}+\xi^{\mu}(x)$ is (with $\xi_{\mu}:=f_{\mu\nu}\xi^{\nu}$)

$$h^{\text{new}}_{00}\simeq h^{\text{old}}_{00}-2\xi_{0,0}\ , \tag{2.24a}$$

$$h^{\text{new}}_{0i}\simeq h^{\text{old}}_{0i}-\xi_{0,i}-\xi_{i,0}\ , \tag{2.24b}$$

$$h^{\text{new}}_{ij}\simeq h^{\text{old}}_{ij}-\xi_{i,j}-\xi_{j,i}\ . \tag{2.24c}$$

It is immediately clear from equations (2.24a) and (2.24b) that, by simple quadratures in time, we can construct a ξ_μ such that the "new" h^{wave} satisfies

$$h_{00}^{\text{wave}} = 0, \tag{2.25a}$$
$$h_{0i}^{\text{wave}} = 0, \tag{2.25a}$$

and that the "new" h_{ij}^{wave} is still zero before some time in the past. Moreover, in this "temporal gauge" one can easily express h_{ij}^{wave} in terms of its associated (linearized) curvature

$$R_{\mu\nu\rho\sigma}^{\text{wave}} = \tfrac{1}{2}(h_{\mu\sigma,\nu\rho}^{\text{wave}} + h_{\nu\rho,\mu\sigma}^{\text{wave}} - h_{\mu\rho,\nu\sigma}^{\text{wave}} - h_{\nu\sigma,\mu\rho}^{\text{wave}}). \tag{2.26}$$

Indeed one sees immediately that

$$R_{0i0j}^{\text{wave}} = -\tfrac{1}{2} h_{ij,00}^{\text{wave}} , \tag{2.27}$$

so that h_{ij}^{wave} is uniquely obtained from R_{0i0j}^{wave} by two time quadratures:

$$h_{ij}^{\text{wave}}(x^0, x^k) = -2 \int_{-\infty}^{x^0} dx'^0 \int_{-\infty}^{x'^0} dx''^0 R_{0i0j}^{\text{wave}}(x''^0, x^k). \tag{2.28}$$

As $R_{\mu\nu\rho\sigma}^{\text{wave}}$ satisfies the vacuum Einstein equations (and the Bianchi identities) it satisfies equations (2.20) and (2.21). In particular, one has

$$R_{0i0j,j}^{\text{wave}} \simeq 0, \tag{2.29a}$$
$$\Box_f R_{0i0j}^{\text{wave}} \simeq 0, \tag{2.29b}$$

and also

$$R_{0j0j}^{\text{wave}} \simeq R_{00}^{\text{wave}} = 0. \tag{2.29c}$$

By integration over the time (see equation (2.28)) one deduces from equations (2.29) that

$$\Box_f h_{ij}^{\text{wave}} \simeq 0, \tag{2.30a}$$
$$h_{ij,j}^{\text{wave}} \simeq 0, \tag{2.30b}$$
$$h_{jj}^{\text{wave}} \simeq 0. \tag{2.30c}$$

Reciprocally it is easily checked that any solution of the equations (2.30), with (2.25), satisfies the vacuum Einstein equations. Therefore the system of equations (2.25), (2.30) describes the most general linearized gravitational wave (in the sense defined above). The "gauge" used in the description is usually called "transverse" because of equation (2.30b) and "traceless" because of equation (2.30c). But, contrary to the usual approach, where these conditions (2.30b-c) are enforced by means of suitable coordinate transformations (see e.g. Misner *et al.*[26]), they have been here obtained as necessary consequences of the (more easily enforced) "temporal" (or "synchronous," or "Gaussian") conditions (2.25).

As is well known the general "progressive plane wave" solution of equations (2.30) can be written as

$$h_{ij}^{\text{wave}}(x^0, x^k) = h_+(x^0 - n_k x^k)(p^i p^j - q^i q^j) + h_\times(x^0 - n_k x^k)(p^i q^j + q^i p^j), \tag{2.31}$$

where $\vec{n}$ is the unit spatial vector pointing in the direction of propagation of the wave, and where $\vec{p}$ and $\vec{q}$ are unit spatial vectors making with $\vec{n}$ an orthonormal euclidean triad $(\vec{n}, \vec{p}, \vec{q})$. The structure (2.31), with the two "linear polarisations" h_+ and $h_\times$ "living" in a 2-plane $(\vec{p}, \vec{q})$ transverse to the propagation, corresponds exactly to the general algebraic structure of gravitational shock waves or of high-frequency waves (as detailed in the references given in §§2.1 and 2.2).

On the other hand, when one is interested not so much in a global description of the propagation of a gravitational wave, but on its effect on a detector of size small compared to a typical wavelength, is is convenient to switch from the previous "TT" coordinate system to a coordinate system better adapted to the physical description of the effect of a gravitational wave on material systems. For the construction of such coordinates we refer to the book of Misner, Thorne, and Wheeler,[26] and to the articles of Grishchuk and Polnarev,[42] and of Brillet *et al.*[6] In the general case of an "accelerated" detector (e.g. a detector fixed on Earth, instead of being in free fall) these special coordinates go under the name of "proper reference frame" of the detector. In the special case of a freely falling detector they are called "locally inertial frame" or "Fermi coordinates." The advantage of such coordinates, say (X^0, X^i), is that one can often use a Newtonian description of the motion of the detector (with time $T := X^0/c$ and cartesian coordinates X^i) and consider that the only effect of a gravitational wave is to exert on each mass (or mass element) m, located in X^i, a force given by

$$F_i^{\text{wave}} = -mc^2 R_{0i0j}^{\text{wave}} X^j. \tag{2.32}$$

For a useful pictorial description of the "wave force" (or, more precisely, "radiative tidal force") (2.32) by means of "lines of force" see Misner *et al.*[26] and references therein. For detailed investigations of the effect of gravitational waves on material systems or fields see e.g. Souriau,[43] Grishchuk and Polnarev,[42] Carter[44] and references therein.

In conclusion, the third viewpoint allows one to describe, in more detail than previous approaches, the effect of gravitational waves on material systems. Moreover, it allows one to tackle some situations which cannot be studied with the previous viewpoints: for instance weak gravitational waves propagating in the presence of curvature inhomogeneities of characteristic distance scale comparable to or smaller than the wavelength (indeed in the linearized approximation (2.23) the wavelength of h^{wave} can take any value, independently of the characteristic time or space scales of h^{local}).

2.4 Other viewpoints

Let us only mention, for the sake of completeness, that one can give meaning to the concept of "gravitational wave" by using still several other viewpoints.

For instance, it can be quite instructive to look at some exact solutions of the Einstein equations which can be interpreted as "waves," see e.g. Landau and Lifshitz,[45] Misner *et al.*,[26] Kramer *et al.*[46] This viewpoint is physically limited because of the idealized structure of such exact waves, but it is interesting because it allows one to take into account the nonlinearities of Einstein's theory.

One can try to define gravitational radiation by looking at the asymptotic behaviour of time-dependent gravitational fields far away from their sources, see §3.1 (problem 2) and §4.13 below.

Also, one can look at small perturbations of some given solution, $\hat{g}_{\mu\nu}$, of the Einstein equations, say

$$g_{\mu\nu}^{\text{perturbed}}(x) = \hat{g}_{\mu\nu}(x) + \varepsilon h_{\mu\nu}(x), \tag{2.33}$$

with $\varepsilon \to 0$. One finds that the metric perturbations $h_{\mu\nu}$ satisfy, in vacuum, when using coordinates such that

$$\hat{g}^{\nu\lambda}(\hat{\nabla}_\lambda h_{\mu\nu} - \tfrac{1}{2}\hat{\nabla}_\mu h_{\nu\lambda}) = 0, \tag{2.34}$$

the following equation

$$\Box_{\hat{g}} h_{\mu\nu} + 2\hat{R}^{\alpha}{}_{\mu}{}^{\beta}{}_{\nu} h_{\alpha\beta} = 0, \tag{2.35}$$

where

$$\Box_{\hat{g}} := \hat{g}^{\alpha\beta}\hat{\nabla}_{\alpha}\hat{\nabla}_{\beta} \tag{2.36}$$

is the covariant wave operator in the unperturbed ("background") spacetime. The equation (2.35) can be interpreted as describing the propagation of small "ripples" on the curved "background" $\hat{g}$. This viewpoint generalizes both the second viewpoint (because now $h_{\mu\nu}$ is not limited to be of "high-frequency") and the third viewpoint above (because now the background $\hat{g}$ can be curved). However the price to pay for this generality is that one does not know how to solve in closed form the propagation equation (2.35). For general investigations of wave equations in curved space see e.g. Friedlander,[47] Choquet-Bruhat *et al.*,[48] Carminati and McLenaghan[49] and references therein. One of the most interesting effects which is contained in the general solution of equation (2.35), and which does not appear in the "high-frequency" formal solutions discussed above is the continuous back-scattering of the waves off the curvature of the background. This effect leads to the formation of "tails" beyond any initially sharp wave-pulse. For investigations of these tails see e.g. the works of DeWitt *et al.*, Couch *et al.*, Price, Bardeen and Press, and Unt and Keres quoted in chapter 35 of Misner *et al.*,[26] as well as Waylen[50] and §5 below.

Finally, let us quote the exact "wave equation" which is satisfied *in vacuum* by the curvature tensor (Penrose[51]) (for the propagation of waves within the matter and its associated dispersion or absorption effects, see e.g. Grishchuk and Polnarev,[42] Thorne,[25] and references therein). It is easily obtained by following the method of §2.3 above, but without making any approximation. It reads:

$$\Box_g R_{\mu\nu\rho\sigma} = R_{\mu\nu}{}^{\alpha\beta} R_{\rho\sigma\alpha\beta} + 2(R_{\mu}{}^{\alpha}{}_{\rho}{}^{\beta} R_{\nu\alpha\sigma\beta} - R_{\mu}{}^{\alpha}{}_{\sigma}{}^{\beta} R_{\nu\alpha\rho\beta}). \tag{2.37}$$

This equation is not very useful in practice, but it is conceptually nice because it exhibits clearly the subtle nature of "gravitational waves": *nonlinearly interacting waves of curvature propagating in the wavy curved space itself.*

§3. BASIC PROBLEMS OF GRAVITATIONAL RADIATION THEORY

3.1. A catalogue of problems.

In spite of the fact that the theory of gravitational radiation, in the context of General Relativity, has been the subject of extensive research, especially during the last twenty-five years, it must be admitted that the basic problems of this field of research have not yet received fully satisfactory answers. Without an attempt at completeness, let us outline some of these basic problems by means of the following questions:

Problem 1. "Definition problem"

Given some gravitational field, is there a "good" criterium to decide if it is "radiative" or not?; and, is there a "good" way of splitting it into a "radiative" part and a "non-radiative part"?

In other words, the problem is to know what is the general definition of a "gravitational wave." In the previous section (§2) we have tried to give meaning to the concept of "gravitational wave" by studying several limiting cases where the propagation properties of the Einstein equations were especially evident. But this does not prove that this concept can be meaningfully and/or usefully applied in general. A clear-cut definition of the "wave content" of some given gravitational field might however be useful in several investigations. For instance it might help in deciding whether a gravitational field contains some "incoming" gravitational radiation or not. Also, the need for such a definition arises naturally in the Hamiltonian approaches to the dynamics of the gravitational field, see especially the work of Arnowitt, Deser and Misner[52] where a definition of the "true dynamical degrees of freedom" of the gravitational field is proposed. Another motivation for finding such a definition is to try to quantize the gravitational field. More recently the search for such a definition has received a new motivation in the 3+1 numerical approaches to the evolution of gravitating systems (see e.g. the articles by York, Piran and Nakamura in reference 15). Such a definition is also useful in Schutz's statistical approach[53,54] to gravitational radiation damping. On the other hand, if it is possible to otherwise exclude (for instance, within the "asymptotic problem") the presence of incoming radiation, a definition at all times of the global "wave content" of a gravitational field becomes unnecessary. Indeed it is needed neither for asking nor for answering questions of direct physical significance.

Problem 2. "Asymptotic problem"

What is the asymptotic behaviour, appropriate to isolated systems and consistent with Einstein's field equations, of radiative gravitational fields far away from their sources?

This problem can be split into several sub-problems, that differ by the kind of limiting process, in spacetime, that is considered. For instance one can consider the asymptotic behaviour of the field when $r \to +\infty$, at fixed t (spatial infinity), or when $r \to +\infty$ and $t \to -\infty$, with $t \sim -r/c$ (past null infinity), or when $r \to +\infty$ and $t \to +\infty$, with $t \sim +r/c$ (future null infinity). There exist powerful methods, initiated by Bondi and Penrose, to investigate the previous sub-problems (for reviews see e.g. Newman and Tod,[21] Walker,[22] Ashtekar,[23,55] Geroch,[56] Schmidt[57] and references therein). However, as stressed e.g. by Schmidt,[57] Damour[58] and Blanchet and Damour,[59] the whole Bondi-Penrose approach to asymptotic structure is still unsatisfactory. Indeed, it provides a *definition* of a class of spacetimes that one would like to associate with radiative isolated systems, but neither the global consistency, nor the physical appropriateness of this definition have been proven. Important progress has been made recently by Friedrich[60] towards proving the consistency of the Bondi-Penrose definition, however some perturbation calculations tend to indicate that the back-scatter of gravitational radiation off the curvature of spacetime jeopardizes the asymptotic conditions that are usually assumed in the Bondi-Penrose approach (see references quoted in refs 58–59). However it seems probable that the dominant asymptotic behaviour of the metric and the curvature will not be affected. This may be sufficient both to define the absence of "incoming radiation" and to define the "outgoing radiation field."

Problem 3. "Generation problem"

What is the link between the "outgoing radiation field" and the structure and motion of the sources?

This problem is rendered very difficult by the all pervasive nonlinearity of Einstein's theory. And yet, the present world-wide development of gravitational wave detectors makes it urgent to get, at least approximate, answers to this "generation problem." Many different approaches have been aimed at answering this problem. In the following we shall summarize the main methods and some of their results (see §§4 and 5 below). One possible way of tackling the generation problem is to split it into two sub-problems: on the one hand, one tries to compute (for instance numerically) the gravitational field generated by a source at some finite distance away from the source, and, on the other hand, one must relate the "outgoing radiation field" to the gravitational field computed at a finite distance. The latter sub-problem can be called the:

Problem 4. "Propagation problem".

What is the link between the "outgoing radiation field" and some, possibly non-radiative, component of the gravitational field at a finite radius away from the source?

This problem has been less investigated than the other problems. Its interest comes from the recognition that it might be impossible to solve the "generation problem" by means of only one method. For instance in the case of a strongly self-gravitating source it may be necessary to use a numerical method to compute the motion of the source and its local gravitational field (within the numerical grid). It is then necessary to complete the numerical calculation by an analytical investigation of the structure of the gravitational field in a vacuum region outside the source to give both the appropriate boundary conditions, at the end of the grid, and the transformation between the local gravitational field and the outgoing radiation at infinity. For a treatment of the "propagation problem" in the WKB approximation see refs 20 and 25, for results in the general case see §5.8 below.

Problem 5. "Radiation reaction problem"

What is the back-reaction on the source of the emission of gravitational radiation?

The observation of a secular acceleration of the orbital motion of the binary pulsar PSR 1913+16 has recently spurred a re-examination of this problem. We shall not try to review all the work done (see e.g. the articles by Thorne, Damour, Eardley and Ashtekar in ref. 15, and the reviews of Schutz,[24] Will[61] and Damour[62]). However we shall present below (§§5 and 6) some of the results which have been obtained concerning the "radiation reaction problem."

Finally, we would like to mention, for completeness, the

Problem 6. "Detection problem"

What are the observable effects of gravitational waves on physical systems?, and, how can one measure them?

We have rapidly dicussed the first half of this problem in §2.3 above. For a fuller discussion see the references quoted in the latter sub-section. For introductions to the second half of this problem see e.g. references 5 and 6; for full details see references 7–17 and references therein.

3.2. A catalogue of approximation methods

Most of the problems listed in the previous sub-section are, at present, beyond the reach of rigourous mathematical methods. In fact the situation is even worse than that in the sense that, as far as I know, there still exists no theorem proving the *global* existence of sufficiently generic solutions of the *inhomogeneous* Einstein equations. For clear introductions to the mathematical problems involved see e.g. Choquet-Bruhat[63] (for recent progress see also Friedrich[60] and Noutchegueme[64]). Moreover, in the case where the source of Einstein's equations consists of one or several fluid bodies having spatially compact supports, it has been stressed by B. Schmidt (personal communication) that there exist no mathematical results guaranteeing a physically interesting domain of existence of solutions (as far as I know, solutions are known to exist only in very small "diamond" spacetime domains contained within each body). In view of this somewhat unsatisfactory mathematical situation, the physicists are forced to rely on (mathematically uncontrolled) approximation methods.

The first step towards setting up approximation methods consists in identifying some small (dimensionless) parameters. For instance in the various weak-field methods this parameter can measure the "strength" (or rather the "weakness") of the gravitational field. Usually one considers as weakness parameter a quantity of the type

$$\gamma_i = \frac{Gm}{c^2 L}, \tag{3.1}$$

where m is a characteristic mass, and L a characteristic linear dimension, of the source. The index i in equation (3.1) stands for "internal," because the parameter γ_i measures the maximal strength of the gravitational field inside the source. Sometimes the parameter γ_i is not small but one can still apply the weak-field method in some region outside the source if the following "external" strength parameter,

$$\gamma_e = \frac{Gm}{c^2 D}, \tag{3.2}$$

is small, D being a characteristic minimal distance away from the source.

An often considered dimensionless quantity is the "slowness" parameter,

$$\beta = \frac{v}{c}, \tag{3.3}$$

where v is a characteristic velocity of the source (in some cases one can distinguish between an "internal" β_i and an "external" β_e). The parameter β can be rewritten as

$$\beta \sim \frac{L}{\lambda} = \frac{L/c}{P} \tag{3.4}$$

where L is, as above, a characteristic linear dimension of the source, and λ (respectively P) a characteristic wavelength (resp. period) of the gravitational radiation emitted by the source. The form (3.4) shows that $\beta \ll 1$ means that the source is well within its near-zone, or, in other words, that the retardation effects within the source, due to the propagation of gravity, are small.

Another ratio of interest can be the WKB parameter (notation of equation (2.8) above)

$$\omega^{-1} = \frac{\lambda}{\mathcal{L}}, \tag{3.5}$$

where $\mathcal{L}$ is a characteristic length scale of the gravitational field.

Finally, some situations might involve strong fields, and variations which are neither very slow, nor very fast, but might deviate only very slightly from a known solution. In such cases one might introduce a small dimensionless parameter, say ϵ, measuring the deviation from the known "background" solution.

Associated to the previous, possibly small, parameters there are some corresponding approximation methods. For instance:

Method 1. "Post-Minkowskian Approximation Methods,"

or in short PMA methods, are weak-field expansions (i.e. expansions in powers of γ_i, or γ_e) which can be used when γ_i, or γ_e, is small.

Method 2. "Post-Newtonian Approximation Methods,"

or in short PNA methods, combine a weak-field expansion and a small-retardation (or well-within-the-near-zone) expansion. It usually assumes that both γ_i (or γ_e) and β are small, and that $\beta^2 \sim \gamma_i$ (or γ_e).

Method 3. "W.K.B. Methods,"

use a high-frequency, or short-wave, expansion, assuming that ω is big (see e.g. §2.2 above).

Method 4. "Small Perturbation Methods,"

use an expansion around a "background" solution (see e.g. §2.4 above).

Moreover one can also use

Method 5. "Multipole Expansion Methods,"

which do not, a priori, require a small parameter (because they are generally expected to converge) but which must converge quickly, in the region where they are used, to be of interest.

Method 6. "Numerical Methods,"

which can be considered as expansions in (grid spacing)/(typical length or time scale of variation). For reviews see e.g. Piran,[65] as well as several contributions to references 8, 12, 14, 15, 17 and 18.

We shall not attempt here to review the previous basic classes of approximation methods (see e.g. references 20, 24, 25, 26, 61, 62, as well as Thorne[66] and Damour,[67] and references therein). However, we would like to stress that all of these methods have limitations, not only in the sense that they cannot be applied to arbitrary sources, but also in the sense that when they are applicable, they are liable to have a domain of validity which is limited in space or in time. For instance the post-Newtonian approximation methods make sense only in the near zone of the source (i.e. a region contained within a radius $r \ll \lambda$ around the source). On the other hand, the post-Minkowskian methods a-priori make sense all over the weak-field zone of the source, independently of its position with respect to the near-zone or the wave-zone. However, explicit calculations indicate that the PMA methods run into some problems in the exponentially far wave-zone:

$$r \gtrsim \lambda \exp[c^2\lambda/(4\pi Gm)]. \tag{3.6}$$

Therefore a useful thing to do, when discussing approximation methods, is, following Thorne,[20,25,66] to distinguish several regions of space around a source, namely: strong-field zone / weak-field zone, and, near-zone / transition-zone / local wave-zone / distant wave-zone. Such distinctions, based on rather precise criteria (see references 20,25,66), have the merit of calling attention to the limitations of validity of some approximation methods. They also suggest that, even in cases where none of the methods are everywhere valid, there might be a way out: namely using simultaneously several approximation methods in a complementary way. This possibility has been advocated and/or used by many authors working in General Relativity (for references see e.g. reference 62, §12 and §§5 and 6 below). However, as stressed by Blanchet and Damour[68] and Damour,[62,69] there are still fundamental issues concerning this combination of approximation methods which have not yet been fully clarified. Indeed the power, but also the basic paradox, of this general approach (borrowed essentially from fluid dynamics) lies in the fact that it "matches" two different asymptotic expansions whose domains of a-priori validity become completely disjoint when the small parameter, say ϵ, tends to zero. For instance, in some studies of slow-motion sources, one might wish to match a small-retardation expansion, a priori valid when $r \lesssim L$, to a weak-field-wave-zone expansion, a-priori valid when $r \gtrsim \lambda$. So that, when the "slowness" parameter $\beta = v/c = L/\lambda$ tends to zero, the two domains of a-priori validity become infinitely separated. Now, the literature which deals with precise rules for the "matching of asymptotic expansions" is full of ambiguities and controversies.

It is particularly so in the tricky situation called "weak matching," which, unfortunately, is realized in General Relativity. Practically speaking, this means that when one uses this powerful method of "matching of asymptotic expansions" in General Relativity, one must take care to control in sufficient detail the structure of the higher-order nonlinear corrections, otherwise one can get wrong results.

If we now start from the "basic" classes of methods considered above, and allow ourselves to combine them, as just stated, to make new approximation methods, we see that we can produce a long "menu" of "cocktail" approximation methods which should allow us to tackle many interesting physical problems. In the rest of these lectures (§§4-5-6) we shall consider several such combinations of approximation methods. Among the other possible "cocktails" not discussed here, let us quote only one that is of interest in relation to the "generation problem" for fast-moving, strongly self-gravitating, sources. It consists of trying to harmoniously combine numerical methods (appropriate to the dynamic strongly self-gravitating central region) with analytical methods (appropriate to the weak-field exterior region). For three different attempts at tackling this problem see e.g. Bardeen,[70] Anderson and Hobill[71] and Damour.[72]

§4. QUADRUPOLE MOMENT FORMALISMS

4.1 "Quadrupole laws" versus "quadrupole equations"

As there has been some confusion concerning the status of the so-called "quadrupole formula" in General Relativity, it may be useful to first clarify the quite different meanings that can be associated to this locution. For lucid presentations of the issues involved, see also Walker[73] and Ehlers and Walker.[74] For clarity, we shall here distinguish several kinds of "quadrupole formulae."

First, we shall make a distinction between what we shall call three general *"quadrupole laws"* and the more specific *"quadrupole equations."* This (tentative) terminology is inspired by a corresponding terminology which has been introduced by Havas and Goldberg[82] in order to clarify some conceptual issues in the general relativistic problem of motion. They distinguished the general *"laws of motion,"* i.e. some general mathematical expressions relating some (more or less specified) particle variables to some unspecified force (say, $dp/dt = F$), from the specific *"equations of motion"* which are the same expressions but with the particle variables and the force fully specified in terms of some common configuration variables (say, $d(m\, dx/dt)/dt = -kx$). In practice the specification of, say, the force, can be somewhat implicit but it has to be, in principle, completely determined. Similarly we shall call "quadrupole laws" some general mathematical expressions relating some (more or less specified) field, or dynamical, gravitational variables to some unspecified "quadrupole moment," and we shall call "quadrupole equations" the same expressions but with the gravitational variables and the quadrupole moment fully (even if implicitly) specified. As will be clear in practice the latter full specification can mean, for some quadrupole laws, that the precise functional dependence on some common configuration variables of both the gravitational variables and the quadrupole moment must be given, or, more simply, that the quadrupole moment is completely specified and that the physical meaning of the gravitational variables is fully clarified (so that some specific physical consequences can be effectively drawn from the corresponding "quadrupole equation"). As will be discussed in more detail below, the reason for introducing such a terminology (which may, at first sight, seem pure hair-splitting) is that not only have some authors used the locution "*the* quadrupole formula" when speaking about quite different gravitational effects, e.g. gravitational radiation emission versus gravitational radiation reaction (here referred to as separate quadrupole *laws*), but also some authors have claimed to have confirmed "*the* quadrupole formula" while their result concerned in fact only a specific quadrupole *equation* (corresponding to a particular class of physical systems). Note that what is called by Ehlers and Walker[74] "quadrupole laws" are in fact, in the terminology proposed here, specific "quadrupole equations" corresponding to the particular class of sources having (everywhere) Newtonian internal gravity. Note also that the list of quadrupole laws that we are going to discuss is not limiting; one can also add further quadrupole laws referring for instance to some "loss of angular momentum." Finally the word "law," used here for conceptual clarification, should not mislead the reader as to the value of these "quadrupole laws" which are neither exact nor general, nor fundamental; they are "laws" in the sense, say, of Snell's laws within the framework of Maxwell theory.

4.2 The three "quadrupole laws"

The *first quadrupole law,* or *far-field* quadrupole law, reads

$$h_{ij}^{\text{wave}} = \frac{2G}{c^4} \cdot \frac{1}{r} \cdot P_{ijkl}(\vec{n}) \cdot \frac{d^2 Q_{kl}^{\text{I}}(u)}{du^2}, \tag{4.1a}$$

where

$$P_{ijkl}(\vec{n}) := (\delta_{ik} - n_i n_k)(\delta_{jl} - n_j n_l) - \tfrac{1}{2}(\delta_{ij} - n_i n_j)(\delta_{kl} - n_k n_l), \tag{4.1b}$$

is the projection operator onto the (euclidean) 2-plane orthogonal to the unit direction vector n_i $(n_j n_j = 1)$ which projects any symmetric tensor onto a symmetric-trace-free tensor "living" within this 2-plane, and where $Q_{ij}^{\text{I}}(u)$ is some symmetric trace-free euclidean 2-tensor function of one real variable $(i, j = 1, 2, 3; u \in \mathrm{R}; Q_{ij}^{\text{I}} = Q_{ji}^{\text{I}}; Q_{jj}^{\text{I}} = 0)$. The physical meaning of the first quadrupole law is the following: h_{ij}^{wave} represents the outgoing gravitational radiation field, seen in the far wave-zone $(r \to \infty)$ and at very late times (i.e. at fixed "retarded time" $u \sim t - r/c$), and expressed in a suitable coordinate system. Roughly speaking, h_{ij}^{wave}, in equation (4.1a), can be thought of as the leading asymptotic term (when $r \to \infty$; u fixed) of the quantity denoted the same way in §2.3, and representing a general past-zero weak gravitational wave, propagating on a flat background, seen in "transverse-traceless gauge." A more precise definition of h_{ij}^{wave} will be given below in §4.13. The expression (4.1a) becomes physically interesting only if one provides some independent information on the "first quadrupole moment," or "radiative quadrupole moment," Q_{ij}^{I}.

The *second quadrupole law,* or *energy-loss* quadrupole law, reads

$$\frac{dE^{\text{II}}}{du} = -\frac{G}{5c^5} \left(\frac{d^3 Q_{ij}^{\text{II}}(u)}{du^3} \right)^2 . \tag{4.2}$$

It links the decrease of some quantity $E^{\text{II}}(u)$ (having the dimensions of energy) with the time variation of some symmetric trace-free tensor $Q_{ij}^{\text{II}}(u)$ (a priori different from Q_{ij}^{I}). This expression is even less specific than the first one in the sense that not only Q_{ij}^{II}, but also E^{II}, have to be specified (or independently defined) to give a meaning to (4.2). Then the physical meaning of a corresponding "quadrupole equation of the second kind" will depend on the physical meaning of the quantity E^{II}.

The *third quadrupole law,* or *radiation-reaction* quadrupole law, reads

$$\mathcal{F}_i^{\text{reac}} = -\rho^{\text{III}} \frac{\partial}{\partial x^i} \left[\frac{G}{5c^5} \frac{d^5 Q_{jk}^{\text{III}}(t)}{dt^5} x^j x^k \right] \tag{4.3}$$

where $\mathcal{F}_i^{\text{reac}}$ is supposed to represent a "reaction force density" within the matter distribution, correlated with the emission of gravitational radiation. This expression is the least specific of the trio. In order to transform it into a physically useful result, one must specify not only $Q^{III}_{ij}(t)$, but also the "matter density" $\rho^{III}(t, x^i)$, as well as the precise meaning of $\mathcal{F}^{reac}_i$ (this implies also that the coordinate system in which (4.3) holds is precisely defined, see the discussion in §4.15 below).

The notation used above, with three different quadrupole moments, Q^{I}_{ij}, Q^{II}_{ij}, Q^{III}_{ij}, corresponding to the various quadrupole laws, has not only been introduced for conceptual clarification, but also because these quadrupole moments can indeed be numerically different for one given system (in fact $Q^{I} = Q^{II} \neq Q^{III}$). We shall give an instance of this in §5.10 below.

4.3 General discussion of "quadrupole equations"

A "quadrupole equation" will be obtained by replacing in one of the previous formal "quadrupole laws," the general quantities appearing, explicitly or implicitly, in them (Q^{I}_{ij} and/or Q^{II}_{ij} and/or ...) by some specific functions (defined either explicitly, or implicitly via some functional dependence on other known quantities). This means evidently that there can be many different such "quadrupole equations," either because they go under different basic "quadrupole laws," or because they go under the same basic quadrupole law but they contain different explicit quantities. Then the resulting quadrupole equations must be carefully distinguished. Two examples will help to clarify the issues involved.

First, if somebody comes up with an end result having the form of expression (4.2), then one must carefully examine how precisely the quantities Q^{II}_{ij} and E^{II} are defined. If one of the two quantities (usually E^{II}) is not precisely defined, then no real physical information is contained in this result ((4.2) can just be used to *define* what one will call a total outgoing *flux* of gravitational radiation). On the other hand, if both quantities Q^{II} and E^{II} are precisely defined, then the corresponding "quadrupole equation of the second kind" (or "energy-loss quadrupole equation") contains some physical information. However the amount and content of this information depends drastically on what is known about E^{II}. If E^{II} is only defined as some coefficient in the asymptotic metric (as is the case for the "Bondi mass") then the corresponding energy-loss quadrupole equation will say something about the asymptotic metric, and may (as in the Bondi mass case) say something about some physical difference between a stationary initial state and a stationary final state. However, it will say nothing about the behaviour of the intermediary states; this means, for instance, that such a formula can say nothing about the (observed) medium-term evolution of a binary system (for a discussion of this issue see e.g. the contributions of Thorne, Damour and Ashtekar in ref. 15, as well as in refs. 62 and 75). In order to be able to draw conclusions about the medium-term kinematical evolution of a binary system one would need *at least* to have E^{II} given (explicitly or implicitly) as a functional of the instantaneous kinematical state of the binary system (as is the case, for instance in the quadrupole equations derived by Damour,[76,67] Schäfer[77] or Grishchuk and Kopejkin[78]). However even such a precise kinematical energy-loss quadrupole equation is not quite sufficient to be able to draw conclusions about the observable features of a binary system. One needs, as discussed in §6 below, to know: 1) the complete coordinate equations of motion, 2) their (approximate) solutions, and 3) the link between the coordinate motion and the observed quantities.

As a second example of the distinctions which must be made between superficially identical "quadrupole formulae" let us consider two results having the mathematical structure of the far-field quadrupole law (4.1) but such that, in one quadrupole equation the quantity Q^{I}_{ij} is just the usual quadrupole moment of the mass-energy distribution of a *weakly-self-gravitating* matter distribution, namely

$$Q^{I}_{ij}{}'(u) = \int d^3x\, c^{-2} T^{00}(u, \vec{x})(x^i x^j - \tfrac{1}{3}\vec{x}^2 \delta^{ij}), \tag{4.4a}$$

while the other equation deals with the gravitational radiation emitted, say, by a system of well separated ***strongly-self-gravitating*** bodies, and contains in lieu of (4.4a)

$$Q^{\rm I}_{ij}{}'' = \sum m(z^i z^j - \tfrac{1}{3}\vec{z}^2 \delta^{ij}), \tag{4.4b}$$

where m is the "Schwarzschild mass," and z^i the "centre of field," of each strongly self-gravitating body (see §6 below). Now it is sometimes formally assumed that equation (4.4b) can be considered as a special case of equation (4.4a), but this is incorrect because the "masses" appearing in equation (4.4b) differ from the integral of $c^{-2}T_{00}$ on the volume of the bodies by numerically important gravitational-binding energy contributions. In fact, the domain of validity of the demonstration leading to (4.4a) forbids one to consider the case of strongly self-gravitating matter distributions.

In the following we shall discuss some quadrupole equations, corresponding successively to each of the three quadrupole laws (4.1)–(4.3), and we shall try not only to make precise the definition and/or functional form of the quantities appearing in each, but also the (a-priori) realm of validity for each, by which we mean the (physical) conditions under which the derivation of each equation is valid. As we shall see below, it can happen that a final equation has a wider range of validity than the one indicated by its initial derivation, but that can be checked only by finding a new derivation.

4.4 The "standard," or "Einstein," far-field quadrupole equation

We shall call here the "standard" far-field quadrupole equation the one which results from the linearised approximation of General Relativity. This equation dates back essentially to Einstein.[79] It has the structure of the first quadrupole law (4.1). The coordinate system is supposed to be an approximately Minkowskian coordinate system (i.e. such that equations (2.15)–(2.16) hold). The quantity r denotes $(\delta_{ij}x^i x^j)^{1/2}$, while $u = (x^0 - r)/c = t - r/c$ and $n^i = x^i/r$. The quantity $Q^{\rm I}_{ij}$ is given as a functional of the energy distribution by the following formula:

$$Q^{\rm I\,standard}_{ij}(t) = \int d^3x\, c^{-2} T^{00}(t,\vec{x})(x^i x^j - \tfrac{1}{3}\vec{x}^2\delta^{ij}). \tag{4.5}$$

The main physical conditions of validity of the derivation of this "standard" quadrupole equation are: 1) ***negligible*** self-gravity of the source (the motion of the source being due to non-gravitational forces) and 2) ***slowness*** of motion of the source. Then the equation obtained by replacing (4.5) in (4.1) represents only the lowest order term of a double expansion: a wave zone expansion, in powers of λ/r, and a retardation-within-the-source expansion in $(L/c)/P \sim v/c$ (see equations (3.3)–(3.4)). We have quoted the formula (4.5) for $Q^{\rm I}_{ij}$ as it comes out of the standard calculation. Let us recall that it is often further argued that, under usual conditions, one can replace $c^{-2}T_{00}$ by some "mass density," say μ (which can be either a rest mass density or a proper mass-energy density of a fluid, etc....). But for our purposes it will be simpler to keep considering the original form (4.5).

The domain of validity of the standard quadrupole equation is too restricted to be useful in astrophysics. Indeed, the standard derivation does not allow one to use it to evaluate the radiation emitted, say, by a binary system made of two ordinary stars (where the motion of the source is influenced by gravitational forces). However two different ways of generalizing the quadrupole moment formalism to, at least weakly, self-gravitating sources have been indicated, by Landau and Lifshitz on the one hand, and by Fock on the other hand.

4.5 The "Landau-Lifshitz" quadrupole equation.

As early as 1941 (according to ref. 66) Landau and Lifshitz in the first edition of their classic volume on the *Theory of Fields* have shown that the formula (4.5) *was still applicable to the case of slow motion sources with Newtonian internal gravity* (like a binary star). Strangely enough, this remarkable improvement in the applicability of the quadrupole equation (4.1) & (4.5) is often ignored in spite of its having been clearly explicated in all the editions of the *Theory of Fields* of Landau and Lifshitz and the well-known (but maybe insufficiently well-read) book of Misner *et al.*[26] The main idea of Landau and Lifshitz is to shuffle to the right-hand side of Einstein's equations all the terms nonlinear in $\bar{h}^{\mu\nu} := g^{1/2}g^{\mu\nu} - f^{\mu\nu}$, and to consider that they add up to the material stress-energy tensor $T^{\mu\nu}$ to make a total (material & gravitational) stress-energy tensor of the type:

$$\tau^{\mu\nu} = T^{\mu\nu} + t^{\mu\nu}(\bar{h}). \tag{4.6}$$

For details about this trick and its utility see e.g. Landau and Lifshitz,[45] Misner *et al.*,[26] Weinberg.[80] Now if one formally follows the steps of the standard derivation and assumes that the fact that $\tau_{\mu\nu}$ has no longer a spatially compact support does not matter, one obtains an "improved" far-field quadrupole equation of the form (4.1) with the "Landau-Lifshitz improved" quadrupole moment

$$Q^{\mathrm{I}}_{ij}(t) = \int d^3x\, c^{-2}\tau^{00}(t,\vec{x})(x^i x^j - \tfrac{1}{3}\vec{x}^2\delta^{ij}). \tag{4.7}$$

Now for systems with Newtonian internal gravity (weakly, but not negligibly, self gravitating systems) the spatial components of $t^{\mu\nu}$ (gravitational stresses) are comparable to the spatial components of $T^{\mu\nu}$, but the other components are very small compared to the corresponding components of $T^{\mu\nu}$. Then one expects equation (4.7) to reduce essentially to equation (4.5) so that the gravitational radiation emitted by weakly self-gravitating slow sources can be computed by the "standard" formula (4.5).

The preceding method is however fraught with several difficulties. They come from the fact that $\tau^{\mu\nu}$ does not have a spatially compact support, and in fact has only a very slow fall-off ($\sim r^{-2}$) in some spacetime directions, so that many of the formal steps of the derivation are dubious. In fact even the meaning of the result (4.7) is unclear (before making the approximate replacement $\tau^{00} \approx T^{00}$) because τ^{00} is expected to fall off as r^{-4} in spatial directions, so that the integral in the right-hand side of (4.7) is not (absolutely) convergent!

4.6 The "Fock" quadrupole equation

Fock, in his famous book[81] (first published in Russian in 1955) has proposed a quite different approach to this problem. This approach proved valuable not only in the specific problem at hand, but has also led to powerful generalizations. The main idea of Fock is to separate the problem into two sub-problems. The first sub-problem consists of computing the gravitational field in the "near-zone" of the source, i.e. in a region where the retardation r/c stays small compared to a characteristic period (in other words: $r \ll \lambda$). The second sub-problem consists of studying the structure of a general radiative gravitational field in the wave-zone ($r \gg \lambda$) (with inclusion of the tricky nonlinear effects linked to the slow fall-off of $\tau^{\mu\nu}$). Then the results of both sub-problems are "matched" via an intermediate expression for the gravitational field which is used to bridge the gap between the near-zone and the wave-zone (see equation (87.63) of ref. 81). The outcome of this type of approach is to link the radiative quadrupole moment Q^{I}_{ij} to some intermediate quantity which appears also in the gravitational field at "moderately large" distances away from a weakly self-gravitating source (see the lucid remarks of Fock at the beginning of his §86 and at the end of his §87). Essentially, we can summarize the Fock approach by the (double) equation:

$$Q^{\mathrm{I}}_{ij}(t) = (Q^{\text{intermediate}}_{ij} =) \int d^3x \, c^{-2}(1 + \tfrac{1}{2}U/c^2)T^{00}(t,\vec{x})(x^i x^j - \tfrac{1}{3}\vec{x}^2\delta^{ij}), \qquad (4.8)$$

where U denotes the (positive) Newtonian potential (see equations (85.39), (87.62) and (87.63) of reference 81). The result (4.8) is, at least to lowest order, consistent with the Landau-Lifshitz one (however in Fock's approach the "retarded time" u appearing in equation (4.1) becomes $u = t - r/c - (2GM/c^3)\log(r - 2GM/c^2)$).

4.7 Recent improvements

The original ways of implementing the preceding two approaches are not quite satisfactory. Recently many attempts have been made to improve this state of affairs. For instance some of the defects of the approach à la Landau-Lifshitz seem to be eased by the work of Futamase and Schutz[83] (see however the critical comments concerning this line of work in ref. 74). Moreover Futamase[84] indicates how to use the Landau-Lifshitz approach in the case of a system of well separated strongly self-gravitating objects having a weak mutual gravitational interaction (see also below). Another line of research, due to Epstein and Wagoner[85] and Thorne[66] has tried to improve the accuracy of the Landau-Lifshitz result (4.7) by taking into account higher retardation effects at the quadrupolar level (an also higher multipoles, see below). However their results are rather formal because all those higher order corrections are given by non-convergent integrals.

As for the approach à la Fock, it has been clarified and completed by the work of Thorne,[20,66] Anderson[86] and Walker and Will.[87] In particular the work of Thorne[20,66] has greatly generalized the scope of Fock's method, and has given rise to a new formalism for computing the gravitational radiation generated by slow motion sources (see the next sub-section).

Besides the main approaches discussed above, some recent work by Winicour[88] (see also Persides[89]) has tried to set up a new approach to deriving the quadrupole equation appropriate to slow moving, weakly self-gravitating sources. The result of this line of work is consistent with the results of Landau-Lifshitz or Fock.

4.8 Thorne's generation formalism

We shall discuss in more detail in §5 some developments along the lines of Thorne's work.[20,66] Suffice it to say here that the basic idea of Thorne's generation formalism is, while keeping the split into two sub-problems (with a "matching" between the two), to treat the second sub-problem (the near-zone one) by a new method which does not rely on any assumption of weakness of the gravitational field within the source. This new treatment relies only on the slowness of the evolution of the source and compares the near-zone external gravitational field to the general stationary vacuum gravitational field. As the latter is fully parametrized by an infinite set of multipole moments, this new treatment of the near-zone sub-problem leads to the possibility of reading off the near-zone field, outside a slow motion source, some slowly evolving multipole moments, say $M^{\text{near-zone}}_{i_1 \ldots i_\ell}$, $S^{\text{near-zone}}_{i_1 \ldots i_\ell}$. Then the matching between the near-zone field and the wave-zone field is done, like in Fock's approach, by introducing some (approximate) intermediate expression for the gravitational field in the "transition" zone $r \sim \lambda$ (see equations (9.31) of reference 66). Essentially, we can summarize the result, at the quadrupole approximation, of Thorne's approach in a way parallel to Fock's result (4.8):

$$Q^{\mathrm{I}}_{ij}(t) = (Q^{\text{intermediate}}_{ij}(t) =) \; M^{\text{near-zone}}_{ij}(t). \qquad (4.9)$$

One of the great advantages of Thorne's quadrupole fomalism (4.9) is that its validity is not a-priori limited to weakly self-gravitating sources like Fock's quadrupole (4.8). It can deal with strongly self-gravitating sources as long as they have only slow internal motions. For instance the source can be a slowly rotating neutron star, of a binary system of well separated black holes or neutron stars. However the price paid for this generality is that the equation(s) (4.9) must still be completed by an independent method for obtaining explicitly the near-zone quadrupole moment of a specific system. One astrophysically important example where this has been possible concerns a system of well separated, slow moving, strongly self-gravitating bodies. Indeed, the work of D'Eath, Kates, Damour and others (see references in §6) allows one to compute independently the near-zone quadrupole moment for such a system. At lowest order it is found that:

$$M_{ij}^{\text{near-zone}}(t) = \sum m[z^i(t)z^j(t) - \frac{\delta^{ij}}{3}\vec{z}^2(t)], \tag{4.10}$$

where m denotes the "Schwarzschild mass" of each body, and $z^i(t)$ its "centre of field." At lowest order, the "positions" $z^i(t)$ of the strongly-self gravitating bodies satisfy the usual Newtonian equations of motion of a system of "point masses" $(m, \vec{z}(t))$ (see §6 below). By replacing (4.10) into (4.9) we get an explicit quadrupole *equation* for the gravitational radiation emitted by a system of strongly self-gravitating bodies. For other investigations allowing one to complete Thorne's generation formalism by obtaining some specific near-zone multipole moment see reference 90. On the other hand in §5 below we shall give some recent results, by Blanchet and myself, which improve the result (4.9) by taking into account higher-order terms. This allows one to show explicitly that the "radiative quadrupole moment" $Q_{ij}^{\mathrm{I}}(u)$ depends not only on the state of the source at the retarded time $u = t - r/c$, but also on the behaviour of the source at all times $< t - r/c$ (see equations (5.28)-(5.30) below).

4.9 Tentative conclusion about the "standard" far-field quadrupole equation

To conclude our account of the derivations of a *general quadrupole equation* appropriate to any *slow moving* sources having *Newtonian internal gravity* we might say that the coincidence of the results of several different approaches makes it practically certain that the gravitational wave emission by such systems is given at lowest order by the "standard" equation (4.5). However a really satisfactory derivation of this equation is still missing, although several partial investigations have suggested ways of improving separately all the bad points of existing derivations. For a fuller critical discussion see reference 74.

4.10 Far-field quadrupole equations for some specific problems

To complete this section we should quote some of the work which has been done on *specific problems*. The gravitational bremsstrahlung radiation emitted during the fast encounter of two Newtonian stars has been treated in detail by Thorne and Kovács[91] (see also Westpfahl[92] and references therein). The methods discussed in §6 show that their result is still valid for the fast encounter of strongly self-gravitating objects. The gravitational radiation emitted by a *slow* binary system of well separated, possibly strongly self-gravitating bodies has been treated by Deruelle[93] by a direct post-Minkowskian method: her result agrees with the result obtained by replacing (4.10) into (4.9). Recalling the above quoted derivation à la Landau-Lifshitz of Futamase,[84] this means that three independent approaches lead to the same quadrupole equation,

$$Q_{ij}^{\mathrm{I}}(t) = \sum m[z^i(t)z^j(t) - \frac{\delta^{ij}}{3}\vec{z}^2(t)], \tag{4.11}$$

for the gravitational wave emission from a slow system of well separated, possibly strongly self-gravitating objects. Another specific problem which has received much attention is a certain free-fall problem posed by Cooperstock; see refs. 73 and 74 for references and a critical discussion, and ref. 111.

4.11 The generalized quadrupole equation of Halpern-Desbrandes and Press

Furthermore, I would like to mention a generalization of the linearized quadrupole equation (4.1) & (4.5) which has been derived by Halpern and Desbrandes[94] and Press.[95] It applies to any *negligibly* self-gravitating source, whatever be its rate of internal evolution or motion. It reads

$$h_{ij}^{\text{wave}}(t,\vec{x}) = \frac{2G}{c^6 r} P_{ijkl}(\vec{n}) \frac{\partial^2}{\partial t^2} \int d^3x' [T_R^{00} - 2T_R^{0r} n_r + T_R^{rs} n_r n_s] x'^k x'^l, \tag{4.12}$$

where $r = (x^i x^i)^{1/2}$, $n_i = x^i/r$ and where the subscript R means that $T^{\mu\nu}$ should be evaluted at the *retarded* event: $T_R^{\mu\nu} = T^{\mu\nu}(t - |\vec{x} - \vec{x}'|/c, \vec{x}')$. The equation (4.12) is valid for sources with negligible internal gravity, and arbitrary internal motion, in the limit where r becomes much greater than the radius of the source. Formally speaking, as suggested by Press,[95] one can even generalise (4.12) to a completely general source having arbitrary internal motion and arbitrary internal gravity (excluding however the presence of a black hole) by using the Landau-Lifshitz trick, i.e. by replacing $T^{\mu\nu}$ by $\tau^{\mu\nu}$. However the resulting equation is not mathematically well defined because of the slow fall-off properties of $\tau^{\mu\nu}$.

4.12 The "far-field multipole law"

It should be mentioned that, with a suitable definition (not linked to the source) of some "radiative multipoles," one can consider that the far-field quadrupole law (4.1) is just one piece of an *exact* law corresponding to the decomposition of rh_{ij}^{wave} into algebraic pieces irreducible under the rotation group. This, by definition exact, "far-field multipole law" reads, with $c = G = 1$, (see Thorne[66,25]):

$$r\, h_{i'j'}^{\text{wave}} = P_{i'j'ij} \left\{ \sum_{l=2}^{\infty} \frac{4}{l!} M_{ij\,L-2}^{I(l)} n^{L-2} + \sum_{l=2}^{\infty} \frac{8l}{(l+1)!} \varepsilon_{ab(i} S_{j)a\,L-2}^{I(l)} n^b n^{L-2} \right\}, \tag{4.13}$$

where L denotes the multi-index $i_1 i_2 \dots i_l$, $L-2$ denotes $i_1 \dots i_{l-2}$, $n^L := n^{i_1} \dots n^{i_l}$, ε_{abc} is the Levi-Civita symbol, $A^{(l)} := d^l A/du^l$, and where $M_L^I(u) = M_{i_1 \dots i_l}^I(u)$ and $S_L^I(u) = S_{i_1 \dots i_l}^I(u)$ are some symmetric and trace-free euclidean tensor functions of one variable which represent the "radiative multipoles" (of electric and magnetic type) of the gravitational field. The lowest order electric-type radiative multipole, M_{ij}^I is nothing but the "radiative" quadrupole moment Q_{ij}^I introduced above. The problem of transforming the (formal) *law* (4.13) into a specific "far-field multipole *equation*" has been tackled by several authors (see the references in Thorne[66]). In the case of slow moving sources with Newtonian internal gravity the use of the same kind of methods as discussed above has led to the following results[66] $(c = 1)$

$$M_{i_1 \dots i_l}^I(t) = \left(\int d^3x\, T^{00}(t,\vec{x}) x^{i_1} \dots x^{i_l} \right)^{STF}, \tag{4.14a}$$

$$S_{i_1 \dots i_l}^I(t) = \left(\int d^3x\, \varepsilon^{i_1 jk} x^j T^{0k}(t,\vec{x}) x^{i_2} \dots x^{i_l} \right)^{STF}, \tag{4.14b}$$

where the superscript STF means that one must take only the symmetric trace-free part with respect to the indices $i_1 \dots i_l$. Thorne[66] has also derived some formally exact equations for the M^I's and S^I's, however they involve non-convergent integrals so that the meaning (and validity) of such formulas is unclear.

4.13 On the definition of the asymptotic outgoing radiation field

Before discussing the other quadrupole laws, it should be remarked that we have written the far-field quadrupole, and multipole, laws (4.1), (4.13) in a form which fits well with the various approaches to the definition of a gravitational wave used in §2, and which, as emphasized ih particular by Thorne,[25] seems well suited for astrophysical studies. However we must still define precisely what was denoted h_{ij}^{wave} in equations (4.1a), (4.13) and relate it to the other investigations of the asymptotic behaviour of radiative fields referred to in §3.1 ("Asymptotic problem").

Although nothing is known for sure, it seems safe to assume that the most general physically relevant solution of the Einstein equations corresponding to an isolated system is such that the following property holds. There exists a coordinate system $x^\mu = (ct, x^i)$ ("radiative coordinate system") such that when $r := |\vec{x}| \to \infty$, with

$$u := t - r/c \tag{4.15}$$

staying fixed, the metric coefficients admit an asymptotic expansion of the type

$$g_{\mu\nu}(t,\vec{x}) = f_{\mu\nu} + \frac{A_{\mu\nu}(t - r/c, \theta, \phi)}{r} + O(1/r^2), \tag{4.16}$$

where $n^i := x^i/r =: (\sin\theta\cos\phi, \sin\theta\sin\phi, \cos\theta)$, and where $A_{\mu\nu}(u,\theta,\phi)$ must satisfy

$$(A_{\mu\nu} - \tfrac{1}{2} f_{\mu\nu} f^{\alpha\beta} A_{\alpha\beta}) k^\nu = C_\mu \tag{4.17}$$

with $k^\mu = (1, n^i)$ and where $C_\mu(\theta,\phi)$ is some time-independent vector satisfying $C_\mu k^\mu = 0$.

Then the part of the metric describing the outgoing radiation can be defined as

$$h_{ij}^{\mathrm{wave}}(t,\vec{x}) := r^{-1} P_{ijkl}(\vec{n}) A_{kl}(t - r/c, \theta, \phi), \tag{4.18}$$

where the projection operator P_{ijkl} has been defined in equation (4.1b).

The preceding "cartesian" way of expressing the lowest order asymptotic behaviour of the metric has been investigated by Papapetrou,[96] Madore,[97] Blanchet[98] and others. The aim of such "cartesian" formulations is to bridge the gap between the "pedestrian" formulations used in finite domains by approximation methods and the more abstract formulations used by many authors working within the Bondi-Penrose approach to the asymptotic problem (referred to in §3.1). Now the explicit link between the "radiative part" of the metric (4.18) and some of the objects used in the Bondi-Penrose approach to describe the outgoing radiation is as follows (see e.g. Madore,[97] Thorne[25]): the asymptotic shear of the outgoing null rays is

$$\sigma^0(u,\theta,\phi) = \tfrac{1}{2} r h_{ij}^{\mathrm{wave}} m^i m^j \tag{4.19}$$

where $\vec{m} = 2^{-1/2}(\vec{e}_\theta + i\vec{e}_\phi)$ is a complex vector tangent to the unit sphere, and the "news function" of Bondi, which fully captures the presence or absence of gravitational radiation, is

$$N = \frac{\partial}{\partial u}\sigma^0 = \tfrac{1}{2} r \frac{\partial h_{ij}^{\mathrm{wave}}}{\partial t} m^i m^j . \tag{4.20}$$

As N is known (see ref. 56) to be an invariantly defined geometrical object at future null infinity, the equation (4.20) indicates that, in fact, only $\partial h_{ij}^{\rm wave}/\partial t$, and not $h_{ij}^{\rm wave}$ itself, constitutes an invariant characterization of the outgoing gravitational radiation.

4.14 Energy-loss quadrupole equations

Contrary to the case of the far-field quadrupole equations, where only approximate (or formal) results are available, there exists one exact result which is directly related to the energy-loss quadrupole law (4.2). Indeed, starting with the pioneering work of Bondi *et al.*,[99] it has been shown that one can define a certain function $M_B(u)$, the Bondi mass, which is a functional of the asymptotic gravitational field, and which satisfies ($c = G = 1$)

$$\frac{dM_B(u)}{du} = -\frac{1}{32\pi}\int\left(\frac{\partial h_{ij}^{\rm wave}}{\partial t}\right)^2 r^2\, d\Omega. \tag{4.21}$$

It is then easily checked that, if one assumes that some far-field quadrupole equation is valid, then equation (4.21) implies a corresponding energy-loss quadrupole equation with

$$E^{\rm II}(u) = M_B(u), \tag{4.22a}$$

and

$$Q_{ij}^{\rm II}(u) = Q_{ij}^{\rm I}(u). \tag{4.22b}$$

However, as discussed above, the catch with this result is that, in order to be able to draw, from the energy-loss quadrupole *equation* (4.2) & (4.22), meaningful predictions about the medium-term evolution of the source, one must still relate the Bondi mass to the structure and motion of the source. This means, for instance, expressing $M_B(u)$ as an explicit functional of the dynamical degrees of freedom of the source. And this functional must be computed, or controlled, with a very great accuracy in order to be sure to include all the terms whose rate of change may be greater that or equal to the (usually very small) right-hand side of (4.2) (for a discussion of this issue see the references quoted in §4.3 above, concerning $E^{\rm II}$). As far as I know this has never been done in detail (see however Anderson[100]).

On the other hand, as a side result of some investigations, e.g. by Chandrasekhar-Esposito, Papapetrou-Linet, Kerlick, Damour, Schäfer or Grishchuk-Kopejkin, concerning the radiation-reaction problem, it has been possible to derive directly from the relativistic dynamics of a self-gravitating source some energy-loss quadrupole *equations*, with $Q^{\rm II}$ equal to the corresponding $Q^{\rm I}$, but with $E^{\rm II}(u)$ given as an explicit functional of the instantaneous kinematical state of the source. Such explicit energy-loss quadrupole equations have been obtained by e.g. Chandrasekhar and Esposito[101] in the case of a slow moving weakly self-gravitating fluid source, and by e.g. Damour[76,67] in the case of a binary system of slow moving strongly self-gravitating objects (see other references in the next sub-section).

Finally, let us mention that replacing in the exact Bondi-energy-loss law (4.21) the general (exact by definition) far-field multipole law (4.19) leads to a correspondingly exact "energy-loss multipole law" which has been worked out by Thorne.[20,66]

4.15 Radiation-reaction quadrupole equations à la Burke-Thorne

The issue of computing the back-reaction effects in a source, correlated with the emission of gravitational radiation, has had a long and checkered history. For a review, see e.g. refs. 67 and 102. Several approaches have been developed to tackle this issue. Some can deal with quite general classes of sources, while others can deal only with rather specific types of sources. Among the general methods, the two main ones are two independent approaches which succeeded, in the late sixties, if not in solving, at least in clarifying the problem by obtaining some explicit radiation-reaction quadrupole equations. The first method was initiated by Burke[103] and Thorne,[104] and the second one, the post-Newtonian method, reached maturity in the work of Chandrasekhar and Esposito.[101]

The Burke-Thorne approach has led to results having the form of the radiation-reaction quadrupole law (4.3). We must however discuss the validity and meaning of the specific radiation-reaction quadrupole *equations* they found. The first result of Burke, obtained in 1969 in his PhD thesis, had been derived under the conditions, not only of slow motion, but also of very weak self-gravity within the source, i.e. under the conditions of applicability of the linearized approximation of General Relativity (like the "standard" derivation of the far-field quadrupole equation, see §4.4). Burke's result in this case was an equation of the form (4.3), with $Q^{\rm III}_{ij}(t)$ given by the right-hand-side of equation (4.5) (i.e. $Q^{\rm III}_{ij} = Q^{\rm I\,standard}_{ij}$) and with $\rho^{\rm III}$ being, say, $c^{-2}T^{00}$ or any quasi-Newtonian mass density, say ρ (it does not matter which one, at this level).

However the direct physical meaning of the other elements appearing in (4.3), i.e. $\mathcal{F}^{\rm reac}_i$ and the coordinate system (ct, x^i), is less clear. Indeed the only thing which can be said is that the coordinate system x^μ has been chosen in such a way that the local equations of motion of the source read

$$\rho\left(\frac{\partial v^i}{\partial t} + v^j\frac{\partial v^i}{\partial x^j}\right) = \mathcal{F}^{\rm Newtonian}_i + \mathcal{F}^{\rm even}_i + \mathcal{F}^{\rm reac}_i, \tag{4.23}$$

where $\mathcal{F}^{\rm Newtonian}_i$ denotes the ordinary force density appropriate to the description of the dynamics of the source in the Newtonian framework (including the Newtonian gravitational forces), and where $\mathcal{F}^{\rm even}_i$ contains all the (special and general) relativistic corrections to $\mathcal{F}^{\rm Newtonian}_i$ which are "time-even," i.e. which do not change sign when reversing the direction of the (coordinate) time.

The problem with equation (4.23) is that, in absence of a more detailed knowledge of the structure and effects of $\mathcal{F}^{\rm even}_i$, it is impossible to draw any firm conclusions from the knowledge of $\mathcal{F}^{\rm reac}_i$ (which is the only thing which is explicitly computed in the Burke-Thorne approach). It is often argued that "because" $\mathcal{F}^{\rm reac}_i$ is "time-odd" (i.e. changes sign under time reversal), while $\mathcal{F}^{\rm even}_i$ is "time-even," then "necessarily" all the "irreversible effects" in the motion of the source can come only from $\mathcal{F}^{\rm reac}_i$, and can be separated in the long term from the "reversible effects" due to $\mathcal{F}^{\rm even}_i$ thanks to their secular nature. These statements sound natural, and certainly correspond to something true, but however, I know of no precise mathematical results which substantiate them. One of the obstacles to the existence of such general mathematical results is, for instance, that the fact that some equations of motion are "time-reversible" does not imply that each of their solutions behaves in a time-symmetric way, but only that to each solution may be associated a new solution by time-reversal. Then, as each individual solution may behave in a quite time-asymmetric way, the identification of some extra "irreversible effects," due to additional time-antisymmetric terms in the equations of motion, is a priori unclear. Moreover in real life, all measurements of the motion of a source are done during a finite lapse of time, say τ. Therefore all phenomena happening on a time scale much bigger than τ can be represented as "secular terms," i.e. short polynomials in the time variable, so that it is, for instance, impossible to distinguish a long-term periodicity from a real irreversible behaviour. All the preceding remarks imply that the Burke-Thorne approach to radiation reaction cannot be considered as fully satisfactory until it is completed by a good control of the structure and effects of the "time-even" forces in equation (4.23).

A further problem with equation (4.23) is that the distinction time-even/time-odd refers to the *coordinate time* and is not invariant under general coordinate transformations. In fact the mathematical expression of $\mathcal{F}_i^{\rm reac}$ can be modified nearly at will by means of suitable changes of the coordinate system (for some examples see refs 109, 110, 112, and eqs (4.24)-(4.26) below). This means that one needs also to control the structure of the metric coefficients in the *same* coordinate system in which (4.23) holds. Indeed, only the simultaneous knowledge of the coordinate motion and of the metric coefficients allows one to compute (in principle) all the quantities that can be observed. Note then that one must control the behaviour of the metric everywhere between the source and the observer. See §6.12 below for an implementation of this operational approach.

The radiation-reaction work of Thorne[104] is still subject to the general criticisms just enounced, but its realm of applicability is different from the original Burke equation. Indeed although the final radiation-reaction quadrupole equation of Thorne[104] has the same form and contains the same quantities, $Q^{\rm III} = Q^{\rm standard}$, $\rho^{\rm III} = \rho$, as the Burke one, its derivation is such that it applies now to any slow non-radial pulsation of a *spherical star having Newtonian internal gravity*. This is an indication that, like for the far-field case, the "standard" (i.e. "linearized") radiation-reaction quadrupole equation, obtained by Burke in 1969, is valid not only for very weakly self-gravitating general sources, but also for general sources having Newtonian internal gravity. Further work of Burke[106] attempted to prove just that, however his reasoning is incomplete and flawed (as well as the corresponding argument in ref. 26). Some flaws in this work have been pointed out by Walker and Will[102] and Blanchet and Damour.[68] After correction of the flaws, the "standard" result, however, comes out unscathed.

Later this approach has been extended in various directions. Kates[107] argued that the radiation-reaction quadrupole equation obtained by replacing in the general form (4.3) $Q_{ij}^{\rm III}$ by the right-hand side of equation (4.11) was applicable to a slow system of well separated, possibly strongly self-gravitating objects. He also studied radiation-reaction in slow axially symmetric systems having possibly strong internal gravity.[108] On the other hand, Blanchet and Damour[68] derived a radiation-reaction *multipole* equation for a slow source having Newtonian internal gravity, in the case where the lowest time-dependent multipoles are of higher order than the usual "electric" (or "mass") quadrupole. A more recent work by Blanchet and myself, reported in §5.10 below, has extended the Burke-Thorne approach by taking explicitly into account the influence of the past-behaviour of the source on the radiation-reaction. It is found that the result can still be put in the form (4.3) but with a "reaction quadrupole" $Q^{\rm III}$ which depends on the past-history of the source. Moreover it is found that this past-dependent "reaction" quadrupole $Q^{\rm III}$ differs also from the corresponding past-dependent "radiative" quadrupole $Q^{\rm I}$.

Although all the preceding work does clarify several aspects of the Burke-Thorne approach, and indicate its ability to deal with various types of sources, the reader must be warned that this whole approach to gravitational radiation-reaction is based on a method of "matching of asymptotic expansions" which, as said in §3.2 above, has not yet been given a clear formal setting. Therefore I think that the following cautious sentences written by Thorne[104] in 1969 are, on the whole, still applicable today: "It should be obvious from the above discussion that the theory of wave-zone coupling in post-Newtonian expansions will not be on a completely firm footing until it has been investigated systematically, order by order. The present analysis and that of Burke (1969a,b) serve only to delineate the main outline of the theory and some of the key terms."

4.16 Post-Newtonian radiation-reaction quadrupole equations

The post-Newtonian approach to gravitational radiation reaction has led to more complete results than the Burke-Thorne approach, in the sense that it gives a complete knowledge of all the terms of the local equations of motion (4.23). Therefore, in principle, one does not have to rely on semi-heuristic arguments to draw consequences from (4.23). Moreover, the problem of separating the effect of $\mathcal{F}_i^{\rm reac}$ from the effect of $\mathcal{F}_i^{\rm even}$ is greatly helped by the existence of several explicit functionals of the source variables, which come up naturally in the post-Newtonian approach, and which would be constant if $\mathcal{F}_i^{\rm reac}$ were equal to zero. As for the explicit form of $\mathcal{F}_i^{\rm reac}$, it has been first obtained by Chandrasekhar and Esposito.[101] It has a more complex form than the Burke-Thorne quadrupole law (4.3) but Miller[109] has shown how to reduce it to the form (4.3) by means of a suitable coordinate transformation (see also Schäfer[110]). Much further work has clarified and systematized this post-Newtonian approach to gravitational radiation reaction (see references in 54, 67, 102). All authors basically agree on the end result, of the type (4.23), thereby confirming and completing the Burke-Thorne result. However none of the derivations of these results can be considered as fully satisfactory at present. As in the case of the far-field problem, ways have been suggested to improve the existing derivations but they have not yet been combined and implemented in a fully convincing manner. The problem is in fact more difficult than in the far-field case because the mere definition of what one is looking for, $\mathcal{F}_i^{\rm reac}$, is somewhat obscure.

4.17 Radiation-reaction quadrupole equations in the N-body problem

As for the methods dealing with more specific sources, the results are somewhat more satisfactory, especially in the case of slow moving systems of well separated objects where it has been possible to compute the complete equations of motion of the "centres of mass" of the objects, up to the level where radiation damping effects first appear. For a discussion of the work of Damour and collaborators, Schäfer,[77] and Grishchuk and Kopejkin[78] see e.g. ref. 62. We shall discuss below, in §6, the problem of the motion of two strongly-self-gravitating bodies, where fully explicit results have been obtained. In the latter problem one can separate clearly the effects of the net "radiation reaction force" acting on each body, $F_i^{\rm reac}$, by showing that the equations of motion containing only $F_i^{\rm even}$ can be deduced from an (approximately) Poincaré-invariant variational principle. The result for $F_i^{\rm reac}$ can be written in terms of the "Schwarzschild mass," m, the "centre-of-field," z^i, of each body (and $v^i := dz^i/dt$, $a^i := dv^i/dt$), the mass moment, I_{ij}, the quadrupole moment, $Q_{ij} = I_{ij} - \frac{1}{3}I_{ss}\delta_{ij}$, and two other moments of the system, as follows

$$F_i^{\rm reac} = m\frac{G}{c^5}\{\tfrac{3}{5}z^j Q_{ij}^{(5)} + 2v^j I_{ij}^{(4)} + \tfrac{10}{3}a^i I^{(3)} + \tfrac{1}{5}I_{iss}^{(5)} - J_{iss}^{(4)}\} \tag{4.24}$$

(see §6 below for references, see also equations (6.23) for the precise definition of the various moments, and eq. (6.19) for the explicit value of $F_i^{\rm reac}$). It can be shown[112] that there exists a suitable change of coordinates (within the system), $x^\mu \to x'^\mu$, such that the new reaction force (i.e. the t'-odd term in the equations of motion of $z'^i(t')$) has the Burke-Thorne form:

$$F_i'^{\rm reac} = m\frac{G}{c^5}\left\{-\frac{2}{5}z'^j Q'^{(5)}_{ij}\right\}. \tag{4.25}$$

The coordinate transformation $x^\mu \to x'^\mu$ is given by eq. (13) of ref. 112 where the functions $\psi_k(t)$, left undetermined by Linet, are obtained from eq. (4.24): $\psi_k(t) = Gc^{-5}(\frac{1}{5}I_{iss}^{(3)} - J_{iss}^{(2)})$. This coordinate transformation does not involve any terms that are cumulative in time ("secular terms"), however it contains terms that blow up in space. This may mean that the coordinates x'^μ in which (4.25) holds are valid *only in the near zone*. If this is the case, the equation (4.25), considered by itself, does not allow one to make any firm observational prediction about the effects of radiation reaction, because the observations are usually made far away from the system (remember, to give a different example, that the use of a local rotating coordinate frame can modify at will the coordinate advance of the periastron of a binary system). It is therefore better to study the effects of $F_i^{\rm reac}$ in the initial coordinate system x^μ which is well-behaved both in the near zone and in the far zone. Other well-behaved coordinates are the ADM coordinates x''^μ used by Schäfer[77] in his study of N-body systems. Again, he can clearly separate the effects of $F_i''^{\text{even}}$ which corresponds to a Hamiltonian system, and his result for the reaction force is:

$$F_{ai}''^{\text{reac}} = m_a \frac{G}{c^5} \left\{ \frac{2}{5} Q''^{(3)}_{jk} \sum_{b \neq a} \frac{\partial}{\partial z''^i_a} \left(\frac{G m_b n''^j_{ab} n''^k_{ab}}{r''_{ab}} \right) + \frac{4}{5} \frac{d}{dt''} (v''^j_a Q''^{(3)}_{ij}) \right\}. \tag{4.26}$$

By a suitable change of coordinates (within the system), $x''^\mu \to x'^\mu$, the reaction force (4.26) can be transformed[110] into the form (4.25). The same remarks as above apply to the domain of validity of the x'^μ.

In §6.12 below we shall free ourselves of all the coordinate ambiguities attached to the definition of the "radiation-reaction force" by computing some directly observable quantities in which appear explicitly some of the effects that one would expect, heuristically, to "happen" in reaction to the emission of radiation (see equations (6.30)-(6.41)).

4.18 Conclusion

To summarize our presentation of the quadrupole moment formalism, we can say that a potential user of any of the specific quadrupole equations should be well aware

(i) of the many difficulties that still beset the derivations of these equations,

(ii) of their restricted domains of applicability and,

(iii) of the conceptual subtleties that arise as regards their precise meaning (for further discussion and other viewpoints see e.g. refs. 24, 25, 73, 74, 111).

That warning being made, it is clear that the three quadrupole laws represent a useful set of relations that can probably be rightly applied to many astrophysical situations, and which, even in situations where their use is not a priori justified, will often give the main features of gravitational radiation effects.

At present, the only astrophysical situation where the application of one of the quadrupole equations has led to a definite and successful prediction (concerning the evolution of a gravitationally radiating source) is the binary pulsar PSR 1913+16. However, when it was first done, the application of the quadrupole moment formalism to this case could only be considered as heuristic, because none of the then existing derivations were applicable to a system of strongly-self-gravitating bodies, and also because, as explained above, none of the quadrupole laws are precise enough to allow one to predict the medium-term kinematical evolution of a binary system. And, strangely enough, after several years of intense theoretic activity, developed to clarify these issues, one cannot say that any definite progress has been made towards deriving, in a satisfactory way, *general* quadrupole laws and showing that they can be meaningfully employed to predict the medium-term evolution of the binary pulsar. In my opinion the progress which has led, in a more satisfactory way, to an a posteriori justification of the above mentioned application of the quadrupole laws, has been made in the general relativistic problem of motion of a binary system, which is the real problem posed by the binary pulsar (see §6 below). Let us hope however that we shall not have to wait for a direct detection of gravitational radiation for the theorists to concoct more satisfactory direct justifications of the quadrupole laws, including perhaps some theorems. As indicated by the recent interesting results of Winicour,[88] this may be done not by trying to improve the existing "derivations," but by finding completely new approaches.

§5. MULTIPOLAR-POST-MINKOWSKIAN FORMALISMS

5.1 Introduction

As we said in §3.1 above the main problems of gravitational radiation theory remain still unsolved. Moreover it is clear that they are all interrelated, because, for instance, no convincing answer can be given to the asymptotic problem if one does not control the generation and/or the propagation problems. In this section we shall discuss an approximation method which has the nice feature of being able, in principle, to control analytically the structure of the gravitational field generated by an isolated source in a large domain of spacetime, namely in the region exterior to a spacetime tube, of radius r_0, enclosing the source. The only a priori restriction on r_0, beyond the fact that it must exceed the radius of the source, is that it must be much larger than $(G/c^2)\times$(mass of the source), so that in the domain

$$D := \{(\vec{x}, t) \mid r > r_0\}, \tag{5.1}$$

where $r := (x^i x^i)^{1/2}$, the gravitational field is everywhere weak. But no restriction is placed on the time variability of the field, so that r_0 can have any relation with a typical wavelength of the emitted radiation: $r_0 \ll \lambda$, $r_0 \sim \lambda$, or $r_0 \gg \lambda$.

Essentially the method consists of combining a post-Minkowskian approximation scheme, i.e. an expansion in powers of

$$\gamma_e = \frac{Gm}{c^2 r_0} \tag{5.2}$$

(m being the mass of the source), with a multipole expansion, i.e. the decomposition of each term of the post-Minkowskian expansion into a series of algebraic pieces irreducible under the (coordinate) rotation group.

By itself this method can, in principle, provide a complete answer to the propagation problem, and at least a partial answer to the asymptotic problem. By "complete" we mean here, complete within the considered approximation scheme, and by "partial" we mean that the method can aid in the study of the influence of non-linear and non-local effects on the asymptotic behaviour of the field, but that it must be completed by some further information about the evolution of the gravitational field within the domain $r \leq r_0$, and the link of the latter with the evolution of the source. Therefore, as was said in §3.2, we need to combine this method with another approximation method dealing with the source itself, if we want to be able the tackle the generation problem, and also the radiation-reaction problem.

The idea of the method, and its first implementation, are due to Bonnor.[113] It was later developed by Bonnor and his collaborators[114] (under the name of the "double series method"). Then Thorne[66] clarified the formal structure of the method, extended its scope, and used it as a tool to investigate several aspects of the theory of gravitational radiation: general structure of radiating gravitational fields in the near-zone, wave generation formalism for slow motion sources, ... The method has been used by Thorne and his collaborators[90,116,119] in several investigations of radiative, or stationary, gravitational fields. Blanchet and Damour[115,68,59] have set up a well-defined formal framework allowing one to implement algorithmically the method, and to study analytically the most general Multipolar-Post-Minkowskian expandable vacuum metric. We shall sketch in the sub-sections 5.2–5.6 the latter formal framework and in the sub-sections 5.7-5.10 we shall present some of the recent results which have been obtained by means of Multipolar-Post-Minkowskian methods (for other results see the references in this sub-section, and in the following ones).

A Multipolar-Post Minkowskian expandable metric, or, in short, an MPM metric, is a formal series in powers of Newton's constant G (standing for γ_e of equation (5.2))

$$\mathrm{g}^{\alpha\beta}(x^\mu) := g^{1/2} g^{\alpha\beta} = f^{\alpha\beta} + G h_1^{\alpha\beta} + \ldots + G^n h_n^{\alpha\beta} + \ldots, \tag{5.3}$$

such that each term of the series, $h_n^{\alpha\beta}(x^0, x^i)$, admits a multipolar expansion associated with the $O(3)$ group of rotation of the spatial coordinates, i.e.

$$h_n^{\alpha\beta}(x^\mu) = \sum_{l\geq 0} h_{nL}^{\alpha\beta}(r,t)\hat{n}^L(\theta,\phi). \tag{5.4}$$

We use the following notations: $r := (\delta_{ij}x^i x^j)^{1/2}$, $t := x^0/c$; L denotes the multi-index $i_1 i_2 \ldots i_l$, $n^L := n^{i_1} n^{i_2} \ldots n^{i_l}$ with $n^i := x^i/r$, $\hat{n}^L$ being the symmetric trace-free (in short STF) part of n^L. Beware of a change of notation: $h_1^{\alpha\beta}$ is not the same as the $h_{\alpha\beta}$ of equations (2.15)–(2.16) but rather $h_1^{\alpha\beta} = -f^{\alpha\mu} f^{\beta\nu} h_{\mu\nu} + \frac{1}{2} f^{\alpha\beta} f^{\mu\nu} h_{\mu\nu}$. The sum appearing in the right-hand side of equation (5.4) is equivalent to an expansion in usual spherical harmonics $Y_l^m(\theta,\phi)$ (for detailed discussions of the expansions using STF tensors, including the link between the "orbital" expansion (5.4) and a fully irreducible tensor spherical harmonics expansion see refs. 66 and 59). Strictly speaking the formal implementation of the method is clearly defined only if one imposes that all the multipolar expansions are finite, however at the end of the day one can let them increase ad infinitum, and as a multipolar expansion is convergent under weak assumptions (see ref. 59, appendix B) one expects to reach, by this limit, a sufficiently generic metric.

One then looks for such MPM metrics which satisfy (in the sense of a formal series) the *vacuum* Eistein equations

$$R_{\alpha\beta}(\mathrm{g}^{\mu\nu}(x^\lambda)) = 0, \tag{5.5}$$

which were stationary in the past, i.e. such that there exists a time T such that,

$$\left(t \leq -T\right) \implies \left(\frac{\partial}{\partial t}\mathrm{g}^{\alpha\beta}(x^i,t) = 0\right), \tag{5.6}$$

and which were asymptotically Minkowskian at spatial infinity in the past, in the weak sense that

$$\left(t \leq -T\right) \implies \left(\lim_{r\to\infty} \mathrm{g}^{\alpha\beta}(x^i,t) = f^{\alpha\beta}\right). \tag{5.7}$$

All conditions (5.5)–(5.7) are assumed to hold in some domain D, eq. (5.1), outside the source. The condition of past-stationarity (5.6) is intended to be only preliminary: it is first imposed to permit the construction of a class of metrics which, by any criteria, contain no incoming radiation, but then we shall take (when it exists) the limit $-T \to -\infty$, thereby defining a class of *retarded* metrics that we wish to associated to physical sources that were never stationary in the past (like a N-body system for instance).

5.3 The hierarchy of equations to be solved

Replacing the formal series (5.3) into the vacuum Einstein equations (5.5) leads to the following sequence of equations:

$$\Box h_n^{\alpha\beta} = \partial^\alpha H_n^\beta + \partial^\beta H_n^\alpha - f^{\alpha\beta}\partial_\mu H_n^\mu + N_n^{\alpha\beta}(h_1,\ldots,h_{n-1}), \tag{5.8}$$

where $\Box$ is the flat-space-time wave operator, where H_n^α denotes the "harmonicity" of $h_n^{\alpha\beta}$, i.e.

$$H_n^\alpha := \partial_\beta h_n^{\alpha\beta}, \tag{5.9}$$

and where $N_n^{\alpha\beta}$ is a complicated nonlinear algebraic function of the previous h_m $(m < n)$ and their first and second partial derivatives (with $N_1^{\alpha\beta}$ being identically zero). The indices in equation (5.8) and below are raised by means of the flat metric $f^{\alpha\beta}$.

Let us make now the further restriction that there exist harmonic coordinates all over the domain D, so that we can impose

$$H_n^\alpha = 0. \tag{5.10}$$

This condition of harmonicity is not essential in the method, it can be relaxed afterwards, or it can be replaced by other convenient conditions (as in Blanchet[98]).

We then obtain the following hierarchy of conditions to be satisfied by the $h_n^{\alpha\beta}$ $(n = 1, 2, \ldots)$:

$$\Box h_n^{\alpha\beta} = N_n^{\alpha\beta}(h_1, \ldots, h_{n-1}), \tag{5.11a}$$

$$\partial_\beta h_n^{\alpha\beta} = 0, \tag{5.11b}$$

$$h_n^{\alpha\beta} = \sum_{l\geq 0} h_{nL}^{\alpha\beta}(r,t)\hat{n}^L, \tag{5.11c}$$

$$\left(t \leq -T\right) \implies \left(\frac{\partial h_n^{\alpha\beta}}{\partial t} = 0\right), \tag{5.11d}$$

$$\left(t \leq -T\right) \implies \left(\lim_{r\to\infty} h_n^{\alpha\beta} = 0\right). \tag{5.11e}$$

The aim of the MPM method is to construct, as explicitly as possible, the *most general solution* of the hierarchy of equations (5.11).

5.4 The first step of the hierarchy

The most general solution of the first step of the hierarchy, i.e. the most general past-stationary, asymptotically-Minkowskian, linearized harmonic metric in vacuum has been obtained by several authors (see references 66, 59 and references therein). It can be written as

$$h_1^{\alpha\beta}[\mathcal{M}, \mathcal{W}] = h_{\text{can.}1}^{\alpha\beta}[\mathcal{M}] + \partial^\alpha w^\beta[\mathcal{W}] + \partial^\beta w^\alpha[\mathcal{W}] - f^{\alpha\beta}\partial_\mu w^\mu[\mathcal{W}], \tag{5.12}$$

with

$$h_{\text{can.}1}^{00}[\mathcal{M}] = -\frac{4}{c^2}\sum_{l\geq 0}\frac{(-)^l}{l!}\partial_L\{r^{-1}M_L(t-r/c)\}, \tag{5.13a}$$

$$h_{\text{can.}1}^{0i}[\mathcal{M}] = +\frac{4}{c^3}\sum_{l\geq 1}\frac{(-)^l}{l!}\partial_{L-1}\{r^{-1}M_{iL-1}^{(1)}(t-r/c)\}+$$
$$+\frac{4}{c^3}\sum_{l\geq 1}\frac{(-)^l l}{(l+1)!}\varepsilon_{iab}\partial_{aL-1}\{r^{-1}S_{bL-1}(t-r/c)\}, \tag{5.13b}$$

$$h_{\text{can.}1}^{ij}[\mathcal{M}] = -\frac{4}{c^4}\sum_{l\geq 2}\frac{(-)^l}{l!}\partial_{L-2}\{r^{-1}M_{ijL-2}^{(2)}(t-r/c)\}-$$
$$-\frac{8}{c^4}\sum_{l\geq 2}\frac{(-)^l l}{(l+1)!}\partial_{aL-2}\{r^{-1}\varepsilon_{ab(i}S_{j)bL-2}^{(1)}(t-r/c)\}, \tag{5.13c}$$

and with

$$w^0[\mathcal{W}] = \sum_{l\geq 0} \partial_L\{r^{-1}W_L(t-r/c)\}, \tag{5.14a}$$

$$w^i[\mathcal{W}] = \sum_{l\geq 0} \partial_{iL}\{r^{-1}X_L(t-r/c)\}+$$

$$+\sum_{l\geq 1}\left[\partial_{L-1}\{r^{-1}Y_{iL-1}(t-r/c)\} + \varepsilon_{iab}\partial_{aL-1}\{r^{-1}Z_{bL-1}(t-r/c)\}\right] \tag{5.14b}$$

In equations (5.13)–(5.14) the notation $L-1$ represents the multi-index $i_1i_2\ldots i_{l-1}$, the superscript in parenthesis denotes a differentiation with respect to the (retarded) time, and $T_{(ij)} := \frac{1}{2}(T_{ij}+T_{ji})$. All the spatial tensors ($M_L := M_{i_1\ldots i_l}$, $S_L := S_{i_1\ldots i_l}$, W_L, X_L, Y_L, Z_L) which appear in equations (5.13)–(5.14) are (time-dependent) symmetric-trace-free (STF) cartesian tensors. The letter $\mathcal{M}$ (which stands for Mutipole Moments) denotes the set of functions $\{M_L(u), S_L(u)\}$ for $l = 0,1,2,\ldots$, the letter $\mathcal{W}$ denotes the set $\{W_L$, X_L, Y_L, $Z_L\}$, and the square brackets around $\mathcal{M}$ or $\mathcal{W}$ indicate a functional dependence on the sets of functions $\mathcal{M}$ or $\mathcal{W}$. Finally the only constraints that the $\mathcal{M}$ and $\mathcal{W}$ must satisfy is that all the tensors $M_L(u),\ldots,Z_L(u)$ must become constant when $u \leq -T$, and that the M_L and the S_L for $l \leq 1$ are always time-independent.

It is clear from equation (5.12) that the $\mathcal{W}$-terms can be eliminated by the (linearized) coordinate transformation $\delta x^\alpha = Gw^\alpha[\mathcal{W}]$ (which preserves the harmonic property) so that the physical content of the general linearized metric is fully contained in Thorne's[66] "canonical" metric $h^{\alpha\beta}_{\text{can.1}}[\mathcal{M}]$, and therefore in the set of time-varying cartesian tensors $\mathcal{M} = \{M_{i_1\ldots i_l}(u), S_{i_1\ldots i_l}(u)\}$.

5.5 On the meaning of the "algorithmic multipole moments"

At this stage of the algorithm it is important to warn the reader that the quantities $M_L(u)$, $S_L(u)$ have no direct physical meaning. They play the role of functional parameters allowing one to give a convenient mathematical representation of the linearized metric. Later, when a recursive algorithm, generating all the higher-order nonlinear corrections $h_2^{\alpha\beta}, h_3^{\alpha\beta},\ldots$ from $h_1^{\alpha\beta}$, will have been defined, they will still play only the role of *arbitrary functional parameters* in the general fully nonlinear vacuum metric. In other words, this means that the general MPM vacuum metric will be represented as a complicated nonlinear functional of an arbitrary set of time-varying tensors $M_L(u)$, $S_L(u)$. Therefore the latter quantities have only an indirect physical meaning, but they are useful technical intermediaries within an algorithm which will take care of all the nonlinear and nonlocal propagation effects of the gravitational field. We shall call $\mathcal{M} = \{M_L(u), S_L(u)\}$ the set of the *algorithmic multipole moments.*

This point having been clarified, it can now be said that the coefficients in equations (5.13) have been chosen (by K.S. Thorne[66]) to be such that when the source happens to be slowly varying and weakly self-gravitating then the algorithmic multipole moments turn out to be approximately equal to the usual Newtonian multipole moments of the source, i.e. the right-hand sides of equations (4.14). We shall report below some recent work which improves this result by taking into account higher order terms. It should also be mentioned that in the case of a stationary gravitational field the algorithmic multipole moments coincide with the geometrically well-defined asymptotic multipole moments of Geroch-Hansen and Thorne (see e.g. refs. 66, 116 and references therein).

Let us outline the construction of the most general solution of the hierarchy (5.11). The general solution of the first step h_1 is given by equations (5.13)–(5.14) where $\mathcal{M}$ and $\mathcal{W}$ are arbitrary sets of tensors, submitted to the sole constraints that $M_L(u), \ldots, Z_L(u)$ become constant when $u \leq -T$, and that the M_L's and the S_L's for $l \leq 1$, i.e. M, M_i, and S_i are always time independent (there is no S for $l = 0$, see equation (5.13b)).

Knowing h_1, one computes the effective nonlinear source $N_2(h_1)$ for h_2. It is then clear that if we can construct a *particular* solution of the equations

$$\Box h_2^{\alpha\beta} = N_2^{\alpha\beta}(h_1), \tag{5.15}$$

$$\partial_\beta h_2^{\alpha\beta} = 0, \tag{5.16}$$

which satisfies also the conditions (5.11c)–(5.11e), the most general h_2 will be obtained by adding the general homogeneous solution of (5.15), (5.16), (5.11c)–(5.11e), which is nothing but the general solution of the first step of the hierarchy discussed in §5.4. Now in order to find a particular h_2, let us first try to solve equation (5.15). The "source term" $N_2(h_1)$, when written explicitly, is a (finite) sum of the type

$$N_2(h_1) = \sum \hat{n}^L \cdot r^{-k} \cdot F(t - r/c), \tag{5.17}$$

where $F(u)$ is a quadratic local-in-time functional of the algorithmic moments $\mathcal{M}$ and $\mathcal{W}$, and is therefore constant when $u \leq -T$ (by *local in time* functional we mean that $F(u)$ is of the type, say, $M_{L_1}^{(p_1)}(u) \times M_{L_2}^{(p_2)}(u)$, with $M_L^{(p)}(u) := d^p M_L(u)/du^p$, in contrast to a *retarded* functional, say, $G(u) = \int_{-\infty}^{u} dv M_{L_1}^{(p_1)}(v) M_{L_2}^{(p_2)}(v)$).

We cannot solve the inhomogeneous wave equation (5.15), subject to the constraint of past-stationarity (5.11d), by means of the ordinary retarded integral

$$(\Box_R^{-1}(N_2))(\vec{x}, t) := -\frac{1}{4\pi} \int_{\mathbf{R}^3} d^3x' \, |\vec{x} - \vec{x}'|^{-1} N_2(\vec{x}', t - |\vec{x} - \vec{x}'|/c), \tag{5.18}$$

because the domain of definition of (5.15) is the open domain $D = \{(\vec{x}, t) \,|\, r > r_0\}$ instead of the full $\mathbf{R}^4$. We could solve (5.15) by means of a truncated retarded integral (with the cut-off $r > r_0$ in the right-hand side of (5.18)), however it is mathematically more convenient, and it will yield a more powerful algorithm, to use the following auxiliary mathematical technique. Let us first remark that the right-hand side of equation (5.17) allows one to extend the definition of N_2 from D to $\mathbf{R}^3_* \times \mathbf{R}$, i.e. to $\mathbf{R}^4$ deprived only of the time axis, $r = 0$. However N_2 tends to infinity as some negative power of r when $r \to 0$ so that we still cannot use the ordinary retarded integral. However if B is a complex number whose real part, $\Re e(B)$, is large enough, the product $(r/b)^B \times N_2$, where b is some fixed length scale, will define a continuous function on the whole $\mathbf{R}^4$. Therefore the ordinary retarded integral of $(r/b)^B N_2$, $\Box_R^{-1}((r/b)^B N_2)$, is well defined if $\Re e(B)$ is large enough (actually one must also be careful about the behaviour for $r \to \infty$; this is achieved by separating N_2 into a constant part and a past-zero one, see ref. 59 for details). Now it is easily proven that the structure of the right-hand side of equation (5.17) guarantees, if the functions $F(u)$ are sufficiently smooth, that, for a fixed "field point" $(\vec{x}, t)$ in the retarded integral (5.18), the function of B,

$$I(B) := \Box_R^{-1}((r/b)^B N_2), \tag{5.19}$$

(originally defined only for $\Re e(B)$ large enough) is a complex analytic function of the complex variable B, which can be analytically continued all over $\mathbf{C}' := \mathbf{C} - \mathbf{Z}$ (with simple poles near some points of $\mathbf{Z}$). In particular we can consider the *finite part* (zeroth power of B) of the Laurent expansion of $I(B)$ near $B=0$. Let us call it $p_2^{\alpha\beta}$:

$$p_2^{\alpha\beta} := \underset{B=0}{\text{Finite Part}} \, \Box_R^{-1}\left((r/b)^B N_2^{\alpha\beta}(h_1)\right). \tag{5.20}$$

One shows that $p_2^{\alpha\beta}$ is a particular solution of equation (5.15), and is stationary in the past. But $p_2^{\alpha\beta}$ does not satisfy (5.16). However the contracted Bianchi identities (which imply $\partial_\beta N_2^{\alpha\beta}=0$) can be used to show that the "harmonicity" of $p_2^{\alpha\beta}$, $\partial_\beta p_2^{\alpha\beta}$, has the same structure as w^α, equation (5.14), with some $\mathcal{W}$ algorithmically constructed from the $F(u)$'s of (5.17) (and therefore from the algorithmic multipoles of h_1). Then it is easy to construct a complementary term $q_2^{\alpha\beta}$ (with a structure similar to, but more general than, the one of $h_{\text{can}.1}$, equation (5.13)) such that

$$\Box q_2^{\alpha\beta} = 0, \tag{5.21}$$

$$\partial_\beta q_2^{\alpha\beta} = -\,\partial_\beta p_2^{\alpha\beta}, \tag{5.22}$$

and that $q_2^{\alpha\beta}$ is stationary in the past (see refs. 115 and 59 for the explicit construction of q_2). Let us mention that q_2 has the structure of the right-hand side of equation (5.17) but with, now, some of the $F(u)$'s being *retarded* functionals of the algorithmic multipole moment, of the type $G(u)=\int_{-\infty}^{u} dv\, M_{L_1}^{(p_1)}(v) M_{L_1}^{(p_1)}(v)$, and even of the type $H(u)=\int_{-\infty}^{u} dv\, G(v)$.

Then the algorithm consists of defining

$$h_{\text{can}.2}^{\alpha\beta} := p_2^{\alpha\beta} + q_2^{\alpha\beta}\,, \tag{5.23}$$

which is indeed a particular solution of (5.15)–(5.16) satisfying the constraints (5.11c)–(5.11e).

This algorithmic construction can be generalized to all the higher orders, $n=3,4,\ldots$, with the only difference being that the simple structure (5.17) for the "source terms" N_n has to be replaced by a structure which is more subtle (appearance of logarithms) and less precise (presence of a remainder term). Also one must recursively prove that if N_n possesses this generalized structure (expressible by saying that N_n belongs to a special class of functions L^{n-2}) then $N_{n+1}(h_1,\ldots,h_n)$ (with h_n constructed from N_n) belongs to the class L^{n-1}. For details of this algorithm see Blanchet and Damour.[59]

The final result is that, given an arbitrary set of time-varying "algorithmic multipoles" $\mathcal{M}=\{M_L(u),S_L(u)\}$ (constant if $l\le 1$), one can algorithmically construct from $\mathcal{M}$ a Multipolar-Post-Minkowskian "canonical" metric,

$$\begin{aligned} \mathbf{g}_{\text{can}.}^{\alpha\beta}[\mathcal{M}] = f^{\alpha\beta} &+ G h_{\text{can}.1}^{\alpha\beta}[\mathcal{M}] + G^2 h_{\text{can}.2}^{\alpha\beta}[\mathcal{M}] + \ldots + \\ &+ G^n h_{\text{can}.n}^{\alpha\beta}[\mathcal{M}] + \ldots \end{aligned} \tag{5.24}$$

which satisfies the equations (5.5)–(5.7) (and the harmonicity condition, $\partial_\beta \mathbf{g}^{\alpha\beta}=0$) in $\mathbf{R}_*^3\times\mathbf{R}$. Moreover one proves that this "canonical" metric contains the general solution of the problem, in the sense that the general MPM (harmonic) vacuum metric can be obtained from some canonical metric by a (harmonicity preserving) coordinate transformation.

One can mention that the utility of the algorithm comes not only from the flexibility of the method of complex analytic continuation, but also from the fact that it has been possible to derive a remarkably simple formula for the retarded integral of a general multipolar extended source (see §6 of ref. 59). The combination of these two results allows one to get a fully explicit expression for $h_2^{\alpha\beta}$, and also for most of the terms in $h_3^{\alpha\beta}, h_4^{\alpha\beta}, \ldots$. However, although one has a good analytical control of the construction of each $h_n^{\alpha\beta}$, the metric $\mathbf{g}_{\text{can.}}[\mathcal{M}]$ is a complicated retarded functional of the algorithmic multipoles. Therefore it is not a trivial matter to study the dependence of $\mathbf{g}_{\text{can.}}$ on the spacetime variables $(\vec{x}, t)$, and/or the conditions under which the limit $-T \to -\infty$ can be taken and what class of metrics it leads to. In principle, one can use $\mathbf{g}_{\text{can.}}[\mathcal{M}]$ to tackle (in the sense of perturbation theory) many of the basic problems of gravitational radiation theory, but in practice this is a difficult task. In the following sub-sections we shall present some partial or preliminary answers to these problems.

5.7 *Partial results on the asymptotic problem*

As recalled in §3.1, there are several different interesting asymptotic regions of spacetime for a metric corresponding to an isolated source. Two of them are of special interest in gravitational radiation theory: past null infinity where one hopes to define (and exclude) any incoming radiation, and future null infinity where one wants to study outgoing radiation. The previous algorithm, if considered before taking the limit $-T \to -\infty$, gives rise to an uninteresting asymptotic structure at past null infinity because the metric is stationary there, so that any sensible definition of the "incoming radiation" will lead to the result that the general MPM metric does not contain any incoming radiation. It would be very interesting to investigate what happens if the limit $-T \to -\infty$ is taken before the limit of going to past null infinity. This is a task for future work.

As for the asymptotic structure at future null infinity, it is not conveniently investigated with the form of the general MPM metric generated by the algorithm sketched in §5.6. Indeed it is well known that, *in harmonic coordinates,* the asymptotic structure of the metric must involve logarithms (see e.g. refs. 81, 97). This is indeed what is found.[59] These logarithms, although negligible in practical considerations, become a nuisance in theoretical investigations because they come from several different places (the logarithmic deviation of curved light-cones with respect to flat ones, and also the genuine Fock-type logarithms), and because they numerically dominate the asymptotic expansion of the metric in the exponentially far wave-zone, equation (3.6). The cure to these problems is, in principle, well-known. One must try to go from harmonic coordinates to "radiative coordinates"[96,97,98] of the type of the coordinates used in §4.13; i.e., more precisely, a (non harmonic) system of coordinates,

$$(\hat{x}^\mu) = (c\hat{t}, \hat{x}^i) = (c\hat{u} + \hat{r}, \hat{r}\sin\hat{\theta}\cos\hat{\phi}, \hat{r}\sin\hat{\theta}\sin\hat{\phi}, \hat{r}\cos\hat{\theta}), \tag{5.25}$$

such that the related coordinates $(\hat{u}, \hat{\Omega}, \hat{\theta}, \hat{\phi})$, with $\hat{\Omega} := \hat{r}^{-1}$, are good Penrose-type coordinates.[117] This means that the conformally rescaled metric,

$$d\tilde{s}^2 = \hat{\Omega}^2 \hat{g}_{\alpha\beta}\, d\hat{x}^\alpha d\hat{x}^\beta, \tag{5.26}$$

is regular, when expressed in terms of $\hat{u}, \hat{\Omega}, \hat{\theta}, \hat{\phi}$, including at $\hat{\Omega} = 0$ i.e. at future null infinity. By regularity we mean here, as in our discussion of the Lichnerowicz conditions in §2.1 above, that the metric coefficients are p-times continuously differentiable in the *closed* half-space $\hat{\Omega} \geq 0$, for some p (as said in §2.1 this condition can also be expressed in other ways).

It has been shown by Blanchet[98] that, for the most general MPM metric generated by the algorithm of §5.6, there existed a coordinate transformation, $x^\mu \to \hat{x}^\mu$, going from the original harmonic coordinates to radiative coordinates, and that the corresponding conformally rescaled metric (5.26) was *smooth* at future null infinity (i.e. of class C^∞ in the half-space $\hat{\Omega} \geq 0$). Actually he found it more convenient to proceed in an indirect manner, by, first, defining a new algorithm which automatically constructs a general "radiative" MPM metric, and by proving only afterwards that each new radiative MPM metric is diffeomorphic to some old harmonic one. This result is interesting because it bridges the gap between an approximation method which can be used in a region quite near to a general source, and the elegant geometrical investigations à la Bondi-Penrose, of the asymptotic gravitational field. Before this work, only partial results had been obtained by Bonnor and coworkers.[113,114]

However, it must be emphasized that the preceding result has been obtained under the strong assumption of stationarity of the metric before some finite date $(-T)$ in the past, which means physically that the source itself was stationary prior to $-T$. Now, several authors (Bardeen-Press, Schmidt-Stewart, Walker-Will, Porrill-Stewart, and myself, see refs. in ref. 58) have stressed that the behaviour of the source in the remote past is very important for the regularity (or lack of regularity) of the asymptotic behaviour of the metric. This means that the result found by Blanchet of the smoothness (infinite differentiability) of the conformal structure at future null infinity may be completely modified if the limit $-T \to -\infty$ is taken first. In fact, several calculations, within various approximation methods, suggest that the order of differentiability of the conformal structure may be quite low. In particular, a recent calculation[58] concludes to a generic violation of the "peeling property" at future null infinity, which means that the differentiability class must be strictly less than C^3 there. In anticipation of such a drastic restriction on the regularity at future null infinity, we have characterized in §4.13 the asymptotic behaviour of the metric only through very weak requirements, see equations (4.16)–(4.17).

5.8 Preliminary results on the propagation problem

The "propagation problem" has not yet received a clear formulation, but the idea behind it is to investigate the link between the asymptotic radiation field and some other quantities which encode a knowledge of the gravitational field at a finite radius away from the source. For instance, it has been suggested recently[72] that, given some null tetrad, l^μ, n^μ, m^μ,$\bar{m}^\mu$ (see e.g. refs. 19 or 21), the following complex component of the Weyl tensor,

$$\Psi_0 = -C_{\alpha\beta\gamma\delta} l^\alpha m^\beta l^\gamma m^\delta, \tag{5.27}$$

which is usually thought of as capturing the "incoming gravitational radiation," may be better thought of as encoding the "non-radiative" part of the curvature, and may be used to define some "non-radiative multipole moments" of the source, as seen at a finite radius. Then the propagation problem associated to this definition consists in relating the "radiating multipole moments," as seen in the asymptotic outgoing gravitational radiation, i.e. the quantities denoted $M_L^I(u)$ and $S_L^I(u)$ in equation (4.13), with the "non-radiative multipole moments" measured at some moving radius $r_0(u)$ near the source (see ref. 72 for a precise definition of the non-radiative quadrupole, and for a discussion of why this problem is of interest in relation with the numerical calculations of the emission of gravitational waves by a collapsing star). An approximate answer to this problem has been obtained by a kind of Multipolar-Post-Minkowskian method, assuming that the main physical effect is the coupling between weak quadrupolar gravitational waves and the Schwarzschild-like curvature associated to the total mass of the source. It reads:

$$\begin{aligned} Q_{ij}^I(u) = Q_{ij}^{\text{non rad}}(u) - \frac{GM}{2c^3}\frac{d}{du}Q_{ij}^{\text{non rad}}(u) + \\ + \frac{2GM}{c^3}\int_{-T}^{u}\frac{du'}{(u-u')^2}\left\{\frac{1+6\frac{c(u-u')}{2r_0(u)}}{\left(1+\frac{c(u-u')}{2r_0(u)}\right)^6} - 1\right\} Q_{ij}^{\text{non rad}}(u') + \\ + O\left((GM/c^3)^2\right). \end{aligned} \tag{5.28}$$

The main interest of the formula (5.28) is that it exhibits clearly the fact that the radiation observed very far from the system at some retarded time $u(= t - r/c - 2(GM/c^3)\log(r - 2GM/c^2))$ depends not only on the state of the gravitational field near the source at the retarded time u, but also on the full-past-behaviour of the latter near-field. This is an effect which was not present in the three viewpoints of gravitational waves discussed in §2, but which is a direct consequence of the formation of wave *"tails"* during the propagation of a wave on a curved background, as mentioned in §2.4 (see references there, as well as in ref. 114).

If we go back to the general Multipolar-Post-Minkowskian metrics constructed by the algorithm of §5.6, one can prove for them a propagation result similar to the equation (5.28) in the case of sources which vary slowly enough for the weak-field domain D ($r \gg GM/c^2$) to overlap with the near-zone ($r \ll \lambda$). Then it can be shown[118] that in the near-zone the canonical metric (5.24), which is in general a *retarded* functional of the algorithmic moments, is, with an excellent approximation (of formal order $O(\beta^7)$ included in a post-Newtonian expansion in $\beta \sim v/c \sim L/\lambda$) an *instantaneous* functional of them. This means that the algorithmic moments themselves can be taken as quantities encoding the instantaneous state of the near-zone field, or in other words that they can be identified with the "near-zone multipole moments" appearing in Thorne's[66] generation formalism.

Therefore a useful propagation problem consists in linking the radiative multipole moments of the equations (4.1), (4.13) to the algorithmic moments themselves. For instance it is found for the quadrupole:

$$Q^{\mathrm{I}}_{ij}(u) = M_{ij}(u) + \frac{2GM}{c^3}\int_0^{+\infty} dy \left[\log\left(\frac{cy}{2b}\right) + \frac{11}{12}\right] M^{(2)}_{ij}(u-y) + O(1/c^5) \qquad (5.29)$$

(where b is the length scale introduced in equation (5.20)). The result corresponding to keeping only the first term in the right-hand side of equation (5.29) has been obtained by Thorne.[66] The explicit nonlinear correction of order $1/c^3$ (second term) has been computed only recently.[118] It can be thought, like the integral in equation (5.28), as due to the "tails" generated by the backscatter of the gravitational radiation off the curvature associated with the monopole M (which is just the constant Arnowitt-Deser-Misner[52] mass of the spacetime). A different aspect of these tails has been investigated by Bonnor and coworkers.[114] Let us conclude by mentioning that the conjecture put forward in refs. 114, namely that quadratic tails arise only in the monopole $\times$ multipole nonlinear coupling and not in the other multipole $\times$ multipole coupling, has been proven[119] to hold true.

5.9 Partial results on the generation problem

The Multipolar-Post-Minkowskian method, by itself, cannot solve the generation problem because it gives information only about the gravitational field outside the source. However it can help in solving the generation problem via the answers that it can give to the various propagation problems. One needs then to complete it by another method giving some information about the link beween the gravitational field at a finite radius away from the source and the source itself. In the general case of a fast moving, strongly self-gravitating source the only known way of finding answers to the latter question is to resort to numerical calculations. The problem of using them to feed the propagation problem has been tackled in e.g. refs. 70-72.

The case of slow moving, possibly strongly self-gravitating, sources has been considered by Thorne.[66] His generation formalism is based on equation (5.29) (taken at lowest order) and on the fact that for a slow moving source the near-zone gravitational field is, within a good approximation, a local-in-time functional of the algorithmic moments, and in fact, at lowest order in $1/c$, a function of the instantaneous value of the algorithmic moments themselves. This means that, at lowest order, the near-zone metric can be considered as an adiabatically changing *stationary* metric depending on the algorithmic moments $M_L(t)$ and $S_L(t)$, so that the latter moments can be, approximately, read off the near-zone metric as some "near-zone multipole moments." Hence, for instance, the equation (4.9) and similar equations for higher multipoles (of electric or magnetic type). Evidently, in our terminology, this "generation formalism" stays a "propagation formalism" until some independent way of getting the near-zone moments is found. Several ways doing that have been indicated in references 66, 25, 90, and in §4.8 above.

If we further restrict ourselves to considering slow moving *weakly* self-gravitating sources then there is a general analytical way of completing the MPM method. It consists of expanding the near-zone field, in and outside the source, by a Post-Newtonian Approximation method, and then in matching somehow the approximate Post-Newtonian field to the general Multipolar-Post-Minkowskian metric. This procedure was implicit in Fock's book, and has been explicitly introduced and partially formalized by Burke and Thorne.[103–106] It has been developed by several authors, e.g. Kates,[107] Anderson *et al.*,[120] Blanchet and Damour[68,118] (see other references in ref. 62, §12). As repeatedly said above this approach is not yet fully understood. However it is extremely powerful. To exhibit its technical power let us quote the equation, recently derived[118] by it means, linking the radiative quadrupole moment to the structure and motion of the source:

$$\begin{aligned} Q^{\mathrm{I}}_{ij}(t) =& I_{ij}(t) + \frac{1}{c^2}(\ldots)+ \\ &+\frac{1}{c^3}\{2GI(t)\int_0^{+\infty} dy\ \log\left(\frac{cy}{2b}\right) I^{(2)}_{ij}(t-y) + \ldots\}+ \\ &+ O(1/c^4). \end{aligned} \tag{5.30}$$

In equation (5.30), $I_{ij}(t)$ denotes (beware of the change of notation with respect to eqs (4.24) and (6.23a))

$$I_{ij}(t) := \int d^3x\ c^{-2}T^{00}(\vec{x},t)(x^i x^j - \tfrac{1}{3}\vec{x}^2\delta^{ij}), \tag{5.31}$$

while $I(t)$ denotes

$$I(t) := \int d^3x\ c^{-2}T^{00}(\vec{x},t), \tag{5.32}$$

b is the length-scale introduced in equation (5.20), and the ... denote some functionals of the *instantaneous* state of the source (i.e. some integral expressions of the type of (5.31)–(5.32), but corresponding to post-Newtonian effects) which have not yet been explicitly calculated (but which should be calculable, if the matching procedure is meaningful). In absence of an explicit expression for the latter missing terms, equation (5.30) gives only a partial answer to the generation problem, but it has the merit of exhibiting explicitly the *influence of the past history of the source on the outgoing radiation* which is fully described, at order $1/c^3$, by the integral appearing in eqn. (5.30).

5.10 Preliminary results on the radiation-reaction problem

Like in the case of the generation problem, the MPM method can help in solving the radiation-reaction problem only if it is completed by another method dealing with the source itself. For instance, in the case of a slow motion weakly self-gravitating source one can apply the combined MPM+PN method mentioned in the last sub-section (for widely separated strongly self-gravitating objects see e.g. Kates[107] and §6 below). This method leads, at lowest order, to the result already discussed in §4.15, namely

$$\mathcal{F}_i^{\mathrm{reac}} = -\rho \frac{\partial}{\partial x^i} V_{BT}^{\mathrm{reac}}(\vec{x},t) + O(1/c^7) \tag{5.33}$$

where we have posed

$$\rho := c^{-2} T^{00}, \tag{5.34}$$

and where the Burke-Thorne reaction potential, corresponding to the quadrupole approximation, reads

$$V_{BT}^{\mathrm{reac}}(\vec{x},t) = \frac{G}{5c^5} I_{jk}^{(5)}(t) x^j x^k. \tag{5.35}$$

In eqn (5.35) $I_{jk}(t)$ denotes the Newtonian quadrupole moment of the source, as defined by equation (5.31). Note that, at this level, any other reasonable definition of a "Newtonian mass density," differing from (5.34) only by post-Newtonian corrections, could be used in defining I_{ij}.

When the source quadrupole moment $I_{ij}(t)$, (5.31), is independent of time the radiation reaction force is of higher order in $1/c$. More generally, if among the mass multipole moments ("electric type") of the source

$$I_{i_1 \ldots i_n}(t) := \left(\int d^3x \; c^{-2} T^{00}(\vec{x},t) x^{i_1} \ldots x^{i_n} \right)^{STF}, \tag{5.36a}$$

and the current multipole moments ("magnetic type") of the source

$$J_{i_1 \ldots i_n}(t) := \left(\int d^3x \varepsilon^{i_1 jk} x^j c^{-1} T^{0k}(\vec{x},t) x^{i_2} \ldots x^{i_n} \right)^{STF}, \tag{5.36b}$$

the low order moments are stationary, and the first time-varying ones are $I_{i_1 \ldots i_l}(t)$ and/or $J_{i_1 \ldots i_{l-1}}(t)$, then the radiation reaction force is of order $1/c^{2l+1}$, instead of the usual $1/c^5$. In this case the explicit expression of $\mathcal{F}_i^{\mathrm{reac}}$ has been obtained by Blanchet and Damour[68] by a combined MPM+PN method. It can be written in the electromagnetic-like form:

$$\vec{\mathcal{F}}^{\mathrm{reac}} = \rho(\vec{E}^{\mathrm{reac}} + \vec{v} \times \vec{B}^{\mathrm{reac}}), \tag{5.37a}$$

with,

$$\vec{E}^{\mathrm{reac}} = -\vec{\nabla} V^{\mathrm{reac}} - \partial_t \vec{A}^{\mathrm{reac}}, \tag{5.37b}$$

$$\vec{B}^{\mathrm{reac}} = \vec{\nabla} \times \vec{A}^{\mathrm{reac}}, \tag{5.37c}$$

and,

$$V^{\mathrm{reac}}(\vec{x},t) = \frac{G}{c^{2l+1}} \frac{(-2)^l (l+1)(l+2)}{l(l-1)(2l+1)!} x^{i_1} \ldots x^{i_l} I_{i_1 \ldots i_l}^{(2l+1)}(t), \tag{5.38a}$$

$$[\vec{A}^{\mathrm{reac}}(\vec{x},t)]^i = \frac{G}{c^{2l+1}} \frac{(-2)^{l+1}(l-1)(l+1)}{l(l-2)(2l-1)!} \varepsilon^{iab} x^a x^{i_1} \ldots x^{i_{l-2}} J_{b\, i_1 \ldots i_{l-2}}^{(2l-1)}(t). \tag{5.38b}$$

Finally, going back to the usual quadrupolar case, $l = 2$, a recent investigation[118] has led to an explicit analytical control of *the influence of the past-behaviour of the source on the radiation reaction.* The result is partial, like equation (5.30), however it completely determines the leading functional dependence on the past. It has been found that the first past-dependent correction to the reaction force could be put in the form

$$\delta^{\text{past}}(\vec{\mathcal{F}}^{\text{reac}}) = -\rho \vec{\nabla}[\delta^{\text{past}} V^{\text{reac}}], \tag{5.39a}$$

with

$$\delta^{\text{past}} V^{\text{reac}} = \frac{4G^2}{5c^8} x^i x^j I(t) \int_0^{+\infty} dy \log\left(\frac{cy}{2b}\right) I_{ij}^{(7)}(t-y). \tag{5.39b}$$

It is remarkable that the result (5.39), added to the lowest-order reaction force (5.33)–(5.35), is of the form of the "third quadrupole law" (4.3), with a *past-dependent "reaction quadrupole moment"*

$$Q_{ij}^{\text{III}}(t) = I_{ij}(t) + \frac{1}{c^3}\{4GI(t) \int_0^{+\infty} dy \log\left(\frac{cy}{2b}\right) I_{ij}^{(2)}(t-y)\}. \tag{5.40}$$

The result (5.40) is still partial, and like in equation (5.30), some ... should be added to represent the *instantaneous* terms which have not yet been calculated. It is interesting to note that although the past-dependent part of Q_{ij}^{III}, eqn (5.40), is of the same functional form as the past-dependent part of Q_{ij}^{I}, eqn (5.30), there is a difference of a factor 2 between them (see ref. 118 for a discussion, and confirmation of this factor 2). This implies in particular that

$$Q_{ij}^{\text{III}}(t) \neq Q_{ij}^{\text{I}}(t) \tag{5.41}$$

as announced in §4.2.

§6. GRAVITATIONAL RADIATION AND BINARY SYSTEMS OF CONDENSED OBJECTS

6.1 One method for two questions

We shall call an object "gravitationally condensed," or simply *"condensed,"* if its characteristic linear dimension, L, is of the order of its gravitational radius, $G(\text{mass})/c^2$. In the notation of §3.2, this means that the parameter measuring the strength of the internal gravitational field,

$$\gamma_i = \frac{Gm}{c^2 L} \sim 1. \tag{6.1}$$

In other words such objects are strongly self-gravitating. Relativistic astrophysics has predicted the existence of two families of condensed objects: the neutron stars and the black holes. It has now been convincingly shown that neutron stars exist in the real world, and, in particular, that pulsars are very probably rotating neutron stars emitting pulses by an electromagnetic "lighthouse effect." The direct experimental evidence for the existence of black holes is not as solid, but many reasons, and some indirect evidence, make it virtually certain that they do exist in the real world (see the references at the end of section 1 of Carter's lectures in this volume). Moreover many observations, of e.g. X-ray sources or pulsars, show that there exist binary systems containing at least one condensed object. We shall concentrate in the following on the case of binary systems made of two condensed objects. Actually most of the results presented below are probably valid also under a weaker condition of wide separation and absence of mass transfer (and thereby weak coupling) between the two members of a binary system, but we shall discuss only the doubly-condensed case where one is in a better position for controlling, at least formally, the terms neglected in the analysis of the system.

There are two main questions that one would like to answer concerning the link between a two-condensed-body system and gravitational radiation.

The *first question* is: what is the precise form, amount, polarization, ..., of gravitational radiation emitted by such a system?

The *second question* being: what are the kinematical effects, in the binary motion, correlated with the emission of gravitational radiation?

Several of the methods that we have quickly reviewed in §4 can help in answering the first question. However, as discussed in §§4.3, 4.4, 4.15 and 4.17 (and in references quoted there) most of the usual methods cannot give a justified answer to the second question, especially if one is interested in the medium-term kinematical effects. These methods can, and did, suggest, on heuristic grounds, some answers to the second question, but one does need a new method to prove that the latter tentative answer is the correct one. In fact, as the problem is of purely kinematical nature, what is needed is a complete relativistic celestial mechanical description of a binary system made of two condensed objects. By "complete," we mean here complete up to, and including, the appearance of radiation reaction effects, and also complete in the sense that it allows one to go all the way to predicting explicit formulae ready to be compared with the raw observational data.

In the following we shall sketch one method which has been especially devised to answer, as completely as needed, the second question above, and which, as a bonus, also answers the first question. For other methods, giving less satisfactory or less complete answers to the second question, see the works of Schäfer[77] and Grishchuk and Kopejkin[78] (for a critical review see e.g. ref. 62).

6.2 The two-condensed-body problem in General Relativity

The gravitational interaction of two condensed bodies poses a problem in General Relativity which cannot be directly tackled by the usual post-Newtonian or post-Minkowskian methods which have been used to deal with the gravitational interactions of the Sun and of the planets. Indeed each body being strongly self-gravitating, eqn (6.1), we are not in a situation of an everywhere weak gravitational field. In fact there is a mixing of two strong-gravitational-field regions (near and in each body) and of a weak-field-region (obtained by cutting out two balls, each enclosing one body.) This is however a case where it has been possible to devise a suitable "cocktail" of approximation methods, combining an asymptotic expansion appropriate to the common weak field domain ("external region") with two asymptotic expansions appropriate to the strong field domains of each body ("internal regions"). The "mixing" of the "cocktail" is then performed by some matching procedure which allows one to propagate the information between the (a-priori disjoint) domains of validity of the various asymptotic expansions. Various "recipes," and various implementations, of such cocktails have been devised, e.g. by Manasse,[121] Demianski & Grishchuk,[122] D'Eath,[123] Kates,[124] Damour,[125] and Thorne & Hartle.[126] In the following we shall briefly sketch the method of ref. 125, and the results obtained by its means (for a general review of the problem of motion in General Relativity, see e.g. ref. 62).

6.3 The internal problems

Intuitively the physical fact which permits one to develop an approximation scheme in each strong-field internal region is that there should exist a frame of reference, linked to each condensed body, in which the influence of the other body will only cause small corrections to the (supposedly known) equilibrium configuration of one *isolated* condensed body. Mathematically this leads to assuming, around each body, the existence of a coordinate chart $X^\mu = (cT, X^i)$ in which

$$\mathrm{g}^{\mu\nu}(X^\lambda) = \mathrm{g}^{\mu\nu}_{\text{isolated}}(\hat{X}^i) + \sum_n \gamma_e^n \mathrm{g}^{\mu\nu}_{(n)}(\hat{X}^i, T), \tag{6.2}$$

where $\mathrm{g}^{\mu\nu}_{\text{isolated}}$ denotes the stationary metric of an isolated condensed body, where $\hat{X}^i := X^i/L$ denote the spatial coordinates a-dimensionalized by a characteristic internal linear dimension, and where

$$\gamma_e := \frac{Gm}{c^2 R} \tag{6.3}$$

is the small *gravitational coupling parameter* between the bodies (m, being a characteristic mass of the bodies, and R, a characteristic separation between the bodies). Note that in the present problem the parameter measuring the strength of the *internal* gravitational field, γ_i, defined by equation (3.1), is not small. In fact, by definition, γ_i is of order unity; see equation (6.1). However the parameter γ_e defined by equation (6.3), which measures the strength of the gravitational field *external* to each body (see equation (3.2)), can still be very small, and is available to set up a meaningful approximation scheme.

A detailed investigation of the structure of the hierarchy of equations satisfied by the $\mathrm{g}^{\mu\nu}_{(n)}$ (see ref. 67, §5) shows that one can choose each internal coordinate system X^μ so as to *"efface"* the influence of the other body up to the order γ_e^3, in the sense that the summation in the right-hand side of equation (6.2) starts at $n = 3$.

6.4 The matching: internal → external

The information contained in the structure of the expansion (6.2) can be propagated to the external region by assuming that the transformation between each internal coordinate system, X^μ, and the common external coordinate system, x^μ, can be expanded, near each body, as:

$$x^\mu = z^\mu(T) + e_i^\mu(T)X^i - \tfrac{1}{2} f_{ij}^\mu(T) X^i X^j + O(|\vec{X}|^3). \tag{6.4}$$

This leads to a partial knowledge of the structure of the external gravitational field near each body, which says essentially that it looks like a boosted Schwarzschild solution associated with the "centre of field" world-line $z^\mu(T)$ (terminology of ref. 67, which avoids the over-used name of "centre of mass"), and containing the *"Schwarzschild mass"* of each isolated body (i.e. the mass that appears in $\mathrm{g}^{\mu\nu}_{\text{isolated}}(\vec{X})$ when $|\vec{X}| \to \infty$).

6.5 The external problem

In the external region, using harmonic external coordinates $x^\mu = (ct, x^i)$, the gravitational field,

$$h^{\mu\nu}(x^\lambda) := \mathrm{g}^{\mu\nu}(x^\lambda) - f^{\mu\nu}, \tag{6.5}$$

must satisfy the following equations

$$\Box h^{\mu\nu} = N^{\mu\nu}(h) = \mathrm{II}^{\mu\nu}(h,h) + \mathrm{III}^{\mu\nu}(h,h,h) + \dots, \tag{6.6}$$

$$\partial_\nu h^{\mu\nu} = 0, \tag{6.7}$$

where $\mathrm{II}^{\mu\nu}$ is quadratic in $h^{\mu\nu}$, $\mathrm{III}^{\mu\nu}$ is cubic in $h^{\mu\nu}$, etc. .... Moreover $h^{\mu\nu}(x^\lambda)$ must satisfy some asymptotic boundary conditions, deduced from the matching: internal →

external, which give a partial knowledge of the structure of $h^{\mu\nu}(x^\lambda)$ near the world-lines $x^\lambda = z^\lambda$. Finally one assumes that $h^{\mu\nu}(x^\lambda)$ satisfies Fock's "no incoming radiation condition" at (Minkowski) past null infinity (see ref. 81, §92).

In the case where the two condensed objects would be spherically symmetric and non rotating if they were isolated, the asymptotic boundary conditions deduced from the matching are called "Dominant Schwarzschild conditions," and involve, near each body, only the centre of field world-line z^μ and the (constant) Schwarzschild mass, m. Then it is possible to show that these Dominant Schwarzschild conditions, together with the Fock no incoming radiation conditions, are sufficient, given two world-lines $z^\mu(s)$, $z'^\mu(s')$ and two constant parameters, m, m', to characterize, up to order γ_e^3 included, at most one solution of (6.6)–(6.7) which admit a (formal) expansion in powers of γ_e (or of G). The last requirement means

$$h^{\mu\nu}(x^\lambda) = \gamma_e h_1^{\mu\nu}(\hat{x}^i, t) + \gamma_e^2 h_2^{\mu\nu}(\hat{x}^i, t) + \gamma_e^3 h_3^{\mu\nu}(\hat{x}^i, t) + \dots, \tag{6.8}$$

where $\hat{x}^i := x^i/R$ denote the spatial coordinates a-dimensionalized by the characteristic separation between the bodies. Moreover the world-lines $z^\mu(s)$ (s being, say, the Minkowskian proper time) cannot be freely given, they must fulfill some integro-differential constraints ("equations of motion") for a solution to exist. And these equations of motion will contain, up to order γ_e^5 included, only the parameters m and m'.

It is in fact a remarkable property of General Relativity that, in the case just considered, the equations of motion of two condensed bodies depend, up to a very good accuracy, only on two constant parameters, the Schwarzschild masses m and m'. This means that all the internal structure of a particular condensed body is "effaced" when seen in the external problem. For instance a neutron star could be replaced by a black hole of the same mass without changing either the motion of the binary system or the gravitational field in the external region. Other investigations,[127,128] within alternative theories of gravity, show that this property of *effacement of the internal structure* does not hold in most other relativistic theories of gravity, in the sense that the equations of motion contain, already at lowest order, some quantities which depend on the internal structure (e.g. equation of state) of the bodies.

6.6 A convenient auxiliary mathematical technique

The uniqueness result quoted in the last sub-section allows one to use any convenient mathematical technique to compute the third order external gravitational field and equations of motion of two condensed bodies. It has been found convenient to introduce, in the external problem, a fictitious energy-momentum tensor whose role is to incorporate the Dominant Schwarzschild conditions.

A priori one would think of representing each body, as seen in the external region, by a Dirac distribution supported on the centre-of-field worldline. But this is meaningless because of the nonlinearity of the Einstein equations. However, following Marcel Riesz,[129] we can replace each Dirac *distribution* on $\mathbf{R}^4$,

$$\delta^4(x^\mu - z^\mu), \tag{6.9}$$

by the following *function* on $\mathbf{R}^4$

$$A^2[-f_{\mu\nu}(x^\mu - z^\mu)(x^\nu - z^\nu)]^{(A-4)/2}\, Y(-f_{\mu\nu}(x^\mu - z^\mu)(x^\nu - z^\nu))Y(x^0 - z^0), \tag{6.10}$$

where A is a complex number, and where Y denotes Heaviside's step function. Then it makes sense to use the functions (6.10) in the right-hand side of the Einstein equations, i.e. to add two terms constructed from (6.10), corresponding to the two condensed objects

(z^μ, z'^μ), in the right-hand side of equation (6.6). Using complex analytic continuation in A, one can then define the first three terms of an expansion of the type (6.8) by iterating some integrals involving the functions (6.10). At the end of the calculation one makes the analytic continuation of A down to zero, so that the functions (6.10) tend, pointwise, to *zero* (although, in the sense of distributions, they tend to (6.9)). This procedure generates a solution to the (harmonically relaxed) *vacuum* Einstein equations (6.6), outside two world-lines, which satisfies the third order Dominant Schwarzschild conditions. A refinement of the method, based on the imposition of the harmonicity condition (6.7), gives also the third order equations of motion of the worldlines. For a pedagogical discussion of the link between this analytic continuation method and the surface-integral formulations of the equations of motion à la Einstein-Infeld-Hoffman see ref. 62 §14.

Note that the preceding technique of analytic continuation plays no privileged role. It is used only because it is well defined, and because it permits the explicit construction of a (perturbative) solution of the Einstein equations satisfying the required boundary conditions. The uniqueness result quoted above then guarantees that the solution so constructed is the looked for gravitational field exterior to two condensed bodies.

6.7 Answer to the first question of §6.1 (generation)

It has been possible to implement, in a fully explicit way, the previous method at the order γ_e^2, i.e. to get the explicit expression of the second order gravitational field everywhere in the external region.[130] By looking at the asymptotic behaviour of the gravitational field far away from the two objects one can also compute the outgoing radiation field h_{ij}^{wave}. In the case of fast moving objects one can then get explicit results for the gravitational Bremsstrahlung radiation emitted during the fly-by of two objects[91,92] (consistently with the above mentioned effacement of internal structure the result is the same for Newtonian stars or condensed objects). In the case of slow moving, gravitationally bound objects the above mentioned explicit second order post-Minkowskian external gravitational field leads,[93] at lowest order, to an outgoing radiation having the form of the far-field quadrupole law (4.1) with

$$\begin{aligned} Q^{\mathrm{I}}_{ij}(t) =& m\{z^i(t)z^j(t) - \frac{\delta^{ij}}{3}\vec{z}^{\,2}(t)\}+ \\ &+ m'\{z'^i(t)z'^j(t) - \frac{\delta^{ij}}{3}\vec{z}'^{\,2}(t)\}. \end{aligned} \tag{6.11}$$

In eqn (6.11) m and m' are the Schwarzschild masses of the condensed objects, and $z^i(t)$, $z'^i(t)$ the positions of their centres-of-field. The latter positions must satisfy the equations of motion (6.17), which reduce, at lowest order, to Newton's equations. For other methods leading to the result (6.11) see §4.10. For partial results on the third order post-Minkowskian gravitational field see ref. 75.

6.8 Equations of motion of a binary system of condensed bodies

The method outlined above leads to equations of motion for two condensed bodies which are expanded in powers of γ_e, and which have the structure of retarded integro-differential equations: the hyperbolic structure of Einstein's equations, i.e. the finite velocity of propagation of gravity, causes the acceleration of (the centre of field of) each body to be a functional of the past history of the motions of the two bodies. Fortunately it is possible[75] to write many terms in the equations of motion of, say, the first body as a function of the "*retarded* kinematical state" of the second body, i.e. its position, velocity, ... considered at time

$$t'_R = t - \frac{1}{c}|z^i(t) - z'^i(t'_R)|. \tag{6.12}$$

If we assume that the relative motion is slow, we find that the retardation appearing in equation (6.12) is small compared to a characteristic period. Indeed

$$\frac{\text{retardation}}{\text{period}} \sim \frac{R/c}{P} \sim \frac{v}{c} = \beta \ll 1. \tag{6.13}$$

This makes it possible to expand all the retarded quantities in an asymptotic series of powers of β, with coefficients depending only on the *instantaneous* positions, velocities, accelerations, ... of the second body (Lagrange expansion theorem[131]). As for the other terms in the equations of motion which cannot be written simply as a function of retarded quantities, they are, either directly expanded in a truncated formal asymptotic series of β, or shown to formally contribute only at order β^6. As, moreover, we shall have, in the case of gravitationally bound systems,

$$\gamma_e \sim \beta^2, \tag{6.14}$$

we see that the original *post-Minkowskian* equations of motion deduced from the method outlined above, symbolically (with $ds^2 = -f_{\mu\nu}\, dz^\mu\, dz^\nu$)

$$\begin{aligned} \frac{d^2 z^\mu}{ds^2} = \sum_{n=1}^{3} \gamma_e^n \{&\text{function of the retarded state of the companion } + \\ &+ \text{functional of the past of the binary system}\} + O(\gamma_e^4), \end{aligned} \tag{6.15}$$

imply some *post-Newtonian* equations of motion of the type

$$\frac{d^2 z^i}{dt^2} = \sum_{n=0}^{5} \beta^n \{\text{function of the instantaneous state of the binary system}\} + O(\beta^6), \tag{6.16}$$

where $O(\beta^6)$ hides many complicated terms, some of which depend in an irreducible manner on the full past history of the motion of the system. It seems plausible, from the work reported in §5.10, that the dominant irreducible contribution of the past-history to the equations of motion is of order $O(\beta^8 \log \beta)$, see eqn (5.39). This would mean that the dominant error term in $O(\beta^6)$, eqn (6.16), can also be written as some function of the instantaneous state of the binary system, but this needs to be confirmed by further work. Now, the (Lagrange) expansion procedure outlined above leads in eqn (6.16) to functions of the instantaneous positions, $z^i(t)$, $z'^i(t)$, velocities $v^i(t) := dz^i/dt$, $v'^i := dz'^i/dt$, accelerations $a^i(t) = dv^i/dt = d^2z^i/dt^2, \ldots$, over-accelerations, $d^3z^i/dt^3, \ldots$, etc.... However by an iterated use of the lower order equations of motion, it is possible to reduce the dependence of the right-hand side of eqn (6.16) to only the instantaneous positions and velocities (for a discussion of this "order reduction" see ref. 67 and references therein, for equations of motions explicitly containing higher order derivatives see ref. 78). This leads finally to a quasi-Newtonian system of *ordinary differential equations* expanded in powers of β (or equivalently of $1/c$) first obtained by Damour and Deruelle[132] and Damour,[76] and confirmed by some post-Newtonian calculations of Grishchuk and Kopejkin[78] (see also Schäfer[77]) (for more references and a discussion of the realm of validity of these calculations see e.g. refs. 62 and 67).

The acceleration of, say, the first body of a binary system of non-rotating condensed bodies is found to satisfy the following equation:

$$\vec{a} = \vec{A}_{\text{conservative}}(\vec{z} - \vec{z}\,', \vec{v}, \vec{v}\,') + \vec{A}_{\text{dissipative}}\,(\vec{z} - \vec{z}\,', \vec{v} - \vec{v}\,') + O(1/c^6), \tag{6.17}$$

where all quantities are taken at the same (coordinate) time $t = z^0/c = z'^0/c$, and where [introducing the notation $RN^i := z^i(t) - z'^i(t)$ (with $N^iN^i = 1$), $V^i := v^i(t) - v'^i(t)$, $(Nv) := N^iv^i$, (etc....), $v^2 := v^iv^i$ (etc....)] one has put

$$A^i_{\text{conservative}} = A^i_0 + c^{-2}A^i_2 + c^{-4}A^i_4, \tag{6.18a}$$

with

$$A^i_0 = -Gm'R^{-2}N^i, \tag{6.18b}$$

$$\begin{aligned} A^i_2 = Gm'R^{-2}\{N^i[-v^2 - 2v'^2 + 4(vv') + \tfrac{3}{2}(Nv')^2 + 5(Gm/R) + 4(Gm'/R)]+ \\ + (v^i - v'^i)[4(Nv) - 3(Nv')]\}, \end{aligned} \tag{6.18c}$$

$$\begin{aligned} A^i_4 = Gm'R^{-2}\{N^i[&-2v'^4 + 4v'^2(vv') - 2(vv')^2 + \tfrac{3}{2}v^2(Nv')^2 + \\ &+ \tfrac{9}{2}v'^2(Nv')^2 - 6(vv')(Nv')^2 - \tfrac{15}{8}(Nv')^4 + \\ &+ (Gm/R)(-\tfrac{15}{4}v^2 + \tfrac{5}{4}v'^2 - \tfrac{5}{2}(vv') + \\ &\quad + \tfrac{39}{2}(Nv)^2 - 39(Nv)(Nv') + \tfrac{17}{2}(Nv')^2) + \\ &+ (Gm'/R)(4v'^2 - 8(vv') + 2(Nv)^2 - 4(Nv)(Nv') - 6(Nv')^2) + \\ &+ (G^2/R^2)(-\tfrac{57}{4}m^2 - 9m'^2 - \tfrac{69}{2}mm')] + \\ &+ (v^i - v'^i)[v^2(Nv') + 4v'^2(Nv) - 5v'^2(Nv') - 4(vv')(Nv) + \\ &\quad + 4(vv')(Nv') - 6(Nv)(Nv')^2 + \tfrac{9}{2}(Nv')^3 + \\ &\quad + (Gm/R)(-\tfrac{63}{4}(Nv) + \tfrac{55}{4}(Nv')) + \\ &\quad + (Gm'/R)(-2(Nv) - 2(Nv'))]\}, \end{aligned} \tag{6.18d}$$

and

$$\begin{aligned} A^i_{\text{dissipative}} = &\tfrac{4}{5}G^2mm'R^{-3}\{V^i[-V^2 + 2(Gm/R) - 8(Gm'/R)] + \\ &+ N^i(NV)[3V^2 - 6(Gm/R) + \tfrac{52}{3}(Gm'/R)]\}. \end{aligned} \tag{6.19}$$

In the case of slowly rotating condensed bodies one must add[76] to the right-hand side of equation (6.17) a supplementary term $\vec{A}_{\text{spin-orbit}}$. This term is very small ($O(1/c^4)$) and its effect can just be added to the effects considered below. It is then found not to interfere with the "radiation reaction effects" considered below (see refs. 76, 62 and references therein).

6.9 Radiation-reaction force versus the relativistic Laplace effect

The main feature which allows one to control all the short-term and medium-term kinematical effects contained in the solution, for gravitationally bound motion, of the equations (6.17)–(6.19), is that the following system of differential equations

$$\begin{aligned} \frac{d^2z^i}{dt^2} &= A^i_{\text{conservative}}(\vec{z}\,(t) - \vec{z}\,'(t), \vec{v}\,(t), \vec{v}\,'(t)), \\ \frac{d^2z'^i}{dt^2} &= A'^i_{\text{conservative}}(\vec{z}\,(t) - \vec{z}\,'(t), \vec{v}\,(t), \vec{v}\,'(t)), \end{aligned} \tag{6.20}$$

has been explicitly shown,[133,76] to be (approximately) deducible from an (approximately) *Poincaré-invariant variational principle,*

$$\delta \int dt\ L(z(t), z'(t), v(t), v'(t), a(t), a'(t)) = 0. \tag{6.21}$$

For the explicit expression of the *acceleration-dependent Lagrangian* L see refs. 133, 76 and 67 (note that one must "order-reduce" the Euler-Lagrange equations of L to get (6.20)). For a discussion of the effect of coordinate transformations on L, the link with previous work, and references to later work see e.g. refs. 135 and 62 and references therein.

The (approximate) Poincaré invariance of (6.21) allows one to construct[134,67] *ten Noetherian quantities,* which are functions of the instantaneous state of the binary system, and which are conserved by the evolution (6.20). If we go back to the full equations of motion (6.17) we now find that the preceding ten Noetherian quantities are no longer conserved but are slowly changing because of the presence of the term $\vec{A}_{\text{diss.}}$. In order to express the changes of the ten Noetherian quantities in terms of objects familiar from §4, it is convenient to rewrite $\vec{A}_{\text{diss.}}$ in the following form:

$$A^i_{\text{diss.}} = \frac{G}{c^5}\{\tfrac{3}{5}z^j Q^{(5)}_{ij} + 2v^j I^{(4)}_{ij} + \tfrac{10}{3}a^i I^{(3)} + \tfrac{1}{5}I^{(5)}_{iss} - J^{(4)}_{iss}\}, \tag{6.22}$$

with ($\sum$ denoting the summation over the two bodies)

$$I_{ij} := \sum m z^i z^j, \tag{6.23a}$$

$$I := I_{ss}, \tag{6.23b}$$

$$Q_{ij} := I_{ij} - \tfrac{1}{3}\delta^{ij} I, \tag{6.23c}$$

$$I_{iss} := \sum m z^i z^2, \tag{6.23d}$$

$$J_{iss} := \sum m v^i z^2. \tag{6.23e}$$

Then it is easy to prove the following results for the time derivatives of the ten Noetherian quantities E_{Noet}, P^i_{Noet}, J^{ij}_{Noet}, and $K^i_{\text{Noet}} := G^i_{\text{Noet}} - tP^i_{\text{Noet}}$

$$\frac{d}{dt}E_{\text{Noet}}(z,v) = -\frac{1}{c^5}\frac{d}{dt}E_5(z,v) - \frac{G}{5c^5}Q^{(3)}_{ij}Q^{(3)}_{ij} + O(1/c^6), \tag{6.24a}$$

$$\frac{d}{dt}P^i_{\text{Noet}}(z,v) = -\frac{1}{c^5}\frac{d}{dt}P^i_5(z,v) + \quad 0 \quad + O(1/c^6), \tag{6.24b}$$

$$\frac{d}{dt}J^{ij}_{\text{Noet}}(z,v) = -\frac{1}{c^5}\frac{d}{dt}J^{ij}_5(z,v) - \frac{2G}{5c^5}(Q^{(2)}_{is}Q^{(3)}_{js} - Q^{(2)}_{js}Q^{(3)}_{is}) + O(1/c^6), \tag{6.24c}$$

$$\frac{d}{dt}[G^i_{\text{Noet}}(z,v) - tP^i_{\text{Noet}}(z,v)] = -\frac{1}{c^5}\frac{d}{dt}[G^i_5(z,v) - tP^i_5(z,v)] + 0 + O(1/c^6). \tag{6.24d}$$

As indicated by the notation all the quantities $E_{\text{Noet}}, E_5, \ldots$ are functions of the instantaneous kinematical state of the binary system: $z(t), z'(t), v(t), v'(t)$ (higher derivatives having been reduced to z and v only). This means in particular that equation (6.24a) provides us with a *fully kinematical energy-loss quadrupole equation.* As discussed in §4.3 this is the type of equation needed to be able to say something about the medium-term kinematical effects in the binary system.

In summary, the separation (6.17) of the equations of motion in "conservative" ($\vec{A}_{\text{cons.}}$) and "dissipative" ($\vec{A}_{\text{diss.}}$) terms, and the result (6.24a) linking dE_{Noet}/dt (which equals $\sum m\vec{A}_{\rm diss.}\cdot\vec{v}$, i.e. the "work" of $m\vec{A}_{\rm diss.}$) to the square of the third time derivative of the quadrupole (6.23c) (which happens to be equal to the radiative quadrupole (6.11)) are sufficient justifications to give to

$$\vec{F}_{\rm reac} = m\vec{A}_{\rm diss.} \tag{6.25}$$

the name of "gravitational-radiation-reaction force" acting on the centre-of-mass motion of the first object. Let us mention also that by a suitable change of coordinates in the near-zone of the binary system one can transform the expression (6.25) for $\vec{F}_{\rm reac}$ into the Burke-Thorne form (4.25). However it is not clear whether the new coordinate system can, like the one used above, be used also in the wave-zone. This issue will become important in §6.12 where one needs a coordinate system which is regular far away from the binary system (see the discussion in §§4.15 and 4.17 above).

On the other hand it must be recalled that the equations of motion (6.17) were obtained by a retardation expansion of retarded equations of motion (of the type (6.15)) which were a direct consequence of the propagation and nonlinear properties of the Einstein equations. Therefore from the point of view of the local dynamics of the binary system it seems more appropriate to think of $\vec{A}_{\rm diss.}$ as the relativistic descendant of an effect predicted by Laplace[136] as early as 1773. Indeed Laplace studied the influence on the motion of the planets or comets of a finite velocity of propagation of gravity, say c_g, ("Sur les altérations que le mouvement des planètes et des comètes peut éprouver [...] par la transmission successive de la pesanteur"). Laplace thought of the gravitational attraction as due to the impulse imparted by some fluid converging towards the attracting body. Then, if the attracted body, at $\vec{z}$, is moving with velocity $\vec{V} = \vec{v} - \vec{v}'$ relative to the attracting one, $\vec{z}'$, and the "gravity fluid" with velocity $-c_g\vec{N}$, the aberration effect will modify Newton's law $\vec{A}_0$ (eqn (6.18b)) to give the following equations of motion:

$$\vec{a} = \vec{A}_0 + \vec{A}_{\rm Laplace}, \tag{6.26}$$

with

$$\vec{A}_{\rm Laplace} = -\frac{Gm'}{R^2}\frac{\vec{V}}{c_g}. \tag{6.27}$$

If we compare the explicit expression of $\vec{A}_{\rm diss.}$ in harmonic coordinates, eqn (6.19) with $\vec{A}_{\rm Lapl.}$, eqn (6.27), we see that $\vec{A}_{\rm diss.}$ contains several terms having the same structure as $\vec{A}_{\rm Lapl.}$, namely the structure of tangential forces, opposite to the velocity. In particular, in the case of (coordinate) circular motion $\vec{A}_{\rm diss.}$ has exactly the structure (6.27). Moreover the physical origin of these tangential components of the force is essentially the same, it is due to the lag caused by the finite velocity of propagation. The main difference is that Laplace predicted an effect at the first order in V/c_g, while the relativistic effect $\vec{A}_{\rm diss.}$, is of fifth order in V/c. In other words the naive first order effect is compensated but, following Eddington,[41] we can think of $\vec{A}_{\rm diss.}$ as a "residual Laplace effect."

6.10 Poincaré on gravitational waves

In connection with the comments of the previous sub-section on the double-headed nature of the reaction/Laplace term $\vec{A}_{\rm diss.}$, it may be of interest to note that, to my knowledge, the concept of gravitational wave (*"onde gravifique"*) has been introduced by Poincaré,[137] as early as June 1905, precisely in relation with a discussion of Laplace's ideas on the effect of a finite velocity of propagation of the gravitational interaction.

Indeed Poincaré, in his pioneering 5 June, 1905 article on Special Relativity (which preceded Einstein's first paper on relativity, received on 30 June, 1905) assumed the gravitational interaction to be described by some Poincaré-invariant retarded action-at-a-distance force law. He notices that this means that gravity propagates with the speed of light, so that there seems to be a contradiction with the well-known result of Laplace[136] concluding that the velocity of propagation of gravity must exceed seven million times the velocity of light. He then resolves the contradiction by proving that the first-order effect predicted by Laplace, eqn (6.27), is compensated, and that the first modification of Newton's law brought about by all Poincaré-invariant retarded laws are of second order in v/c (see equations (6.18)). He concludes by saying that the latter modifications will cause probably only very small effects in the dynamics of the solar system, but that a thorough discussion of these effects is called for.

In a later popular article of 1908 (see ref. 137) he points out that the effects of the $(v/c)^2$ terms will be mainly visible in the motion of the perihelion of Mercury. The same article contains also a remarkable prediction which shows the good physical insight of Poincaré and his clear understanding of the various aspects of gravitational waves. Indeed Poincaré says that the planetary system should emit some outgoing gravitational waves ("onde d'accélération"), and that this emission should dissipate the energy of the planetary system. He then predicts that this energy-loss should cause a secular acceleration of the mean orbital motion of the planets: "Il en résulterait que les moyens mouvements des astres iraient constamment en s'accélérant, comme si ces astres se mouvaient dans un milieu résistant." This is the effect of $\vec{A}_{\rm diss.}$ which will be studied in the next sub-section (see eqn (6.29)). He adds that this effect is much too weak to be detected, which is true for the planetary system. However we shall see below that the qualitative prediction of Poincaré has been recently substantiated by the observations of the binary pulsar, thereby strikingly confirming both the existence of gravitational waves and their double nature: time lag within the system and outgoing radiation.

6.11 Answer to the second question of §6.1 (radiation reaction)

The equations (6.24) are useful to investigate some consequences of the equations of motion (6.17), but they are not quite sufficient to fully control the short-term and medium-term kinematical behaviour of a binary system. A satisfactory answer to the latter question (which was the second equation asked in §6.1) can be obtained in two steps[138] (following essentially the method used by Laplace in integrating his equation (6.26)). The first step consists of solving, with sufficient accuracy, the truncated equations of motion (6.20). This is done by using the ten Noetherian conserved quantities. The end result is that one can write (in principle) explicitly the solution of (6.20) by means of some hyperelliptic functions of the time. For instance the coordinate radial motion of the first body, in a suitable centre-of-mass frame for the binary system, is given (in the gravitationally bound case) by:

$$|\vec{z}(t)| =: r(t) = S(\ell(t), c_1, c_2), \tag{6.28}$$

where S is a hyperelliptic function of its first variable, ℓ, with period 2π and where c_1 and c_2 are two constants linked to $E_{\rm Noet}$ and $|\vec{J}_{\rm Noet}|$. In eqn (6.28) $\ell(t)$ is a linear function of t, which can be thought of as a relativistic mean anomaly. $\ell(t)$ takes the value $2\pi N$ when the first body returns for the Nth time to its coordinate periastron.

Then the second step consists in applying the method of the variation of arbitrary constants, considering $\vec{A}_{\rm diss.}$, in (6.17), as a perturbative force added to the conservative system (6.20). For details see refs. 138. Let us only mention that, in this second step, one needs to vary more "constants" than in equations (6.24). Of particular interest for the following is the radial motion of the first body. It is given again by equation (6.28) with the constants c_1 and c_2 being replaced by slowly varying quantities $c_1(t)$, $c_2(t)$ and with the relativistic mean anomaly, ℓ, being now of the form:

$$\frac{\ell(t)}{2\pi} = \frac{C_0}{2\pi} + \frac{t}{P_0} - \frac{1}{2}\frac{\dot{P}_0}{P_0^2}t^2 + \text{periodic terms}, \tag{6.29}$$

where the periodic terms have only a *short* period $\simeq P_0$ ("orbital period"). This allows one to clearly separate the secular effect $\propto t^2$ in eqn (6.29) on any time scales greater than P_0 and smaller than $P_0/\dot{P}_0$. The coefficient $\dot{P}_0$ of the latter secular effect measures the secular shortening of the coordinate time of return to the coordinate periastron, or, in other words, the secular acceleration of the mean orbital motion (it is often called the "period derivative" but this is a somewhat misleading name, the three dimensional motion not being periodic because of the periastron advance). The method of variation of constants gives for $\dot{P}_0$ an expression of the form

$$\dot{P}_0 = -\tfrac{3}{2}P_0 E_{\text{Noet}}^{-1}\langle \tfrac{d}{dt}E_{\text{Noet}}\rangle, \tag{6.30}$$

where the angular brackets denote the average over the explicit dependence on the relativistic mean anomaly:

$$\langle F(\ell(t), c(t))\rangle := \frac{1}{2\pi}\int_0^{2\pi} d\ell\, F(\ell, c(t)). \tag{6.31}$$

The result for $\dot{P}_0$ can be put in the form:

$$\dot{P}_0 = -\frac{192\pi}{5c^5}\left(\frac{2\pi G}{P_0}\right)^{5/3}\frac{mm'}{(m+m')^{1/3}}(1 + \tfrac{73}{24}e^2 + \tfrac{37}{96}e^4)(1-e^2)^{-7/2}, \tag{6.32}$$

where e denotes a relativistic eccentricity which reduces to the usual Keplerian eccentricity when $\frac{v}{c} \to 0$. For details of the meaning and proofs of equations (6.28)–(6.32) see ref. 138. Earlier arguments[139–141] had led to the prediction of the same end result (6.32). However they can be considered only as heuristic, because they were based on an ill-specified energy-loss quadrupole law (see the discussion in §4.3 and references therein), and because also, as stressed above (and in ref. 138) even a fully kinematical energy-loss equation, like (6.24a), is insufficient, taken by itself, to derive (6.29).

Besides the secular acceleration of the mean orbital motion, $\dot{P}_0$, there are several other important secular kinematical effects induced by $\vec{A}_{\text{diss.}}$: notably a shrinkage and a circularization of the orbit.[140] All these effects were already predicted by Laplace[136] as consequences of a finite velocity of propagation of gravity (by integrating his equation (6.26)). For other, purely relativistic, secular effects see ref. 138.

6.12 Application to the binary pulsar PSR 1913+16

In 1974 Hulse and Taylor[4] discovered a pulsar member of a binary system: the "binary pulsar" PSR 1913+16. This discovery of a very precise clock (the pulsar) rapidly moving in a highly elliptic orbit around the centre of mass of a binary system opened up a new testing field for General Relativity (see e.g. ref. 128 and references therein). In particular it immediately raised the hope to observe secular effects due to gravitational radiation reaction.[141] Indeed the measurement of the arrival times on Earth of the radio pulses emitted by the pulsar allows one to continuously track its orbital motion with a remarkable precision (now reaching $\pm$ 6 km) and without any loss of information over more than ten years.[142] Therefore even very small damping forces in the system will become observable by their secular effects on the orbital motion.

In order to be able to compare the predictions of General Relativity with the observations of the binary pulsar, one must first choose a theoretical model of the binary pulsar and then compute, via a general relativistic treatment of the latter model, the quantities which are directly observed. The most plausible model for the binary pulsar is a clean system of two condensed bodies, one of them (the pulsar) emitting electromagnetic radiation like a rotating beacon (for discussions of the plausibility of this model, and for some tests of it, see e.g. refs. 142-144 and references therein). We have seen above, (6.17)–(6.19), what the equations of motion are for such a system, and how they can be solved (in principle) explicitly, with proper account taken of all short-term and medium-term effects (by short-term we mean of the order of one "orbital period," $P_0 \sim$ 8 hours, and by medium-term we mean on time scales much longer than P_0 but much smaller than the radiation damping time scale $P_0/\dot{P}_0 \sim 400$ million years). However all the results quoted above, including $\dot{P}_0$, concern only the coordinate motion, as seen in the exterior (harmonic) coordinate system of §6.5. What is needed now is to transform this coordinate knowledge into an invariant knowledge by computing the (necessarily invariant) quantities which are observed.

This means computing the so-called "timing formula,"

$$\tau_N^{\text{Earth}} = F(N, \xi^a), \tag{6.33}$$

where τ_N^{Earth} denotes the proper time of arrival on Earth of the Nth electromagnetic pulse, and where F is a function of the integer N and of some parameters ξ^a ($a = 1, 2, \ldots$) describing the theoretical model. The "timing formula" (6.33) has been derived, in an increasingly accurate manner, by Blandford and Teukolsky, Epstein, Haugan, and Damour and Deruelle (see references in ref. 143).

This "timing" problem poses both theoretical and pragmatical problems. The theoretical problems are due (like in §6.2) to the mixing of strong-field and weak-field effects: indeed the electromagnetic pulse originates from the vicinity of the pulsar (strong-field region) and travels afterwards in weak-field regions (between the condensed objects, and in the solar system). The pragmatical problems come from the fact that the theoretical "timing formula" $F(N, \xi^a)$ must be, at the same time, very accurate, fully explicit, and simple enough to allow one to understand analytically the influence of each physical effect in the final result. These aims seem to be met in the latest derivation of the timing formula.[143] The formula is obtained by a ladder of relations linking, first, N to some intermediate time T, then T to τ_N^{bary} (proper time of arrival at the barycenter of the solar system, in absence of all solar-system-gravitational fields) and finally τ_N^{bary} to τ_N^{Earth}. The first two steps of the ladder read as follows:

$$N = N_0 + \nu T + \tfrac{1}{2}\dot{\nu} T^2 + \tfrac{1}{6}\ddot{\nu} T^3, \tag{6.34}$$

$$D\tau_N^{\text{bary}} = T + \Delta_R(T) + \Delta_E(T) + \Delta_S(T) + \Delta_A(T), \tag{6.35}$$

where we have distinguished:

the Römer effect (time of flight across the binary system)

$$\Delta_R(T) = \frac{a_r \sin i}{c}\{\sin\omega(\cos u - e_r) + (1 - e_\theta^2)^{1/2}\cos\omega\,\sin u\}, \tag{6.36}$$

the Einstein effect (integrated gravitational redshift + second order Doppler effect on the pulsar-clock)

$$\Delta_E(T) = \gamma \sin u, \tag{6.37}$$

the Shapiro effect (weak-field gravitational time delay across the binary system)

$$\Delta_S(T) = -\frac{2Gm'}{c^3}\log\{1 - e\cos u - \sin i[\sin\omega(\cos u - e) + (1-e^2)^{1/2}\cos\omega\,\sin u]\}, \tag{6.38}$$

and the aberration effects (caused by the interplay between the rotation and translation of the pulsar-beacon)

$$\Delta_A(T) = A\{\sin(\omega + A_e(u)) + e\sin\omega\} + B\{\cos(\omega + A_e(u)) + e\cos\omega\}. \tag{6.39}$$

The quantity $\omega(T)$ (slowly-precessing relativistic argument of the periastron) is given by

$$\omega = \omega_0 + 2k\arctan\left[\left(\frac{1+e}{1-e}\right)^{1/2}\tan\tfrac{1}{2}u\right], \tag{6.40}$$

where $u(T)$ (modified relativistic eccentric anomaly) is implicitly given by the following "Kepler equation":

$$C_0 + 2\pi\left[\frac{T}{P_0} - \frac{1}{2}\frac{\dot{P}_0}{P_0^2}T^2\right] = u - e_T\sin u. \tag{6.41}$$

For the meaning of the parameters appearing in equations (6.34)–(6.41) see ref. 143 and references therein. The alert reader will have recognized in the left-hand side of equation (6.41) the secular part of the relativistic mean anomaly (6.29) which contains the secular acceleration term $\dot{P}_0$.

For detailed discussions of the comparison between the theoretical timing formulae and the observational results see references 142 and 144. Here we wish only to quote the fact that the parameter $\dot{P}_0$ (which, because of equation (6.41), appears as one of the "observables" ξ^a in the timing formula (6.33)) has been directly measured with a value:[142,144]

$$\dot{P}_0^{\text{observed}} = (-2.40 \pm 0.09) \times 10^{-12}. \tag{6.42}$$

On the other hand, the direct measure of the parameters P_0, m, m' and e, allows one to compute a "theoretical" value of $\dot{P}_0$, as predicted by the equation (6.32). One finds[142]

$$\dot{P}_0^{\text{theory}} = (-2.403 \pm 0.002) \times 10^{-12}. \tag{6.43}$$

6.13 Conclusion

The 4% agreement between (6.42) and (6.43) provides an impressive confirmation of the structure (6.17) of the equations of motion and of the general relativistic value (6.19) for the "dissipative" part of the equations of motion, $\vec{A}_{\text{diss.}}$. This means that both the fact that the gravitational interaction propagates with the velocity c between two condensed objects ($\vec{A}_{\text{diss.}}$ as a residual Laplace effect), and the fact that there is a balance between the local loss of mechanical energy and the outgoing flux of gravitational radiation ($\vec{A}_{\text{diss.}}$ as a radiation-reaction force) seem to be well confirmed by the observations of the binary pulsar. This result is important because it is the first confirmation of General Relativity which involves both strong gravitational fields and propagation effects. This makes us confident that Einstein's theory correctly describes the behaviour of strong and/or rapidly varying fields, and therefore that the general relativistic investigations in the theory of gravitational radiation reviewed in these lectures constitute an appropriate tool to prepare the ground for the future "Gravitational Wave Astronomy."

ACKNOWLEDGEMENTS

I thank the organizer, M. Lévy, the directors, J. Audouze and H. Reeves, and the coordinators, B. Carter and J.B. Hartle, of this summer school for planning and running such a stimulating meeting. For helpful comments and discussions, during or after the school, I am very grateful to J.D. Barrow, J. Bičák, L. Blanchet, Ch.J. Bordé, T. Dray, J.B. Hartle, G. t'Hooft, G. Schäfer, P. Spindel, W.M. Suen, and, last but certainly not least, K.S. Thorne.

Special thanks go to B.P. Jensen for his suggestions on wording, and his careful help with a difficult manuscript.

SOME BOOKS FULLY DEVOTED TO GRAVITATIONAL RADIATION (See also references 9–18 below)

J. Bičák and V.N. Rudenko, "Gravitational radiation in General Relativity and the methods of its detection," (in Russian) Moscow State University Press, Moscow (to appear in 1987).

Centre National de la Recherche Scientifique, colloque international n° 220, "Ondes et Radiations gravitationnelles," Editions du C.N.R.S., Paris (1974).

N. Deruelle and T. Piran, editors, "Gravitational Radiation," North-Holland, Amsterdam (1983).

V. De Sabbata and J. Weber, editors, "Topics in Theoretical and Experimental Gravitation Physics," Plenum Press, London (1977).

L.L. Smarr, editor, "Sources of Gravitational Radiation," Cambridge University Press, Cambridge (1979).

H.J. Treder, "Gravitative Stosswellen, nichtanalytische Wellenlösungen der Einsteinschen Gravitationsgleichungen," Akademie-Verlag, Berlin (1962).

J. Weber, "General Relativity and Gravitational Waves," Interscience, N.Y. (1961).

BIBLIOGRAPHICAL REFERENCES

1. F.A.E. Pirani, Introduction to Gravitational Radiation Theory, *in:* "Lectures on General Relativity," A. Trautman, F.A.E. Pirani and H. Bondi, editors (Brandeis, 1964), pp 249-273, Prentice-Hall, Englewood Cliffs (1965).

2. F.J. Dyson, *in:* "Interstellar Communication," A.G.W. Cameron, editor, p. 118, Benjamin, New York (1963).

3. J. Weber, Opening Remarks, *in:* "Topics in Theoretical and Experimental Gravitation Physics," V. De Sabbata and J. Weber, editors, pp ix-xi, Plenum Press, London (1977).

4. R.A. Hulse and J.H. Taylor, *Astrophys. J. (Letters)* **195**, L51-L53 (1975).

5. K.S. Thorne, *Rev. Mod. Phys.* **52**, 285-297 (1980). See also in the same issue the article by Caves *et al.*.

6. A. Brillet, T. Damour and Ph. Tourrenc, *Ann. Phys. Fr.***10**, 201-218 (1985). See also in the same issue the articles by Brillet, Reynaud and Heidmann, Tourrenc and Deruelle, Vinet, and Teissier du Cros.

7. V. De Sabbata and J. Weber, editors, "Topics in Theoretical and Experimental Gravitation Physics," Plenum Press, London (1977).

8. L.L. Smarr, editor, "Sources of Gravitational Radiation," Cambridge University Press, Cambridge (1979).

9. S.W. Hawking and W. Israel, editors, "General Relativity: An Einstein Centenary Survey," Cambridge University Press, Cambridge (1979).

10. C. Edwards, editor, "Gravitational Radiation, Collapsed Objects and Exact Solutions," Springer-Verlag, Berlin (1980).

11. A. Held, editor, "General Relativity and Gravitation," 2 volumes, Plenum Press, London (1980).

12. R. Ruffini, editor, "Proceedings of the Second Marcel Grossmann Meeting," North-Holland, Amsterdam (1982).

13. P. Meystre and M.O. Scully, editors, "Quantum Optics, Experimental Gravitation, and Measurement Theory," Plenum Press, London (1983).

14. Hu Ning, editor, "Proceedings of the Third Marcel Grossmann Meeting," North-Holland, Amsterdam (1983).

15. N. Deruelle and T. Piran, "Gravitational Radiation," North-Holland, Amsterdam (1983).

16. Laboratoire "Gravitation et Cosmologie Relativistes," editor, "Gravitation, Geometry and Relativistic Physics," Lecture Notes in Physics, no. 212, Springer-Verlag, Berlin (1984).

17. R. Ruffini, editor, "Proceedings of the Fourth Marcel Grossmann Meeting," North-Holland, Amsterdam, in press.

18. H. Sato and T. Nakamura, editors, "Proceedings of the XIV Yamada Conference on Gravitational Collapse and Relativity," World Scientific, Singapore, in press .

19. R.K. Sachs, Gravitational Radiation, *in:* "Relativity, Groups and Topology," C. and B. De Witt, editors, pp 521-562, Gordon and Breach, New York (1964).

20. K.S. Thorne, The Generation of Gravitational Waves: A Review of Computational Techniques, *in:* reference 7, pp 1-61.

21. E.T. Newman and P. Tod, Asymptotically Flat Space-Times, *in:* reference 11, volume 2, pp 1-36.

22. M. Walker, Isolated Systems in Relativistic Gravity, *in:* "Relativistic Astrophysics and Cosmology," X. Fustero and E. Verdaguer, editors, World Scientific, Singapore (1983).

23. A. Ashtekar, Asymptotic Properties of Isolated Systems: Recent Developments, *in:* "General Relativity and Gravitation," B. Bertotti *et al.*, editors, Reidel, Dordrecht (1984).

24. B.F. Schutz, Motion and Radiation in General Relativity, *in:* "Relativity, Supersymmetry and Cosmology," O. Bressan *et al.*, editors, pp 3-80, World Scientific, Singapore (1985).

25. K.S. Thorne, The Theory of Gravitational Radiation: An Introductory Review, *in:* reference 15, pp 1-57.

26. C.W. Misner, K.S. Thorne and J.A. Wheeler, "Gravitation," Freeman, San Francisco (1973).

27. A. Lichnerowicz, "Théories Relativistes de la Gravitation et de l'Electromagnétisme," Masson, Paris (1955).

28. A. Papapetrou, Shock Waves in General Relativity, *in:* reference 7, pp 83-102.

29. A. Lichnerowicz, Sur les ondes de choc gravitationnelles et electromagnétiques, *in:* "Ondes et Radiations Gravitationnelles," pp 47-56, Editions du C.N.R.S., Paris (1974) and references therein.

30. Y. (Choquet-)Bruhat, *C.R. Acad. Sc. Paris*, **258**, 3809-3812 (1964).

31. R.A. Isaacson, *Phys. Rev.* **166**, 1263-1271 and 1272-1280 (1968).

32. J.A. Wheeler, "Geometrodynamics," Academic Press, New York (1962).

33. D. Brill and J.B. Hartle, *Phys. Rev.* **135**, B271 (1964).

34. Y. Choquet-Bruhat, *Commun. math. Phys.* **12**, 16-35 (1969); Y. Choquet-Bruhat, Couplage d'ondes gravitationnelles et électromagnétiques à haute fréquence, *in:* "Ondes et Radiations Gravitationnelles," pp 85-100, Editions du C.N.R.S., Paris (1974); Y. Choquet-Bruhat, Coupling of high frequency gravitational and electromagnetic waves, *in:* "Proceedings of the First Marcel Grossmann Meeting," R. Ruffini, editor, pp 415-428, North-Holland, Amsterdam (1977).

35. J. Madore, *Commun. math. Phys.*, **27**, 291 (1972) and **30**, 355 (1973); J. Madore, *Gen. Rel. Grav.* **5**, 169 (1974).

36. A. Taub, Variational methods and gravitational waves, *in:* "Ondes et Radiations Gravitationnelles," pp 57-71, Editions du C.N.R.S., Paris (1974); A. Taub, *in:* reference 11, volume 1, pp 539-555.

37. U.N. Gerlach, *Phys. Rev. Lett.* **32**, 1023 (1974).

38. M.E. de Araujo, *Gen. Rel. Grav.* **18**, 219-233 (1986); C.D. Ciubotariu, Nonlinear wave interactions in General Relativity, *in:* "Abstracts of contributed papers to the 11th international conference on General Relativity and Gravitation," volume II, p 472 (Stockholm, Sweden, July 6-12, 1986).

39. J. Leray, *Cahiers de Physique*, **15**, 373-381 (1961).

40. L. Gårding, T. Kotake and J. Leray, *Bull. Soc. math. France*, **92**, pp 263-361 (1964).

41. A.S. Eddington, "The mathematical theory of relativity," Cambridge University Press, Cambridge (1965). See note 7 to §57 (page 246) for the propagation of curvature. See also note 8 to §57 (page 248) for the problem of radiation reaction within the source, and a discussion of the "residual Laplace effect" mentioned in §6 of these lectures.

42. L.P. Grishchuk and A.G. Polnarev, Gravitational Waves and Their Interaction with Matter and Fields, *in:* ref. 11, volume 2, pp 393-434.

43. J.M. Souriau, Le milieu élastique soumis aux ondes gravitationnelles, *in:* "Ondes et Radiations Gravitationnelles," pp 243-256, Editions du C.N.R.S., Paris (1974).

44. B. Carter, Interaction of gravitational waves with an elastic solid medium, *in:* ref. 15, pp 455-464.

45. L.D. Landau and E.M. Lifšits, "Teoria dei campi," sixth edition, Mir, Mocow and Editori Riuniti, Rome (1976).

46. D. Kramer, H. Stephani, M. MacCallum and E. Herlt, "Exact solutions of Einstein's field equations," Cambridge University Press, Cambridge (1980).

47. F.G. Friedlander, "The wave equation on a curved space-time," Cambridge University Press, Cambridge (1975).

48. Y. Choquet-Bruhat, D. Christodoulou and M. Francaviglia, *Ann. Inst. Henri Poincaré, A, Physique Théorique,* **31**, 399-414 (1979).

49. J. Carminati and R.G. McLenaghan, *Ann. Inst. Henri Poincaré, Physique Théorique,* **44**, 115-153 (1986), and references therein on the "Huygens' principle".

50. P.C. Waylen, *Proc. R. Soc. Lond. A.*, **321**, 397-408 (1971).

51. R. Penrose, *Ann. Phys.* **10**, 171-201 (1960).

52. R. Arnowitt, S. Deser and C.W. Misner, The Dynamics of General Relativity, *in:* "Gravitation: an introduction to current reserch," L. Witten, editor, pp 227-265, Wiley, N.Y. (1962) and references therein.

53. B.F. Schutz, *Phys. Rev.* **D22**, 249-259 (1980).

54. T. Futamase and B.F. Schutz, *Phys. Rev.* **D28**, 2363-2372 (1983); T. Futamase, *Phys. Rev.* **D28**, 2373-2381 (1983).

55. A. Ashtekar, Asymptotic structure of the gravitational field at spatial infinity, *in:* ref. 11, volume 2, pp 37-69.

56. R. Geroch, *in:* "Asymptotic Structure of Space-Time," F.P. Esposito and L. Witten, editors, pp 1-105, Plenum Press, N.Y. (1977).

57. B.G. Schmidt, Asymptotic Structure of Isolated Systems, *in:* "Isolated Gravitating Systems in General Relativity," J. Ehlers, editor, pp 11-49, North-Holland, Amsterdam (1979).

58. T. Damour, Analytical calculations of gravitational radiation, *in:* "Proceedings of the Fourth Marcel Grossman Meeting," R. Ruffini, editor, North-Holland, Amsterdam, in press.

59. L. Blanchet and T. Damour, *Phil. Trans. R. Soc. Lond. A* **320**, 379-430 (1986).

60. H. Friedrich, *Commun. Math. Phys.* **100**, 525-543 (1985) and references therein.

61. C.M. Will, *Can. J. Phys.* **64**, 140-145 (1986).

62. T. Damour, The problem of motion in Newtonian and Einsteinian gravity, *in:* "300 Years of Gravitation," S.W. Hawking and W. Israel, editors, Cambridge University Press, Cambridge (1987).

63. Y. (Choquet-)Bruhat, The Cauchy problem, *in:* "Gravitation: an introduction to current research," L. Witten, editor, pp 130-168, Wiley, N.Y. (1962): Y. Choquet-Bruhat and J.W. York, The Cauchy problem, *in:* ref. 11, volume 1, pp 99-172; Y. Choquet-Bruhat, Mathematical problems in General Relativity, *in:* "Proceedings of the International School 'Cosmology and Gravitation'," M. Novello, editor, Rio de Janeiro, Brazil (in press), see also Usp. Mat. Nauk. **40**, no. 6, 1-40 (1985) (in Russian).

64. N. Noutchegueme, Solutions semi-globales asymptotiquement minkowskiennes pour les équations d'Einstein, *Ann. Inst. Henri Poincaré, Physique Théorique,* in press.

65. T. Piran, Methods of Numerical Relativity, *in:* ref. 15, pp 203-256.

66. K.S. Thorne, Multipole Expansions of Gravitational Radiation, *Rev. Mod. Phys.* **52**, 299-339 (1980).

67. T. Damour, Gravitational Radiation and the Motion of Compact Bodies, *in:* ref. 15, pp 59-144.

68. L. Blanchet and T. Damour, *Phys. Lett.* **104A**, 82-86 (1984).

69. T. Damour, Sur les nouvelles méthodes d'approximation en relativité générale, *in:* "Géométrie et Physique, Journées Relativistes SMF 1985," Y. Choquet, B. Coll, R. Kerner and A. Lichnerowicz, editors, Hermann, Paris (série "Travaux en Cours") (1986).

70. J.M. Bardeen, Gauge and Radiation Conditions in Numerical Relativity, *in:* ref. 15, pp 433-441.

71. J.L. Anderson and D.W. Hobill, A study of nonlinear radiation damping by matching analytic and numerical solutions, preprint Stevens Institute of Technology, (1986).

72. T. Damour, On the propagation problem in gravitational radiation theory, *in:* "Proceedings of the XIV Yamada Conference on Gravitational Collapse and Relativity," H. Sato and T. Nakamura, editors, World Scientific, Singapore, in press.

73. M. Walker, The quadrupole approximation to gravitational radiation, *in:* "General Relativity and Gravitation," B. Bertotti *et al.*, editors, pp 107-123, Reidel, Dordrecht (1984).

74. J. Ehlers and M. Walker, Gravitational radiation and the "quadrupole" formula, Report of workshop A1, *in:* "General Relativity and Gravitation," B. Bertotti *et al.*, editors, pp 125-137, Reidel, Dordrecht (1984).

75. T. Damour, Radiation damping in General Relativity, *in:* ref. 14, pp 583-597.

76. T. Damour, *C. R. Acad. Sc. Paris, série II,* **294**, 1355-1357 (1982).

77. G. Schäfer, *Ann. Phys. (N.Y.)* **161**, 81-100 (1985).

78. L.P. Grishchuk and S.M. Kopejkin, Equations of motion for isolated bodies with relativistic corrections including the radiation reaction force, *in:* "Relativity in Celestial Mechanics and Astrometry," J. Kovalevsky and V.A. Brumberg, editors, pp 19-34, (114th Symposium of IAU) Reidel, Dordrecht (1986); S.M. Kopejkin, *Astron. Zh.* **62**, 889-904 (1985); and references therein.

79. A. Einstein, *König. Preuss. Akad. der Wissenschaften, Sitzungsberichte*, Erster Halbband, p 154 (1918).

80. S. Weinberg, "Gravitation and Cosmology," Wiley, N.Y. (1972).

81. V. Fock, "The Theory of Space Time and Gravitation," Pergamon Press, London (1959).

82. P. Havas and J.N. Goldberg, *Phys. Rev.* **128**, 398-414 (1962).

83. T. Futamase and B.F. Schutz, *Phys. Rev.* **D32**, 2557-2565 (1985).

84. T. Futamase *Phys. Rev.* **D32**, 2566-2574 (1985).

85. R. Epstein and R.V. Wagoner, *Astrophys. J.* **197**, 717-723 (1975); R.V. Wagoner and C.M. Will, *Astrophys. J.* **210**, 764-775 (1976); see also N. Spyrou and D. Papadopoulos, *Gen. Rel. Grav.* **17**, 1059-1067 (1985).

86. J.L. Anderson, *Phys. Rev. Lett.* **45**, 1745-1748 (1980).

87. M. Walker and C.M. Will, *Phys. Rev. Lett.* **45**, 1741-1744 (1980).

88. J. Winicour, The quadrupole radiation formula, preprint Max Planck Institut für Physik und Astrophysik 253, August 1986, and references therein; R.A. Isaacson, J.S. Welling and J. Winicour, *Phys. Rev. Lett.* **53**, 1870-1872 (1984).

89. S. Persides, A new derivation of the quadrupole formula for gravitational radiation, preprint University of Thessaloniki (1986).

90. B.L. Schumaker and K.S. Thorne, *Mon. Not. Roy. Astron. Soc.* **203**, 457-489 (1983); K.S. Thorne and Y. Gürsel, *Mon. Not. Roy. Astron. Soc.* **205**, 809-817 (1983); L.S. Finn, G-modes of non-radially pulsating relativistic stars: the slow-motion formalism, submitted to *Mon. Not. Roy. Astron. Soc.* and contribution to ref. 17; see also M. Zimmermann, *Phys. Rev.* **D21**, 891 (1980).

91. K.S. Thorne and S.J. Kovács, *Astrophys. J.* **200**, 245-262 (1975); R.J. Crowley and K.S. Thorne, *Astrophys. J.* **215**, 624-635 (1977); S.J. Kovács and K.S. Thorne, *Astrophys. J.* **217**, 252-280 (1977) and **224**, 62-85 (1978).

92. K. Westpfahl, *Fortschritte der Physik* **33**, 417-493 (1985) and references therein.

93. N. Deruelle, Thèse de Doctorat d'Etat, Paris 6 (1982), unpublished; N. Deruelle, Gravitational radiation from two compact bodies: the radiative losses at the post linear approximation, *in:* ref. 14, part B, pp 955-958.

94. L.E. Halpern and R. Desbrandes, *Ann. Inst. Henri Poincaré A, Physique Théorique,* **11**, 309-329 (1969).

95. W.H. Press, *Phys. Rev.* **D15**, 965-968 (1977).

96. A.Papapetrou, *Ann. Inst. Henri Poincaré A, Physique Théorique,* **11**, 251-275 (1969).

97. J. Madore, *Ann. Inst. Henri Poincaré A, Physique Théorique,* **12**, 285-305 (1970); and **12**, 365-392 (1970).

98. L. Blanchet, Radiative gravitational fields in General Relativity II. Asymptotic behaviour at future null infinity, *Proc. R. Soc. Lond. A,* in press.

99. H. Bondi, M.G.J. van der Burg and A.W.K. Metzner, *Proc. R. Soc. Lond. A* **269**, 21-52 (1962).

100. J.L. Anderson, Gravitational radiation damping of systems with compact sources, *in:* "Abstracts of contributed papers to the 11th international conference on General Relativity and Gravitation," volume II, p 468 (Stockholm, Sweden, July 6-12, 1986).

101. S. Chandrasekhar and F.P. Esposito, *Astrophys. J.* **160**, 153-180 (1970).

102. M. Walker and C.M. Will, *Astrophys. J. Lett.* **242**, L129-L133 (1980).

103. W.L. Burke, The coupling of gravitational radiation to nonrelativistic sources, Ph.D. Thesis, Caltech (1969), unpublished.

104. K.S. Thorne, *Astrophys. J.* **158**, 997-1019 (1969).

105. W.L. Burke and K.S. Thorne, Gravitational radiation damping, *in:* "Relativity," M. Carmeli *et al.*, editors, pp 209-228, Plenum Press, N.Y. (1970).

106. W.L. Burke, *J. math. Phys.* **12**, 401-418 (1971).

107. R.E. Kates, *Phys. Rev.* **D22**, 1871-1878 (1980).

108. R.E. Kates, *Gen. Rel. Grav.* **18**, 235 (1986).

109. B.D. Miller, *Astrophys. J.* **187**, 609-620 (1974).

110. G. Schäfer, *Lett. Nuovo Cim.* **36**, 105-108 (1983).

111. F.I. Cooperstock and D.W. Hobill, *Gen. Rel. Grav.* **14**, 361-378 (1982) and references therein; F.I. Cooperstock and P.H. Lim, *Can. J. Phys.* **64**, 134-139 (1986); and *Astrophys. J.* **304**, 671-681 (1986); and references therein.

112. B. Linet, *C.R. Acad. Sc. Paris, série II* **292**, 1425-1427 (1981).

113. W.B. Bonnor, *Phil. Trans. R. Soc. Lond. A* **251**, 233-271 (1959).

114. W.B. Bonnor and M.A. Rotenberg, *Proc. R. Soc. Lond. A* **289**, 247-274 (1966); A.J. Hunter and M.A. Rotenberg, *J. Phys. A* **2**, 34-49 (1969); W.B. Bonnor, Gravitational wave tails, *in:* "Ondes et Radiations Gravitationelles," pp 73-81, Editions du C.N.R.S., Paris (1974).

115. L. Blanchet and T. Damour, *C.R. Acad. Sc. Paris, sérié II* **298**, 431-434 (1984).

116. Y. Gürsel, *Gen. Rel. Grav.* **15**, 737-754 (1983).

117. R. Penrose, *Proc. R. Soc. Lond. A* **284**, 159-203 (1965).

118. L. Blanchet and T. Damour, On the influence of the past in gravitationally radiating systems, in preparation.

119. W.M. Suen, Multipole moments for stationary non-asymptotically flat systems in general relativity, *Phys. Rev. D*, in press; X.H. Zhang, Multipole expansions of the general relativistic field of the external universe, *Phys. Rev. D*, in press.

120. J.L. Anderson, R.E. Kates, L.S. Kegeles and R.G. Madonna, *Phys. Rev.* **D25**, 2038-2048 (1982); J.L. Anderson and R.G. Madonna, *Gen. Rel. Grav.* **15**, 1121-1129 (1983).

121. F.K. Manasse, *J. math. Phys.* **4**, 746-761 (1963).

122. M. Demianski and L.P. Grishchuk, *Gen. Rel. Grav.* **5**, 673 (1974).

123. P.D. D'Eath, *Phys. Rev.* **D11** 1387-1403 and 2183-2199 (1975).

124. R.E. Kates, *Phys. Rev.* **D22**, 1853-1870 and 1871-1878 (1980).

125. T. Damour, ref. 67 and ref. 69.

126. K.S. Thorne and J.B. Hartle, *Phys. Rev.* **D31**, 1815-1837 (1985).

127. D.M. Eardley, *Astrophys. J.* **196**, L59-L62 (1975); C.M. Will and D.M. Eardley, *Astrophys. J.* **212**, L91-L94 (1977).

128. C.M. Will, "Theory and experiment in gravitational physics," Cambridge University Press, Cambridge (1981); and *Phys. Rep.* **113**, 345-422 (1984).

129. M. Riesz, *Acta Mathematica* **81**, 1-223 (1949).

130. L. Bel, T. Damour, N. Deruelle, J. Ibañez and J. Martin, *Gen. Rel. Grav.* **13**, 963-1004 (1981).

131. L. Lagrange, Nouvelle méthode pour résoudre les équations littérales par le moyen des séries, *Mémoires de l'Académie Royale des Sciences et Belles Lettres de Berlin*,

tome 24 (1770); see also E.T. Whittaker and G.N. Watson, "A course of modern analysis," 4th ed., p 132, Cambridge University Press, Cambridge (1978) .

132. T. Damour and N. Deruelle, *Phys. Lett.* **87A**, 81-84 (1981).

133. T. Damour and N. Deruelle, *C.R. Acad. Sc. Paris, série II*, **293**, 537-540 (1981).

134. T. Damour and N. Deruelle, *C.R. Acad. Sc. Paris, série II*, **293**, 877-880 (1981).

135. T. Damour and G. Schäfer, *Gen. Rel. Grav.* **17**, 879-905 (1985).

136. P.S. Laplace, Sur le principe de la gravitation universelle, *Mémoires de l'Académie Royale des Sciences de Paris (Savants Etrangers) année 1773*, t. VII (1776), reprinted *in:* "Oeuvres complètes de Laplace," t. 8, pp 201-275 (especially pp 219-234), Gauthier-Villars, Paris (1891); see also P.S. Laplace, "Traité de Mécanique Céleste," t. 4, seconde partie, livre X, Chapitre VII, reprinted *in:* "Oeuvres complètes de Laplace," t. 4, pp 314-327 (especially pp 326-327), Gauthier-Villars, Paris (1880).

137. Henri Poincaré, *C.R. Acad. Sc. Paris* **140**, 1504-1508, 5 June, 1905; *Rendiconti del Circolo matematico di Palermo*, **21**, 129-176 (1906) (written in July 1905); *Revue générale des sciences pures et appliquées*, **19**, 386-402 (1908). These articles are reprinted *in:* "Oeuvres de Henri Poincaré," t. IX, pp 489-493, 494-550, and 551-586, Gauthier-Villars, Paris (1954).

138. T. Damour, *Phys. Rev. Lett.* **51**, 1019-1021 (1983); T. Damour, Un nouveau test de la relativité générale, *in:* "Proceedings of Journées Relativistes 1983," S. Benenti, M. Ferraris and M. Francaviglia, editors, pp 89-110, Pitagora Editrice, Bologna (1985).

139. P.C. Peters and J. Mathews, *Phys. Rev.* **131**, 435-440 (1963).

140. P.C. Peters, *Phys. Rev.* **136**, B1224-B1232 (1964).

141. L.W. Esposito and E.R. Harrison, *Astrophys J.* **196**, L1-L2 (1975); R.V. Wagoner, *Astrophys. J.* **196**, L63-L65.

142. J.H. Taylor and J.M. Weisberg, *Astrophys. J.* **253**, 908-920 (1982); J.M. Weisberg and J.H. Taylor, *Phys. Rev. Lett.* **52**, 1348-1350 (1984).

143. T. Damour and N. Deruelle, *Ann. Inst. Henri Poincaré, Physique Théorique* **43**, 107-132 (1985) (paper I, The post-Newtonian motion); and **44**, 263-292 (1986) (paper II, The post-Newtonian timing formula).

144. J.H. Taylor, Astronomical and Space Experiments to test Relativity, invited lecture given at the 11th International Conference on General Relativity and Gravitation, (Stockholm, Sweden, July 6-12, 1986) to be published in the Proceedings.

MATHEMATICAL FOUNDATIONS OF THE THEORY OF RELATIVISTIC STELLAR AND BLACK HOLE CONFIGURATIONS

Brandon Carter

Groupe d'Astrophysique Relativiste - D.A.R.C.
CNRS - Observatoire de Paris
92195 Meudon Principal Cedex, France

§1 INTRODUCTION

1.1 Background

Late in the year 1783, Benjamin Franklin (then U.S. representative in France) wrote from Paris to his regular London scientific correspondent Sir Joseph Banks (then president of the Royal Society) that the most exciting recent development in France was the breakthrough in ballooning resulting from the use by Charles of hydrogen (as an alternative to the hot air technique that had just been tried out with only moderate success by the Montgolfiers). After the (manifestly reluctant) admission that "our friends on your side of the water" were more advanced in the competition to achieve *practical* flying, and the declaration that his own side nevertheless claimed credit for the fundamental research on which it was based (thinking particularly of the hydrogen bubbles blown by Cavendish, who had recently carried out the first serious study of the chemical and physical properties of the lightest element), Banks went on to reply (9th Dec. 1783) to the effect that the most interesting recent scientific event in London had been the presentation at the Royal Society of a "very curious paper" on the influence of gravity on light. In the paper in question, which was published soon after,[1] John Michell (1724-1783) foreshadowed modern black hole theory by evaluating the critical radius given in modern notation as

$$r = \frac{2GM}{c^2} \tag{1.1}$$

within which a body of mass M would become invisible because the Newtonian gravitational potential

$$\varphi = -\frac{GM}{r} \tag{1.2}$$

would exceed the specific (Newtonian) kinetic energy $\frac{1}{2}c^2$ of a projectile moving with the speed c of light. Michell's line of reasoning was later taken over almost word for word – but without acknowledgement – by Laplace in his 1796 edition of the "Exposition du Système du Monde".[2]

Laplace's omission of any reference to Michell's prior discovery of the formula for what is now known as the "Schwarzschild" radius (despite the fact that, as the above correspondence shows, Michell's paper attracted wide attention at the time) is not in itself sufficient to explain the historical oblivion to which Michell succumbed until his role was brought to light by work of Hardin,[3] McCormack[4] and in the present context by Schaffer.[5] Nor can this oblivion be explained just by the fact that the "Black Hole Paradigm" (as it would now be termed) fell out of fashion altogether (even Laplace dropped the issue in later editions of his book) soon after the general adoption early in the nineteenth century of the *wave theory* of light (which, prior to Einstein, was perceived as being inconsistent with it subjection to direct gravitational influence). Michell has in fact an at least equally great claim to fame among gravitational physicists as the inventor and builder of the prototype for the "Cavendish" experiment [i.e. the measurement of Newton's G by the *torsion balance,* whose modern descendents, as developed by Eötvös, Dicke and others are still responsible for the most precise and probing experimental results in the discipline], having been fully recognised[6] by Cavendish as the initiator of the programme. Perhaps the reason for his historical eclipse was simply that Michell was outshone by his colleagues, of whom even Cavendish himself was not the most famous: before his semi-retirement as a Yorkshire clergyman, Michell had in fact been a very active and versatile Cambridge don as Queens College lecturer in Classics and Mathematics, and later as University (Woodwardian) Professor of Geology, with a particularly strong interest in the field of Astronomy, where he made a very important contribution to the early development of the greatest astronomer of the age, William Herschel. Not only did he strongly influence Herschel's speculations about the significance of his discovery of the extragalactic nebulae,[7] Michell also contributed to the discovery in the first place by teaching Herschel very down-to-earth experimental techniques for the construction of the first high quality reflecting telescopes. In his own right Michell made one of the most important early studies of stellar parallax,[8] and of the possibility of detecting stellar binary systems, envisaging in particular the possibility of "black hole" companions, of whose existence, in his own words[1,5] "we could have no information from sight, yet if any other luminous body should happen to revolve around them we might still perhaps from the motion of these revolving bodies infer the existence of the central ones." Michell's speculations (referred to by Priestly[9] long before their official publication,[8,1]) were made two centuries before the discovery of Cygnus X-1.

Be that as it may, interest in the "Black Hole Paradigm" as such was not resuscitated immediately even on the occasion of Einstein's General Relativity Theory (which revalidated Michell's supposition that light is subject to the effect of gravity) nor even by Schwarzschild's rediscovery,[10] in the relativistic context, of the Michell radius, (which was at first misinterpreted as the locus of a mathematical breakdown of the theory rather than a regular physical phenomenon).

The question of gravitational trapping of light (and hence, in a relativistic context, of everything else) began to be taken seriously again only after the introduction of quantum theory had for the first time provided a basic understanding of the microscopic structure and the nature of the internal equilibrium of ordinary matter, along lines I have reviewed elsewhere,[11] thus making possible Chandrasekhar's epoch-making discovery[12] of the upper limit on the mass of a star in a cold static equilibrium state, which posed the question of the ultimate gravitational collapse of any larger mass when its thermal energy is exhausted.

Even without any detailed understanding of the quantum structure of matter one could in principle have predicted in advance that for some mass short of the Michell (and Laplace) limit, – based on extrapolation at normal densities comparable with that of water – cold self-gravitating equilibrium states would cease to be possible simply because of the special relativistic causality requirement that the speed of light c should be an upper bound on the speed of sound $(dP/d\rho_m)^{1/2}$ where P is the pressure as a function of the density ρ_m of the cold matter, which implies that whatever the micro-structure of matter at high densities, the pressure should be subject to the order of magnitude limit

$$P \lesssim \rho_m c^2. \tag{1.3}$$

Now the requirement that the pressure should be sufficient (without reinforcement by rotational or other effects) to prevent or stop a gravitational collapse, i.e. that the internal pressure gradient given in order of magnitude by P/r should balance the gravitational force density similarly given in order of magnitude by $\rho_m|\varphi|/r$ where φ is the gravitational scalar potential, is expressible as

$$P \gtrsim \rho_m|\varphi| \tag{1.4}$$

where equality corresponds to the virial theorem condition for self gravitating equilibrium. It is immediately evident that these conditions are compatible only if we have

$$|\varphi| \lesssim c^2. \tag{1.5}$$

As soon as one enters the Michell regime characterised by the violation of (1.5),i.e. by the *gravitational trapping of light,* it will become impossible to satisfy (1.4) which implies that runaway collapse will inevitably ensue, leading directly to the formation of a *singularity* unless the nature of the problem is modified by the build up of rotational or other effects.

If this line of reasoning is crude, it has the corresponding advantage of being robust, not even depending on the precise inverse square law on which (1.2) is based. The scenario of spherical collapse to a singularity in the context of General Relativity was first worked out in a particular case by Oppenheimer and Snyder[13] (see the accompanying lectures of Eardley for an account of more modern and complete work on this subject). The first attempt to show rigourously that the light trapping phenomenon is inevitably followed by the formation of some sort of singularity in General Relativity theory, whether or not there is spherical symmetry, was made by Penrose,[14] a more complete treatment having been given more recently in the treatise of Hawking and Ellis.[15]

1.2 Purpose and Plan

Much of the understanding of such phenomena that has been attained in recent years, at least in the spherically symmetric case, is available by now in standard textbooks. To earlier examples such as those of Zeldovitch and Novikov,[16] Weinberg,[17] and particularly the monumental work of Misner, Thorne, and Wheeler[18] we can now add more up to date (and for many purposes more easily readable, albeit not necessarily more complete) textbooks such as the excellent and mutually complementary works of Demianski[19] and Wald.[20] In view of the availability of such texts, the written lectures to be given below will differ somewhat from those actually delivered in the summer school at Cargèse, in so much as I shall leave out the introductory material concerned with the widely familiar spherical case in order to be able to give more attention to less readily available (and in some cases new) results concerning more general configurations. The objective here is essentially the same as I set myself in the 1972 Les Houches lectures[21] namely to give a reasonably self-contained and as far as possible up-to-date account of the solid mathematical foundations (in so far as they exist) of the classical relativistic theory of *self gravitating stellar, and more particularly black hole, configurations.* Although ultimate astrophysical relevance has been a guiding criterion in deciding what topics were to be included in the limited space available (for example I have made no attempt to discuss the Penrose type singularities predicted to occur *inside* the black hole horizon) the spirit of these lectures is essentially mathematical. For discussions of more physically and astrophysically motivated applications the reader is referred to works such as those of Shapiro and Teukolsky,[22] or Thorne, Price and Macdonald,[23] and to the following courses by Thorne, Lynden-Bell and Kuperus in the present volume. Further mathematical developments concerning the very important question of stability of equilibrium configurations will be developed in the course by Schutz, while the question of singularities has been discussed in the lectures by Eardley.

The plan of the following lectures is as follows. Section 2 will be devoted to the basic ideas (mainly due to Penrose and Hawking) of dynamic black hole theory. Section 3 will be concerned with general principle of stationary equilibrium states, with in particular a new demonstration that in the *non-rotating* case such states must have the *staticity* property (on which the Israel type sphericity theorems are based) except perhaps in extreme circumstances such as might result from electric fields of the order of 10^{27} volts. Section 4 will be concerned with *rotating* equilibrium states, which Hawking has shown to be necessarily *axisymmetric*. Finally Section 5 will present the essential steps in the (recently completed) demonstration that if there are no external sources the solution must belong to the Kerr (or charged Kerr-Newman) family, whose special algebraic properties will be presented as arising from the presence of a Killing-Maxwell-Yano system.

The signature, normalisation and other notation conventions used here are different from those of Hawking and Ellis[15] and of the 1972 Les Houches Lectures,[21, 24] but are based on those of Misner, Thorne and Wheeler,[18] with the speed of light c (but not Newton's constant G) set equal to unity, and with the intention of being as far as possible consistent with my more recent review[25] (concerned particularly with electromagnetic effects) and with my previous Cargèse Lecture notes on "Underlying Mathematical Structures of Classical Gravitation Theory."[26]

§2 NOTIONS OF GENERAL (DYNAMIC) BLACK HOLE THEORY

2.1 Definition

The understanding of the comparitively simple spherical collapse scenario[27] that had been attained by the mid-1960's suggested the utility of defining a *black hole* in an asymptotically flat spacetime (we shall not attempt to consider more general, perhaps cosmologically relevant, cases here) as the part of spacetime from which *no future directed timelike or null line can escape to arbitrarily large distance* into the asymptotically flat outer region.[28] Although it is less obviously relevant to contempory astrophysical (as opposed to cosmological) contents, it is also useful to the analogous but *time reversed* concept (as specified by a definition that is identical except for the replacement of "future" by "past") for which the term *white hole* is commonly employed,[29] since such regions turn up automatically in the analytic extension to the past of the stationary spacetimes in which we shall be interested as asymptotic equilibrium states (as we have already seen in the case of the Kruskal extension of the Schwarzschild solution).

What we shall mostly be concerned with in practice is the region outside, i.e. the part of spacetime belonging neither to the black hole nor the white hole region, for which I use the term *Domain of Outer Communication* (D.O.C. for short) in recognition of the fact that it can be defined in a more positive manner as the set of points in the asymptotically flat spacetime from which it is possible to construct both a future and a past directed timelike line to arbitrarily large distance into the outer region.[30]

Before considering applications to black holes as such, we shall recall some basic properties[14,15,25] of timelike and null geodesic congruences, particularly in the case when the latter are generators of characteristic hypersurfaces.

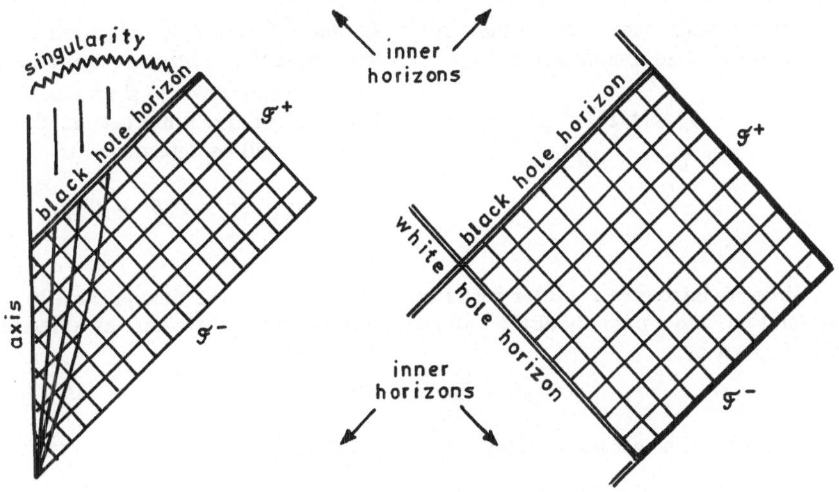

Figure : Cross hatched area indicates extent of D.O.C. in conformal spacetime diagram for:
(a) realistic gravitational collapse (matter lines shown),
(b) analytic extension of stationary equilibrium limit.

2.2 Kinematics of Characteristic (Null) Boundaries

The normal covector $\underline{\ell}$ to any hypersurface must satisfy the irrotationality condition $\underline{\ell} \wedge \partial \underline{\ell} = 0$, i.e.

$$3\ell_{[\mu}\partial_\nu \ell_{\rho]} = 0 \tag{2.1}$$

In the case of a *null* hypersurface, for which $\vec{\ell} \cdot \underline{\ell} = 0$, i.e.

$$\ell^\mu \ell_\mu = 0, \tag{2.2}$$

one will therefore have $\underline{\ell} \wedge (\vec{\ell} \cdot \partial \underline{\ell}) = 0$ and hence for *any* extension off the hypersurface of the vector field $\vec{\ell}$, we shall have $\vec{\ell} \wedge (\vec{\ell} \mathcal{L} \underline{\ell}) = 0$ where

$$\vec{\ell} \mathcal{L} \ell_\mu = \ell^\nu \partial_\nu \ell_\mu + \partial(\ell^\mu \ell_\mu) = \dot{\ell}_\mu + \tfrac{1}{2}\partial_\mu(\ell^\nu \ell_\nu) \tag{2.3}$$

using the notation

$$\dot{\ell}_\mu = \ell_{\mu;\nu}\ell^\nu . \tag{2.4}$$

Thus one sees that the null generator of a null hypersurface satisfies $\underline{\ell} \wedge \dot{\underline{\ell}} = 0$, i.e.

$$\dot{\ell}_{[\mu}\ell_{\nu]} = 0 \tag{2.5}$$

which is the equation of a geodesic with not necessarily affine parametrisation.

2.3 Generalised Raychaudhuri Equation for Timelike and Null Geodesic Congruences

In order to analyze the behaviour of a congruence generated by a vector $\ell^\mu = \frac{dx^\mu}{dt}$ say, of either null geodesic or ordinary timelike nature let us choose a 1-form $\underline{\beta}$ cutting across the congruence, and an associated projection tensor γ with components β_μ and $\gamma^\mu{}_\nu$, in such a way as to satisfy the normalisation and orthogonality conditions

$$\begin{aligned} \beta_\mu \ell^\mu &= -1, \qquad & \gamma^\mu{}_\nu \ell^\nu &= 0, \qquad & \ell_\nu \gamma^\nu{}_\mu &= 0 \\ \gamma^\nu{}_\rho \gamma^\rho{}_\mu &= \gamma^\nu{}_\mu, \qquad & \beta_\nu \gamma^\nu{}_\mu &= 0, \qquad & \gamma^\mu{}_\nu \beta^\nu &= 0 \end{aligned} \tag{2.6}$$

In the timelike case $\underline{\ell}\cdot\vec{\ell}<0$ these requirements lead unambigously to

$$\beta_\mu = (\ell_\nu \ell^\nu)^{-1}\ell_\mu$$
$$\gamma^\mu{}_\nu = g^\mu{}_\nu + \beta^\mu \ell_\nu.$$

In the null case $\underline{\ell}\cdot\vec{\ell}=0$, even when we impose the additional requirements $\vec{\beta}\cdot\underline{\beta}=0$ the choice of $\vec{\beta}$ remains ambiguous, but once it has been made the projector will be determined as

$$\gamma^\mu{}_\nu = g^\mu{}_\nu + \ell^\mu\beta_\nu + \beta^\mu\ell_\nu.$$

Here we introduce a dot for derivation with respect to the time parameter t under consideration, and a possibly distinct affine parameter τ so that we have an affine tangent vector $\vec{u}$ given by

$$u^\mu = \frac{dx^\mu}{d\tau}, \qquad \ell^\mu = \dot{\tau} u^\mu.$$

The coordinate time acceleration for this vector will be given by

$$\underline{\dot{u}}^\mu = u^\mu{}_{;\nu}\ell^\nu \tag{2.7}$$

– it automatically satisfies the orthogonality condition

$$\dot{u}^\mu \ell_\mu = 0. \tag{2.8}$$

If we now define

$$a_\mu = \beta^\nu(\nabla_\nu \ell_\mu + \beta_\mu \ell^\rho \nabla_\nu \ell_\rho) \tag{2.9}$$

then we can see that:

In the *timelike* case $\underline{\ell}\cdot\vec{\ell}<0$, with proper time normalisation, $\underline{u}\cdot\vec{u}=-1$, then $\underline{a}$ will be the *ordinary proper time acceleration.*

In the *null* case $\underline{\ell}\cdot\vec{\ell}=0$ such an interpretation is not possible; even if we impose the restriction that the congruence be geodesic $\dot{u}=0$, the form $\underline{a}$ need not vanish.

It will however be true in both the timelike and null geodesic cases that $\dot{\tau}\underline{\dot{u}} = -(\underline{\ell}\cdot\vec{\ell}\,)\underline{a}$ and hence that we shall have

$$\dot{\ell}_\mu = \kappa \ell_\mu - (\ell_\nu \ell^\nu) a_\mu \tag{2.10}$$

where the coefficient κ is a measure of non-affinity

$$\kappa = (\ln\dot{\tau})^{\cdot} = \ddot{\tau}/\dot{\tau} = -\beta^\mu \dot{\ell}_\mu \tag{2.11}$$

which will have an important role in what follows. One can now decompose the gradient of $\underline{\ell}$ in the form

$$\nabla_\mu \ell_\nu = v_{\mu\nu} - \ell_\nu(\kappa\beta_\mu + \gamma^\kappa_\mu \beta^\lambda \nabla_\kappa \ell_\lambda) - \ell_\mu a_\nu \tag{2.12}$$

where the orthogonal projection $v_{\mu\nu} = \gamma_\mu{}^\kappa \gamma_\nu{}^\lambda \nabla_\kappa \ell_\lambda$ can itself be decomposed in symmetric and antisymmetric parts

$$v_{\mu\nu} = \theta_{\mu\nu} + \omega_{\mu\nu}; \qquad \theta_{[\mu\nu]} = 0, \qquad \omega_{(\mu\nu)} = 0. \tag{2.13}$$

The Frobenius irrotationality condition (2.1) is equivalent to $\omega_{\mu\nu}=0$. For infinitesimal neighboring members of the congruence separated by a relative displacement vector $d\vec{x}$ orthogonal to $\vec{\ell}$, the rate of change of the squared distance will be given by

$\frac{1}{2}(ds^2)^{\cdot} = \theta_{\mu\nu}dx^\mu dx^\nu$ which means that $\theta_{\mu\nu}$ is interpretable as the expansion (strain) rate tensor, its trace $\theta = \theta_\mu{}^\mu = \gamma^{\mu\nu}\nabla_\mu \ell^\nu$ being the divergence of the congruence, which is interpretable as the fractional rate of expansion of the volume (in the timelike case) or area (in the null case).

When the parametrisation is non-affine, the vector field divergence is not the same as the expansion rate of the congruence: there is an extra term

$$\ell^\mu{}_{;\mu} = \theta + \kappa. \tag{2.14}$$

The *shear rate* is the trace free part of the strain rate tensor,

$$\sigma_{\mu\nu} = \theta_{\mu\nu} = (\gamma^\kappa_\kappa)^{-1}\theta\, \gamma_{\mu\nu}. \tag{2.15}$$

Defining scalars

$$\omega^2 = \tfrac{1}{2}\omega_{\mu\nu}\omega^{\mu\nu},, \qquad \sigma^2 = 2\sigma_{\mu\nu}\sigma^{\mu\nu}, \tag{2.16}$$

one will have

$$\ell_{\mu;\nu}\ell^{\mu;\nu} = \kappa^2 + (\gamma^\kappa_\kappa)^{-1}\theta^2 + \tfrac{1}{2}\sigma^2 - 2\omega^2. \tag{2.17}$$

Now if we take the contraction of the Ricci identity

$$(\nabla_\mu\nabla_\nu - \nabla_\nu\nabla_\mu)\ell^\kappa = R_{\mu\nu}{}^\kappa{}_\lambda \ell^\lambda \tag{2.18}$$

that can be considered as a definition of the Riemann curvature tensor, we obtain the identity

$$(\ell^\nu{}_{;\nu})_{;\mu}\ell^\mu = (\ell^\mu{}_{;\nu}\ell^\nu)_{;\mu} - \ell^{\nu;\mu}\ell_{\mu;\nu} - R_{\mu\nu}\ell^\mu\ell^\nu \tag{2.19}$$

(where the Ricci tensor is defined by $R_{\mu\nu} = R_{\kappa\mu}{}^\kappa{}_\nu$) which holds for an arbitrary vector field $\vec{\ell}$. We thus obtain the generalised Raychaudhuri equation

$$\dot\theta - \kappa\theta = -(\gamma_\kappa{}^\kappa)^{-1}\theta^2 - \tfrac{1}{2}\sigma^2 + 2\omega^2 + (\ell_\kappa \ell^\kappa \alpha^\mu)_{;\mu} - R_{\mu\nu}\ell^\mu\ell^\nu \tag{2.20}$$

in a form that is valid for arbitrarily parametrised timelike or null geodesic congruences.

In the case with which we are principally concerned here, that of the null geodesic generators of a characteristic hypersurface in four dimensions, for which $\ell_\mu\ell^\mu = 0$, $\omega_{\mu\nu} = 0$, $\gamma_\kappa{}^\kappa = 2$, this reduces to the form

$$\dot\theta - \kappa\theta = -\tfrac{1}{2}(\theta^2 + \sigma^2) - R_{\mu\nu}\ell^\mu\ell^\nu \tag{2.21}$$

where the left hand side is expressible in terms of the expansion rate $\theta^{(0)}$ say with respect to an affine parameter τ, i.e. $\theta = \dot\tau\theta^{(0)}$, by $\dot\theta - \kappa\theta = \dot\tau\dot\theta^{(0)}$.

2.4 The Horizon of a Black Hole

Much of our interest will be concentrated on the boundary, $\mathcal{H}^+$, of the black hole region, which includes the part of the boundary of the Domain of Outer Communication that lies in its future. (The past boundary of the D.O.C. would belong to the boundary

$\mathcal{H}^-$ of the white hole region – if present.) The boundary $\mathcal{H}^+$ which is known as the *horizon* of the black hole is an example of a *past event horizon*, meaning a surface specified as the boundary or the past of some given set of events, excluding the closure of the set itself. Such an event horizon would be an ordinary null hypersurface wherever it is smooth, but as was pointed out by Penrose and Hawking,[14, 31] it can be characterised more generally (even in the presence of intersections and caustics) by the properties of being achronal (i.e. no two points on it can be connected by a timelike curve) but nevertheless such that through any given point there is (at least) one future directed null geodesic that never leaves the horizon when extended to the future (unless it reaches the defining set).

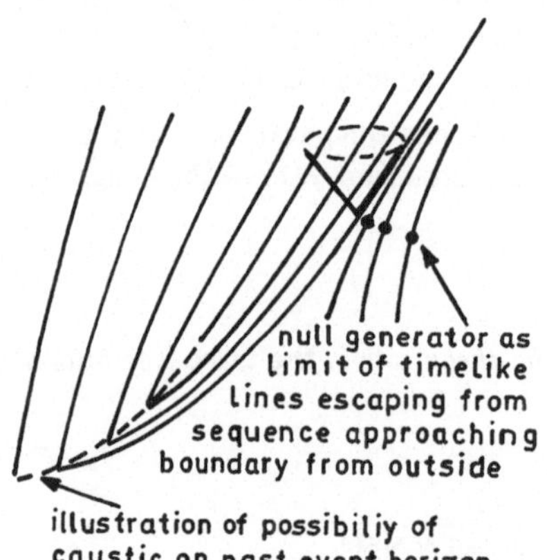

On any past event horizon one can apply the Penrose null version of the Raychaudhuri divergence equation described in §2 to deduce that (except on intersections and caustics) wherever the energy inequality $R_{\mu\nu}\ell^\mu\ell^\nu \geq 0$ on the contraction of the Ricci tensor with the null tangent generator ℓ^μ at a point on the horizon is satisfied, it will follow that the divergence $\theta^{(0)}$ of the null generator will evolve subject to the inequality

$$\frac{d\theta^{(0)}}{d\tau} \leq -\tfrac{1}{2}(\theta^{(0)})^2 \tag{2.22}$$

with respect to any affine time parametrization τ. It follows that if $\theta^{(0)}$ becomes negative at any point on a horizon (i.e. if there is convergence) then the null generator can continue in the horizon for at most a finite affine distance before reaching a point at which $\theta^{(0)} \to -\infty$, i.e. a point of infinite convergence representing a caustic beyond which the generators intersect and hence leave the horizon (since otherwise achronality would be violated). Now if one applies this lemma (due to Penrose) to the particular case of the horizon $\mathcal{H}^+$ of a black hole, which can be considered as the boundary of the past of a region at large distance from the horizon, so that (unless the D.O.C. is incomplete) its null generators can be extended infinitely to the future, then one can deduce (following Hawking) that they cannot have negative divergence $\theta^{(0)}$.

Now whenever a horizon is smooth, an infinitesimal 2-surface element transported by the null generators will have a corresponding area $d\mathcal{S}$ that evolves according to

$$\frac{d}{dt}(d\mathcal{S}) = \theta\, d\mathcal{S} \tag{2.23}$$

where t is any time parameter and θ is the corresponding divergence which will be related to the affine divergence $\theta^{(0)}$ defined with respect to the affine parameter τ by

$$\theta = \dot{\tau}\theta^{(0)}, \qquad \dot{\tau} = d\tau/dt. \tag{2.24}$$

Since (at points where the horizon is not smooth) new null generators may begin but old ones cannot end, the inequality derived above shows that the total area

$$\mathcal{A} = \oint d\mathcal{S} \tag{2.25}$$

of any compact spacelike section through a black hole horizon must evolve subject to the inequalities[32]

$$\frac{d\mathcal{A}}{dt} \geq \oint \theta \, dS \geq 0. \tag{2.26}$$

In particular if two black holes with areas $\mathcal{A}_1$ and $\mathcal{A}_2$ undergo a merger, then the area $\mathcal{A}_3$ say of the combined black hole must satisfy

$$\mathcal{A}_3 > \mathcal{A}_1 + \mathcal{A}_2. \tag{2.27}$$

In view of the obvious analogy between the black hole sectional area and the entropy in ordinary thermodynamics, this famous result due to Hawking was baptised as the "Second Law of Black Hole Mechanics" almost immediately after it was discovered. It was not until several years later that Bekenstein[33] guessed and Hawking[34] subsequently proved (on the basis of the quantum particle creation mechanism) that there is not merely an analogy but an actual proportionality between the area and an associated black hole entropy.

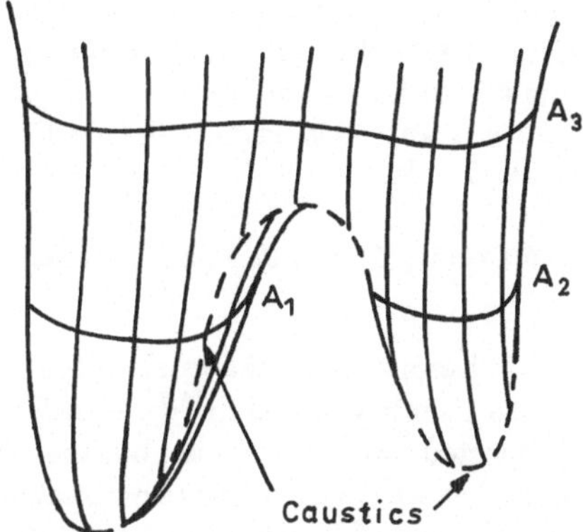

2.5 Asymptotic Predictability, Closed Trapped Surfaces, and Apparent Horizons

Except in strictly stationary (time independent) situations, the black hole and its horizon $\mathcal{H}^+$ have the inconvenient property of being defined teleologically in the sense of being specified in terms of what will happen in the (in practice unknown) future. It is therefore often useful to work with the related but contemporally defined concepts of *trapped regions*[14,15] and *apparent horizons*[32,15] specified with respect to a timelike hypersurface having the property of being a *Cauchy initial data surface,* at least for the part of spacetime at large radius in the asymptotically flat region meaning that any past directed timelike or null line from the future of the hypersurface at large radius can always be extended back until it intersects the hypersurface somewhere. The existence of such a Cauchy surface for the future of the outer asymptotically flat part of spacetime is the condition that has been dubbed as *asymptotic predictability* by Hawking.[24,15] The famous cosmic censorship hypothesis of Penrose[28] is interpretable as the conjecture that this condition will be satisfied in any "physically realistic" mode. This conjecture arose out of Penrose's pioneering study[14] of the implications of the existence of a *closed trapped surface,* meaning a compact spacelike 2-surface such that not only its ingoing null normals but even the outgoing ones have negative divergence, $\theta < 0$ towards the future. Penrose pointed out that the boundary of its future would be a future event horizon (the time reversed analogue of a past event horizon as defined above) and thus generated by past directed null geodesics that can end only on the closed trapped surface itself. Moreover if it lies in the asymptotically predictable future of a Cauchy hypersurface, then all such null generators will extend back to the closed trapped surface. This enabled Penrose to deduce that the boundary of its future would be *compact* by introducing the inequality (2.22) which tells us that from a point at which $\theta < 0$ the null generator can remain on the boundary for only a finite distance to the future. Since it is intuitively inconceivable (and can be shown to be mathematically impossible) for such a future boundary surface to be compact if spacetime is *well behaved* and *complete* Penrose thereby obtained the first *"singularity" theorem.* Hawking[32,24,15] subsequently pointed out that it also followed that the future of the closed trapped surface could not extend to the outer flat region (since otherwise its boundary would obviously not be compact) and hence that it must lie entirely *within* the black hole region. He introduced the term *apparent horizon* for the limiting trapped surface (with outer null normals satisfying $\theta = 0$) for the boundary of the region guaranteed in advance to lie within a black hole.

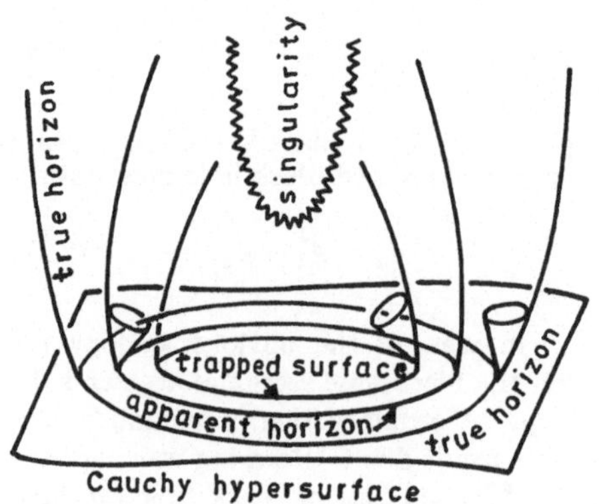

2.6 Cosmic Censorship and the Existence of an Asymptotic Equilibrium

Despite of all the work in the many intervening years the general validity of Penrose's cosmic censorship[28,29] is still a subject of controversy.[36] There have been many purported counterexamples, but they are widely judged to be too specialised (e.g. depending on exactly zero pressure or exact spherical symmetry) to satisfy the rather subjective requirement of "physical realism." Even if physically plausible counterexamples ("naked singularities") were one day to be discovered after all, the theorems based on asymptotic predicability would very likely remain relevant for a considerable range of physical circumstances nevertheless.

Subject to asymptotic predictability, considerations of the conservation of asymptotically radiated gravitational and other forms of energy, and the positivity of mass (which has recently been established on a rigorous basis by work of Schoen, Yau, Witten and others[37,38]) makes plausible the further conjecture that an isolated system will ultimately settle down towards a stationary state, and nearly all recent progress in black hole theory has been concerned with the study of the approximately or exactly stationary equilibrium states that one expects to obtain in this way.

In the asymptotically stationary case, Hawking's black hole area increase theorem[32] can be extended[39,25] to more detailed results concerning the rate of area increase, starting from the general Raychaudhuri equation (obtained in §2) for the null geodesic generators in the horizon which will be expressible in the form $\dot{\tau}\dot{\theta}^{(0)} = -8\pi GD$ where in terms of an arbitrary (non-affine) time t and expansion rate θ we shall have $\dot{\tau}\theta^{(0)} = \dot{\theta} - \kappa\theta$ (where τ is affine and $\theta^{(0)}$ is the corresponding affine expansion rate) while the "effective dissipation rate," D, is given by the expression

$$D = \frac{1}{16\pi G}(\sigma^2 + \theta^2 + 2R_{\mu\nu}\ell^\mu\ell^\nu) \tag{2.28}$$

Subject to the asymptotic stationarity condition, which implies $\theta \to 0$ as $t \to \infty$, the Penrose-Raychaudhuri differential equation for θ can be cast into integral form as

$$\theta = \int_t^\infty e^{\kappa t - \kappa' t'} \left(8\pi GD' + \dot{\kappa}'\theta' t'\right) dt' \tag{2.29}$$

which implies that the change in the area dS of a transported 2 dimensional cross sectional element between, times t_0 and t_1 say, will be given by

$$\ln\left(\frac{dS_1}{dS_0}\right) = \int_{t_0}^{t_1} \theta\, dt$$
$$= \left\{\int_{t_0}^{t_1} dt' \int_{t_0}^{t_1} dt + \int_{t_1}^{\infty} dt' \int_{t_0}^{t_1} dt\right\} e^{\kappa t - \kappa t'}(8\pi G D' + \dot{\kappa}'\theta' t') \tag{2.30}$$

This expression would of course appear much simpler if the parametrisation were chosen to be affine, so as to set $\kappa = 0$, but for the purpose of studying the stationary limit it is usually more convenient to choose the parametrisation given by the action of the stationary symmetry group, with respect to which κ will in general be a non-zero constant.

2.7 Approximate Equilibrium

As we shall show explicitly in the next section, when it is defined with respect to the group action parameter of a stationary solution, the parameter κ will be constant not only in time but also spatially over the horizon. Supposing that it has a value $\kappa \doteq \kappa_1 = \text{constant} > 0$ during an interval $0 = t_0 < t < t_1$ the preceeding expression gives[25]

$$\ln\left(\frac{dS_1}{dS_0}\right) \doteq \frac{8\pi G}{\kappa_1} \int_{0}^{t_1} dt\, (1 - e^{-\kappa_1 t})D + \frac{e^{\kappa_1 t_1} - 1}{\kappa_1} \int_{t_1}^{\infty} dt\, e^{-\kappa t}(8\pi G D + \dot{\kappa}\theta t) \tag{2.31}$$

in which we shall be able to neglect the second (teleological) term if D and $\dot{\kappa}$ remain zero for a sufficiently long time (compared with the natural timescale ${\kappa_1}^{-1}$) after the period t_0 to t_1. If we set the origin t_0 sufficiently far in advance of the first perturbing contribution D, which we suppose due to an infall of matter or radiation, then the first term simplifies also, leaving

$$\ln\left(\frac{dS_1}{dS_0}\right) \doteq \frac{8\pi G}{\kappa_1} \int_{t_0}^{t_1} D\, dt. \tag{2.32}$$

Substituting from Einstein's equations,

$$R_{\mu\nu} - \tfrac{1}{2} R g_{\mu\nu} = 8\pi G T_{\mu\nu} \tag{2.33}$$

and neglecting the θ^2 contribution (since θ is already necessarily small near equilibrium) Hartle and Hawking[39] used the foregoing formula to obtain the area increase formula

$$d\mathcal{A} = \oint \frac{8\pi G}{\kappa}\, dS \int dt\, \{\frac{\sigma^2}{16\pi G} + T_{\mu\nu}\ell^\mu \ell^\nu\}. \tag{2.34}$$

The first term inside the curly brackets above is interpretable as a viscous dissipation contribution if the viscosity coefficient is taken to be $1/16\pi G$. The second term is the energy flux across the horizon.

In the thermodynamic analogy[35] first suggested by Bekenstein,[33] in which $\mathcal{A}/\alpha G$ is taken to correspond to entropy for some proportionality constant α, the foregoing formula for the effect of a dissipative energy input suggested that $\alpha\kappa/8\pi$ should be taken to correspond to temperature. (Hawking's quantum particle creation mechanism has shown that the appropriate value is $\alpha = 4$ in units such that $\hbar = c = k = 1$.)

3. STATIONARY AND STATIC EQUILIBRIUM

3.1 Overview

For the remainder of this course we shall restrict our attention to *exactly stationary* equilibrium states (such as those to which the systems discussed in the previous section were supposed to be converging asymptotically) leaving the important question of the stability of such systems to be discussed in the following course by Schutz. [The drift of our by no means complete present day understanding is that the pure vacuum black hole equilibrium configurations are effectively stable even when rotating, whereas for material e.g. stellar type – equilibrium configurations tend to be beset by instabilities – though not necessarily on practically relevant timescales – except in the non-rotating limit.]

For nearly all of the following work we shall limit our attention to configurations that are not only stationary but also axisymmetric. In so far as complicated (solid as opposed to fluid) equations of state are excluded, this probably does not imply any serious loss of generality for isolated stable systems, since it is hard to imagine physically realistic deviations from axisymmetry that would not entail departure also from stationarity as a result of gravitational radiation. (Such execeptions as the Dedekind type internally shearing stationary ellipsoidal perfect fluid configurations are in effect special limit cases whose existence is unstable against inclusion of any viscosity, and as such do not constitute relevant counterexamples.) For ordinary stellar type fluid configurations I do not know of any significant mathematical progress towards rigorous proofs (as opposed to heuristic justifications) but in the case of pure vacuum black holes (whose physical intuition on the basis of practical experience is entirely lacking) an important result by Hawking,[32,24,15] which I refer to as the *Strong Rigidity Theorem* goes a large part of the way.

The point of departure for this result is the Penrose-Raychaudhuri equation (2.21) used in deriving the area increase theorem and more particularly the Hartle-Hawking formula in the previous section. It can be seen that for any material coupling satisfying the usual condition that $R_{\mu\nu}\ell^\mu\ell^\nu$ be positive definite for any timelike or, more pertinently, null vector ℓ^μ, a steady increase in the sectional area $\mathcal{A}$ of the horizon of the black hole can be avoided only if the null congruence generating the horizon satisfies

$$R_{\mu\nu}\ell^\mu\ell^\nu = 0, \qquad \theta = 0, \qquad \sigma = 0. \tag{3.1}$$

The absence of expansion and shear means that the field ℓ^μ generates an isometry of the degenerate, effectively 2-dimensional, metric induced on the horizon. This important *rigidity* property of the horizon results from the effective viscosity described in the previous section (and precludes any possibility of Dedekind type behaviour). A weak rigidity theorem which I had obtained previously[42,78] (by a simple argument recapitulated in §4.5) in the axisymmetric case, shows that the evolution of the horizon is rigid not only in having no intrinsic distortion but in the extrinsic sense that ℓ^μ must coincide with a Killing field combination of the postulated stationarity and axisymmetry generators. Hawking's *strong rigidity* theorem (whose proof[15] is too long to recapitulate here) goes much farther, in that on the basis only of the assumption of existence of the stationary symmetry, with asymptotically timelike Killing vector k^μ say

$$k_{(\mu;\nu)} = 0, \qquad V = -k_\mu k^\mu \implies V^\infty = 1 \tag{3.2}$$

(but *without* any assumption of axisymmetry) it establishes that the null tangent vector ℓ^μ of the horizon can be normalised so as to coincide with a Killing vector congruence, as characterised by

$$\ell_{(\mu;\nu)} = 0 \tag{3.3}$$

over any finite neighbourhood of the horizon on which the geometry is analytic, a property which is guaranteed[40] by the Einstein or Einstein-Maxwell equation for the *vacuum* (or ordinary material influx would violate the condition that $R_{\mu\nu}\ell^\mu\ell^\nu$ must vanish) surrounding the horizon, by the elliptical nature of these field equations wherever k^μ is *timelike*.

Unfortunately both the applicability of this theorem and the strength of the conclusions that can be drawn from it are still severely limited by the difficulty of excluding the possiblity that this latter condition might be violated in relevant applications, i.e. that there might be an *ergo-region* (so called because the specific energy $u^\mu k_\mu$ of a particle on an orbit with tangent vector u^μ might become negative there) on which k^μ becomes negative, i.e. where

$$V < 0. \tag{3.4}$$

Indeed as shown by the lemma given immediately below, V cannot remain strictly positive everywhere unless no black hole is present at all. Moreover in what we refer to as the *rotating-case*, i.e. the case for which the Killing vector ℓ^μ is *distinct* from the stationary generator k^μ – which implies the existence of an *axisymmetry* generated by some combination of k^μ and ℓ^μ – it is possible for an ergo region to occur in very rapidly rotating (unstable) fluid configurations even when there is no black hole at all, and the existence of an ergo sphere extending outside the horizon even in the pure vacuum case is familiar from the example of the Kerr black holes. On the other hand, for the alternative possibility, which we refer to as the *non-rotating case*, i.e. the case in which ℓ^μ is the same as the original Killing vector field k^μ (so that we cannot immediately infer the existence of axisymmetry) with the implication that V vanishes on the horizon which therefore lies in the boundary at the ergosphere, it can at least be not implausibly conjectured that in the pure vacuum (Einstein or Einstein-Maxwell) case, i.e. in the absence of external material structures such as a pair of rapidly counter-rotating matter rings) the condition that V should remain positive everywhere outside the horizon might be demonstrable after all.

It was in fact established long ago by work of Vishveshwara[41] and myself[42,21,43] on the basis of the two lemmas given in §3.5 and §3.8 that V must indeed satisfy the positivity requirement everywhere outside the horizon in the strictly *static* case for which the Killing vector satisfies the Frobenius irrotationality (hypersurface orthogonality) condition

$$k_{[\mu}k_{\nu;\rho]} = 0 \tag{3.5}$$

However to prove the *staticity theorem* (generalising a result proved before by Lichnerowicz[44] for the simpler case in which no black hole is present at all) to the effect that this irrotationality condition will in fact hold for a *non-rotating* black hole in the pure vacuum case, Hawking[32,15] was again obliged to invoke the positivity condition on V as an assumed axiom. Actually Hajicek[45] has attempted to give a mathematical justification, but only at the expense of other litigious assumptions, while Hawking and Ellis[15] have attempted to provide a heuristic justification, ruling out the existence of an ergosphere detached from the horizon on the grounds that the possibility of extracting energy by stacking particles there in negative energy orbits –from which they would not be able to escape either to infinity or over the horizon – could ultimately lead to violation of the total mass positivity condition – which has since been placed on a rigorous basis by work of Schoen and Yau,[37] Witten,[38] and others. However I find such arguments quite implausible, because such detached ergosphere regions are well known to be possible – albeit unstable as discussed in the accompanying lectures of Schutz – in very rapidly rotating stellar configurations. There is no danger of arriving at the paradoxical situation envisaged by Hawking and Ellis, because although their angular velocities must be prograde, the negative energy orbits will have *retrograde angular momentum* relative to the sense of the bulk motion responsible for the frame dragging that gives rise to the ergo region. Feeding more particles into such orbits will therefore cause a positive readjustment of the energy of the particles already there – naturally at the expense of the energy of matter in positive energy orbits – while the extent of the ergoregion diminishes. Before too much energy can have been extracted, the ergosphere will have simply faded out of existence.

In the extension of Hawking's staticity theorem to cover the electromagnetic (Einstein-Maxwell) case, as described in the following section, I have found it necessary to impose not only that V be everywhere positive but that it should satisy the even stronger condition (which is interpretable as a requirement of exclusion of a certain kind of generalised ergosphere of the class defined by Denardo and Ruffini[46,47])

$$V > 4G(\Phi - \Phi^H)^2 \tag{3.6}$$

where Φ is the electrostatic potential as defined in any stationary gauge (i.e. one that is invariant under the action of the stationary action generated by k^μ) by

$$\Phi = A_\mu k^\mu \tag{3.7}$$

and where Φ^H is its value on the black hole horizon, which as I had first noted in the 1972 Les Houches Lectures,[21] and will show explicitly in §3.3, must necessarily be uniform over the horizon, consistently with the conducting membrane analogy described in the accompanying lecture notes of Thorne. (If no black hole is present Φ^H could be considered as an arbitrary constant and adjusted so as to minimise $(\Phi - \Phi^H)^2$ thereby facilitating the satisfaction of the condition.) I have taken the trouble of inserting Newton's constant G explicitly (instead of just setting it to unity) because in my original discussion of this subject in the 1972 Les Houches lecture notes I overhastily concluded that the condition $V > 0$ was sufficient even in the electromagnetic case, as a result of a sign mistake in the gravitational constant that was pointed out to me only very recently by Gibbons who had noticed it in the course of a collaboration with Breitenlohner and Maison in which they had hoped to use the staticity theorem in conjunction with positive mass theorem[37,38] to show that Minkowski space is the only possibility for a stationary source-free solution without any black hole present. Gibbons has noticed more recently that this purpose can be achieved (as described in §3.5) by using (instead of the staticity theorem) the lemma in the derivation of the generalised Smarr formula[21] (to be described in §4.6). My earlier conclusion should now be considered as applying only to the (physically uninteresting) "anti-Einstein-Maxwell" theory in which the gravitational coupling is taken to be repulsive. When the sign is taken correctly so as to correspond to ordinary Einstein-Maxwell theory, the same method of derivation will be shown to lead the more severe requirement expressed above. This requirement can easily be violated by very strong electric fields, (i.e. fields of the order of the Planck potential, 10^{27} volts, which is attained on the surface of an extreme Reissner-Nordstrom black hole) for which the theorem as given below does not apply. Whether it can be violated even in the non-electromagnetic case remains an open question: I know of no (regular, non-rotating) counterexample but on the other hand I know of no convincing mathematical or heuristic argument either. It is also an open question whether the conclusion (staticity) remains valid even when (3.6) *is* violated.

I have insisted on the problem of excluding a (generalised) ergosphere on this occasion because it constitutes the most important outstanding gap in the programme of classifying all vacuum Einstein or Einstein-Maxwell black hole equilibrium states. As far as *rotating* axisymmetric black holes are concerned, the programme that I initiated in 1971[30] along lines described in detail in the 1972 Les Houches lectures[21] has now been virtually completed – confirming that the only possibilities are those of the Kerr-Newman family – by the work of Robinson,[48,49] Bunting[50,51] and Mazur[52] in the manner to be described in the next section, the main remaining lacuna being that no one has yet rigorously excluded the possibility of two distinct holes in a (no doubt highly unstable and thus astrophysically irrelevant) equilibrium on the same axis, with corotational repulsion balancing gravitational attraction. In so far as the non-rotating black hole solutions may be considered to be *static*, in has also been virtually established by work of Müller-zum-Hagen, Robinson, and Seifert,[53,54] Robinson[55] and Bunting[50] using foundations laid previously by Israel,[56,57] that there are no possibilities other than the static (Schwarzschild and Reissner Nordstrom) limits of the Kerr-Newman family, provided that, as was necessary in the rotating case also, the possibility of *multiple* black hole solutions is excluded from consideration. In so far as this latter eventuality is concerned there here has been recent progress by Bunting and Massood-ul-Alam[58] who (generalising an earlier result obtained by Gibbons[59] for the static axisymmetric case) have exploited the mass positivity theorem in demonstrating the non-existence of multiple black holes in the pure vacuum case. It may be plausibly be conjectured that this conclusion will extend to the electrovac case in general except in the degenerate limit, with electrostatic repulsion exactly balancing gravitational attraction, for which multiple, not necessarily axisymmetric, black hole configurations are known in fact[60] to be possible. [Although the marginally stable Majumdar-Papapetrou solutions[61] in question are of no obvious astrophysical interest, they have recently attracted much attention in the context of pure mathematical physics due to the recognition that they exhibit the special property of supersymmetry.[62]]

3.2 Elementary Local Properties of Killing Horizons

The existence (as predicted by the strong or weak rigidity theorem) of a vector field having the Killing antisymmetry property

$$\ell_{\mu;\nu} = \ell_{[\mu;\nu]} \tag{3.8}$$

and coinciding with the null tangent vector of a null (characteristic) hypersurface, characterises the latter as what I have termed a local *Killing horizon*, a geometric structure whose intrinsic properties were first studied explicitly by R.H. Boyer[63] (who was the original discoverer of the black hole character of the Kerr solutions with $a \leq M$).[64] On the Killing horizon itself, as characterised by

$$\ell_\mu \ell^\mu = 0 \tag{3.9}$$

the vector field must also satisfy the Frobenius hypersurface orthogonality condition

$$\ell_{[\mu}\ell_{\nu;\rho]} = 0 \tag{3.10}$$

which implies as we have already seen for the case of more general (non-Killing) null hypersurfaces, that it must also satisfy the geodesic equation

$$\ell_{\mu;\nu}\ell^\nu = \kappa \ell_\mu \tag{3.11}$$

where κ is a coefficient measuring the extent to which the parametrisation is non affine, and which can be seen (using the Killing equations) to be definable alternatively by

$$V^\dagger{}_{,\mu} = 2\kappa \ell_\mu, \qquad V^\dagger = -\ell_\mu \ell^\mu \tag{3.12}$$

Since construction of the irrotationality condition with the gradient 2-form gives

$$(\ell^{\rho;\nu}\ell_{\nu;\rho})\ell_\mu = \kappa(\ell_\nu \ell^\nu)_{,\mu} \tag{3.13}$$

we obtain the explicit expression

$$\kappa^2 = \tfrac{1}{2}\ell_{\mu;\nu}\ell^{\mu;\nu} = \lim_{V^\dagger \to 0} \frac{V^\dagger{}_{,\mu} V^{\dagger,\mu}}{4V^\dagger} \,. \tag{3.14}$$

It follows from the original definition in terms of the non-affine null geodesic equation that this quantity κ is interpretable as expressing the relation between an ignorable *symmetry* group parameter t, as characterised by $\ell^\mu \tau,_\mu = 1$, and an *affine* parameter τ along the null geodesics, in accordance with

$$\kappa = \frac{d}{dt}\left(\ln \frac{d\tau}{dt}\right) \tag{3.15}$$

and from the preceding limit relation one sees that it is also interpretable as an *improper* gravitational acceleration of a circular orbit: if outside the horizon one defines an effectively corotating unit vector u^μ by

$$\ell^\mu = \dot\tau u^\mu, \qquad \dot\tau \equiv \frac{d\tau}{dt} = (V^\dagger)^{1/2} \tag{3.16}$$

the improper acceleration defined by

$$\dot u_\mu = u_{\mu;\nu}\ell^\nu = \dot\tau a_\mu \tag{3.17}$$

(where a_μ is the ordinary proper acceleration) will satisfy

$$\dot u_\mu \dot u^\mu \to \kappa^2 \tag{3.18}$$

on the horizon.

3.3 Uniformity of the Corotating Potential on a Killing Horizon

In the presence of electromagnetic field

$$F_{\mu\nu} = 2A_{[\nu;\mu]} \tag{3.19}$$

the Killing vector ℓ^μ will determine corresponding electric and magnetic field components which we distinguish by a dagger† from those defined analogously with respect to k^μ in the rotating case (for which the latter Killing vector is distinct) given by

$$E^\dagger_\mu = F_{\mu\nu}\ell^\nu, \qquad B_\mu = \tfrac{1}{2}\varepsilon_{\mu\nu\rho\sigma}\ell^\nu F^{\rho\sigma} \,. \tag{3.20}$$

Assuming that the field is invariant under the action generated by ℓ^μ, and that the gauge is chosen so as to exhibit this symmetry, i.e.

$$A_{\mu;\nu}\ell^\nu + A_\nu \ell^\nu{}_{;\mu} = 0 \tag{3.21}$$

the electric component will be expressible as

$$E^\dagger_\mu = \Phi^\dagger{}_{,\mu}\,, \qquad \Phi^\dagger = A_\mu \ell^\mu. \tag{3.22}$$

The observation[21] that this corotating potential $\Phi^\dagger$ must necessarily be *uniform* over the horizon was the first indication of the strong analogy with an ordinary electrically conducting membrane that was developed by Znajek[65] Damour[66] and others and of which the details are discussed in the accompanying lectures of Thorne. To derive this result it suffices to substitute the Maxwell energy momentum contribution

$$T_F{}^{\mu\nu} = \frac{1}{4\pi}(F^\mu{}_\rho F^{\nu\rho} - \frac{1}{4}F_{\sigma\rho}F^{\sigma\rho}g^{\mu\nu}) \tag{3.23}$$

which entails

$$T_F{}^{\mu\nu}\ell_\mu\ell_\nu = \frac{1}{8\pi}(E^\dagger_\mu E^{\dagger\mu} + B^\dagger_\mu B^{\dagger\mu}) \tag{3.24}$$

in the total energy-momentum restriction

$$T^{\mu\nu}\ell_\mu\ell_\nu = 0 \tag{3.25}$$

where

$$T^{\mu\nu} = T_m{}^{\mu\nu} + T_F{}^{\mu\nu} \tag{3.26}$$

that must hold on a stationary horizon by Einstein's equations, as a consequence of the vanishing of the Ricci contractioin $R_{\mu\nu}\ell^\mu\ell^\nu$ that was derived from the Penrose-Raychaudhuri equation at the outset of this discussion. The positivity of the material contribution $T_m{}^{\mu\nu}\ell_\mu\ell_\nu$ for normal matter implies that the neighbourhood of a stationary horizon must be a vacuum in which the electromagnetic field contribution must also vanish, i.e.

$$E^\dagger_\mu E^{\dagger\mu} + B^\dagger_\mu B^{\dagger\,\mu} = 0. \tag{3.27}$$

Since $E^\dagger_\mu$ and $B^\dagger_\mu$ are everywhere orthogonal to ℓ^μ by construction, and therefore nowhere timelike where ℓ^μ is timelike or null, it follows that each must be null and therefore, by the orthogonality, also parallel to ℓ_μ on the horizon, i.e.

$$E^\dagger_{[\mu}\ell_{\nu]} = 0, \qquad B^\dagger_{[\mu}\ell_{\nu]} = 0 \tag{3.28}$$

there. Since an *arbitrary* (spacelike or null) tangent vector ξ^μ lying in the horizon is characterised by $\ell_\mu\xi^\mu = 0$, it follows that the potential satifies

$$\Phi^\dagger{}_{,\mu}\xi^\mu = 0 \tag{3.29}$$

which is the required uniformity condition, establishing that there is a well defined limit, Φ^H say, for the value of $\Phi^\dagger$ on the horizon.

3.4 Uniformity of κ (the "zeroth law") on a Killing Horizon

By a somewhat analogous but technically more complicated argument one can establish the even more fundamental condition that the decay parameter κ must be uniform over a Killing horizon.[21,15,35] To derive this and other important properties it is convenient to start by using the fact that the Frobenius orthogonality condition on the horizon is equivalent to the condition that $\ell_{\mu;\nu}$ (which is known to be antisymmetric by the Killing property) be expressible in the form

$$\ell_{\mu;\nu} = 2\ell_{[\mu}q_{\nu]} \tag{3.30}$$

for some covector q_ν on the horizon, where q_ν can be fixed uniquely with respect to the transverse null vector β^μ introduced in the discussion of general (dynamic) horizon properties by the normalisation condition $q_\mu\beta^\mu = 1$. It is easy to see in terms of this covector that for any fields ξ^μ, η^μ in the horizon, i.e. such that $\ell_\mu\xi^\mu = \ell_\mu\eta^\mu = 0$, we shall get

$$\xi^\mu\eta^\nu\ell_{\nu;\mu} = \xi^\mu\eta^\nu(q_\mu\ell_\nu - q_\nu\ell_\mu) = 0 \tag{3.31}$$

and hence, (differentiating the orthogonality relation $\eta^\mu \ell_\mu = 0$) that

$$\ell^\nu \xi^\mu \eta_{\nu;\mu} = 0 \tag{3.32}$$

which shows that the horizon is *extrinsically flat* (and hence geodesically generated) since it implies that the vector $\eta_{\mu;\nu}\xi^\nu$ necessarily lies in the horizon. Now a further derivation gives

$$\ell_{\nu;\mu;\rho}\xi^\mu \eta^\nu = -\xi_{\nu;\mu}(\xi^\mu \eta^\nu{}_{;\rho} + \eta^\nu \xi^\mu{}_{;\rho}) = 0 \tag{3.33}$$

(using $\xi^\mu{}_{;\rho}\ell_\mu = \xi^\nu q_\nu \ell_\rho$) and hence since the Killing vector property implies

$$\ell_{\nu;\mu;\rho} = R_{\mu\nu\rho}{}^\tau \ell_\tau \tag{3.34}$$

one deduces that for any vectors ξ^μ, η^ν on the horizon we shall have

$$R_{\mu\nu\rho\tau}\ell^\mu \xi^\rho \eta^\tau = 0. \tag{3.35}$$

Contracting this with the metric in the form $g_{\mu\nu} = \gamma_{\mu\nu} - \beta_\mu \ell_\nu - \ell_\mu \beta_\nu$ leads to

$$R_{\mu\nu\rho\tau}\xi^\mu \ell^\nu \beta^\rho \ell^\tau = R_{\mu\nu}\xi^\mu \ell^\nu. \tag{3.36}$$

Now since the original definition of κ implies that it can be expressed as

$$\kappa = -\beta^\mu \ell^\nu \ell_{\mu;\nu} = \ell^\mu q_\mu \tag{3.37}$$

we deduce that its gradient in the direction ξ^μ is given by

$$\kappa,_\rho \xi^\rho = -\xi^\rho(\ell_{\mu;\nu;\rho}\beta^\mu \ell^\nu + \beta^\mu \ell^\nu{}_{;\rho}\ell_{\mu;\nu} + \kappa \ell_\mu \beta^\mu{}_{;\rho}). \tag{3.38}$$

Since the last two terms cancel (being respectively equal to $\kappa q_\rho \xi^\rho$) one is left with the simple final expression

$$\kappa,_\rho \xi^\rho = -R_{\mu\nu}\ell^\nu \xi^\mu \tag{3.39}$$

from which our original condition, that $R_{\mu\nu}\ell^\mu \ell^\nu$ vanishes, can be recovered as the special case when ξ^ρ is set equal to ℓ^ρ. The required result

$$\kappa,_\rho \xi^\rho = 0 \tag{3.40}$$

is now immediately obvious in the pure Einstein vacuum case, while for the Einstein-Maxwell vacuum as described in the preceding subsection it can be seen that it still holds, because although the contraction $R_{\mu\nu}\ell^\nu$ is no longer zero, it is proportional to the null cotangent ℓ_μ which is sufficient.

3.5 The Mass of a Stationary System

In general the definition of conserved mass and angular momentum can be difficult in the absence of corresponding spacetime symmetries. Asymptotic definitions are possible if the metric tends sufficiently rapidly to flatness at large distances $r \to 0$, the requirements for the definition of angular momentum being more severe than those for mass. However there is no difficulty in extending the applicablilty of the definitions to finite distances if appropriate symmetries are present. The existence of the *stationary symmetry* (that we have postulated throughout the present section) generated by a Killing vector k^μ whose trajectories are timelike at least at sufficiently large asymptotic distance r allows a straitforward definition of the asymptotic mass M by means of the *Komar integral*[67]

$$M = -\frac{1}{4\pi G}\oint_{\infty} k^{\mu;\nu}\, dS_{\mu\nu} \tag{3.41}$$

taken over a spacelike topological 2-sphere in the limit of large radius r, while a corresponding Komar integral for angular momentum can be defined under conditions of axial symmetry (as postulated in the next section) in the manner to be described in §4.7.

The Killing (metric invariance) equations

$$k_{(\mu;\nu)} = 0 \tag{3.42}$$

give corresponding Ricci curvature conditions

$$k^{\mu;\nu}{}_{;\nu} = -R^{\mu}{}_{\nu}k^{\nu}. \tag{3.43}$$

If the distant 2-sphere is filled in by a spacelike hypersurface Σ, with 3-ball topology if no black hole is present, and otherwise bounded in the interior by a 2-surface on the horizon, then (3.41) can be used with Green's theorem to obtain

$$M = \frac{1}{4\pi G}\int_{\Sigma} R^{\mu}{}_{\nu}k^{\nu} d\Sigma_{\mu} + \frac{1}{4\pi G}\oint_{H} k^{\mu;\nu}\, dS_{\mu\nu} \tag{3.44}$$

where the 2-surface integral is taken over the inner boundary H of Σ on the horizon. In the *non rotating* case, when k^{μ} actually coincides with the null generator in the horizon, the 2-surface integral can be evaluated as being proportional to a product of the decay parameter κ and the surface area $\mathcal{A}$, according to an expression obtainable as the non-rotating limit of the formula to be derived for the general axisymmetric case in §4.6.

In evaluating the Ricci curvature contribution to (3.44) using the Einstein equations (2.33) it is convenient to use the decomposition (3.26) of the local energy momentum contribution $T^{\mu}{}_{\nu}k^{\nu}$ into a material source contribution $T_m{}^{\mu}{}_{\nu}k^{\nu}$ and an electromagnetic field contribution $T_F{}^{\mu}{}_{\nu}k^{\nu}$, and to use the stationary invariance property

$$F_{\mu\nu;\rho}k^{\rho} = 2\, F_{[\rho\mu}k^{\rho}{}_{;\nu]} \tag{3.45}$$

together with the Maxwell-Faraday integrability condition

$$F_{[\mu\nu;\rho]} = 0 \tag{3.46}$$

and the Ampère-Maxwell source equation

$$F^{\mu\nu}{}_{;\nu} = 4\pi\, j^{\mu} \tag{3.47}$$

to decompose the electromagnetic field contribution (specified by (3.24)) into a current source contribution and a bivector divergence term, in the form[21,25]

$$T_F{}^{\mu}{}_{\nu}k^{\nu} = (A_{\nu}k^{\nu})j^{\mu} - \frac{1}{2}(A_{\nu}j^{\nu})k^{\mu} + \frac{1}{4\pi}\{(A_{\nu}k^{\nu})F^{\mu\rho} + A_{\nu}F^{\nu[\rho}k^{\mu]}\}_{;\rho} \tag{3.48}$$

where we have taken advantage of (3.46) and (3.45) so as to express the field in the form (3.19) with a 4-potential A_{μ}, chosen so as to be subject to the same stationary symmetry property as the field itself, i.e. so that

$$A_{\mu;\nu}k^{\nu} + A_{\nu}k^{\nu}{}_{;\mu} = 0. \tag{3.49}$$

Provided that the magnetic monopole moment

$$P = \frac{1}{8\pi} \oint_{\infty} \varepsilon^{\mu\nu\rho\sigma} F_{\mu\nu} \, dS_{\rho\sigma} \tag{3.50}$$

(which is automatically conserved, by (3.46)) is postulated, on physical grounds, to vanish, i.e.

$$P = 0 \tag{3.51}$$

the further postulate that the electromagnetic field dies off at large distance can be used to impose a corresponding requirement on the potential, i.e.

$$A_\mu \to 0 \qquad \text{as} \qquad r \to \infty. \tag{3.52}$$

Substituting this in (3.44) using the Einstein equations and Green's theorem gives

$$\begin{aligned} M = \int_\Sigma \{2(T_m{}^\mu{}_\nu + j^\mu A_\nu)k^\nu - (T_m{}^\nu{}_\nu + j^\nu A_\nu)k^\mu\} d\Sigma_\mu \\ + \oint_H \Big\{ \frac{1}{4\pi G} k^{\mu;\rho} + 2A_\nu (k^\nu F^{\mu\rho} + F^{\nu\rho} k^\mu) \Big\} \, dS_{\mu\rho} \end{aligned} \tag{3.53}$$

where we have got rid of a 2-surface boundary term at large distance by using (3.52).

In the *non-rotating* case when k^μ actually coincides with the null generator in the horizon, the black hole surface integral contributions in (3.53) can be evaluated in term of the decay parameter κ, the surface area $\mathcal{A}$, the effective black hole charge Q_H and the limiting value of the potential $A_\mu k^\mu$ (which by the work of the preceding section is constant over the horizon) in accordance with the non-rotating limit of the formula to be derived for the general axisymmetric case[21,25] in §4.6.

When there is *no black hole* present at all, it can be seen that (3.53) expresses M entirely in terms of the material source and current contributions, and that when they are absent, i.e. when the *source-free Einstein-Maxwell equations* are satisfied, it will follow that the *mass, M,* must *vanish.* It has been pointed out by Gibbons[68] that taken in conjunction with the mass positivity theorem[37,38] this establishes (subject to the asymptotic flatness requirements) that a source free Einstein-Maxwell solution without any inner black hole can only be a field free Minkowski space (the purpose for which, in collaboration with Breitenlohner and Maison, he had previously thought of applying the Staticity theorem, thereby coming to notice the sign error that will be rectified in §3.7).

3.6 Globally Bradyonic Character of generators of Stationary D.O.C.

We now wish to point out a useful property[21,43] characterising the trajectories of the asymptotically timelike symmetry generator k^μ in the *Domain of Outer Communications,* namely that of forming a *maximal connected set with the property of being globally bradyonic* in the following sense. We described an oriented curve in a time oriented Lorentz signature pseudo-Riemannian manifold as being *globally bradyonic* if the future of any point $\mathbf{X}_0$ contains the entire curve forward of some point $\mathbf{X}^+$ say, and if the past of $\mathbf{X}_0$ contains the entire curve backward of some point $\mathbf{X}_0^-$ say. This concept generalises that of an ordinary timelike (locally bradyonic) curve for which $\mathbf{X}^+$ and $\mathbf{X}^-$ may be taken to coincide with $\mathbf{X}_0$ itself. It is therefore obvious that the trajectories of k^μ are bradyonic in both the local the global sense wherever V is positive. To see that they remain bradyonic in the weaker global sense throughout the Domain of Outer Communications (even in parts lying in the ergoregion where V is negative) we use the fact that any point $\mathbf{X}_0$ in the D.O.C. must (by definition of the D.O.C.) be simultaneously connectable by respectively a future and a past directed timelike line to the same points x_1, x_2 say in the outer asymptotically flat region where the trajectories have been stipulated to be timelike, and where it is evidently possible to construct a further future directed timelike line from x_1 to some point x_3 on the (timelike) trajectory of k^μ through x_2. Then by using the stationary group action that transports x_2 to x_3 to transport the entire timelike curve from x_2 to x_0 onto a new timelike curve from x_3 to a point x_4, we can complete the construction of a timelike curve from x_0 via x_1 and x_3 back to the point x_4 on the Killing vector trajectory through x_0. Having thus demonstrated the existence of a segment of the trajectory lying in the future of x_0, one can obviously go on (by applying the same construction to points in this segment) to construct more such segments until a stage at which they will overlap and beyond which the trajectory through x_0 will never again leave the future of x_0. Since precisely analogous considerations apply to the extrapolation of the trajectory to the past, the globally bradyonic character of the trajectory is thus established. Since it is fairly obvious[21,43] that in any *connected* region in which all the trajectories are globally bradyonic, each trajectory can be connected to any other one by a timelike line with either orientation one finally arrives at the following useful *lemma:*

> A stationary D.O.C. may be characterised as the maximal connected domain, including the outer asymptotically flat region, such that the trajectories of the stationary action are globally bradyonic.

It follows that in *part* of any black hole region – including the *entire boundary* of the region – the trajectories of k^μ must cease to be globally bradyonic, which implies a fortiori that they are non-timelike there.

3.7 The staticity Theorem for Non-Rotating Electromagnetic Black Holes

Due to lack of space here, the reader is referred to the article of Robinson[55] for a brief introduction to the essential ideas in the work on classification of *static* vacuum black holes by Israel and others to which we referred at the beginning of this section. However before moving on to consider in detail the astrophysically more important *axisymmetric* case – including all rotating configurations – we shall complete the present section by providing a *partial justification* (unavailable elsewhere except in the pure vacuum case first treated by Hawking[32,15]) of the assumption of staticity in the *non-rotating* case for which the null generator ℓ^μ of the horizon *coincides* with the asymptotically timelike Killing vector field k^μ (that we have just shown to be globally bradyonic on the D.O.C.), i.e.

$$\ell^\mu = k^\mu. \tag{3.54}$$

The line of reasoning given below is essentially the same as that which I wrote up hurriedly for inclusion in the 1972 Les Houches Lecture notes,[21] but which included a sign mistake in the gravitational coupling (not to mention several inconsequent transcription and printing errors) that is rectified here.

The purpose of the theorem is to establish staticity both in the electromagnetic sense, meaning that in the decomposition into electric and magnetic parts

$$E_\mu = F_{\mu\nu}k^\nu, \qquad B_\mu = \tfrac{1}{2}\varepsilon_{\mu\nu\rho\sigma}k^\nu F^{\rho\sigma} \tag{3.55}$$

only the former is present, and in the geometric sense meaning that the rotation vector

$$\omega_\mu = \tfrac{1}{2}\varepsilon_{\mu\nu\rho\sigma}k^\nu k^{\sigma;\rho} \tag{3.56}$$

is also zero, which is equivalent to the Frobenius hypersurface orthogonality condition stated above. In addition to the present necessity (that one might hope to be able to dispense with in some more powerful future version of the theorem) of postulating a certain *non-negative lower bound* on the magnitude

$$V = -k_\mu k^\mu \tag{3.57}$$

the theorem is based on the obviously necessary requirement that any electromagnetic or material sources present should themselves be static in the sense of having no flow relative to the local rest frame of the Killing vector, which means explicitly that any electromagnetic current j^μ should satisfy

$$j^{[\mu} k^{\nu]} = 0 \tag{3.58}$$

while any non-electromagnetic contribution $T_m{}^{\mu\nu}$ say to the energy-momentum should satisfy

$$k^\rho T_{m\,\rho}{}^{[\mu} k^{\nu]} = 0, \tag{3.59}$$

(these requirements being satisfied automatically in the source free vacuum case with which the applications referred to above are concerned).

Under these conditions, and subject of course to the stationary invariance requirements, (3.42) and (3.45) the Maxwell-Faraday integrability conditions (3.46) reduce to the conditions

$$V B_\mu{}^{;\mu} = V,_\mu B^\mu - 2E_\mu \omega^\mu\,, \qquad E_{[\mu;\nu]} = 0 \tag{3.60}$$

while the source equation (3.47) gives

$$V E_\mu{}^{;\mu} = V,_\mu E^\mu + 2E_\mu \omega^\mu - 4\pi k_\mu j^\mu\,, \qquad B_{[\mu;\nu]} = 0. \tag{3.61}$$

Finally using the Einstein equation

$$\begin{aligned} R^\mu{}_\nu = {} & 2G(F^\mu{}_\rho F_\nu{}^\rho - \tfrac{1}{4} F_{\rho\sigma} F^{\rho\sigma} \delta^\mu_\nu) \\ & + 8\pi G(T_m{}^\mu{}_\nu - \tfrac{1}{2} T_m{}^\rho{}_\rho \delta^\mu_\nu) \end{aligned} \tag{3.62}$$

together with the Killing vector properties

$$k^{\mu;\nu}{}_{;\nu} = -R^\mu{}_\nu k^\nu\,, \qquad \{k^{[\mu} k^{\nu;\rho]}\}_{;\rho} = \tfrac{2}{3} k^\rho R_\rho{}^{[\mu} k^{\nu]}, \tag{3.63}$$

one deduces that the rotation vector will satisfy

$$V \omega_\mu{}^{;\nu} = 2\, V,_\mu \omega^\mu, \qquad \omega_{[\mu;\nu]} = 2G\; E_{[\mu} B_{\nu]}\;. \tag{3.64}$$

The conditions on the exterior derivatives of E_μ, B_μ, and ω_μ imply the existence (globally over the D.O.C. provided it is assumed to be simply connected) of three corresponding scalar potentials, Φ, Ψ, U say, such that

$$E_\mu = \Phi,_\mu\,, \qquad B_\mu = \Psi,_\mu \tag{3.65}$$

$$\omega_\mu = U,_\mu + G(\Psi\Phi,_\mu - \Phi\Psi,_\mu) \tag{3.66}$$

while the divergence conditions allow us to construct two quantities that are divergence free, since we shall have

$$\left\{\frac{\omega_\mu}{V^2}\right\}^{;\mu} = 0, \tag{3.67}$$

$$\left\{\frac{\tilde{B}_\mu}{V}\right\}^{;\mu} = 0, \tag{3.68}$$

where we introduce the abbreviation

$$\tilde{B}_\mu = B_\mu + \frac{2\Phi}{V}\,\omega_\mu\,. \tag{3.69}$$

The idea now is to use these vanishing divergence conditions to construct a combined convergence that will be a positive definite function of the quantities ω_μ and B_μ (or equivalently $\tilde{B}_\mu$) that one wishes to show to be zero. Two obviously promising possibilities, depending on a choice of sign $\pm$, are given by

$$\left\{\frac{(U+G\Phi\Psi)\omega_\mu}{V^2} \pm \frac{G\Psi}{V}\tilde{B}_\mu\right\}^{;\mu} = \frac{(\omega^\mu + 2G\Phi B^\mu)\omega_\mu}{V^2} \pm \frac{GB^\mu\tilde{B}_\mu}{V}\,. \tag{3.70}$$

The simplest form for the right hand side is obtained by choosing the $\pm$ sign to be negative, which gives

$$\left\{\frac{(U-G\Phi\Psi)\omega_\mu}{V^2} - \frac{G\Psi}{V}B_\mu\right\}^{;\mu} = \frac{\omega^\mu\omega_\mu}{V^2} - \frac{GB^\mu B_\mu}{V}\,. \tag{3.71}$$

This relation, which would have the required form for a negative (repulsive) gravitational coupling, is the one derived in the 1972 Les Houches Notes, in which I had inadvertently set $G = -1$. For the normal attractive gravity with a positive value for Newton's G, we need rather to choose the $\pm$ sign to be positive, which leads to the slightly more complicated relation

$$\left\{\frac{(U+G\Phi\Psi)\omega_\mu}{V^2} + \frac{G\Psi}{V}\tilde{B}_\mu\right\}^{;\mu} = \left(1 - \frac{4G\Phi^2}{V}\right)\frac{\omega^\mu\omega_\mu}{V^2} + \frac{G\tilde{B}^\mu\tilde{B}_\mu}{V} \tag{3.72}$$

in which the right hand side has the required positive definite form provided we have

$$V > 4G\Phi^2 \tag{3.73}$$

(since, being orthogonal to k^μ by construction, ω^μ, B^μ, and $\tilde{B}^\mu$ will all be spacelike.) In order to maximise the chance of satisfying this condition, and also in order to get rid of surface integrals obtained by the application of Green's theorem to this relation, we are free to adjust the arbitrary constant of integration in the potential Φ in the most favorable manner. In the non-rotating case this potential may be identified with the corotating potential, $\Phi^\dagger$ which was shown to have a constant value, let us say Φ^H on the horizon of the black hole. In order to ensure that the required inequality is satisfied in the neighbourhood of the horizon where V tends to zero, we shall evidently be obliged to choose the gauge constant so as to obtain

$$\Phi^H = 0. \tag{3.74}$$

[If we wished to use a potential gauge invariant formulation, we could replace Φ by $\Phi - \Phi^H$ throughout the preceding work, which leads to the expression for the lower bound on V given at the beginning of this section.] In the demonstration that $\Phi^\dagger$ is constant on the horizon we saw that $B^\dagger_\mu$ shares with $E^\dagger_\mu$ the property of lying along the horizon tangent vector ℓ_μ. It therefore follows that in the non-rotating case and subject to the current staticity condition that is necessary for the existence of the potential Ψ this value will also have a constant value Ψ^H on the horizon which may also be set equal to zero by choice of gauge, i.e.

$$\Psi^H = 0 \tag{3.75}$$

and since the rotation vector ω^μ itself must vanish on the horizon (whose normal lies along k_μ) the same applies to the rotation potential U, i.e. it necessarily has a uniform value U^H on the horizon and can therefore be gauge adjusted so as to satisfy

$$U^H = 0.$$

Subject to these gauge conditions it can be seen that the coefficients of ω_μ and B_μ or $\tilde{B}_\mu$ in the differands of (3.71) and (3.72) will remain bounded on the horizon despite the fact that V tends to zero there.

We can convert a 4-dimensional divergence of the kind just obtained into a form suitable for application of Green's theorem[26] on a 3-dimensional hypersurface whenever the flux n^μ say in question is invariant under the action of a Killing vector k^μ, i.e. such that

$$n^\mu{}_{;\nu} k^\nu - k^\mu{}_{;\nu} n^\nu = 0$$

by using the fact that the Green theorem for the 2-dimensional boundary S say of a 3-dimensional hypersurface Σ gives

$$\begin{aligned} \oint_S n^\mu k^\nu \, dS_{\mu\nu} &= \int_\Sigma (n^{[\mu} k^{\nu]})_{;\nu} \, d\Sigma_\mu \\ &= -\tfrac{1}{2} \int n^\mu{}_{;\mu} k^\nu \, d\Sigma_\nu \, . \end{aligned} \tag{3.76}$$

If we apply this to a spacelike hypersurface Σ extending from the horizon out to infinite radial distance r in the asymptotically flat region, we see that the surface inner contribution will vanish provided we have

$$n^\mu k_\mu = 0 \tag{3.77}$$

on the horizon, which will automatically be satisfied when we take

$$n^\mu = \frac{U + G\Phi\Psi}{V^2} \omega^\mu \pm G\Psi \frac{\tilde{B}^\mu}{V} \tag{3.78}$$

subject to the preceding gauge conditions, since ω^μ, B^μ, and hence also $\tilde{B}^\mu$ are orthogonal to k^μ by construction. For a bounded source distribution the usual conditions of asymptotic flatness and the requirement that the electromagnetic field contribution should decay to zero at large distance implies with the usual normalisation for k^μ that we shall have

$$V \to 1 \qquad U \to U^\infty \tag{3.79}$$

$$\Phi \to \Phi^\infty \qquad \Psi \to \Psi^\infty \tag{3.80}$$

as $r \to \infty$, where U^∞, Φ^∞, Ψ^∞ are constants. Therefore we see that the result of substituting the above expression for n^μ in the Green theorem will be expressible as

$$\int (-k^\mu \, d\Sigma_\mu) \left\{ \left(1 - \frac{4G\Phi^2}{V}\right) \frac{\omega^\mu \omega_\mu}{V^2} + G \frac{\tilde{B}^\mu \tilde{B}_\mu}{V} \right\} = 4\pi G \Psi_\infty P \, . \tag{3.81}$$

Hence provided we make the physically reasonable postulate (3.51) that the monopole moment vanishes, the positive definite nature of the functional dependence in the integrand ensures that we must have

$$\tilde{B}^\mu = 0 \tag{3.82}$$

$$\omega^\mu = 0 \tag{3.83}$$

and hence also

$$B^{\mu} = 0 \tag{3.84}$$

throughout the D.O.C. in all cases for which the lower bound (3.73) on V is satisfied. Whether this sufficient lower bound (which could be violated only in rather extreme circumstances) is actually mathematically necessary remains an open question. (In normal astrophysical circumstances with fields small compared with 10^{27} volts, no danger of its violation will arise.)

3.8 Local Characterisation of Simply Connected Static Domain of Outer Communications

When it has been established (on the basis of the preceding theorem or otherwise) that the stationary Killing vector k^{μ} of a D.O.C. has the staticity property (3.83) which will be expressible in expanded form as

$$2k_{\rho;[\mu}k_{\nu]} = k_{\rho}k_{[\mu;\nu]} \tag{3.85}$$

then, provided it is simply connected, the D.O.C. will be characterisable locally[41,42] as the (maximal) connected outer region, S say, on which the Killing vector is timelike, i.e. on which

$$V > 0. \tag{3.86}$$

To see this we start by noting that the normal to the *outer ergosurface* boundary, $S^{\bullet}$, of S will in general be expressible in the form

$$\lambda_{\mu} = (V^{1/(n+1)})_{,\mu} \tag{3.87}$$

where n is the order of the zero of V as a local analytic function of the space position (analyticity being presumably ensured by the elliptic nature of the stationary field equations), so that in particular one will have $n = 0$ in the usual non-degenerate case. We then contract (3.85) with k^{ρ} so as to obtain

$$(n+1)(V^{1/(n+1)})_{,[\mu}k_{\nu]} = V^{1/(n+1)}k_{\mu;\nu}\,. \tag{3.88}$$

Applying this on the ergosurface boundary where

$$V = 0 \tag{3.89}$$

we obtain

$$\lambda_{[\mu}k_{\nu]} = 0 \tag{3.90}$$

which means that λ_{μ} is proportional to k_{μ} and thus that (as originally remarked by Vishveshwara[41] and myself[42]) such a static ergosurface boundary is *necessarily null*, except on fixed points of the action whre k^{μ} is not merely null but actually zero.

We now invoke the lemma[43] that was described in Section 3.5, to the effect that the D.O.C. is the maximal outer connected set on which the trajectories of k^{μ} are globally bradyonic. This implies that the D.O.C. *includes* the set S defined above on which the trajectories are not just globally bradyonic but locally timelike, and also that the D.O.C. *excludes* all fixed points on which k^{μ} becomes zero (which is obviously not compatible with the bradyonic property that the trajectory through a point enters the future of that point). It follows that any part of $S^{\bullet}$ lying *within* (rather that on the boundary of) the D.O.C. would have to consist of discrete null hypersurface segments whose outgoing normals would in each case have a uniform (future or past directed) *time orientation* (since the time orientation could change only on a Kruskal type crossover at a fixed point of the stationary action, which, we have seen, can occur only on the boundary of the

D.O.C., but not within it). Now provided that the D.O.C. is simply connected, it will follow that any connected component D say of any part of the complement of S in the D.O.C. would have a boundary $D^{\bullet}$ in the D.O.C. that would itself have to be connected, and hence have to be a null hypersurface segment with uniform time orientation. This implies that D could not be connected to S by both future *and* past timelike lines, as is nevertheless necessary for D to belong to the D.O.C. The only consistent conclusion is thus that any such D is empty and hence that S coincides with the entire D.O.C. which is the required result.

The simple connectivity requirement assumed above is also sufficient to ensure that the orthogonal hypersurfaces whose local existence is ensured by the staticity condition (equivalently (3.5), (3.83), or (3.85)) can also be constructed *globally,* and hence can be used to define a canonical time coordinate t say that will be *well behaved over the entire D.O.C.*, whose metric will therefore be expressible by

$$ds^2 = g_{ij} dx^i \, dx^j - V \, dt^2 \tag{3.91}$$

where V and g_{ij}, $i = 1, 2, 3$ are functions only of the space coordinates x^i not of t, g_{ij} being a positive-definite (Euclidean signature) metric on each of the orthogonal hypersurfaces, and V being a positive scalar tending to 1 at large distance and to zero on the boundary (i.e. the black hole horizon) of the D.O.C. This form is the starting point for the derivation of the Israel type theorems referred to above.[53,54,55,56,57,58]

§4 AXISYMMETRIC EQUILIBRIUM STATES

4.1 Mass and Angular Momentum of Stationary Axisymmetric Systems

The present section will be devoted to the case in which, in addition to being invariant under a stationary action generated by an asymptotically timelike Killing vector k^μ, the system is also invariant under the action of a Killing vector m^μ with *closed circular* orbits, which are necessarily *spacelike* if causality violation is excluded, except on the rotation axis where m^μ is zero.

Under these circumstances instead of just one Komar integral for the mass M we shall have a pair of Komar integrals for the mass M and angular momentum J, whose forms are given respectively by

$$M = -\frac{1}{4\pi G} \oint_\infty k^{\mu;\nu} dS_{\mu\nu} \qquad\qquad J = \frac{1}{8\pi G} \oint_\infty m^{\mu;\nu} dS_{\mu\nu} \tag{4.1}$$

where the integration is taken over a topological spacelike 2-spheres in the limit of large radius r. The Killing (metric invariance) equations

$$k_{(\mu;\nu)} = 0 \qquad\qquad m_{(\mu;\nu)} = 0 \tag{4.2}$$

give corresponding Ricci curvature conditions

$$k^{\mu;\nu}{}_{;\nu} = -R^\mu{}_\nu k^\nu \qquad\qquad m^{\mu;\nu}{}_{;\nu} = -R^\mu{}_\nu m^\mu \tag{4.3}$$

which can be used, if the distant 2-sphere can be filled in by a spacelike 3-surface Σ with 3-ball topology, to obtain

$$M = M_{GS}, \qquad\qquad J = J_{GS} \tag{4.4}$$

where the gravitational source contributions are defined by

$$M_{GS} = \frac{1}{4\pi G}\int_\Sigma R^\mu{}_\nu k^\nu \, d\Sigma_\mu \qquad J_{GS} = -\frac{1}{8\pi G}\int_\Sigma R^\mu{}_\nu m^\nu \, d\Sigma_\mu. \tag{4.5}$$

(The extra terms required for non trivial topology resulting from presence of a black hole will be described later.)

In the special case of a rigidly rotating body there will be a unit flow vector u^μ which can be related to the mean redshift factor $\dot{r}$ to construct the *co-rotating Killing vector* $\bar{u}^\mu = \dot{r}u^\mu$ of the form $\bar{u}^\mu = k^\mu + \Omega m^\mu$ where Ω is the *uniform* angular velocity. For a perfect fluid energy momentum tensor

$$T_m{}^{\mu\nu} = (\rho_m + P)u^\mu u^\nu + Pg^{\mu\nu} \tag{4.6}$$

the Einstein equations give the relation

$$M_{GS} - 2\Omega J_{GS} = \int_\Sigma (\rho_m + 3P)\dot{r}u^\mu \, d\Sigma_\mu. \tag{4.7}$$

Under these conditions a change $d\Omega$ of the rigid angular velocity can be shown[69,70] to give corresponding mass and angular momentum variations related by

$$dM = \Omega \, dJ. \tag{4.8}$$

This variation can also be shown[71] to apply to the more general case of *solid* matter (such as that forming the crust of a neutron star) provided its deformation under the change of rigid angular velocity is *elastic*.

4.2 Circularity Theorem for Stationary Axisymmetric Systems

We shall now show that, under appropriate conditions, of which the archetypical example is the case of a perfect fluid whose flow is *circular* in the sense that its unit tangent is a combination of commuting, stationary, and axisymmetric Killing vectors,

$$\dot{r}u^\mu = k^\mu + \Omega m^\mu \tag{4.9}$$

(i.e. no radial or azimuthal component) where (unlike what was assumed for the preceding paragraph) the angular velocity, Ω, is not required to be uniform, we can be assured of the possibility that the spacetime coordinates can be chosen in such a way that (wherever the 2-surface elements generated by the Killing vectors is non-degenerate) the metric takes the standard (Papapetrou)[72] form

$$ds^2 = g_{00}dt^2 + 2g_{03}dtd\varphi + g_{33}d\varphi^2 + g_{ij}dx^i dx^j \qquad (i,j = 1,2) \tag{4.10}$$

where all components are independent of t and φ which are the ignorable coordinates corresponding to the symmetries:

$$k^\mu \longleftrightarrow \frac{\partial}{\partial t} \qquad m^\mu \longleftrightarrow \frac{\partial}{\partial \varphi}. \tag{4.11}$$

The non trivial part[73,74,42] of this *circularity theorem*, i.e. the eliminability of all cross components of the type g_{0i} or g_{3i} $(i = 1,2)$ is equivalent to establishing the integrability of 2-surface elements orthogonal to the Killing vectors, which by the standard Frobenius theorem is equivalent to

$$k_{[\mu;\nu}k_\rho m_{\sigma]}=0 \qquad m_{[\mu;\nu}k_\rho m_{\sigma]}=0. \tag{4.12}$$

The condition that the Killing vectors commute (so that t and φ can be simultaneously ignorable) i.e. $k^\mu{}_{;\nu}m^\nu = m^\mu{}_{;\nu}k^\nu$ together with the consequences of the Killing equations,

$$k_{[\mu;\nu}k_\rho m_{\sigma]}{}^{;\sigma}=-\tfrac{1}{2}k^\sigma R_{\sigma[\mu}k_\nu m_{\rho]}, \qquad m_{[\mu;\nu}m_\rho k_{\sigma]}{}^{;\sigma}=-\tfrac{1}{2}m^\sigma R_{\sigma[\mu}k_\nu m_{\rho]}, \tag{4.13}$$

establishes immediately that the Froebenius orthogonality condition requires what we shall refer to as the Ricci circularity condition namely

$$k^\sigma R_{\sigma[\mu}k_\nu m_{\rho]}=0, \qquad m^\sigma R_{\sigma[\mu}k_\nu m_{\rho]}=0, \tag{4.14}$$

which themselves are obvious consequences of the flow circularity condition assumed at the outset provided the energy momentum tensor in the Einstein equations has the perfect fluid form. To obtain the required result one needs to show that these Ricci circularity conditions are not just necessary but also sufficient for the Frobenius orthogonality conditions to hold. Defining "twist forms"

$$\omega_\mu=\tfrac{1}{2}\varepsilon_{\mu\nu\rho\sigma}k^\nu k^{\rho;\sigma} \qquad \psi_\mu=\tfrac{1}{2}\varepsilon_{\mu\nu\rho\sigma}m^\nu m^{\rho;\sigma} \tag{4.15}$$

one sees that they will satisfy

$$(m^\sigma\omega_\sigma)_{,\mu}=\tfrac{1}{2}\varepsilon_{\mu\nu\rho\sigma}k^\nu m^\rho R^\sigma{}_\tau k^\tau, \qquad (k^\sigma\psi_\sigma)_{,\mu}=\tfrac{1}{2}\varepsilon_{\mu\nu\rho\sigma}k^\nu m^\rho R^\sigma{}_\tau m^\tau. \tag{4.16}$$

Thus the Ricci circularity conditions imply that the scalars $m^\sigma\omega_\sigma$ and $k^\sigma\psi_\sigma$ are uniform. Since it is obvious that they must satisfy

$$m^\sigma\omega_\sigma=0, \qquad k^\sigma\psi_\sigma=0 \tag{4.17}$$

on the rotation axis where m^σ itself vanishes it follows that (4.15) will hold *everywhere*, which is equivalent to the required orthogonal transitivity requirement (4.10), provided that postulated circularity condition (4.12) also holds *everywhere*. If (4.12) was only satisfied locally in a region (say the interior of a smoke ring-like configuration) not including the rotation axis then the conclusion (4.7) could not be established directly at a purely local level but the converse, that (4.7) implies (4.12), is derivable.

In the prototype version of this result, Papapetrou[73] only considered the pure vacuum case in which (4.12) is satisfied trivially, but it can be seen that it will be ensured by the Einstein equation if the material contribution $T_m^{\mu\nu}$ to the energy momentum tensor satisfies the analogous circularity condition

$$k_\sigma T_m{}^\sigma_{[\mu}k_\nu m_{\rho]}=0, \qquad m_\sigma T_m{}^\sigma_{[\mu}k_\nu m_{\rho]}=0 \tag{4.18}$$

which will be the case in particular for a perfect fluid, as defined by (4.6) provided that the flow is circular in the literal sense of satisfying the condition (4.8) stated at the outset and provided that any electromagnetic contribution is provided by a field that is itself axisymmetric and stationary, in the sense of satisfying

$$F_{\mu\nu;\rho}m^\rho-2F_{\rho[\mu}m^\rho{}_{;\nu]}=0 \tag{4.19}$$

as well as the analogous condition (3.46) with respect to k^μ, and the its source current j^μ has the same circularity property as we postulated at the outset for the fluid flow, meaning that it is a (locally variable) combination of Killing vectors, i.e.

$$j^{[\mu}k^\nu m^{\rho]} = 0. \tag{4.20}$$

To establish this electromagnetic part of the result[42] we shall show that (4.20) is (globally) a sufficient condition (as well as obviously being a locally necessary condition) for the electromagnetic field circularity condition

$$F_{\mu\nu}k^\mu m^\mu = 0, \qquad F^{[\mu\nu}k^\rho m^{\sigma]} = 0 \tag{4.21}$$

which is itself obviously sufficient to ensure that the Maxwell energy-momentum tensor $T_F{}^{\mu\nu}$ (as defined by (3.22)) will satisfy conditions analogous to (4.18). We start by pointing out that subject to the stationarity and axisymmetry conditions (3.45) and (4.19), the scalars

$$F_{\mu\nu}k^\mu m^\nu = E_\rho m^\rho, \qquad \tfrac{1}{2}\varepsilon_{\mu\nu\rho\sigma}F^{\mu\nu}m^\rho k^\sigma = B_\rho m^\sigma \tag{4.22}$$

[whose vanishing is evidently equivalent to (4.20)] will in general satisfy

$$(E_\rho m^\rho)_{,\mu} = 0, \qquad (B_\rho m^\rho)_{,\mu} = 4\pi\varepsilon_{\mu\nu\rho\sigma}m^\nu k^\rho j^\sigma \tag{4.23}$$

in consequence of the Maxwell equations (3.46) and (3.47) respectively. It follows that the circularity condition (4.20) suffices to ensure that these scalars will *both* be uniform. As in the gravitational case, it is enough to notice that they automatically satisfy

$$E_\rho m^\rho = 0, \qquad B_\rho m^\rho = 0 \tag{4.24}$$

on the rotation axis, where m^ρ is zero, in order to see that they will vanish everywhere, as required to establish (4.21), provided that (4.20) also holds everywhere.

It is to be remarked, as a corollary that this result implies that the field $F_{\mu\nu}$ will be derivable (in accordance with (3.19)) from a 4-potential that not only has the same stationarity and axisymmetry invariance properties as the field itself, i.e.

$$A_{\mu;\nu}k^\nu + A_\nu k^\nu{}_{;\mu} = 0, \qquad A_{\mu;\nu}m^\nu + A_\nu m^\nu{}_{;\mu} = 0 \tag{4.25}$$

but also has a circularity property analogous to that postulated for the current, i.e.

$$A^{[\mu}k^\nu m^{\rho]} = 0 \tag{4.26}$$

which ensures incidentally that it will automatically satisfy the Lorentz gauge condition

$$A^\rho{}_{;\rho} = 0. \tag{4.27}$$

We emphasize that although (subject to stationarity and axisymmetry) (4.20) is a *local* consequence of (4.25), the possibility of obtaining (4.26) depends on (4.20) holding *globally* over a domain including the rotation axis. The classic counterexample to the local sufficiency of (4.20) is provided by the familiar flat space configuration where one has a current spiralling (and thus violating the circularity condition (4.20)) round a toroidal solenoid, but no sources elsewhere: *outside* the solenoid the circularity theorem will be applicable consistently with the well known fact that there will actually be no field there at all; however *inside* the solenoid there will of course be a magnetic field in the φ direction, which is contrary to what would be allowed by (4.24) or (4.26). A failure to understand the gravitational analogue of this situation in the early days after the discovery by Boyer and Lindquist[67] that the Kerr vacuum solution could be cast in the t, φ reversal invariant form (4.9) above, led to speculation that any interior matter source for the Kerr solution would have to exhibit the same dicrete symmetry. What the Papapetrou's theorem showed was that the *exterior* field *must* posess such a discrete symmetry *regardless* of whether or not it is possessed by the source.

4.3 The Ergoregion and ZAMOs in Papapetrou Coordinates

Before drawing consequences of the orthogonal transitivity property established by the preceding Circularity Theorem, it is to be emphasized that it would fail to hold inside a convecting star for which one would no longer be able to set $g_{0i} = 0$, $g_{3i} = 0$. It would still hold in the external vacuum but not in a conceivable enclosed vacuum with a hollow (smoke-ring type) torus.

The ignorable coordinate coefficients in any manifestly stationary and axisymmetric metric form are expressible invariantly as (say),

$$g_{00} = k^\mu k_\mu = -V, \qquad g_{03} = k^\mu m_\mu = W, \qquad g_{33} = X. \tag{4.28}$$

The quantity X must be everywhere positive by the causality requirement (no closed timelike or null lines) whereas the quantity V is only required to be positive at large distance r, but may become negative in an ergoregion where $\vec{k}$ becomes timelike, so that the energy,

$$-k^\mu u_\mu = V\frac{dt}{d\tau} - W\frac{d\varphi}{d\tau} - g_{0i}\frac{dx^i}{d\tau} \tag{4.29}$$

per unit mass (of a particle with 4-velocity u^μ, proper time τ) may become negative there. Throughout space, the angular momentum

$$m^\mu u_\mu = X\frac{d\varphi}{d\tau} + W\frac{dt}{d\tau} + g_{3i}\frac{dx^i}{d\tau} \tag{4.30}$$

may have either sign. One often refers to a preferred congruence of trajectories known as ZAMOs,[76] which I interpret as standing for "zero angular momentum orbiter," meaning particles moving in *circular* orbits with $m^\mu u_\mu = 0$. The required angular velocity, $\frac{d\varphi}{dt} = \omega$, say, will be given by

$$\omega = -\frac{g_{03}}{g_{33}} = -\frac{W}{X}. \tag{4.31}$$

This formula would still give angular velocity corresponding to zero angular momentum even for non circular (e.g. infalling) trajectory subject to the requirement for the circularity theorem $g_{3i} = 0$ but would fail more generally. When the circularity conditions hold, ZAMOs are characterised (but by no means uniquely) by the property of irrotationality, since they are definable as the congruence orthogonal to hypersurfaces of constant t, x^1, x^2.

It can be seen that in order for the ZAMO worldlines to be *timelike* it is necessary and sufficient that the quantity

$$\rho^2 = W^2 + VX = (g_{03})^2 - g_{00}g_{03} \tag{4.32}$$

should be positive everywhere except on the rotation axis (where both W and X vanish, so that one only need V to be positive). In terms of this quantity

$$ds^2 = g_{ij}dx^i dx^j + X\{(d\varphi - \omega dt)^2 - \rho^2 dt^2\} \tag{4.33}$$

Apart from the familiar symmetry axis singularity where $X = 0$, this latter metric form can be shown to be valid not just locally but even globally in many important cases so long as one restricts oneself to the domain of outer communications as defined in the previous chapter. To this effect, a key lemma, to be demonstrated in §4.8, states that as (suggested by the notation) the discriminant ρ^2 is positive throughout the domain of outer communications (within which, it vanishes only on the symmetry axis where $X = 0$). One consequence is to ensure that the ZAMOs have well behaved timelike trajectories throughout the domain of outer communications (whereas, as we shall see, they become null on the horizon of any black hole that may be present, and are blatantly tachyonic inside the hole).

4.4 Slowly Rotating (but Strongly Gravitating) Stellar Equilibrium Configurations

In most applications to observed strongly gavitating systems (I am thinking of the neutron star models for pulsars, etc. in which rotation velocity is small compared with orbital velocity at the equator) a slow rotation perturbation of spherical equilibrium is sufficient. Moreover although one of the most important quantities, the rotation energy $E(\Omega) = M(\Omega) - M(0)$ is a quadratic function of the angular velocity Ω which we shall take to be uniform (I shall not consider non-rigid rotation) it can be obtained from a treatment linear in Ω by using the Hartle-Sharp-Bardeen variation principle[69,70] quoted above (which I have shown[71] to be valid even for a partially solid star as long as it is elastic) which implies

$$E = \tfrac{1}{2} I(0)\Omega^2 + O(\Omega^4) \tag{4.34}$$

where we have chosen to generalise the definition of the moment of inertia, I, from flat to curved space by taking $I = J/\Omega$ (which in Newtonian mechanics is a theorem), the required spherical limit value $I(0)$ being read out from the relation

$$J = I(0)\Omega + O(\Omega^3) \tag{4.35}$$

whose use does not necessitate going beyond the first order in Ω. To this order the specification of the perturbed spherical metric requires the introduction of only one function, namely the ZAMO angular velocity, ω, in addition to the unperturbed spherical quantities ϕ, μ (whose evaluation is described below) in the corresponding metric expression is

$$ds^2 = e^{2\mu}\, dr^2 + r^2\, d\theta^2 + r^2 \sin^2\theta\, (d\varphi - \omega\, dt)^2 - e^{-2\phi}\, dt^2 . \tag{4.36}$$

At first order in ω (which must be of the same order as the fluid angular velocity Ω) the only one of the perfect fluid Einstein equations that is non trivial has been shown by Hartle[77] to be reducible to

$$\frac{1}{r^4}\frac{\partial}{\partial r}\left(r^4 e^{-(\mu+\phi)}\frac{\partial \omega}{\partial r}\right) + \frac{e^{(\mu-\omega)}}{r^2\sin^3\theta}\frac{\partial}{\partial\theta}\left(\sin^3\theta\frac{\partial\omega}{\partial\theta}\right) = \frac{4(\Omega-\omega)}{r}\frac{\partial}{\partial r}e^{-(\mu+\phi)}. \tag{4.37}$$

A Legendre expansion of the form

$$\omega(r,\theta) = \sum_l \omega_l(r)\frac{dP_l(\cos\theta)}{d(\cos\theta)} \tag{4.38}$$

gives the ordinary differential equation

$$\frac{1}{r^4}\frac{d}{dr}\left(r^4 e^{-(\mu+\phi)}\frac{d\omega_l}{dr}\right) + \frac{e^{-(\mu-\phi)}}{r^2}[l(l+1)-2]\omega_l = \frac{4(\Omega_l-\omega_l)}{r}\frac{d}{dr}e^{-(\mu+\phi)}. \tag{4.39}$$

Our rigidity assumption implies $\Omega = \Omega_1$, $\Omega_l = 0$, $(l \neq 1)$, i.e., only the $l = 1$ equation has a source term. Hence we deduce that $\omega_l = 0$ for $l \neq 1$. The remaining Legendre coefficient ω_1 will be fixed by boundary conditions of finiteness at the origin $r \to 0$, and by $\omega_1 \sim 2GJ/r^3$ as $r \to \infty$. In practice (starting at the regular origin) we take *any* solution of the homogeneous equation

$$\frac{d}{dr}\left(r^4 e^{-(\mu+\phi)}\frac{d\bar{\omega}}{dr}\right) + \frac{e^{-(\mu-\phi)}}{r^2}\frac{d\bar{\omega}}{dr} = 0 \tag{4.40}$$

for the combination

$$\bar{\omega} = \omega - \Omega, \tag{4.41}$$

and read off the required ratios $I(0) = J/\Omega$ from the coefficients in the asymptotic form

$$\bar{\omega} \sim \Omega - 2\frac{GJ}{r^3}. \tag{4.42}$$

To carry out this first order part of the calculation we need of course to have solved the zero order perfect fluid equations for μ and ϕ, which are familiar from the long well known theory of spherical equilibrium, which gives

$$\frac{dM}{dr} = 4\pi r^2 \rho_m \qquad \frac{dP}{dr} = (\rho_m + P)\frac{d\phi}{dr} \tag{4.43}$$

$$\frac{d\phi}{dr} = \frac{G(M + 4\pi r^3 P)}{r(r - 2GM)} \tag{4.44}$$

where the effective mass function $M(r)$ is given in terms of μ by the standard definition

$$e^{-2\mu} = 1 - 2GM/r. \tag{4.45}$$

4.5 Killing Horizon (Nullity and Rigidity) Property of Locus Where ZAMOs Go Null

The Frobenius orthogonality conditions established by the preceding Circularity Theorem (whose requirements are always satisfied in vacuum) can be written out as

$$\begin{aligned} k_{\mu;[\nu}\rho_{\sigma\tau]} &= -k_\mu k_{\nu;\sigma} m_{\tau]} + m_\mu k_{\nu;\sigma} k_{\tau]} \\ m_{\mu;[\nu}\rho_{\sigma\tau]} &= m_\mu m_{\nu;\sigma} k_{\tau]} - k_\mu m_{\nu;\sigma} m_{\tau]} \end{aligned} \tag{4.46}$$

where $\underline{\rho} = \underline{k} \wedge \underline{m}$ i.e.

$$\rho_{\sigma\tau} = k_{[\sigma} m_{\tau]} \tag{4.47}$$

is a 2-form that is non-zero where and only where the Killing vectors are linearly independent, so we can deduce that this 2-form will satisfy

$$2\rho_{\kappa\mu;[\nu}\rho_{\sigma\tau]} = \rho_{\kappa\mu}\rho_{\sigma\tau;\nu}. \tag{4.48}$$

Contracting with $\rho^{\kappa\mu}$ and noting that the discriminant ρ^2 that has been introduced above is given by

$$\rho^2 = -\tfrac{1}{2}\rho_{\mu\nu}\rho^{\mu\nu} \tag{4.49}$$

we obtain[42] the relation

$$(\rho^2)_{,[\nu}\rho_{\sigma\tau]} = \rho^2 \rho_{[\sigma\tau;\nu]} \tag{4.50}$$

which tells us that wherever ρ^2 vanishes, its gradient $(\rho^2)_{,\mu}$ must be linearly dependent on k_μ and m_μ. One familiar situation in which this occurs is on the axis of symmetry where m_μ itself vanishes and where the gradient of ρ^2 must obviously vanish too. However the *non-degenerate* possibility, meaning $(\rho^2)_{,\mu} \neq 0$ where $\rho^2 = 0$, occurs where the 2-element generated by k^μ and m^μ is *null* so that it contains a unique combination of the form

$$\ell^\mu = k^\mu + \Omega^H m^\mu \tag{4.51}$$

with

$$\ell^\mu \ell_\mu = 0. \tag{4.52}$$

Since ρ^2 is invariant its gradient must be orthogonal to both vectors k^μ and m^μ. The only way it can not be a combination of the covectors **k** and **m** is for it to be proportional to the unique null combination, i.e. it must satisfy

$$(\rho^2)_{,\mu} = 2\kappa(m_\nu m^\nu)\ell, \tag{4.53}$$

for some proportionality factor κ which can be identified with the decay parameter introduced in section 3. The nullity of the gradient of ρ^2 implies that (when non-degenerate) the locus on which ρ^2 vanishes is a null hypersurface. Since ℓ_μ is orthogonal to any vector within this null surface, and in particular to the Killing vectors k_μ and m_μ, contraction of the preceding equation with the latter gives

$$k^\mu m_\mu + \Omega^H m^\mu m_\mu = 0 \tag{4.54}$$

which shows that the effective angular velocity Ω^H of the local horizon where $\rho^2 \to 0$ is just the limit there of the ZAMO angular velocity introduced above, i.e. we have $\omega \to \Omega^H$ there.

The locus bounding the outer region in which ZAMOs are timelike is thus a horizon of the type under consideration. Such a horizon has the important property of being rigid[78] in the sense that the angular velocity Ω^H of its null generators (i.e. the limiting ZAMO angular velocity) is uniform over it. To see this we can use the fact that (by the Killing equations and the commutativity condition) the gradient of the ZAMO angular velocity $\omega = -W/X$ will be given by

$$\begin{aligned} X^2\omega_{,\mu} &= WX_{,\mu} - XW_{,\mu} \\ &= 2(Wm^\nu - Xk^\nu)m_{\nu;\mu} \end{aligned} \tag{4.55}$$

so that by the Frobenius orthogonality condition we obtain

$$X^2\rho_{[\mu\nu}\omega_{,\sigma]} = 2\rho^2 m_{[\sigma}m_{\nu;\mu]}. \tag{4.56}$$

Hence on the horizon where $\rho^2 = 0$, the quantity ω has the same property as ρ^2 of having its gradient both orthogonal to and linearly dependent on the Killing vectors which means that it is also normal to the null hypersurface, i.e. $\omega_{,\mu} = \lambda\ell_\mu$ for some proportionality constant λ, which establishes that it has a value $\omega = \Omega^H$ that is *constant* over the horizon. This rigidity property means that we can *globally* define a combination ℓ^μ of the form given above (4.48) taking Ω^H to be constant throughout space,

$$\Omega^H{}_{,\mu} = 0, \tag{4.57}$$

so that the Killing vector property $\ell_{(\mu;\nu)} = 0$ is satisfied everywhere, while the nullity property $\ell^\mu\ell_\mu = 0$ holds only on the horizon. This completes our demonstration that the (non-degenerate part) of the locus bounding the region where ZAMOs are timelike has the property of being a *Killing horizon* in the technical sense[42] of being a null hypersurface whose null generators are also trajectories of a Killing vector field, namely ℓ^μ as it has just been defined.

4.6 Black Hole Mass and Angular Momentum and their Variations

The existence of a well definined Killing vector combination $\ell^\mu = k^\mu + \Omega^H m^\mu$ where Ω^H is the constant angular velocity of the black hole horizon enables us to obtain a useful formula relating the mass and angular momentum of the hole.

The previously introduced 2-surface integral definitions of the asymptotic mass M and angular momentum J suggest corresponding definitions[21,25] for the hole contributions in terms of two surface integrals on the horizon in the form

$$M_H = \frac{1}{8\pi G}\oint_H k^{\mu;\nu}\, dS_{\nu\mu}, \qquad J_H = -\frac{1}{16\pi G}\oint_H m^{\mu;\nu}\, dS_{\nu\mu} \tag{4.58}$$

so that by Green's theorem one will have

$$M = M_{GS} + M_H, \qquad J = J_{GS} + J_H \tag{4.59}$$

where the gravitational source contributions are given by (4.5). The black hole contributions introduced in this way satisfy

$$M_H - 2\Omega^H J_H = \frac{1}{8\pi G}\int \ell^{\mu;\nu}\, dS_{\nu\mu}. \tag{4.60}$$

Since the surface element on the horizon can be expressed in terms of the null tangent ℓ^μ and the transverse null vector β^μ in the form

$$dS_{\mu\nu} = 2\beta_{[\mu}\ell_{\nu]}dS \tag{4.61}$$

where dS is the 2-surface measure) the expression $\kappa = -\ell^\mu \ell_{\mu;\nu}\beta^\nu$ for the time decay parameter κ leads us to the result

$$M_H = 2\Omega^H J_H + \frac{\kappa \mathcal{A}}{4\pi G} \tag{4.62}$$

which generalises[21,35] a relation first observed by Smarr[79] in the Kerr vacuum case. (One can obtain a more detailed relation explicitly distinguishing a possible electromagnetic contribution.)

The relation (3.48) suggests the utility of introducing a modified (material plus current) energy momentum tensor

$$\tilde{T}^\mu_{\ \nu} = T_m{}^\mu{}_\nu + j^\mu A_\nu \tag{4.63}$$

in terms of which we can define combined matter and current contributions to the mass and angular momentum by

$$M_{MC} = 2\int (\tilde{T}^\mu_{\ \nu} - \tfrac{1}{2}\tilde{T}^\rho_{\ \rho} g^\mu_{\ \nu})k^\nu\, d\Sigma_\mu, \qquad J_{MC} = -\int \tilde{T}^\mu_{\ \nu} m^\nu\, d\Sigma_\mu. \tag{4.64}$$

Hence applying Green's theorem to the analogue of (3.48) with ℓ^μ in place of k^μ, and dealing with the surface contributions on the horizon by using (4.61) together with the boundary conditions (3.28), we obtain[21,35]

$$M_{GS} - 2\Omega^H J_{GS} = M_{MC} - 2\Omega^H J_{MC} + \Phi^H Q_H \tag{4.65}$$

where the charge Q_H of the black hole is given in terms of the total charge moment Q by

$$Q = \frac{1}{4\pi}\int_{\infty} F^{\mu\nu}\, dS_{\mu\nu} = Q_H + Q_C \tag{4.66}$$

where

$$Q_H = \frac{1}{4\pi}\oint_H F^{\mu\nu}\, dS_{\mu\nu}\,, \qquad Q_C = -\int j^{\mu}\, d\Sigma_{\mu} \tag{4.67}$$

Combining this with (4.62), we obtain

$$M - 2\Omega^H J - \Phi^H Q - \frac{\kappa \mathcal{A}}{4\pi G} = M_{MC} - 2\Omega^H J_{MC} - \Phi^H Q_C. \tag{4.68}$$

When there is no external material or current contribution (so that M_{MC}, J_{MC}, Q_C vanish) the right hand side will drop out and we shall be left with the relation first observed by Smarr[79] in explicit case of the Kerr-Newman solutions.

Let us now consider a variation between two neighbouring equilibrium states, adjusting the mapping between the two corresponding manifolds (i.e. choosing the gauge, as we are free to do) in such a way as to conserve the basic stationarity and axisymmetry Killing fields, k^{μ} and m^{μ}, so that

$$\delta k^{\mu} = 0 \qquad \delta m^{\mu} = 0 \tag{4.69}$$

which implies for the corotating Killing vector $\delta\ell^{\mu} = m^{\mu}\delta\Omega^H$ and hence

$$\delta\ell_{\mu} = h_{\mu\nu}\ell^{\nu} + m_{\mu}\delta\Omega^H \tag{4.70}$$

where $\delta\Omega^H$ is the angular momentum variation and $h_{\mu\nu}$ denotes the metric variation, i.e.

$$h_{\mu\nu} = \delta g_{\mu\nu}. \tag{4.71}$$

Since $\underline{\ell}$ remains orthogonal to the horizon, which we also suppose to remain fixed (by choice of gauge), its direction will remain unaffected so that we shall have $\underline{\ell}\wedge\delta\underline{\ell} = 0$. Combining this with the group invariance property $\vec{\ell}\mathcal{L}\delta\underline{\ell} = 0$ one can deduce that the variation of κ will be given by

$$\delta\kappa = \tfrac{1}{2}\ell^{\mu}h_{\mu\nu}{}^{;\nu} - \ell_{\mu}\beta_{\nu}m^{\nu;\mu}\delta\Omega^H \tag{4.72}$$

which gives

$$\begin{aligned} \mathcal{A}\,\delta\kappa &= \oint_{\kappa} (\delta\kappa)\; dS \\ &= \oint_H h^{[\mu}{}_{\nu}{}^{;\nu]}\ell^{\rho}\, dS_{\mu\rho} - 8\pi G J_H \delta\Omega_H \end{aligned} \tag{4.73}$$

since the asymptotic metric variation will be related to that of the mass by

$$h_{ab} = 2G\frac{\delta M}{r}\delta_{ab} + O(1/r^2) \tag{4.74}$$

which gives

$$4\pi G\,\delta M = \oint_{\infty} h^{[\mu}{}_{\nu}{}^{;\nu]}\kappa^{\rho}\, dS_{\mu\rho}. \tag{4.75}$$

Hence, using the postulate that the surface be invariant under the rotation symmetry, which gives $m^\nu dS_{\mu\nu} = 0$, one obtains

$$\oint_H \ell^\rho h^{[\mu}{}_\nu{}^{;\nu]}\, dS_{\mu\rho} = \int_\Sigma k^\mu h^{[\nu}{}_\nu{}^{;\rho]}{}_{;\rho}\, d\Sigma_\mu - 4\pi G\,\delta M \tag{4.76}$$

and thus

$$4\pi G\delta M + 8\pi G J_H \delta\Omega^H + A\delta\kappa = \int_\Sigma k^\mu h^{[\nu}{}_\nu{}^{;\rho]}{}_{;\rho}\, d\Sigma_\mu. \tag{4.77}$$

The integrand on the right occurs in the derivation of the Einstein equations from the variational principle based on the Hilbert action $R(\det g)^{1/2}$ which satisfies

$$\tfrac{1}{2}(\det g)^{-1/2}\delta(R(\det g)^{1/2}) = -\tfrac{1}{2}(R^{\mu\nu} - \tfrac{1}{2}Rg^{\mu\nu})h_{\mu\nu} - h^{[\nu}{}_\nu{}^{;\rho]}{}_{;\rho}\,. \tag{4.78}$$

Since the last term is a pure divergence, the requirement that this should vanish for arbitrary $h_{\mu\nu}$ gives the vacuum condition that $R^{\mu\nu} - \frac{1}{2}Rg^{\mu\nu}$ should vanish. For our present purpose, where matter is supposed to be present, we use this identity with the expression

$$\delta(d\Sigma_\mu) = (\det g)^{-1/2}\delta(\det g)^{1/2} = \tfrac{1}{2}h^\nu{}_\nu d\Sigma_\mu \tag{4.79}$$

to obtain

$$\begin{aligned}4\pi G\delta M + 8\pi G J_H \delta\Omega^H + A\delta\kappa = &-\tfrac{1}{2}(R^{\mu\nu} - \tfrac{1}{2}Rg^{\mu\nu})h_{\mu\nu}k^\sigma d\Sigma_\sigma \\ &- \tfrac{1}{2}\delta\int Rk^\mu d\Sigma_\mu.\end{aligned} \tag{4.80}$$

The right hand side here has the same form as that occuring in the ordinary Hartle-Sharp type variation principle that applies in the absence of a black hole. Its evaluation depends on the kind of variation envisaged, but for a perfectly elastic variation of a perfect fluid (or solid) star with rigid angular velocity Ω and material angular momentum J_M say, it can be shown[71] to reduce simply to $\Omega\, dJ_M$. Additional contributions arising from electromagnetic effects are discussed elsewhere.[21,25]

4.7 Superradiance

The flux rate per unit of the time t associated with a Killing vector $\ell^\mu = dx^\mu/dt$ across the 2-surface S of the mass energy current $T^\mu{}_\nu k^\nu$ or the angular momentum current $T^\mu{}_\nu m^\nu$ associated with any radiation field with energy momentum $T^\mu{}_\nu$ will be given[35] respectively by

$$\dot{M} = \oint_S T^\mu{}_\nu k^\nu \ell^\rho\, dS_{\mu\rho} \tag{4.81}$$

$$\dot{J} = -\oint_S T^\mu{}_\nu m^\nu \ell^\rho\, dS_{\mu\rho}. \tag{4.82}$$

Applying this to the particular case of flux across the horizon, [using (4.48) and (4.59)] one deduces that on the horizon of any black hole in axisymmetric equilibrium we shall have

$$\begin{aligned}\dot{M} - \Omega^H \dot{J} = \oint_H T_{\mu\nu}\ell^\mu\ell^\nu\, dS = \frac{\kappa}{8\pi G}\dot{A} \\ \geq 0.\end{aligned} \tag{4.83}$$

In the case of a periodic wave, proportional to the real part of $e^{i(\sigma t - m\phi)}$ the average energy and angular momentum fluxes $\langle\dot{M}\rangle$ and $\langle\dot{J}\rangle$ will be conserved (independent of the 2-surface) and respectively proportional to the time and angular frequency quantum numbers σ and m, i.e. we shall have

$$\langle \dot{M} \rangle / \sigma = \langle \dot{J} \rangle / m. \tag{4.84}$$

The preceding inequality therefore gives

$$\langle \dot{M} \rangle (1 - \frac{m}{\sigma} \Omega^H) \geq 0. \tag{4.85}$$

For a non rotating hole, $\Omega^H = 0$, this simply tells us that $\langle \dot{M} \rangle$ must be positive, i.e. energy can only flow into the hole. However for non-zero rotation velocity, let us say $\Omega^H > 0$ by choice of parity conservation, there will be a range[21,47,43] of *super-radiant modes* characterised by $0 < \sigma/m < \Omega^H$ for which $\langle \dot{M} \rangle$ must be *negative*, i.e. incoming radiation from infinity will be *amplified* as it is reflected by the black hole.

4.8 Local Characterisation of Simply Connected Stationary (Circularity) Axisymmetric D.O.C.

We have already seen in §3.3 that a simply connected static D.O.C. can be completely characterised as the maximal outer connected set satisfying the local condition that k^μ be timelike, i.e. that V be positive. We shall now show that, provided it has the *circularity* property discussed in §4.2, a non-static but *stationary axisymmetric* D.O.C. can analogously be completely characterised by a purely local condition as being the maximal outer connected set, Z say, on which the Killing bivector $k_{[\mu}m_{\nu]}$ is timelike, which except of the rotation axis is equivalent to the condition that the quantity

$$\rho^2 = (k^\nu m_\nu)^2 - (k^\mu k_\mu)(m^\rho m_\rho) \tag{4.86}$$

introduced in section 4.3 satisfies the positivity requirement

$$\rho^2 > 0. \tag{4.87}$$

In short such a D.O.C. is the *maximal outer connected set, Z, on which the ZAMO trajectories are timelike.*

To see this we start by noting that the set Z so defined must evidently lie *within* the D.O.C.,[43] since by a construction of the type described already in §3.6, any connecting curve within it wil be continuously deformable along the local direction of the ZAMO trajectories so as to become timelike (with future or past orientation accorking to choice).

Since the global bradyonicity lemma in section 3.6 ensures that the stationarity action cannot have any fixed points or closed trajectories (it being implicitly taken for granted throughout this work that we are dealing with a time oriented manifold that is *causal* in the sense that there are no closed timelike curves) it follows that the vector k^μ cannot coincide with m^μ anywhere in the D.O.C. It follows that the Killing bivector $k^{[\mu}m^{\nu]}$ must be *non-degenerate* everywhere in the D.O.C. except of the rotation axis where m^μ itself is zero. It therefore follows from the lemma of §4.5 that the boundary, $Z^\bullet$, of the region where this bivector, and therefore also the ZAMO direction, is timelike will have the Killing horizon property of being locally a null hypersurface segment whose null generator is a Killing vector combination, with a locally uniform time orientation.

We can now complete the argument in exactly the same way as we did for the analogous demonstration in the static case of §3.8. Subject to the assumption (which has not been invoked up to this point in the reasoning) that the D.O.C. is *simply* connected, any connected component D say of any part of the complement of Z in the D.O.C. would have a boundary $\mathcal{D}^{\bullet}$ in the D.O.C. that would itself have to be connected, and hence have to be a null hypersurface segment with uniform time orientation. This implies that $\mathcal{D}$ could not be connected to $\mathcal{Z}$ by both future and past timelike lines, as would be necessary for $\mathcal{D}$, if it exists, to belong to the D.O.C. It is therefore evident that $\mathcal{D}$ must be empty, which completes the required demonstration that $\mathcal{Z}$ must not just lie within the D.O.C. but must cover the whole of it.

As in the analogous static case, the condition of simple connectivity implies the *global* not just local regularity of the 2-surfaces orthogonal to the Killing vectors whose existence is implied by the circularity theorems of §4.2. The timelike nature of the Killing bivector that we have shown to apply throughout the D.O.C. entails correspondingly that the orthogonal 2-surfaces will be spacelike, with positive definite induced metric. Since any such surface deviates from flatness only by a locally variable conformal factor Σ say, it follows that the space coordinates x^i $(i = 1, 2)$ in the general Papapetrou form given above may be chosen more specifically to be cylindrical type coordinates ϖ, z say, in such a way that the metric will be expressible in the form

$$ds^2 = \Sigma(d\varpi^2 + dz^2) + X\{(d\varphi - \omega dt)^2 - \rho^2 dt^2\} \tag{4.88}$$

which will be *globally valid over the entire D.O.C.* except for the familiar degeneracy on the symmetry axis where X vanishes, ρ^2 and X being positive elsewhere (while the factor Σ is positive everywhere, including the axis).

In so far as the electromagnetic field is concerned, just as we have already seen in §3.6 how the stationarity condition (3.46) implies the existence of an electric potential Φ such that

$$F_{\mu\nu}k^{\nu} = \Phi,_{\mu} \tag{4.89}$$

so analogously the axisymmetry condition (4.19) implies the existence of a corresponding magnetic potential B (not to confused with the potential Ψ introduced above in the non-rotatating case) such that

$$F_{\mu\nu}m^{\nu} = B,_{\mu} \tag{4.90}$$

where such a B must obviously have a value that remains constant along any rotation axis segment. The magnetic monopole moment associated with any 2-surface S generated by the axisymmetry action on any curve s beginning and ending on the symmetry axis will be expressible as

$$P = \frac{1}{4\pi}\oint_S F_{\mu\nu} dx^{[\mu} ds^{\nu]} \tag{4.91}$$

where we may take $dx^{\mu} = m^{\mu} d\varphi$. This shows that P determines the difference $[B]$ between the values of B and the two ends of s, i.e.

$$[B] = \int B_{\mu} ds^{\mu} \tag{4.92}$$

and hence that if as before we make the usual physical postulate that such magnetic monopole moments must always be zero, we can deduce that B must take the *same* constant value on *all* segments of the rotation axis, which enables us to specify it uniquely by imposing the requirement that we should have

$$B \to 0 \tag{4.93}$$

on each segment of the rotation axis. Now provided that the current circularity condition (4.20) is satisfied everywhere, we can apply the theorem of §4.2 to conclude that the electromagnetic 4-potential will be expressible in the gauge characterised by (4.25) and (4.26) by

$$A_\rho dx^\rho = \Phi dt + B d\varphi \tag{4.94}$$

where Φ and B are independent of the t and φ and where the hypothesis of simple connectivity of the D.O.C. is sufficient to ensure that they are well defined globally. In order for A_μ to be not only globally well defined but also *globally well behaved*, even on the rotation axis where $d\varphi$ is singular, it is necessary and sufficient to use the gauge specified as above by (4.93), which as we have seen, is possible only on condition that magnetic monopoles are excluded.

§5 THE SOURCE FREE EQUILIBRIUM STATE PROBLEM FOR AXISYMMETRIC BLACK HOLES

5.1 Canonical Global Coordinate System for the D.O.C. of a Stationary Axisymmetric Black Hole

Whereas the previous section was concerned with material stellar type equilibrium configurations in which the possibility of external orbiting matter rings was explicitly taken into account, we shall now restrict our attention to the case of *source-free* (axisymmetric, asymptotically flat) *pure or electromagnetic vacuum* black hole equilibrium states, with the aim of describing the main steps (though not all the technical details) of the line of reasoning that leads (with by now *almost* complete rigour) to the conclusion that the only possibilities are those belonging to the well known family of Kerr vacuum solutions[83] and their electromagnetic Kerr Newman[84] generalisation. Before proceeding, it is to be mentioned that it may be plausibly conjectured[85] that the more general class of solutions for the source-free equations with cosmological Λ term[21,86] which I originally discovered on the basis of their special separability properties[87,88] may similarly include the only source free solutions representing axisymmetric black hole equilibrium states in an asymptotically De Sitter or anti-De Sitter background, but all attempts to prove this have failed so far because the *essential step* to be described in the present subsection works only when the Λ *term* is set exactly to *zero.*

The step in question is based on the fact that for a stationary axisymmetric metric of the orthogonally transitive (circular) type (4.7) derived in section 4.2, the trace of the projection of the Ricci tensor in the plane of the Killing vectors takes the simple form

$$R^{00}g_{00} + 2R^{03}g_{03} + R^{33}g_{33} = -\frac{1}{\Sigma\rho}\nabla^2\rho \tag{5.1}$$

where the quantities Σ and ρ are those appearing in the metric form (4.82), and the operator ∇^2 is the Laplacian of the *two-dimensional flat* metric $d\varpi^2 + dz^2$ that is introduced there. Subject to the source free Einstein-Maxwell equations

$$R^{\mu\nu} = 8\pi G T_F^{\mu\nu} + \Lambda g^{\mu\nu} \tag{5.2}$$

where $T_F^{\mu\nu}$ is the (trace free) Maxwell energy momentum tensor specified in section 3.2 and Λ is the cosmological term, we see that we can rewrite the right hand side of (5.1) in the form

$$\rho^2(VR_{33} + 2WR_{03} - XR_{33}) = 2\Lambda - 2GF_{03}F^{03}. \tag{5.3}$$

Furthermore, subject to the source free Maxwell equations, we can apply the circularity theorem of section 4.2 which by equation (4.19) implies that we shall have

$$F_{03} = 0 \tag{5.4}$$

so that (5.1) will reduce to the form

$$\nabla^2 \rho = -2\Lambda\Sigma\rho. \tag{5.5}$$

If we now restrict ourselves again to the non-cosmological case

$$\Lambda = 0 \tag{5.6}$$

with which we have been primarily concerned throughout this work, we see that this implies that ρ will be a harmonic function, which implies (as was noticed long ago by Papapetrou[72] and other workers on axisymmetric systems) the possibility of choosing ρ itself as one of the cylindrical type coordinates in the conformally flat form of the metric on the spacelike 2-surfaces orthogonal to the Killing vectors, i.e. one may set

$$\varpi = \rho \tag{5.7}$$

in the form (4.80) which thus becomes

$$ds^2 = \Sigma(d\rho^2 + dz^2) + X\{(d\varphi - \omega\, dt)^2 - \rho^2 dt^2\}. \tag{5.8}$$

In the case of a black hole with non-trivial, e.g. toroidal, topology for which the Domain of Outer Communications would not be simply connected the quantity ρ that we have thus introduced as a coordinate would not be globally regular (in particular in the toroidal case it would have a minimax somewhere between the inner side of the torus and the symmetry axis). However for a black hole with simple spherical topology (or even in the still not rigourously excluded though physically implausible case of a pair of corotating black holes balanced on a common symmetry axis) for which the Domain of Outer Communications would be simply connected, I have been able to establish[21] that there is no possibility for ρ to have any critical points at all, by applying the strongly restrictive type of Morse theory that applies to harmonic functions.[89] [Ordinary Morse theory establishes that the number of maxima plus the number of minima minus the number of minimaxes is a topological invariant subject to appropriate boundary conditions and subject to the assumption that no degenerate critical points occur. In the harmonic case minima and maxima cannot occur at all, and degenerate critical points become easy to deal with by labelling them with an appropriate positive degeneracy index, the theorem being that the index weighted sum over all critical points *including* possible degenerate ones is a topological invariant which in the case in question can be seen – e.g. by considering any simple example such as the Schwarzschild solution – to be zero.]

Since Hawking has been able to demonstrate[32,15] by a rather intricate geometrical argument that (subject to an energy inequality guaranteed automatically by the source free Einstein Maxwell equations without Λ term that we are using here) the 2-dimensional cross section through the horizon of any connected black hole segment must have strictly positive Euler number and hence *can only have the topology of a 2-sphere,* it follows that the simple connectivity requirement on which the preceding argument is based must in fact apply.

We are therefore able to conclude that (in this source free case *without* Λ term) the metric from (5.8) is valid *globally over the whole of the D.O.C.* except for the usual trivial degeneracy on the symmetry axis where X vanishes.

5.2 Reduction to a 2-dimensional Boundary Problem

Having established the *global* regularity of the form (5.8) over the entire D.O.C., we are now in a position to reduce the *4-dimensional* geometric problem under consideration to a boundary problem for a (non-linear) differential system on the *2-dimensional* background space with *known* flat background metric given by $d\rho^2 + dz^2$ over the half plane $\rho > 0$. This half plane is interpretable as representing one of the 2-surfaces on which the ignorable coordinates t and φ each take some fixed value, e.g. $t = 0$, $\varphi = 0$, the inner boundary where $\rho \to 0$ being interpretable as consisting partly of segments of the rotation axis and partly of the intersection of the 2-surface with the black hole horizon.

In addition to the possible appearance of the two electromagnetic scalars Φ and B that appear in the 4-potential form (4.94) there remain just two independent geometrical scalars, X and ω say, in the coupled differential system that is finally obtained on this half plane. The variable ρ has effectively been eliminated as an unknown by being *promoted* to the status of one of the "known" coordinates in terms of which position on the 2-surface is specified, while the only other variable Σ in the metric form (4.88) turns out to *decouple* from the source free Einstein or Einstein Maxwell equations for the more specialised form (5.8) (obtained in the case when Λ is absent) as has been well known since early work on stationary axisymmetric systems in other contexts.[72,90] After the system has been solved for the independent varialbe X, ω and (in the electromagnetic case) Φ, B, the remaining variable Σ is determined by a simple quadrature operation up to an arbitrary multiplicative constant of integration that represents the freedom to adjust the relative scale of the non-ignorable coordinates ρ, z with respect to that of the ignorable coordinates t, φ, this arbitrariness being removed when we impose the standard normalisation in the asymptotically flat region by requiring that $\Sigma \to 1$ at large distance.

In order to specify the boundary conditions on the independent variables X, ω, Φ, B it is necessary to distinguish between the *rotation axis* segments in the D.O.C. and the *horizon* bounding the D.O.C., which both form part of the boundary $\rho \to 0$ of the half plane under consideration. In the case when the black hole has just one *single* topologically spherical component, to which we now restrict ourselves (while recalling that the physically implausible possibility of a pair of black holes balanced on a common rotation axis has not yet been mathematically excluded) the boundary conditions can be most conveniently specified by replacing the cylindrical type coordinates ρ, z by ellipsoidal type coordinates λ, μ chosen in such a way that the horizon boundary occurs for a constant value, c say, of λ while the two rotation axis segments branching off to the "north" and "south" of the horizon are given by the respective limits $\mu = \pm 1$. The most natural way to do this is to choose c to be half the difference between the values of z at which the north and south pole intersect the horizon, which can be seen to be equivalent to taking

$$\begin{aligned} c =& \frac{\mathcal{A}\kappa}{4\pi} \\ =& M - \Phi^H A - 2\Omega^H J \end{aligned} \tag{5.9}$$

where the last line is obtained from (4.68) by setting the source contributions to zero. By adjusting the origin of z so that the "equatorial" value $z^\bullet = 0$ lies midway between these polar values, which thus become

$$z = \pm c^\bullet \tag{5.10}$$

we can arrange for a suitable transformation to ellipsoidal coordinates to be given by

$$z = \lambda\mu, \qquad \rho^2 = (\lambda^2 - c^2)(1 - \mu^2). \tag{5.11}$$

In this system the metric takes the form

$$ds^2 = \Xi d\hat{s}_{II}{}^2 + X\, d\varphi^2 + 2W\, d\varphi\, dt - V\, dt^2 \tag{5.12}$$

where

$$d\hat{s}_{II}{}^2 = \frac{d\lambda^2}{\lambda^2 - c^2} + \frac{d\mu^2}{1 - \mu^2} \tag{5.13}$$

and the modified conformal factor Ξ is given by

$$\Xi = (\lambda^2 - c^2\mu^2)\Sigma. \tag{5.14}$$

We recall that in terms of X and ω, V and W are given by

$$W = X\omega, \qquad V = X^{-1}(\rho^2 - W^2). \tag{5.15}$$

In terms of the 2-dimensional covariant differentiation operator $\hat{\nabla}$ defined with respect to the conformally flat metric (5.13) the system of independent source free Einstein Maxwell equations reduces to the form of two Maxwell equations,

$$\hat{\nabla}\left\{\frac{X}{\rho}(\hat{\nabla}\Phi - \omega\hat{\nabla}B)\right\} = 0 \tag{5.16}$$

$$\hat{\nabla}\left\{\frac{\rho}{X}\hat{\nabla}B + \frac{\omega}{\rho}(\hat{\nabla}\Phi - \omega\hat{\nabla}B)\right\} \tag{5.17}$$

and two Einstein equations

$$\hat{\nabla}\left\{\frac{X^2}{\rho}\hat{\nabla}\omega + \frac{4GB}{\rho}(\hat{\nabla}\Phi - \omega\hat{\nabla}B)\right\} = 0 \tag{5.18}$$

$$\hat{\nabla}\left\{\frac{\rho}{X}\hat{\nabla}X\right\} + \frac{|\hat{\nabla}\omega|^2}{\rho} + \frac{2GX}{\rho}|\hat{\nabla}\Phi - \omega\hat{\nabla}B|^2 + \frac{2G\rho}{X}|\hat{\nabla}B|^2 = 0 \tag{5.19}$$

Although this system is singular wherever ρ or X is zero, and hence on the boundary of the half plane $\rho > 0$ under consideration, it is guaranteed to be regular everywhere *within* the half plane by the causality condition (no closed timelike or null lines) as applied to the axisymmetry generators, which implies that we must have

$$X > 0 \tag{5.20}$$

everywhere except on the rotation axis, and hence everywhere *within* the half plane $\rho > 0$.

This is one of the principle motivations for our having chosen to work with a system based on the variables X and ω (derived essentially from the axial Killing vector m^μ) as principle geometric unknowns. The more traditional approach giving preference to the stationarity Killing vector k^μ which leads to the use of V and $V^{-1}W$ as principle unknowns, has the serious disadvantage for the present purpose that the corresponding system becomes singular on the surface of the ergosphere, where $V = 0$, which in general occurs in the *interior* of the domain under consideration. In the context of the traditional system (based on V and $V^{-1}W$) it was pointed out by Ernst[90] that the system admits very useful transformation of variables, which of course can be carried over to the present system (based on X and $X^{-1}W$) which considerably simplifies the field equations, and which I found[91,25] to have the additional advantage in the present context of considerably simplifying the boundary conditions on the horizon. I also found that the transformed system has the additional important advantage of being derivable from a *Lagrangian*[21] *that is positive definite* (when based on X, though not when based on V) a property whose full significance has been made clear by the recent work of Bunting[50,51] and Mazur.[52]

The Ernst type[90] transformation (analogous to our introduction of the variables Ψ and U in the staticity theorem) by which this result is acheived, makes use of the Maxwell equation (5.16) to justify the introduction of an electric (stream function type) potential E given by

$$X\left(\frac{\partial\Phi}{\partial\lambda}-\omega\frac{\partial B}{\partial\lambda}\right)=(1-\mu^2)\frac{\partial E}{\partial\mu}, \tag{5.21}$$

$$X\left(\frac{\partial\Phi}{\partial\mu}-\omega\frac{\partial B}{\partial\mu}\right)=-(\lambda^2-c^2)\frac{\partial E}{\partial\lambda} \tag{5.22}$$

in replacement of Φ, and of the Einstein equation (5.18) to justify the introduction of an analogous rotation potential Y given by

$$\begin{aligned} X^2\frac{\partial\omega}{\partial\lambda}&=(1-\mu^2)\left\{\frac{\partial Y}{\partial\mu}+2G\left(E\frac{\partial B}{\partial\mu}-B\frac{\partial E}{\partial\mu}\right)\right\}\\ X^2\frac{\partial\omega}{\partial\mu}&=-(\lambda^2-c^2)\left\{\frac{\partial Y}{\partial\lambda}+2G\left(E\frac{\partial B}{\partial\lambda}-B\frac{\partial E}{\partial\lambda}\right)\right\}\end{aligned} \tag{5.23}$$

in replacement of ω. In terms of the new set of variables E, B, X, Y the Maxwell equations (5.16), (5.17) are to be replaced by

$$\hat\nabla\left\{\frac{\rho}{X}\hat\nabla B\right\}+\frac{\rho}{X^2}\{\hat\nabla Y+2G(E\hat\nabla B-B\hat\nabla E)\}\cdot\hat\nabla E=0 \tag{5.24}$$

$$\hat\nabla\left\{\frac{\rho}{X}\hat\nabla E\right\}-\frac{\rho}{X^2}\{\hat\nabla Y+2G(E\hat\nabla B-B\hat\nabla E)\}\cdot\hat\nabla E=0 \tag{5.25}$$

while the Einstein equations (5.18), (5.19) are to be replaced by

$$\hat\nabla\left\{\frac{\rho}{X^2}(\hat\nabla Y+2G(E\hat\nabla B-B\hat\nabla E)\right\}=0 \tag{5.26}$$

$$\hat\nabla\left\{\frac{\rho}{X^2}\hat\nabla X\right\}+\frac{\rho}{X^3}\left\{|\hat\nabla X|^2+|\hat\nabla Y+2G(E\hat\nabla B-B\hat\nabla E|^2\right\} +\frac{2G\rho}{X^2}(|\hat\nabla E|^2+|\hat\nabla B|^2)=0 \tag{5.27}$$

When I first noticed that this Ernst type system is obtainable from the application of a variational principle to a Lagrangian integral

$$I=\int\mathcal{L}\,d\lambda_\mu \tag{5.28}$$

where the function $\mathcal{L}$ is given[21] by

$$\mathcal{L}=\frac{1}{2X^2}\{|\hat\nabla X|^2+|\hat\nabla Y+2G(E\hat\nabla B-B\hat\nabla E)|^2\}+\frac{2G}{X}\{|\hat\nabla E|^2+|\hat\nabla B|^2\} \tag{5.29}$$

(whose form generalises that of an expression given previously by Ernst himself[90] in the non-electromagnetic case) I immediately suspected that its obvious property of being *positive definite* (wherever (5.20) holds) must be relevant to the existence of the divergence relations with which it was found possible[91,48,49] to obtain successively more complete no-hair and uniqueness theorems, but this property was not actually exploited explicitly until the more recent and fully complete results obtained (using the "harmonic mapping" property expressed by the fact that the Lagrangian is homogeneously quadratic in the gradients) by Bunting,[50,51] and using a rather different method (depending on more specialised properties), to be described in the next section, by Mazur.[52]

Before describing this method, I shall complete the specification of the boundary conditions that I found[91,21] to be not only *necessary* but also (as may be relevant for future work on systems with annular material source distributions) *sufficient* for asymptotic flatness in the limit $\lambda \to \infty$ and for regularity on the symmetry axis, $\mu \to \pm 1$, and the horizon $\lambda \to c$.

The asymptotic boundary conditions (subject to our requirement that the magnetic monopole moment P vanishes) are given by the requirement that E, B, Y, and $\lambda^{-2}X$ be well behaved functions of λ, μ as $\lambda^{-1} \to 0$, with

$$\begin{aligned} E &= -Q\mu + O(\lambda^{-1}), \qquad B = O(\lambda^{-1}) \\ Y &= 2J\mu(3-\mu^2) + O(\lambda^{-1}) \\ \lambda^{-2}X &= (1-\mu^2) + O(\lambda^{-1}). \end{aligned} \tag{5.30}$$

The mass parameter does not appear explicitly at this order, but it comes into the problem indirectly, via (5.9) in the specification of the length parameter c that sets the overall scale.

The most delicate parts of the boundary conditions are those that apply the north and south polar segments, $\mu \to 1$ and $\mu \to -1$, of the rotation axis where $X \to 0$ so that the system becomes singular. It is required that as $\mu^2 \to 1$ the functions E, B, Y, X should be regular functions of λ, μ such that

$$\begin{aligned} \frac{\partial E}{\partial \lambda} &= O(1-\mu^2), \qquad \frac{\partial B}{\partial \lambda} = O(1-\mu^2) \\ \frac{\partial Y}{\partial \lambda} &= O(1-\mu^2), \qquad \frac{\partial Y}{\partial \mu} + 2G\Big(E\frac{\partial B}{\partial \mu} - B\frac{\partial E}{\partial \mu}\Big) = O(1-\mu^2) \\ X &= O(1-\mu^2), \qquad \frac{(\mu^2-1)}{2X}\frac{\partial X}{\partial \mu} = 1 + O(1-\mu^2) \end{aligned} \tag{5.31}$$

where the last equation is an expression that although X itself must vanish on the rotation axis, its derivative with respect to μ must remain finite.

One of the advantages of formulating the system in the present form is that the part of the boundary that is the least familiar (from the point of view of everyday geometric experience) is also the one on which the final form[91,21] of the boundary conditions turns out to be simplest. What one obtains as a necessary and sufficient condition for regularity of the horizon is merely the requirement that E, B, X, Y should be regular as functions of λ, μ in the limit as $\lambda \to c$.

5.3 The Final Step in the Uniqueness Theorem

At the time when I first succeeded in reducing the black hole problem to the 2-dimensional boundary problem, (in the form that has just been derived) the best I could do with it was to establish a pure vacuum "no hair" (or in more formal language "no-bifurcation") theorem.[91] As subsequently extended by Robinson[48] to include the general electromagnetic case, the no-hair theorem stated that the solutions came in continuous non-intersecting families parametrised completely by three independent quantities appearing in the boundary conditions, namely Q, J, and c (or equivalently as Q, J and M), the only such family with an uncharged non-rotating limit being that of the Schwarzschild solution, i.e. the already known family of Kerr Newman. While the possibility that there might exist some other family without any uncharged limit never seemed at all plausible, it was not excluded by any rigorous mathematical proof until the much more recent work of Bunting[50,51] and Mazur,[52] the possibility of uncharged families without any non-rotating limit having been excluded in the meantime by Robinson.[49]

The basic method used in all these works consists of equating a *divergence* such that its integral can be eliminated using the boundary conditions (by Green's theorem) to a *positive definite* functional of the *difference* between hypothetically distinct solutions (for the full non-linear uniqueness theorems[49,50,51,52]) or of a *linearised perturbation* [for the weaker no-hair theorems[91,48]]. The earlier divergence relations of this type were based on a trial and error method that Robinson was able to push to an amazing degree of complexity only after Bunting and Mazur had found out how to obtain divergence relations of the required type in a more systematic manner.

The very powerful method used by Bunting (whose details were originally available only in an unpublished Ph.D. thesis[50]) is outlined in a discussion that I have provided elsewhere.[51] On the present occasion I shall describe only the more specialised (but in some ways more elegant) method due to Mazur,[52] which is based on the discovery essentially due to the work of Kinnersley[92] and Geroch[92] that the Ernst type systems belong to a class of non-linear σ-models whose field equations are equivalent to a (partially redundant) set of conservation laws of the form

$$\hat{\nabla}\mathbf{J} = 0 \tag{5.32}$$

where $\mathbf{J}$ is a matrix vector constructed in terms of a *positive definite* hermitian matrix function $\mathbf{\Phi}$ say of the field components according to a formula of the form

$$\mathbf{J} = \rho^{-1}\mathbf{\Phi}^{-1}\cdot\hat{\nabla}\mathbf{\Phi} \tag{5.33}$$

where ρ is a known positive scalar weight factor (which actually does not appear in the traditional flat space σ-models, but whose presence here does not introduce any significant complication). In the present application the matrix has the form

$$\mathbf{\Phi}_{ab} = \eta_{\dot{a}b} + 2\bar{v}_{\dot{a}}v_b \tag{5.34}$$

(using the standard convention that a dot distinguishes antilinear from linear tensor indices) where $\eta_{\dot{a}b}$ is just a fixed Lorentz signature (Minkowski type) metric on a 3-dimensional complex vector space (2 dimensions being sufficient in the non electromagnetic case) and v^a is a complex vector with components given by the Kinnersley transformation

$$\begin{pmatrix} v_0 \\ v_1 \\ v_2 \end{pmatrix} = \frac{1}{2\sqrt{X}}\begin{pmatrix} -1 - E^2 - B^2 - X + iY \\ 1 - E^2 - B^2 - X + iY \\ E + iB \end{pmatrix} \tag{5.35}$$

which is such that it makes v^a automatically "timelike" (with respect to the abstract complex 3-space) with unit normalisation property

$$\eta^{a\dot{b}}v_a\bar{v}_{\dot{b}} = -1 \tag{5.36}$$

which is sufficient to guarantee the positive definiteness of the matrix defined by (5.35). The SU(2,1) invariance of the system (under the action leaving the metric $\eta_{\dot{a}b}$ invariant) that Kinnersley thus made evident can in fact be seen to extend "off shell" in the sense that it applies also to our Lagrangian (5.29) which will be transformed to the form

$$\begin{aligned} \mathcal{L} &= 2|\eta^{a\dot{b}}v_a\hat{\nabla}\bar{v}_{\dot{b}}|^2 - \eta^{a\dot{b}}\hat{\nabla}v_a\hat{\nabla}\bar{v}_{\dot{b}} \\ &= \tfrac{1}{2}\hat{g}_{ij}\mathrm{tr}\,\{\mathbf{J}^i\cdot\mathbf{J}^j\}. \end{aligned} \tag{5.37}$$

The Mazur method of establishing uniqueness of solutions for such a system depends on positive definiteness both of the hermitian matrix $\mathbf{\Phi}$ and of the base space matrix $\hat{g}_{ij}$ say, where in our case i, j run over the values 1,2 ($x^1 = \lambda$, $x^2 = \mu$). One wishes to prove the vanishing of the difference

$$\overset{\circledcirc}{\mathbf{\Phi}} = \mathbf{\Phi}_{[1]} - \mathbf{\Phi}_{[0]} \tag{5.38}$$

between any pair of matrices representing solutions for some appropriate set of boundary conditions, which is equivalent to the vanishing of what I shall refer to as the deviation matrix,

$$\mathbf{\Delta} = \mathbf{\Phi}_{[1]} \cdot \mathbf{\Phi}_{[0]}^{-1} - \mathbf{1} = \overset{\circledcirc}{\mathbf{\Phi}} \cdot \mathbf{\Phi}_{[0]}^{-1} \tag{5.39}$$

where $\mathbf{1}$ is the unit matrix (in the 3-dimensional complex space of our application). The gradient of this deviation matrix will be expressible by

$$\rho \nabla \mathbf{\Delta} = \mathbf{\Phi}_{[1]} \cdot \overset{\circledcirc}{\mathbf{J}} \cdot \mathbf{\Phi}_{[0]}^{-1} \tag{5.40}$$

where

$$\overset{\circledcirc}{\mathbf{J}} = \mathbf{J}_{[1]} - \mathbf{J}_{[0]} \tag{5.41}$$

is the difference between the corresponding current vectors. Taking the divergence we obtain

$$\nabla(\rho\nabla\mathbf{\Delta}) = \mathbf{\Phi}_{[1]} \cdot \{\nabla\overset{\circledcirc}{\mathbf{J}} + \rho^{-1}\hat{g}_{ij}(\mathbf{J}^i_{[1]} \cdot \mathbf{J}^j_{[1]} - 2\mathbf{J}^i_{[1]} \cdot \mathbf{J}^j_{[0]} + \mathbf{J}^i_{[0]} \cdot \mathbf{J}^j_{[0]})\} \cdot \mathbf{\Phi}_{[0]}^{-1}. \tag{5.42}$$

Now since the hermicity property

$$\mathbf{\Phi} = \mathbf{\Phi}^* \tag{5.43}$$

(where the star denotes the complex conjugate of the transpose) implies

$$\mathbf{J}^* = \mathbf{\Phi} \cdot \mathbf{J} \cdot \mathbf{\Phi}^{-1} \tag{5.44}$$

the quadratic terms above can be rewritten as

$$\mathbf{\Phi}_{[1]}(\mathbf{J}^j_{[1]} \cdot \overset{\circledcirc}{\mathbf{J}}_j - \overset{\circledcirc}{\mathbf{J}}_j \cdot \mathbf{J}^j_{[0]})\mathbf{\Phi}_{[0]}^{-1} = \mathbf{J}^{*\,j}_{[1]} \cdot \mathbf{\Phi}_{[1]} \cdot \overset{\circledcirc}{\mathbf{J}}_j \cdot \mathbf{\Phi}_{[0]}^{-1} - \mathbf{\Phi}_{[1]} \cdot \overset{\circledcirc}{\mathbf{J}}_j \cdot \mathbf{\Phi}_{[0]}^{-1} \cdot \mathbf{J}^{*\,j}_{[0]}. \tag{5.45}$$

Hence taking the trace one obtains the scalar identity

$$\nabla[\rho\nabla \mathrm{tr}\,(\mathbf{\Delta})] - \mathrm{tr}\,\{\mathbf{\Phi}_{[0]}^{-1} \cdot \mathbf{\Phi}_{[1]} \cdot \overset{\circledcirc}{\mathbf{J}}\} = \rho^{-1} g_{ij} \mathrm{tr}\,\{\mathbf{\Phi}_{[0]}^{-1} \cdot \overset{\circledcirc}{\mathbf{J}}^{*\,i} \cdot \mathbf{\Phi}_{[1]} \cdot \overset{\circledcirc}{\mathbf{J}}^j\} \tag{5.46}$$

which is the identity of Mazur[52] (and which includes earlier identities given by Robinson[48,49] and in the linearised limit by myself [91] as special cases).

Subject to the field equations (5.33), which imply

$$\nabla\overset{\circledcirc}{\mathbf{J}} = 0 \tag{5.47}$$

the left hand side of (5.46) reduces to a divergence whose integral can be converted to a surface contribution which will vanish, i.e.

$$\oint dS_i\, \rho \hat{g}^{ij} \hat{\nabla}_j (\mathrm{tr}\,\mathbf{\Delta}) \to 0 \tag{5.48}$$

subject to boundary conditions such as those that hold in the application under consideration here, for which using the notation convention[25]

$$\langle X\rangle = (X_{[0]}X_{[1]})^{1/2}, \qquad \tilde{X} = X_{[1]} + X_{[0]}, \qquad \overset{\odot}{X} = X_{[1]} - X_{[0]}, \tag{5.49}$$

we shall have

$$\mathrm{tr}\,(\boldsymbol{\Delta}) = \frac{\overset{\odot}{X}{}^2 + 2\tilde{X}(\overset{\odot}{E}{}^2 + \overset{\odot}{B}{}^2)^2 + (\overset{\odot}{Y} + \tilde{E}\overset{\odot}{B} - \tilde{B}\overset{\odot}{E})^2}{\langle X\rangle} \tag{5.50}$$

which by the conditions stated in the previous section will satisfy

$$\mathrm{tr}\,(\boldsymbol{\Delta}) = O(\lambda^{-2}), \qquad \frac{\partial \mathrm{tr}\,(\boldsymbol{\Delta})}{\partial\lambda} = O(\lambda^{-3}) \tag{5.51}$$

as $\lambda \to \infty$, where $\rho^2 = O(\lambda^2)$,

$$\mathrm{tr}\,(\boldsymbol{\Delta}) = O(1), \qquad \frac{\partial \mathrm{tr}\,(\boldsymbol{\Delta})}{\partial\mu} = O(1) \tag{5.52}$$

(despite the vanishing of the denominator $\langle X\rangle$) on the axis, as $\mu^2 \to 1$, where $\rho^2 = O(1-\mu^2)$, and

$$\mathrm{tr}\,(\boldsymbol{\Delta}) = O(1), \qquad \frac{\partial \mathrm{tr}\,(\boldsymbol{\Delta})}{\partial\lambda} = O(1) \tag{5.53}$$

on the horizon, as $\lambda \to c$, where $\rho^2 = O(\lambda^2 - c^2)$. Under such conditions we can deduce that the integral of the right hand side of the Mazur identity (5.46) must vanish, and since the positive definiteness property (resulting from (5.34) and (5.36)) of the matrices $\boldsymbol{\Phi}_{[1]}$, $\boldsymbol{\Phi}_{[0]}$ and hence also $\boldsymbol{\Phi}_{[0]}^{-1}$ ensures that this right hand side is a positive definite function of the current difference $\overset{\odot}{\mathbf{J}}$, we can conclude that the latter must vanish everywhere

$$\overset{\odot}{\mathbf{J}} = 0 \tag{5.54}$$

in the half plane $\rho > 0$ under consideration. It is now immediately obvious from (5.40) that this implies

$$\boldsymbol{\Delta} = \mathbf{C} \tag{5.55}$$

where $\mathbf{C}$ is some constant matrix, and since the boundary conditions ensure that $\boldsymbol{\Delta} \to 0$ as $\lambda \to \infty$, this constant must be zero,

$$\mathbf{C} = 0 \tag{5.56}$$

which by (5.39) is equivalent to the required result, namely that

$$\boldsymbol{\Phi}_{[1]} = \boldsymbol{\Phi}_{[0]} \tag{5.57}$$

throughout the domain.

The actual solutions to the problem belong to the Kerr-Newman family, which is traditionally specified in terms of the Kerr rotation parameter

$$a = J/M \tag{5.58}$$

and the Kerr coordinates r, θ which are related to λ, μ by

$$\lambda = r - M, \qquad \mu = a\cos\theta \tag{5.59}$$

the metric being given by[21]

$$X = \left\{ r^2 + a^2 + \frac{(2Mr - P^2 - Q^2)a^2 \sin^2\theta}{r^2 + a^2\cos^2\theta} \right\} \sin^2\theta \tag{5.60}$$

$$Y = \left\{ M(2 + \sin^2\theta) - \frac{\sin^2\theta[(Q^2 + P^2)r - Ma^2\sin^2\theta]}{r^2 + a^2\cos^2\theta} \right\} 2a\cos\theta \tag{5.61}$$

which leads to

$$W = \frac{(2Mr - P^2 - Q^2)a^2\sin^2\theta}{r^2 + a^2\cos^2\theta} \tag{5.62}$$

$$V = 1 - \frac{2Mr - Q^2 - P^2}{r^2 + a^2\cos^2\theta} \tag{5.63}$$

$$\Xi = r^2 + a^2\cos^2\theta \tag{5.64}$$

while the electromagnetic field is given by

$$E = \frac{Q(r^2 + a^2)\cos\theta - Pra\sin^2\theta}{r^2 + a^2\cos^2\theta} \tag{5.65}$$

$$B = \frac{P(r^2 + a^2)\cos\theta - Qra\sin^2\theta}{r^2 + a^2\cos^2\theta} \tag{5.66}$$

which leads to

$$\Phi = \frac{Qr - Pa\cos\theta}{r^2 + a^2\cos^2\theta}. \tag{5.67}$$

In order to satisfy the boundary conditions of the problem as posed above we must restrict the parameters so that the horizon scale parameter c as given by

$$c^2 = M^2 - a^2 - P^2 - Q^2 \tag{5.68}$$

is real, i.e. we must have

$$c^2 > 0 \tag{5.69}$$

and so that there is no magnetic monopole moment, i.e. we must have

$$P = 0. \tag{5.70}$$

It is to be remarked by the way[93] that there is a magnetic dipole Qa corresponding to a value of the *gyromagnetic ratio* that is exactly the *same* as that for the simple Dirac equation.

Apart from the desirable feature of being expressed in terms of rational algebraic functions, the solution appears at first sight rather a mess when expressed in the general framework used here. However as soon as one tries to solve explicit problems, starting with the integration of the geodesic equations,[93] one finds that the local algebraic properties are in fact (when expressed in the right framework) amazingly convenient –indeed by far the best that could have been hoped for on the basis of purely local desiderata– as will be discussed in the next section.

Another fascinating aspect[75,93] which however will *not* be dealt with here (because of its lack of direct astrophysical relevance) is the analytic extrapolation of the solutions *within* the horizon, i.e. to the region $\lambda < c$, where many weird features such as the existence of closed timelike lines[43] occur. In practice, most of these features would not even be relevant for an explorer bold enough to fly *into* the hole, since there is strong evidence[94] that the inside (unlike the outside) is *unstable.*

5.4 Killing-Maxwell-Yano system

The Kerr[83] and Kerr-Newman[84] solutions to which one is led by the foregoing global boundary problem turn out to belong to a very special class of metrics (including asymptotically DeSitter or anti-DeSitter solutions[21,86] in which a cosmological Λ term is present) originally discovered by the present author[87,88] which are characterised by an extremely high degree of local "hidden symmetry" which in particular allows one to obtain explicit solutions by separation of variables of a very large class of associated differential systems ranging in simplicity from the simple scalar Dalembertian wave equation[88] to the Dirac equation for a massive charged particle,[95] of which perhaps the most physically important examples are those for the massless electromagnetic and gravitational perturbation equations,[96] which have been used to provide a very convincing (although, for the rotating case, not yet absolutely rigorous) demonstration that the amplification factor for superadiant perturbation modes remains bounded, at least in the uncharged case, and hence that the corresponding black hole equilibrium states are stable.[97,98,99]

The technical details of the way in which many of these results were obtained are described at length in a recent treatise by Chandrasekhar,[100] in which even more elaborate and miraculous cancellations and integrability properties are revealed one by one in successive chapters. However although such a treatment has the advantage of maintaining a high level of excitement in what might otherwise have been a rather tedious labour, nevertheless for anyone who is not particularly gifted at heavy algebra (particularly if he shares my propensity for sign errors) there is a tremendous advantage in sacrificing the dramatic seduction of such an epic approach for a more pedestrian and deductive (though by no means necessarily less æsthetic) treatment in which one strives systematically to keep as much symmetry as possible in view at all stages, following the example of the Debever school.[88,101,102,103,104]

What is in a certain sense the most elementary structure underlying the hidden symmetries (and even the manifest symmetries generated by the Killing vectors k^μ and m^μ) of the solutions under consideration was brought to light by the work of Penrose and several of his collaborators[105,106,107,108,109] when they discovered the presence of a certain kind of Killing-Yano tensor $\gamma_{\mu\nu}$ whose existence is equivalent to what I like to refer to as a Killing-Maxwell system[110] in which the most fundamental element of all from which the Killing-Yano tensor and the ordinary Killing vectors are obtainable by purely differential operations, is a covector $\hat{A}_\mu$ satisfying the basic defining equation

$$\hat{A}_{\sigma;\rho;(\mu}\varepsilon_{\nu)\lambda}{}^{\sigma\rho} = 0 \tag{5.71}$$

for what I shall refer to as a Killing-Maxwell 1-form. The Kerr and Kerr-Newman solutions are in fact distinguished among all asymptotically flat vacuum and source-free Einstein Maxwell solutions (without any prior restriction of stationarity of axisymmetry of the existence of a regular black hole horizon) merely by the property that they admit a solution to the preceding equation that is non-trivial in the sense that the corresponding Maxwell type field

$$\hat{F}_{\mu\nu} = 2\hat{A}_{[\nu;\mu]} \tag{5.72}$$

is non-zero. The existence of such a 1-form is sufficient and necessary for the corresponding dual tensor

$$f_{\mu\nu} = \tfrac{1}{2}\varepsilon_{\mu\nu\rho\sigma}\hat{F}^{\rho\sigma} \tag{5.73}$$

(discovered in the case of the Kerr solutions by Penrose and Floyd[109]) to satisfy a Killing Yano type equation which in this case is equivalent to the condition that the covariant derivative by totally antisymmetric

$$f_{\mu\nu;\rho} = f_{[\mu\nu;\rho]} \tag{5.74}$$

while for the dual field we shall have

$$F_{\mu\nu;\rho} = k_{[\mu}g_{\nu]\rho} \tag{5.75}$$

where k^{μ} is evidently proportional to the associated current $\hat{j}^{\mu}$, i.e. we have

$$F^{\rho\mu}{}_{;\mu} = 4\pi\hat{j}^{\mu} \,, \qquad \hat{j}^{\mu} = \frac{3}{4\pi}k^{\mu} \,. \tag{5.76}$$

Using the ensuing integrability conditions[111]

$$f_{\mu\nu;\rho;\sigma} = \tfrac{3}{2} f_{\tau[\nu}R_{\mu\rho]\sigma}{}^{\tau} \tag{5.77a}$$

$$R_{\mu\nu[\sigma}{}^{\tau}f_{\rho]\tau} + R_{\sigma\rho[\mu}{}^{\tau}f_{\nu]\tau} = 0 \tag{5.77b}$$

we respectively obtain

$$k_{\mu;\nu} = \tfrac{1}{2}R^{\rho}{}_{\nu}\hat{F}_{\mu\rho} + \tfrac{1}{4}R_{\mu\nu}{}^{\rho\sigma}\hat{F}_{\rho\sigma} \tag{5.78}$$

and

$$f^{\rho}{}_{(\mu}R_{\nu)\rho} = 0, \qquad \hat{F}^{\rho}_{(\mu}R_{\nu)\rho} = 0 \tag{5.79}$$

from which it can be seen that the vector field k^{μ}, which is expressible explicitly as

$$k^{\mu} = \tfrac{1}{3!}\varepsilon^{\mu\nu\rho\sigma}f_{\nu\rho;\sigma} \tag{5.80}$$

necessarily[112] satisfies the Killing equations, i.e.

$$k_{(\mu;\nu)} = 0 \tag{5.81}$$

and thus generates an "unhidden" symmetry. From this *primary* Killing vector one can proceed algebraically[112] to construct a *secondary* Killing vector

$$h^{\mu} = a^{\mu}{}_{\nu}k^{\nu} \tag{5.82}$$

generating another such "unhidden" symmetry where the symmetric tensor

$$a_{\mu\nu} = f_{\mu}{}^{\rho}f_{\rho\nu} \tag{5.83}$$

will satisfy the ordinary Killing tensor equation

$$a_{(\mu\nu;\rho)} = 0 \tag{5.84}$$

which entails

$$h_{(\mu;\nu)} = 0 \tag{5.85}$$

as a consequence.

We recall that the preceding conditions are obviously necessary and sufficient for the linear and quadratic combinations $k_{\rho}u^{\rho}$, $a_{\rho\sigma}u^{\rho}u^{\sigma}$, to be constants of an affine geodesic motion $u^\mu{}_{;\nu}u^\nu = 0$ and that they are also necessary[113] for the corresponding self adjoint scalar operators

$$K = ik^\mu \nabla_\mu, \qquad L^2 = \nabla_\mu a^{\mu\nu} \nabla_\nu \tag{5.86}$$

to commute with the ordinary Dalembertian

$$\Box = \nabla_\mu \nabla^\mu : \qquad [K, \Box] = 0 = [L^2, \Box] \tag{5.87}$$

a sufficient condition for such commutation being that we should have

$$a^\rho{}_{[\mu} R_{\nu]\rho} = 0 \tag{5.88}$$

which will be satisfied automatically in the present case by the above integrability conditions. The very special condition that $a_{\mu\nu}$ should have a square root satisfying the Killing-Yano equation, implies more particularly that the uncontracted (generalised angular momentum) vector $f_{\mu\nu}u^\nu$ will be parallel propagated, and that we can construct not only an ordinary energy-type 4-spinor operator[119]

$$K = i(k^\mu \nabla_\mu - \tfrac{1}{4}\gamma^\mu \gamma^\nu k_{\mu;\nu}) \tag{5.89}$$

but also[111] a generalised (unsquared) angular momentum operator

$$L = i\gamma_\mu(\gamma^5 f_\mu{}^\nu \nabla_\nu - k_\mu) \tag{5.90}$$

that commute with the Dirac operator

$$D = \gamma^\mu \nabla_\mu : \qquad [K, D] = 0 = [L, D]. \tag{5.91}$$

From the properties established above, it also follows directly that k^μ and h^μ generate symmetries not only of the metric but also of the system itself. For the secondary Killing vector we have

$$f_{\mu\nu;\rho}h^\rho + 2\, f_{\rho[\mu}h_{\nu]}{}^{;\rho} = 0 \tag{5.92}$$

(i.e. the corresponding Lie derivative of $f_{\mu\nu}$ vanishes) while for the primary condition we shall have the two separate conditions

$$f_{\mu\nu;\rho}k^\rho = 0, \qquad f_{\rho[\mu}k_{\nu]}{}^{;\rho} = 0 \tag{5.93}$$

which add up to an analogous result, which implies in particular that the two Killing vectors *commute*, i.e.

$$k^\mu{}_{;\rho}h^\rho - h^\mu{}_{;\rho}k^\rho = 0. \tag{5.94}$$

Finally it can be seen that the system will have both geometric and (although this is less directly important since the field distinguished by a hat symbol ^ is to be thought of as just a mathematical construct not a physical Maxwell field) also electromagnetic *circularity* properties of the same kind as were postulated and proved from global physical considerations in the generalised Papapetrou theorem described above, i.e. we shall have

$$k^\rho h^\sigma f_{\rho\sigma} = 0 \qquad k^\rho h^\sigma \hat{F}_{\rho\sigma} = 0 \tag{5.95}$$

and

$$h_{[\lambda;\mu}k_\rho h_{\sigma]} = 0 \qquad k_{[\lambda;\mu}k_\rho h_{\sigma]} = 0. \tag{5.96}$$

Relative to the primary Killing vector both the electric and magnetic parts of the field have the form of gradients given explicitly by

$$\hat{F}_{\mu\rho}k^{\rho} = \tfrac{1}{4}(\hat{F}_{\rho\sigma}\hat{F}^{\rho\sigma})_{,\mu} \tag{5.97}$$

$$f_{\mu\rho}k^{\rho} = \tfrac{1}{4}(f_{\rho\sigma}\hat{F}^{\rho\sigma})_{,\mu} \tag{5.98}$$

while for the secondary Killing vector (which in general rotates relative to the current $\hat{j}^{\mu}$) we shall have an analogous property only for the electric part:

$$F_{\mu\rho}h^{\rho} = \tfrac{1}{32}\{(f_{\rho\sigma}\hat{F}^{\rho\sigma})^2\}_{,\mu} \tag{5.99}$$

The foregoing properties allow us to introduce a canonically preferred coordinate system in which $\tilde{t}$ and $\tilde{\varphi}$ are *ignorable* coordinates such that the Killing vectors have the form

$$k^{\mu} \leftrightarrow \frac{\partial}{\partial \tilde{t}}, \qquad h^{\mu} \leftrightarrow \frac{\partial}{\partial \tilde{\varphi}} \tag{5.100}$$

while r and q say are non-ignorable coordinates defined in terms of the local Killing-Maxwell invariants by

$$r^2 - q^2 = \tfrac{1}{2}\hat{F}_{\rho\sigma}\hat{F}^{\rho\sigma}, \qquad 2rq = \tfrac{1}{2}f_{\rho\sigma}\hat{F}^{\rho\sigma} \tag{5.101}$$

and then to choose a canonical Lorentzian gauge

$$\hat{A}_{\rho}{}^{;\rho} = 0 \tag{5.102}$$

in which the Killing-Maxwell 1-form will be given explicitly by

$$\hat{A}_{\rho}\, dx^{\rho} = \tfrac{1}{2}(q^2 - r^2)d\tilde{t} - \tfrac{1}{2}r^2q^2\, d\tilde{\varphi} \tag{5.103}$$

which implies that the Killing-Yano 2-form will be given by

$$f_{\mu\rho}\, dx^{\mu} \wedge dx^{\nu} = q\, dr \wedge (d\tilde{t} + q^2 d\tilde{\varphi}) + r\, dq \wedge (d\tilde{t} - r^2 d\tilde{\varphi}). \tag{5.104}$$

5.5 The Canonical Tetrad

Since at this point the available space and time no longer allow us to continue a step by step approach we must now simply quote the result that specialisation of the class of spaces described in the previous subsection to the asymptotically flat vacuum or source free Einstein-Maxwell case leads to the same Kerr and Kerr-Newman solutions that resulted (subject to appropriate inequalities on the mass, angular momentum, and charge parameters) from the essentially quite different black hole problem discussed in the earlier sections. In terms of the *algebraically preferred coordinates* of the type originally introduced by the present author,[88,103,100] as set up in the preceding section the metric can be expressed in the form

$$ds^2 = \eta_{\hat{i}\hat{j}}(\theta^{\hat{i}}_{\mu}\, dx^{\mu})(\theta^{\hat{j}}_{\nu}\, dx^{\nu}) \tag{5.105}$$

where $\eta_{\hat{i}\hat{j}}$ is a fixed Minkowski metric and the $\theta^{\hat{i}}$ are 1-forms of the *canonical separable tetrad* discovered long ago by the present author[88] but unfortunately neglected by many more recent calculations[96,100] which is given (in terms of the 1-forms that have already made their appearance in the expression for the Killing Yano tensor) by

$$\begin{aligned} \theta^{\hat{1}}_{\mu}\, dx^{\mu} &= \left(\frac{r^2+q^2}{\Delta_{(+)}}\right)^{\frac{1}{2}} dr & \theta^{\hat{3}}_{\mu}\, dx^{\mu} &= \left(\frac{\Delta_{(-)}}{r^2+q^2}\right)^{\frac{1}{2}} (r^2 d\tilde{\varphi} - d\tilde{t}\,) \\ \theta^{\hat{2}}_{\mu}\, dx^{\mu} &= \left(\frac{r^2+q^2}{\Delta_{(-)}}\right)^{\frac{1}{2}} dq & \theta^{\hat{4}}_{\mu}\, dx^{\mu} &= \left(\frac{\Delta_{(+)}}{r^2+q^2}\right)^{\frac{1}{2}} (q^2 d\tilde{\varphi} + d\tilde{t}\,) \end{aligned} \tag{5.106}$$

where $\Delta_{(+)}$ and $\Delta_{(-)}$ are single variable functions of r and q respectively. One can allow for the presence of a *cosmological* Λ constant, by inclusion of an appropriate *quartic* term,[88,21,115,103] but in the asymptoticallyl flat case with which we are concerned here, the Einstein equations require that these functions have purely quadratic form:

$$\Delta_{(+)} = b^2 - 2Mr + r^2, \qquad \Delta_{(-)} = a^2 - 2Nq - q^2 \tag{5.107}$$

where the constants a, b, M, N are arbitrary but where the imposition of asymptotic flatness in the global not just the local sense (in effect the exclusion of the Newman-Demianski solutions[113]) means that we must ultimately set the "NUT" parameter N to zero, i.e.

$$N = 0. \tag{5.108}$$

[In practice it is often convenient to carry this parameter along as a free variable in intermediate stages of calculation in order to maintain the manifest symmetry (modulo an imaginary factor i) between the coordinates r and q.] The difference between a^2 and b^2 determines the strength of the electromagnetic field contributions:

$$b^2 - a^2 = Q^2 + P^2 \tag{5.109}$$

where the source-free electromagnetic field 1-form A_μ is given by

$$A_\mu = \frac{Qr(q^2 d\tilde{\varphi} + d\tilde{t}) - qP(r^2 d\tilde{\varphi} - d\tilde{t})}{r^2 + q^2} \tag{5.110}$$

Q being the electric charge parameter while P is the magnetic monopole parameter, which we have written in only to demonstrate the $r \leftrightarrow q$ symmetry, but which one normally excludes, setting

$$P = 0 \tag{5.111}$$

on physical grounds similar to those on which the "gravimagnetic" monopole moment N was ruled out.

Although the analytic separability properties can most easily be studied in the (maximally symmetric) *algebraically preferred* coordinates[88] we have been using so far, they are of course preserved by the transformation to the standard *geometrically preferred* Boyer-Lindquist coordinates[75] r, θ, φ, t (or indeed to the even less symmetric coordinates of the type originally used by Kerr[83]). The non-ignorable Boyer-Lindquist coordinates (which are the same as those of Kerr[83]) are obtained from ours simply by setting

$$r \leftrightarrow r, \qquad q \leftrightarrow a\cos\theta \tag{5.112}$$

while the ignorable Boyer-Lindquist coordinates t, φ (which are *not* the same as those of Kerr) are obtained by setting

$$\tilde{\varphi} \leftrightarrow a^{-1}\varphi, \qquad \tilde{t} \leftrightarrow t - a\varphi \tag{5.113}$$

the point being to ensure that the new φ coordinate is *periodic* (with the standard periodicity 2π). This means that the secondary Killing vector h^μ introduced in the preceding subsection will be related to the stationary Killing vector k^μ (which is *itself* the primary one) and the usual axial Killing vector m^μ (which is distinguished by having closed trajectories with period 2π) by

$$h^\mu = a^2 k^\mu + a m^\mu \tag{5.114}$$

from which it can be seen that its trajectories rotate with rigid angular velocity (interpretable as that of the inner ring singularity) given by

$$\Omega = a^{-1}. \tag{5.115}$$

(In the spherical limit, as a tends to zero, the secondary Killing vector tends to m^μ in direction but to zero in magnitude.) This is quite different from the angular velocity Ω^H of the horizon at $r = r_+$, as given by

$$r_+ = M + \sqrt{M^2 - a^2 - Q^2} \tag{5.116}$$

where $\Delta_{(+)}$ vanishes; it is given by

$$\Omega^H = \frac{a}{2Mr_+ + Q^2} \tag{5.117}$$

which is closely analogous to the expression for the potential, as given by

$$\Phi^H = \frac{Qr_+}{2Mr_+ + Q^2}. \tag{5.118}$$

The study of higher spin wave perturbations may be most conveniently carried out in terms of a complex null tetrad formalism,[112,118] the canonical null tetrad being constructed from the orthonormal tetrad given above by relations of the usual form

$$\begin{aligned} \ell_\mu &= \tfrac{1}{\sqrt{2}}(\theta^{\hat{4}}_\mu + \theta^{\hat{1}}_\mu), \qquad & n_\mu = \tfrac{1}{\sqrt{2}}(\theta^{\hat{4}}_\mu - \theta^{\hat{1}}_\mu) \\ m_\mu &= \tfrac{1}{\sqrt{2}}(\theta^{\hat{2}}_\mu + i\theta^{\hat{3}}_\mu). \end{aligned} \tag{5.119}$$

For an ordinary Maxwell field (spin 1) $F_{\mu\nu}$ one can construct tetrad components

$$\Phi_1 = F_{\mu\nu}\ell^\mu m^\nu, \qquad \Phi_{-1} = F_{\mu\nu}\bar{m}^\mu n^\nu \tag{5.120}$$

(for obvious reasons of symmetry and notational convenience below we do not use the traditional designations which would be Φ_0 and Φ_2). Similarly (this time subtracting 2 from the traditional index values) for a gravitational (spin 2) perturbation with Weyl tensor $C_{\mu\nu\rho\sigma}$ one defines

$$\Psi_2 = -C_{\mu\nu\rho\sigma}\ell^\mu m^\nu n^\rho m^\sigma, \qquad \Psi_{-2} = -C_{\mu\nu\rho\sigma}\ell^\mu \bar{m}^\nu n^\rho \bar{m}^\sigma \tag{5.121}$$

Then if Ψ_{*0} is a scalar field, and if for the higher helicities $s = \pm 1$, $s = \pm 2$ we set[103]

$$\Psi_{*s} = (r - iq)^{|s|}\,\Psi_s \tag{5.122}$$

then the corresponding Dalembertian, Maxwellian, or perturbed Einstein wave equations decouple and separate in the simple form

$$\Psi_{*s} = X_s(r)Y_s(q)e^{-i(Et-\tilde{\Phi}\tilde{\varphi})} \tag{5.123}$$

for which the single variable functions respectively satisfy

$$\left\{\frac{d}{dr}\Delta_{(+)}\frac{d}{dr} + \frac{(Er^2 + is(r-M) - \tilde{\Phi})^2}{\Delta_{(+)}} + 4isEr\right\} X_s = \tilde{K}X_s \tag{5.124}$$

and

$$\left\{\frac{d}{dq}\Delta_{(-)}\frac{d}{dq} - \frac{(Eq^2 + sq + \tilde{\Phi})^2}{\Delta_{(-)}} + 4sEq\right\} Y_s = -\tilde{K}Y_s \tag{5.125}$$

where $\tilde{K}$ is the separations constant whose existence expresses the "hidden symmetry", and where by setting

$$E\tilde{t} - \tilde{\Phi}\tilde{\varphi} = Et - \Phi\varphi \tag{5.126}$$

we see that E is the ordinary energy constant (associated with the primary Killing vector k^μ) while the constant $\tilde{\Phi}$ (associated with the secondary Killing vector h^μ) is related to the ordinary axial angular momentum Φ (associated with the axial Killing vector m^μ) by

$$\tilde{\Phi} = a\Phi - a^2E. \tag{5.127}$$

The scalar ($s = 0$) case of this separation was originally discovered by myself[88] in essentially the same form as given here (though with greater generality in so far as the field was allowed to have its own charge e and Klein-Gordon mass m) but the higher spin generalisations were discovered later by Teukolsky[96] using a non-canonical tetrad (of a kind introduced for a different purpose by Kinnersley[119]) related to the canonical one given above by a symmetry violating transformation of the form

$$\ell^\mu \to \left(\frac{2}{\Delta_{(+)}}\right)^{\frac{1}{2}} (r^2+q^2)^{\frac{1}{2}}\ell^\mu, \qquad n^\mu = \left(\frac{2}{\Delta_{(+)}}\right)^{-\frac{1}{2}} (r^2+q^2)^{-\frac{1}{2}}n^\mu, \tag{5.128}$$

$$m^\mu \to (r-iq)(r^2+q^2)^{-\frac{1}{2}}\, m^\mu, \tag{5.129}$$

which not only violates the symmetry under simultaneous time and axial angle inversion (and hence the symmetry between positive and negative values of the helicity parameter s) but it also destroys the symmetry whereby the q equation can be obtained directly from the r equation simply by setting $r \to iq$, $M \to 0$. This loss of symmetry does not destroy the separability property, but taken with the further complication resulting from the replacement of q by $a\cos\theta$ it leads to the replacement of the pair of structurally analogous single line equations written above by the four structurally distinct and individually much longer equations of the standard Teukolsky formulation.[96] The prestige unfortunately conferred on the Kinnersley tetrad, by the fact that this separation worked at all, was such that most American workers[96,100] have adopted it exclusively for all purposes ever since, apparently not realising that there is a much simpler way of doing things using the canonical tetrad, whose advantages become particularly valuable[102,104,120,121] for more complicated problems such as that of dealing with the massive Dirac equation, which separates[95,122] while remaining coupled. In dealing with the even more difficult problems still outstanding (such as that of the coupled electromagnetic and gravitational perturbations in the charged black hole background) general experience (and in particular the example of the Mazur construction in the uniqueness problem) tells us that it will probably be necessary to exploit all available symmetry to the utmost in order to make progress.

ACKNOWLEDGEMENTS

I am grateful to many participants of the Cargèse summer school for comments, questions and suggestions, and particularly to Bruce Jensen for assistance in the preparation of the final manuscript. I also wish particularly to thank Gary Gibbons for the detailed discussions on the staticity problem referred to in section 3. For the historical information in the introduction I am indebted to Jean Eisenstaedt, who first drew my attention to Schaffer's paper on Michell, and to my father, Harold B. Carter, who pointed out the relevant reference in the Banks-Franklin correspondence.

REFERENCES

1. J. Michell, Phil. Trans. Roy. Soc. Lond. **LXXIV**, 35 (1784).
2. P.S. Laplace "Expos. du Système du Monde II," 305, Paris (1796).
3. C.L. Hardin, Annals of Science **XXII**, 27 (1966).
4. R. McCormack, British Journal for History of Science **IV**, 126 (1968).
5. S. Schaffer, J.Hist. Astron.**10**, 42 (1979).
6. H. Cavendish, Phil. Trans. Roy. Soc. Lond. **LXXXVIII**, 469 (1798).
7. W. Herschel, Phil. Trans. Roy. Soc. Lond. **LXXXI**, 71 (1791).
8. J. Michell, Phil. Trans. Roy. Soc. Lond. **LVII**, 234 (1767).
9. J. Priestly, "History and Present State of Discoveries Relating to Vision, Light, and Colour," p. 387-90 and p. 787-98, London (1772).
10. K. Schwarzschild, Sitzber. Deut. Akad. Wiss., Berlin Kl. Math-Phys. Tech., 424 (1916).
11. B. Carter, Journal de Physique C7,**34**, 7 (1973).
12. S. Chandrasekhar, Astrophys. J., **74**, 81 (1931).
13. J.R. Oppenheimer, H. Snyder, Phys. Rev. **56**, 455 (1939).
14. R. Penrose, Phys. Rev. Lett. **14**, 57 (1965).
15. S.W. Hawking, G.F.R. Ellis, "The Large Scale Structure of Space Time," Cambridge University Press, Cambridge (1973).
16. Ya. B. Zeldovitch, P. Novikov " Relativistic Astrophysics, I," University of Chicago Press, Chicago (1971).
17. S. Weinberg, "Gravitation and Cosmology," Wiley, New York (1972).
18. C.W. Misner, K.S. Thorne, J.A. Wheeler, "Gravitation," Freeman, San Francisco (1973).
19. M. Demianski, "Relativistic Astrophysics," Pergamon, Oxford (1985).
20. R.M. Wald, "General Relativity," University of Chicago Press, Chicago (1984).
21. B. Carter, in "Black Holes," ed. C. and B. DeWitt, (1972 Les Houches Summer School Lectures) Gordon and Breach, New York (1973).

22. S.L. Shapiro, S.A. Teukolsky, "Black Holes, White Dwarves, and Neutron Stars; The Physics of Compact Objects," Wiley, New York (1983).

23. K.S. Thorne, R.H. Price, D.A. Macdonald, "Black Holes: the Membrane Paradigm," Yale University Press, New Haven (1986).

24. S.W. Hawking, in "Black Holes," ed. C. and B. DeWitt, (1972 Les Houches Summer School Lectures) Gordon and Breach, New York (1973).

25. B. Carter, in "General Relativity: an Einstein Centenary Survey," ed. S.W. Hawking, W. Israel, Cambridge Univ. Press, Cambridge (1979).

26. B. Carter, in "Recent Developments in Gravitation," (1978 Cargèse Summer School) ed. M. Levy, S. Deser, Plenum, New York (1979).

27. B.K. Harrison, K.S. Thorne, M. Wakano, J.A. Wheeler, "Gravitation Theory and Gravitational Collapse," Univ. of Chicago Press, Chicago (1965).

28. R. Penrose, Riv. Nuovo Cimento I **1**, 252 (1969).

29. R. Penrose, in "General Relativity: an Einstein Centenary Survey," ed. S.W. Hawking, W. Israel, Cambridge Univ. Press, Cambridge (1979).

30. B. Carter, Phys. Rev. Lett. **26**, 233 (1971).

31. S.W. Hawking, R. Penrose, Proc. Roy. Soc. Lond. **A324**, 529 (1970).

32. S.W. Hawking, Commun. Math. Phys. **25**, 152 (1972).

33. J. Bekenstein, Phys. Rev. **D7**, 949 (1973).

34. S.W. Hawking, Commun. Math. Phys. **43**, 199 (1975).

35. J.H. Bardeen, B. Carter, S.W. Hawking, Commun. Math. Phys. **31**, 181 (1973).

36. D. Christodoulou Commun. Math. Phys. **93**, 171 (1984); D.M. Eardley, L. Smarr, Phys. Rev. **D19**, 2239 (1979).

37. R. Schoen, S.T. Yau, Commun. Math. Phys. **65**, 45 (1979).

38. W. Israel, J.M. Nester, Phys. Lett. **85A**, 259 (1981).

39. J.B. Hartle and S.W. Hawking, Commun. Math. Phys. **27**, 283 (1972).

40. H. Müller zum Hagen, Proc. Camb. Phil. Soc. **68**, 199 (1970).

41. C.V. Vishveshwara, J. Math. Phys. **9**, 1319 (1968).

42. B. Carter, J. Math. Phys. **10**, 70 (1969).

43. B. Carter, J. Gen. Rel. Grav. **9**, 437 (1978).

44. A. Lichnerowicz, "Theories Relativistes de la Gravitation et de l'Electromagnetism," Masson, Paris (1955).

45. P. Hajicek, Phys. Rev. **D7**, 2311 (1973); J. Math. Phys. **16**, 518 (1975).

46. G. Denardo, R. Ruffini, Phys. Lett. **45B**, 259 (1973).

47. S.V. Dhurandar, N. Dadhich, Phys. Rev. **D29**, 2712 (1984).

48. D.C. Robinson, Phys. Rev. **D10**, 458 (1974).

49. D.C. Robinson, Phys. Rev. Lett. **34**, 908 (1975).

50. G. Bunting, "Proof of the Uniqueness Conjecture for Black Holes," University of New England, Armidale N.S.W. (1983).

51. B. Carter, Commun. Math. Phys. **99**, 563 (1985).

52. P.O. Mazur, J. Phys. **A15**, 3173 (1982).

53. H. Müller zum Hagen, D.C. Robinson, H.J. Seifert, J. Gen. Rel. Grav. **4**, 53 (1973).

54. H. Müller zum Hagen, D.C. Robinson, H.J. Seifert, J. Gen. Rel. Grav. **5**, 59 (1974).

55. D.C. Robinson, J. Gen. Rel. Grav. **8**, 659 (1977).

56. W. Israel, Phys. Rev. **164**, 1776 (1967).

57. W. Israel, Commun. Math. Phys. **8**, 245 (1968).

58. G. Bunting, A.K.M. Massood-ul-Alam, J. Gen. Rel. Grav. to appear (1986).

59. G.W. Gibbons, Commun. Math. Phys. **27**, 87 (1972).

60. J.B. Hartle, S.W. Hawking, Commun Math. Phys. **26**, 37 (1972).

61. S.D. Majumdar, Phys. Rev. **72**, 930 (1947); A. Papapetrou, Proc. R. Irish Acad. **A51**, 191 (1947) .

62. G.W. Gibbons, C.M. Hull, Phys. Letters **109B**, 190 (1982).

63. R.H. Boyer, Proc. Roy. Soc. Lond. **A311**, 245 (1969).

64. R.H. Boyer, T.G. Price, Proc. Camb. Phil. Soc. **62**, 531 (1965).

65. R.L. Znajek, Mon. Not. R. Astron. Soc. **182**, 639 (1978).

66. T. Damour, Phys. Rev. **D18**, 3598 (1978).

67. A. Komar, Phys.Rev. **113**, 934 (1959).

68. P. Breitenlohner, D. Maison, G. Gibbons, draft preprint (1986).

69. J.B. Hartle and D.H. Sharp, Astrophys. J. **147**, 317 (1967).

70. J. Bardeen, Astrophys. J. **162**, 71 (1970).

71. B. Carter, Commun. Math. Phys. **30**, 261 (1973).

72. A. Papapetrou, Ann. Physik. **12**, 309 (1953).

73. A. Papapetrou, Ann. Inst. H. Poincaré **A4**, 83 (1966).

74. W. Kundt, M. Trumper, A. Physik **192**, 414 (1966).

75. R.H. Boyer, R.W. Lindquist, J. Math. Phys. **8**, 265 (1967).

76. J.M. Bardeen, in "Black Holes," ed. C. and B. DeWitt, (1972 Les Houches Summer School Lectures) Gordon and Breach, New York (1973).

77. J.B. Hartle, Astrophys. J. **150**, 1005 (1967).

78. B. Carter, Nature (Phys. Sci.) **238**, 71 (1973).

79. L. Smarr, Phys. Rev. Lett. **30**, 71 (1973).

80. C.W. Misner, Phys. Rev. Lett. **28**, 994 (1972).

81. W.H. Press, S.A. Teukolsky, Nature **238**, 211 (1972).

82. Ya. B. Zeldovich, J. Exp. Theor. Phys. **62**, 2076 (1972).

83. R.P. Kerr, Phys. Rev. Letters **11**, 238 (1963).

84. E.T. Newman *et al*, J. Math. Phys. **6**, 918 (1965).

85. W. Boucher, G. Gibbons, G.T. Horowicz, Phys. Rev. **D30**, 2447 (1984).

86. G. Gibbons, S.W. Hawking, Phys. Rev. **D15**, 2738 (1976).

87. B. Carter, Phys. Letters **A26**, 399 (1968).

88. B. Carter, Commun. Math. Phys. **10**, 280 (1968).

89. M. Morse, M. Heins, Ann. of Math. **46**, 625 (1945).

90. F.J. Ernst, Phys. Rev. **167**, 1175 (1968).

91. B. Carter, Phys. Rev. Lett. **26**, 331 (1971).

92. W. Kinnersley, J. Math. Phys. **14**, 651 (1973); R. Geroch, J. Math. Phys. **12**, 918 (1971).

93. B. Carter, Phys. Rev. **174**, 1559 (1971).

94. R. Penrose, Nature **236**, 377 (1972); M. Simpson, R. Penrose, Int. J. Theor. Phys. **3**, 183 (1973).

95. S. Chandrasekhar, Proc. Roy. Soc. Lond. **A349**, 571 (1976).

96. S.A. Teukolsky, Astrophys. J. **185**, 283 (1973).

97. W. Press, S.A. Teukolsky, Astrophys. J. **185**, 649 (1973).

98. J.B. Hartle, D.C. Wilkins, Commun. Math. Phys. **38**, 47 (1974).

99. B. Whiting, in preparation (1987).

100. S. Chandrasekhar, "The Mathematical Theory of Black Holes," Clarendon, Oxford (1983).

101. R. Debever, Bull. Soc. Math. Belgique **XXIII**, 360 (1971).

102. R.L. Znajek, Mon. Not. R. Astr. Soc. **179**, 457 (1977).

103. B. Carter, R.G. McLenaghan in "Recent Developments of General Relativity," ed. by R. Ruffini, North Holland, Amsterdam (1982).

104. R. Debever, N. Kamran, R.G. McLenaghan, J. Math. Phys. **25**, 1955 (1984).

105. M. Walker, R. Penrose, Commun. Math. Phys. **18**, 265 (1970).

106. L.P. Hughston, R. Penrose, P. Sommers, M. Walker, Commun. Math. Phys. **27**, 303 (1972).

107. L.P. Hughston, P. Sommers, Commun. Math. Phys. **32**, 147 (1973).

108. L.P. Hughston, P. Sommers, Commun. Math. Phys. **33**, 129 (1973).

109. R. Penrose, Ann. N.Y. Acad. Sci. **224**, 125 (1973).

110. B. Carter, preprint, Obs. Paris-Meudon, (1986).

111. B. Carter, R.G. McLenaghan, Phys. Rev. **C19**, 1093 (1979).

112. W. Dietz, R. Rudiger, Proc. Roy. Soc. **A375**, 361 (1981).

113. B. Carter, Phys. Rev. **D16**, 3414 (1977).

114. Y. Kosman, Ann. di Mat. Pura ed. Appl. **IV.91**, 317 (1972).

115. J.F. Plebanski, M. Demianski, Ann. N.Y. Acad. Sci. **98**, 98 (1976).

116. M. Demianski, E. Newman, Bull. Acad. Pol. **14**, 653 (1966).

117. E.T. Newman, R. Penrose, J. Math. Phys. **3**, 566 (1962).

118. R. Debever, Bull CI. Sci. Acad. Roy. Belg. **LXIII**, 662 (1976).

119. W. Kinnersley, J. Math. Phys. **10**, 1195 (1969).

120. N. Kamran, R.G. McLenaghan, J. Math. Phys. **25**, 1019 (1984).

121. N. Kamran, J.A. Marck, J. Math. Phys. **27**, 1589 (1986).

122. D.N. Page, Phys. Rev. **D14**, 1509 (1976).

RELATIVISTIC GRAVITATIONAL INSTABILITIES

Bernard F. Schutz

Department of Applied Mathematics and Astronomy
University College
Cardiff CF1 1XL, Great Britain

INTRODUCTION

The purpose of these lectures is to review and explain what is known about the stability of relativistic stars and black holes, with particular emphasis on two instabilities which are due entirely to relativistic effects. The first of these is the post-Newtonian pulsational instability discovered independently by Chandrasekhar (1964) and Fowler (1964). This effectively ruled out the then-popular supermassive star model for quasars, and it sets a limit to the central density of white dwarfs. The second instability was also discovered by Chandrasekhar (1970): the gravitational wave induced instability. This sets an upper bound on the rotation rate of neutron stars, which is near that of the millisecond pulsar PSR 1937+214, and which is beginning to constrain the equation of state of neutron matter.

I will follow the notation of Misner, *et al* (1973) and of Schutz (1985): the metric has signature +2; Greek indices run from 0 to 3, Latin from 1 to 3. I set c and G to 1 everywhere. For perfect fluids, my notation is: n is the number density of conserved particles; ρ is the density of total mass-energy; S is the specific entropy; and γ is the adiabatic index, defined by

$$\gamma = \left.\frac{\partial \ln p}{\partial \ln n}\right|_S .$$

All these quantities are defined in the local rest frame of the fluid. In these terms, the first law of thermodynamics (energy conservation) becomes

$$nTdS = d\rho - (\rho+p)dn/n. \tag{1}$$

SPHERICAL PULSATION OF SPHERICAL STARS

Newtonian Stars

Although our subject is relativistic instability, it will help us to get a general feeling for the way instability arises in Newtonian stars before tackling the relativistic case. Not only is the Newtonian case simpler, but also by comparing the Newtonian and relativistic versions of stability criteria we will be able to see exactly which instabilities are attributable to general relativity.

The most convenient way to describe the small-amplitude spherical pulsation of a spherical star is in terms of the displacement ψ of a thin shell of the star from its equilibrium position. In terms of ψ the first-order perturbation in the Euler and continuity equations can be written in the form (Ledoux & Walraven 1958)

$$\rho\psi_{,tt} + C(\psi) = 0, \tag{2}$$

where a subscripted ",t" denotes a partial derivative with respect to time and where C is the operator

$$C(\psi) = -\frac{d}{dr}\left[\frac{\gamma p}{r^2}\frac{d}{dr}(r^2\psi)\right] - \frac{4Gm(r)\rho}{r^3}\psi, \tag{3}$$

in which m(r) is the mass inside radius r. It is easy to see that C is selfadjoint with respect to the L^2 norm weighted by r^2 (*i.e.,* integrating over the volume of the star rather than the 1-dimensional radius), with the boundary condition that ψ vanishes at r=0. It is more relevant to solving Eq.(4) below that $\rho^{-1}C$ is selfadjoint in the density-weighted norm (Eisenfeld 1969).

The stability of the star could be studied directly by showing that all solutions of Eq.(2) are bounded in time if and only if the operator C is positive-definite (Laval, *et al* 1965). But for our later purposes it is useful to introduce here the *normal mode* problem. If we assume that the perturbation ψ has harmonic time dependence, $\psi(t,x^i) = \chi(x^i)\exp(i\omega t)$, then the dynamical equation (2) becomes the eigenvalue problem:

$$C(\chi) = \rho\omega^2\chi. \tag{4}$$

This problem can essentially be put into Sturm-Liouville form for the eigenvalue ω^2 by a change to the variable χ/r. Several consequences follow immediately. (i) There is an ascending series of eigenvalues ω_n^2, (n = 0,1,2,...), which approaches infinity as n does. (ii) The eigenfunction associated with ω_n^2 has n nodes in the radial direction. (iii) The eigenfunctions are complete, so that a star is stable if and only if all the ω_n are real, *i.e.* if and only if all ω_n^2 are positive. From property (i) it follows that the star is stable if and only if ω_0^2 is positive. (iv) This in turn will be true if and only if the integral

$$\int \chi^* C(\chi) r^2 dr > 0 \tag{5}$$

for all χ. Using the explicit form of C gives, after some algebra, the simple stability criterion (Ledoux 1958)

$$\frac{d}{dr}[(\gamma - 4/3)p] < 0 \Rightarrow \text{stability}. \tag{6}$$

Thus, in the linear approximation (small amplitude perturbations), a star with constant γ is stable if $\gamma > 4/3$. If γ is not constant, then the stability is harder to decide. This is a *sufficient* condition for stability.

There is a simple way to understand why 4/3 should be the critical value of the adiabatic index. This is an order-of-magnitude argument based on binding energy (*cf.* Zel'dovich & Novikov 1971). The total binding energy is

$$E = U + W,$$

where U is the internal energy and W the gravitational potential energy. For a star of radius R, mass M, typical pressure p, and typical density ρ, these are

$$U = apR^3 \quad \text{and} \quad W = -bM^2R^{-1},$$

where a and b are constants of order unity. Given that p is proportional to ρ^γ, we have

$$E = kM^\gamma R^{3-3\gamma} - bM^2R^{-1},$$

where k is another constant. The star will be in equilibrium if its binding energy is an extremum against variations of R with M fixed. This gives

$$\left.\frac{\partial E}{\partial R}\right|_M = 0 \quad \Rightarrow \quad E = -b\,\frac{\gamma - 4/3}{\gamma - 1}\,\frac{M^2}{R},$$

so that a star is bound (hence stable) if γ exceeds 4/3 and unbound (unstable) if γ is less than 4/3. The marginally stable case is $\gamma = 4/3$: if such a star is perturbed, it will experience no net restoring force, so its radius will simply increase or decrease linearly with time, at least until nonlinear effects become important.

<u>Relativistic stars</u>

Spherical motions of a star do not radiate gravitational waves, so we might guess that there is no qualitative difference between the evolution of a perturbation in the relativistic case from that in the Newtonian one. This expectation is basically correct, but it is also easy to see that we should expect the Newtonian criterion of $\gamma = 4/3$ to be different in general relativity. This is because the binding-energy argument presented above must be different for relativistic stars, whose gravitational binding energy is larger than that of their Newtonian counterparts (*cf.* Harrison, *et al* 1965). In the first post-Newtonian approximation (that is, taking into account the first relativistic corrections to Newtonian theory), the appropriate criterion was found independently by Chandrasekhar (1964) and Fowler (1964):

$$\text{stability} \iff \gamma > 4/3 + K, \tag{7}$$

where K is a positive constant that depends on the equation of state and which increases with M/R.

In itself, this represents a small correction to the stability criterion, and it would not be remarkable except for the *coincidence* that γ approaches 4/3 as a fluid becomes more relativistic. (I call this a coincidence because I can see no fundamental relation between the way 4/3 is singled out as special in the binding-energy argument above and the fact that 4/3 is also the relativistic limit of the adiabatic index.) One class of stars in which $\gamma \simeq 4/3$ is massive main-sequence stars, where radiation pressure is the dominant support. As Chandrasekhar and Fowler both showed, stars with masses approaching $10^6\ M_\odot$ have γ so close to 4/3 that the small correction K makes them unstable. Thus, stars whose structure is essentially completely Newtonian have their stability decided by effects of general relativity. This happens because the Newtonian forces nearly cancel: the star is almost marginally stable in Newtonian theory, and the issue is decided by tiny relativistic corrections.

Another class of stars where $\gamma \simeq 4/3$ is white dwarfs with large central densities. Consider a sequence of white dwarfs with increasing central density. As the central density gets larger, the electrons providing the pressure need to get more and more relativistic, and γ approaches 4/3. This competes with another destabilizing effect: as central density increases, the *neutronization* reaction in which a proton and an electron combine to form a neutron and a neutrino becomes energetically favorable. This reaction removes the pressure-providing electrons, so that the equation of state softens and γ drops below 4/3. For helium and carbon white dwarfs, general relativity limits the central density; for iron white dwarfs, it is neutronization (*cf.* Shapiro & Teukolsky 1983).

The turning point criterion for white dwarfs and neutron stars

There is an easy way to decide the overall stability of stellar models that are members of a one-parameter family of models, such as one obtains by taking "cold" matter (minimum entropy) for various central densities. This method appears to have been devised by Zel'dovich and by Wheeler, and it is described in Harrison, *et al* (1965). Figure (1) contains a plot of the mass versus the radius of such stars, parametrized by their central density. The curve should be regarded as schematic, since quantitative details will depend upon the particular equation of state chosen. Suppose we follow the curve from the low-mass, low-density objects at the left margin (rocks and planets)

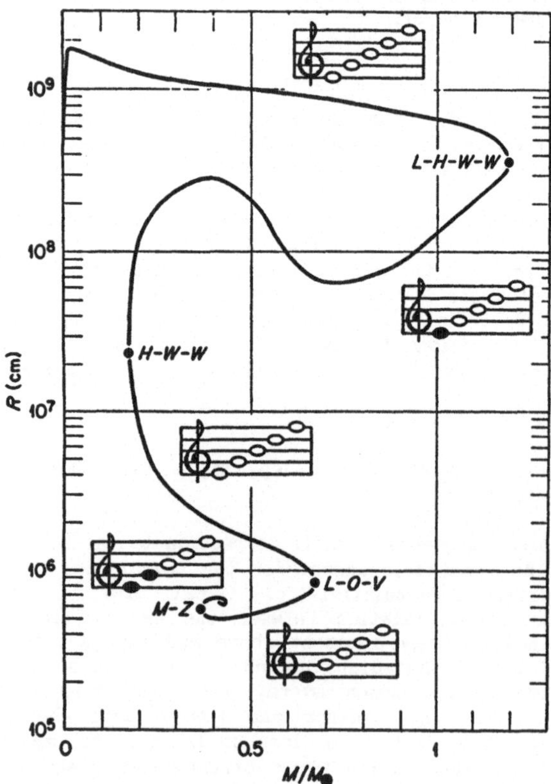

Fig. 1. A typical plot of radius versus mass for stars parametrized by their central densities. The musical notation indicates which modes (overtones of the fundamental) are stable (open circles) or unstable (filled circles). Taken from Harrison, *et al* (1965).

until M reaches its first maximum. This is a place where nearby models have the same masses but different radii. This means that we could perturb the star at the maximum by, say, increasing its radius, and it would remain in equilibrium: there would be no restoring force, and the star would be neutrally stable. Since rocks are stable and this is the first such neutrally stable point on the curve, it is reasonable to expect that the star goes *unstable* at this point. A more careful analysis bears this out and shows, in fact, that if the curve is curling clockwise at an extremum, a mode is going unstable, while if it is curling counterclockwise then an unstable mode is returning to stability. The first maximum of M in Figure (1) is the instability point of the white dwarfs. The slide to the minimum represents a sequence of unstable models, and past the minimum we reach the stable neutron stars. They become unstable (and therefore reach their maximum mass) at the next maximum of M, and after that the continued spiralling of the curve indicates that more and more modes are going unstable. This section of the curve is especially sensitive to the assumptions one makes about the largely unknown properties of high-density matter.

The unstable region between white dwarfs and neutron stars means that neutron stars have a *minimum* mass, a fact first appreciated by Oppenheimer & Serber (1938). White dwarfs have no such minimum: they follow smoothly after smaller objects, like planets. The turning point criterion has been used for axisymmetric perturbations of rotating stars by Hartle & Thorne (1969), and extended by Ipser & Horowitz (1979); it has been generalized to many-parameter families of models by Sorkin (1982).

Star clusters

Clusters made of collisionless particles interacting gravitationally have also been studied extensively, mainly as models for globular clusters and elliptical galaxies in Newtonian theory, and for quasars or quasar precursors in general relativity. There is no room to review that work here, but the reader is referred to Ipser (1969). The subject has been considerably enlivened recently by the numerical calculations of Shapiro & Teukolsky (1985a,b,c) showing how an unstable relativistic cluster can quickly form a black hole containing a considerable fraction of its mass. These calculations have been summarized in Shapiro & Teukolsky (1986).

NONSPHERICAL PULSATION OF SPHERICAL STARS

Newtonian stars

Things get a little more complicated when we consider nonspherical perturbations of stars. A good reference for this subject is Unno, *et al* (1979). Since the problem is a linear one and the unperturbed star is spherically symmetric, we can analyze the perturbations into spherical harmonics. Scalar functions, such as ρ, have perturbations expandable in the usual way:

$$\delta\rho(r,\theta,\phi) = \sum_{lm} \delta\rho_{lm}(r)\, P_l^m(\cos\theta)\, e^{im\phi}. \tag{8}$$

Vectors are expanded in vector spherical harmonics, one version of which is as follows:

$$\chi(r,\theta,\phi) = \sum_{lm} [\chi^r_{lm}(r)\, P_l^m(\cos\theta)\, e^{im\phi}\, e_r + \chi^+_{lm}(r)\, \nabla(P_l^m\, e^{im\phi}) + \chi^-_{lm}(r)\, {}^*\nabla(P_l^m\, e^{im\phi})]. \tag{9}$$

Here the radial component of the displacement vector $\chi(r)$ is expanded as a scalar (because that is how it behaves under rotations), and the tangential components are expanded in terms of the gradient of the spherical harmonic of order (l,m) and of its *dual* gradient in the sphere,

$$*\nabla \equiv e_r \times \nabla \ . \tag{10}$$

The gradient and dual gradient in the sphere produce linearly independent vectors. Since the sphere is two-dimensional, they form a basis for vectors tangent to the sphere, so Eq.(9) is perfectly general.

This representation of vector perturbations is useful because we will now see that the dual-gradient parts of expressions can essentially be ignored. The argument is one of *parity*. Consider a coordinate change $\phi \to -\phi$. Under such a change, and its associated change of basis $e_\phi \to -e_\phi$, true scalars and tensors do not change, but pseudovectors do change sign. Now the spherical harmonics and their gradients are true scalars and vectors, respectively, but the dual gradient is a pseudovector. The unperturbed star is of course *invariant* under this change (not true for rotating stars), so the differential equations for the perturbation will not mix the two classes of perturbations. Since the pressure and gravitational field perturbations are scalars, the pseudovector -- called the *odd-parity* part of the perturbation -- does not elicit any restoring forces, so it is a neutral perturbation, one with zero frequency. An example of an odd-parity perturbation is setting the star into rotation: it simply continues to rotate. The odd-parity normal modes are usually called *toroidal* modes.

The *even-parity* normal modes contain the interesting dynamical information. For a relatively simple star (in a sense to be made clear below), the eigenvalues fall into three classes, called f-, p-, and g-modes. Their typical behavior as a function of l is illustrated in Figure 2, taken from Cox (1980). (Because of the spherical symmetry, the eigenfrequencies do not depend on m.) The p-modes form an ascending sequence of eigenfrequencies, with $\omega \to \infty$. The g-modes form another infinite sequence, but

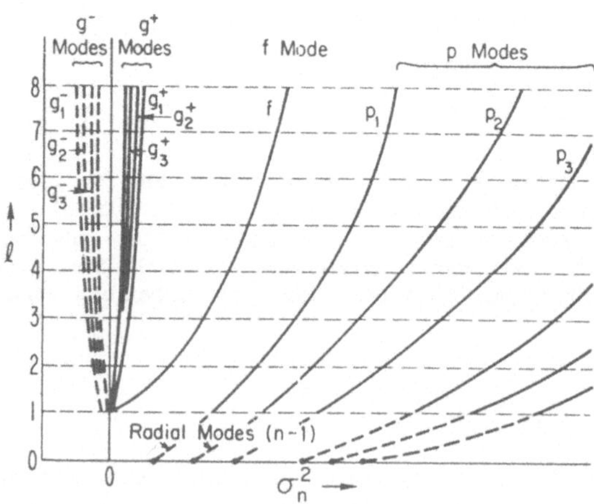

Fig. 2. The qualitative behavior of the even-parity nonradial modes of a typical spherical star. (From Cox 1980.)

with $\omega \to 0$ in the limit. The f-mode is a single mode in between, which shares characteristics of both types. In general, a star may have both stable g-modes (g^+-modes) and unstable (g^-) ones, but the f-mode and p-modes will always be stable. In the limit $\omega \to \infty$, the mode equation approaches a

Sturm-Liouville equation with ω^2 as the eigenvalue, and in the opposite limit of $\omega \to 0$ it approaches a Sturm-Liouville equation with $1/\omega^2$ as the eigenvalue; this explains the asymptotic behavior of the two sequences.

The remarkable feature of the nonradial pulsation problem is the simplicity of its stability criterion (Lebovitz 1966):

$$\text{nonradial stability} \Longleftrightarrow dS/dr > 0. \tag{11}$$

This is usually called the *Schwarzschild criterion*, although it was first derived from a local argument based on convection and bouyancy by Lord Kelvin, as related by Chandrasekhar (1939). For an accessible derivation see Cox (1980). If the star is simple enough that dS/dr is of the same sign everywhere, then the g-modes will be either all stable or all unstable. If the star is more complicated (as when it has different zones of convection, ionization, composition, etc.), then it may have both types of g-modes. Physically, the g-modes are associated with convection; their eigenfunctions are dominated by velocity rather than density perturbations. If one has $dS/dr = 0$ throughout the star, then the g-modes will all have zero frequency; this happens for a polytrope, for example, if one takes γ equal to $1 + 1/n$. The p-modes are associated with sound waves; as their order gets larger, they become just local waves travelling at the speed of sound.

An important concept is the *pattern speed* of a mode, which is its phase angular velocity. Since the perturbation is proportional to $\exp(im\phi + i\omega t)$, surfaces of constant phase at some fixed r and θ will satisfy

$$\omega t + m\phi = \text{const}.$$

Differentiating this with respect to t gives

$$d\phi/dt = -\omega/m \equiv \omega_p, \tag{12}$$

where ω_p is called the pattern speed of the mode. It is clear from Figure 2 that the frequency of p-modes typically increases with l less rapidly than linearly, so that the smallest pattern speed associated with any p-mode (obtained by dividing $-\omega$ by l) decreases towards zero as l increases. Therefore, although the p-modes contribute arbitrarily high frequencies, one can find p-modes with arbitrarily small pattern speed. This will be important to us later when we discuss the gravitational-wave-induced instability in rotating stars.

Relativistic stars

When we turn to relativistic stars, we should expect a qualitative difference from the Newtonian theory because nonradial pulsations can emit gravitational radiation. On the other hand, at least for nearly Newtonian systems, we should also expect the quantitative effect of this to be small. If a mode has real frequency in the Newtonian star, then the energy carried away by gravitational radiation should damp the mode slowly, and this should appear as a small positive imaginary part of the eigenfrequency. So we expect stable relativistic stars to have complex eigenfrequencies with $|\mathrm{Re}(\omega)| \gg \mathrm{Im}(\omega) > 0$. This expectation has been verified by extensive investigations, and it extends essentially unchanged even to highly relativistic stars.

How does the imaginary part of the frequency actually arise in the eigenvalue calculation? It does not come from any qualitative change in the local perturbation equations; rather, it comes form imposing an *outgoing-wave boundary condition* on the eigenfunction. One demands that the energy flux at infinity represent only waves emitted by the star. This is a time-asymmetric condition, and it therefore produces eigenfrequencies which are "biased" -- if ω is an eigenfrequency, then its complex conjugate ω^* is not. (In fact, this ω^* is an eigenfrequency for a mode which satisfies an *incoming* wave boundary condition at infinity.) We shall meet the outgoing-wave boundary condition in the next section and again in Eq.(39) below.

The first calculations of the f- and p-modes of relativistic stars was performed numerically by Thorne (1968, 1969) using the analytic formalism developed by Thorne & Campolattaro (1967). For the l=2 f-mode, he found that the period was typically about a millisecond, and the damping time roughly 10^3 times as long, even for highly relativistic stars. The best calculations of these modes to date have been by Lindblom & Detweiler (1983), using the models that Arnett & Bowers (1977) calculated for a variety of recent equations of state.

The g-modes present special problems for numerical calculations because their very low frequencies and small density perturbations mean they give off extremely small amounts of gravitational radiation. There are two problems because of this: first, a numerical eigenfrequency calculation has to hunt for a tiny imaginary part of the frequency, smaller than the numerical accuracy of typical codes for p-modes; and second, since the wavelength of the waves is very large, the outgoing-wave boundary condition must be imposed very far from the star, where numerical errors can accumulate. Finn (1986) has recently overcome these problems by doing an analytic approximation to the near-zone gravitational wave field of the star and finding the appropriate boundary conditions for numerical computations. He is preparing a further paper with the results of realistic calculations.

Despite the complications of gravitational waves, it seems likely that the Schwarzschild criterion governs the stability of nonradial pulsation in general relativity as well as in Newtonian theory. The argument, basically given by Thorne (1966), is that instability sets in through a zero-frequency mode (established in general by Friedman & Schutz 1975, as we will discuss in detail later in these lectures). But such a mode will not give off gravitational radiation, so the local physics of convection will be essentially the same as in Newtonian theory. Since in Newtonian theory the local criterion for convection is all one needs for the stability of the star, one can conjecture that this will be true as well in general relativity. Chandrasekhar (1965) has extended Lebovitz's Newtonian proofs to post-Newtonian general relativity, and steps toward a rigorous fully relativistic proof have been taken by Thorne (1966) and Islam (1970). But the conjecture has not been fully established.

For completeness, let us recall the odd-parity toroidal modes of the Newtonian case. The argument that they were zero frequency was purely a symmetry one, so we may apply it in general relativity as well. The difference is that the gravitational field is no longer a scalar. As a tensor it can have odd-parity parts as well. But these still do not couple radiation to the star: the pressure and density perturbations are scalars, and velocity perturbations simply cause the star to rotate steadily, without radiating. There are odd-parity gravitational waves that propagate without disturbing the star, and we will consider them in more detail when we study perturbations of spherical black holes.

Strongly damped modes

The above discussion of relativistic modes was motivated by the Newtonian analogy, and all the numerical calculations performed so far have looked only for relativistic modes that may be regarded as small perturbations of Newtonian modes. But gravitational radiation has its own dynamical freedom, and one might ask whether there are any modes associated with this that have no analogy in Newtonian theory. In fact, a family of such modes was discovered by Dyson (1980) in a model problem. An even simpler model problem analyzed recently by Kokkotas & Schutz (1986) will show clearly how they arise. Figure 3 shows the physical system: one finite string of length 2L and one semi-infinite string coupled to it by a massless spring with spring constant k. The strings have the same tension T and wave speed c. The finite string represents the star, while the semi-infinite

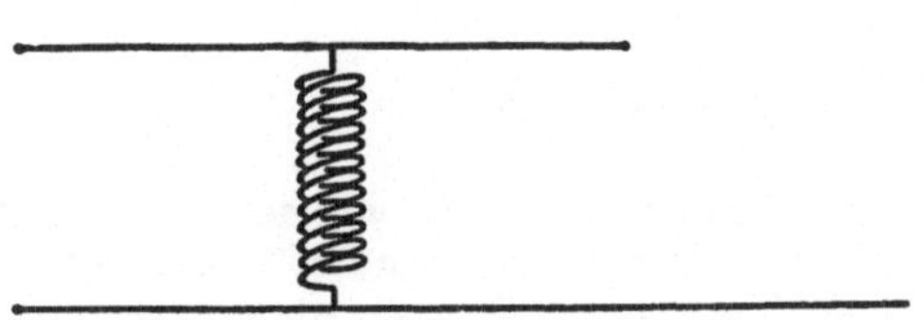

Fig. 3. A simple model system which has both weakly and strongly damped modes.

string is the analogue of the gravitational wave field; the spring couples the two as weakly (the Newtonian case) or strongly (the relativistic one) as we like. The analysis is just elementary mechanics apart from the imposition of an outgoing-wave boundary condition. We want to allow solutions for the amplitude $y(t,x)$ of the semi-infinite string of the form $f(ct-x)$ but to exclude $f(ct+x)$. Therefore we require $y_{,t} + cy_{,x} = 0$, or for a mode with frequency ω:

$$\frac{\partial y}{\partial x} = -i\omega y.$$

The explicit appearance of the frequency in the boundary condition is what makes the eigenvalues complex. Using this, it is not hard to show that the eigenvalue equation is

$$z \cosh z = -K \sinh z \,(2 + e^{-2z}), \tag{13}$$

where z is the dimensionless frequency and K the ratio of the strength of the spring to the tension in the strings:

$$z = i\omega L/c, \qquad K = kL/2T.$$

In the limit of weak coupling ($K \ll 1$), two families of eigenfrequencies emerge. The first has

$$\mathrm{Re}(\omega) \simeq (2n+1)\pi c/2L, \qquad \mathrm{Im}(\omega) = \frac{8K^2 c}{(2n+1)^2\pi^2 L} \tag{14}$$

and the second family has the same $\mathrm{Re}(\omega)$ but $\mathrm{Im}(\omega) = ac/L$, where a solves

$$a = K \exp(2a). \tag{15}$$

The modes of the first sequence are clearly small perturbations of the odd-order modes of the finite string, and are analogous to the modes of the relativistic stars described above. [The even-order modes of the finite string have a node at the attachment point of the spring, so they do not couple to the other string: their eigenfrequencies emerge unchanged from

Eq.(13).] The other modes become more and more strongly damped ($a \to \infty$) as $K \to 0$. The eigenfunctions of these families help us make sense of them. The weakly damped family have their energy primarily in the finite string; it gradually leaks through the spring and is radiated away. The strongly damped modes, on the other hand, have larger amplitude in the semi-infinite string, exciting the finite string only weakly. If we think in terms of the initial-value problem, data that excite the finite string can be represented by the weakly-damped normal modes, but they then have no freedom left to represent any initial excitement of the semi-infinite string. This is the reason for the existence of the strongly damped modes: any initial excitement of the semi-infinite string will be radiated away very quickly, so these modes have strong damping. This physical argument makes it seem plausible to me that strongly damped modes should exist in relativistic stars as well, but so far they have not been seen.

Quadrupole gravitational radiation

In the lectures by Damour in this volume, the reader will find an extensive discussion of the "quadrupole formulas" that describe gravitational radiation in the slow-motion limit. I will simply extract one result from that discussion, namely that a nearly Newtonian system loses energy to gravitational radiation at an average rate given by

$$\frac{dE}{dt} = -\frac{1}{5}\langle \bar{I}^{(3)}_{jk} \bar{I}^{(3)}_{jk} \rangle, \tag{16}$$

where there is an implied sum on j and k, the superscripted "(3)" means three time derivatives, and the reduced quadrupole tensor $\bar{I}_{jk}$ is defined by

$$\bar{I}_{jk} = I_{jk} - \frac{1}{3}\delta_{jk} I^{1}_{1}, \quad I_{jk} = \int T^{00} x_j x_k d^3x. \tag{17}$$

This gives us another method to calculate the normal modes of at least nearly Newtonian stars: calculate the normal modes of a Newtonian star, find its energy radiation from Eq.(16) applied to the eigenfunction, and estimate the damping rate of the relativistic mode from the equation

$$\mathrm{Im}(\omega) = \tfrac{1}{2} E_{,t}/E,$$

where E is the energy of the mode. This method gives a good test of both the validity of the quadrupole formula and the accuracy of numerical p-mode eigenfrequencies for weakly relativistic stars. It has been used by Balbinski, *et al* (1985) to show that the quadrupole formula works surprisingly well even for highly relativistic stars, and to improve the numerical methods used for fully relativistic stars.

NONSPHERICAL PERTURBATIONS OF SPHERICAL BLACK HOLES

The other spherical system that has received a lot of attention in general relativity is the Schwarzschild black hole. Because Birkhoff's theorem (*cf.* Misner, *et al* 1973) excludes any nontrivial spherical perturbations of the hole, we need only study its nonradial stability. The problem was first studied by Regge & Wheeler (1957), but at that time the nature of the black hole was not understood (indeed, Wheeler didn't coin the term "black hole" until a decade later), and they used an inappropriate boundary condition at the horizon. A definitive proof of the stability of the Schwarzschild metric was finally given by Vishveshwara (1970). Nevertheless, the Schwarzschild perturbation problem continues to be interesting, partly because normal mode oscillations of a black hole might be seen by gravitational wave antennas, partly as a guide to the much more difficult problem of perturbations of the Kerr metric (the rotating black hole), and partly because the normal mode problem has some peculiar and challenging features (Chandrasekhar 1983).

Formulation as a scattering problem

The interesting features of the problem are most easily illustrated by studying the so-called odd-parity equation for gravitational waves. Unlike the stellar case, the odd-parity perturbations of the Schwarzschild metric are just as interesting as the even-parity ones. At first it was thought that they obeyed a different equation from the even-parity "Zerilli" wave equation (Zerilli 1970), but it has since been shown that the two equations can in fact be transformed into one another (Chandrasekhar 1975, Chandrasekhar & Detweiler 1975; see Chandrasekhar 1983). The equation has a form which is similar to that of scattering problems in quantum mechanics (Regge & Wheeler 1957):

$$\psi'' + (\omega^2 - V)\,\psi = 0, \qquad V = \left(1 - \frac{2M}{r}\right)\left[\frac{l(l+1)}{r^2} - \frac{6M}{r^3}\right], \tag{18}$$

where M is the mass of the hole and primes (') denote, not derivatives with respect to r, but derivatives with respect to r_*, defined by

$$r_* = r + 2M \ln(r/2M-1),$$

which is an affine parameter on the outgoing null geodesics. The amplitude ψ is a metric component, from a knowledge of which all the odd-parity metric components can be calculated. Since r_* is the fundamental variable, we must ask what its limits are and what V looks like in these limits. As $r \to \infty$, so does r_*, and $V \simeq l(l+1)/r_*^2$. As $r \to 2M$, we have $r_* \to -\infty$ and V falling off exponentially as $\exp(r_*/2M)$. Between these two extremes, the potential is smooth, reaching a maximum at some intermediate point, with no complicated wiggles.

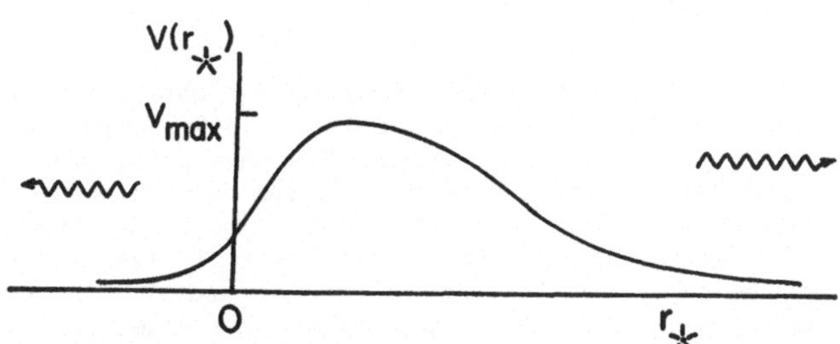

Fig. 4. The qualitative shape of the potential V. The waves show the appropriate boundary conditions for a normal mode.

If this were a standard scattering problem in quantum mechanics, then there would be no difficulty. What makes this problem different are the normal-mode boundary conditions: in order to exclude any outside "forcing" of the hole's oscillations, we demand that the waves in ψ must be outgoing far from the hole and ingoing across the horizon. That is, we are looking for solutions of Eq.(18) which have waves moving *away* from the potential barrier on both sides. Simple flux-conservation arguments of the type one makes for the Schrödinger equation show that it is impossible to satisfy this condition with a purely real frequency ω, and it is not hard to extend them to show that the imaginary part of ω must be positive, so that all the modes are damped away. (We still lack any completeness theorem for the normal modes, however, so that this result cannot be used to infer the stability of the hole.)

In fact, we can go further and use WKB-type arguments to see that the real part of ω^2 must be near the maximum of V. Suppose this were not the case. If ω^2 were larger than the barrier maximum, then a wave outgoing on the right could only match to a wave with a substantial component approaching the barrier from the left, in violation of our boundary condition. If ω^2 is too small, then the left-moving wave on the left of the barrier would "tunnel" through the barrier as a linear combination of two exponentials, one growing and the other dying. Each of these would, in turn, match across the right-hand edge of the barrier to a pair of waves, one ingoing and the other outgoing. For the two ingoing waves to cancel (and thus satisfy the boundary condition), they would have to have the same amplitude; but if the tunneling has been strong, they cannot have the same amplitude. Therefore, the tunnelling has to be essentially negligible: ω^2 has to be near the top of the potential barrier. [A more extended version of this discussion may be found in Schutz (1984).]

Calculations of the normal modes

Numerical calculations bear out this simple argument. The first extensive calculations were by Chandrasekhar & Detweiler (1975). They were able to get the first few modes for each l, but soon lost accuracy. They found that for each l the modes formed a sequence in which the real part of the frequency did not change very much, but the imaginary part increased rapidly (reminiscent of the strongly damped modes of our model string problem, above).

An approximate solution can be sought by replacing the potential V with the simpler *Price* potential V_P,

$$V_P = \begin{cases} l(l+1)/r_*^2 & r_* > a \\ 0 & r_* < a \end{cases}$$

for some a. The solutions for this potential are obtainable in terms of Bessel functions. Remarkably, there are only a finite number of modes for each l. This raises the question of whether the real potential V only has a finite number of modes, as well. It also forces one to wonder about the completeness problem: since only a finite number of eigenfunctions exist, they cannot be a basis for arbitrary initial data. What is the evolution of the data that cannot be expressed in terms of normal modes? The completeness problem for radiating systems has not received the attention it deserves. The first extensive investigation of which I am aware was in the Ph.D. thesis of Dyson (1980), and it led to the discovery of the strongly damped modes. Leaver (1986b) has recently studied the problem for the Schwarzschild metric, showing in particular that there are non-normal-mode contributions that give rise to a radiating, power-of-time falloff in the "tails" of gravitational wave perturbations, eventually becoming the nonradiative decay tails discovered by Price (1972). It seems to me that this is one of the most important and potentially fruitful areas for research in the black-hole normal mode problem.

Another approximation to the Schwarzschild potential has been explored by Blome & Mashhoon (1984) and Ferrari & Mashhoon (1984). Here another analytic form is substituted for the full potential V, permitting an analytic solution. This one has a couple of parameters that can be adjusted so that it is a good fit at the maximum, rather than having the correct behavior at infinity, as does the Price potential. This method yields good approximations for the lowest modes, but it is hard to improve it to do better for higher-order modes.

In another attempt at approximation, Schutz & Will (1985) introduced WKB methods to obtain the eigenfrequencies. This also gave good results only for the lowest modes, but it has the advantage that it can be extended to higher orders. Iyer & Will (1986) and Iyer (1986) have gone up to fifth order, giving vastly improved accuracy. The WKB approach also gives an analytic formula for the behavior of the eigenfrequencies as l gets large:

$$M\omega \simeq [(l+\tfrac{1}{2}) + i(n+\tfrac{1}{2})]/3^{3/2} + O(1/l),$$

where n is the order of the mode. The real part of the frequency gives, for modes with $l=m$, a pattern speed $M\omega_p \simeq 3^{-3/2}$, which is the orbital frequency of a photon in the (unstable) circular photon orbit around Schwarzschild (Goebel 1972). The fact that, for $n=0$, the imaginary part of the frequency approaches the limit .0962, had been noticed by Detweiler (1979), but not shown analytically until this WKB work.

Great progress on the development of suitable numerical techniques for this problem has recently been made by Leaver (1985, 1986a). He has given a very detailed discussion of the spheroidal wave equation, of which Eq. (18) is a special case, and he has adapted approximation methods developed in atomic physics to this problem. Future work on this problem will surely take this work as its starting point. Leaver's methods strongly suggest that there will be an infinite number of normal modes for any l, but they don't quite prove it: this is still one of the most important unsolved problems in this field.

STABILITY OF ROTATING STARS: GENERAL REMARKS

In the preceding sections I have reviewed a subject that is reasonably well understood, and I was able only to highlight some important results. When we turn to the study of the perturbations and stability of rotating stars, we find a very different story: despite considerable interest in the problem, there are few general theoretical results; there have been no extensive calculations of the modes of relativistic, rapidly rotating stars; and even the Newtonian theory is poorly understood. Nevertheless, some remarkable features of the problem have been discovered, including the fact that the emission of gravitational radiation can actually *destabilize* a rotating star; even more, *all* perfect-fluid rotating stars are unstable in this way! It appears that these instabilities may play an important role in the formation and subsequent evolution of neutron stars, and so the main aim of the remainder of these lectures will be to try to understand them.

Before going on, it is well to ask why adding rotation to the star makes the problem so much more difficult. Except in the limit of slow rotation, where the problem is not much harder than the spherical one (Newtonian theory reviewed by Tassoul 1978; relativistic theory treated by a number of authors, *e. g.*: Hartle 1967; Hartle & Thorne 1968, 1969; Hartle, *et al*, 1972; Hartle & Munn 1975; Chandrasekhar & Friedman 1972; Abramowicz & Wagoner 1978), there are two factors that make the normal mode problem harder in the rotating star:

(i) *It is harder to compute results.* When we lose spherical symmetry, the mode problem becomes an elliptic boundary value problem in two dimensions, rather than an ordinary differential equation. This is not an insuperable problem with today's computers, but it has inhibited progress in the past. I suspect that the motivation to tackle the problem has been lacking until recently, but the possibility of gravitational wave observations in the near future may change that.

(ii) *Even in the Newtonian case, the eigenfrequencies are not eigenvalues of a selfadjoint operator.* In fact, this is true of the spherical problem as well, but there we are able to solve the problem in terms of a selfadjoint operator whose eigenvalues are ω^2. But ω itself is the eigenvalue of the square root of this operator, which is not necessarily selfadjoint. It is easy to see why: in order to allow for the possibility of the star being unstable, it must be possible to have complex eigenfrequencies ω; but a selfadjoint operator has real eigenvalues, so we cannot expect ω to be the eigenvalue of such an operator. This argument is as true for the rotating star as for the nonrotating one. But there is no lucky way around it in the rotating case. For example, in terms of the Lagrangian displacement vector field χ, the dynamical equation for the perturbations of a Newtonian perfect fluid star has the general form (Lynden-Bell & Ostriker 1967)

$$\rho\chi_{,tt} + B(\chi_{,t}) + C(\chi) = 0, \tag{19}$$

where B and C are operators. The operator B involves Coriolis-type terms, and is present for nonaxisymmetric modes only when the background star is rotating. In the absence of rotation, when $B = 0$, then there is an associated eigenvalue problem for ω^2 which is selfadjoint, as we have seen. The axisymmetric normal mode problem is also selfadjoint. But in the rotating case, the eigenvalue problem for nonaxisymmetric modes is genuinely quadratic, and there is no associated simple selfadjoint problem. Because the nonaxisymmetric problem is so different from the axisymmetric one, and because it has unique astrophysical implications, I will concentrate on it in what follows. The axisymmetric problem resembles in many ways the spherical one. In the spherical problem, we are lucky. For the nonaxisymmetric modes of rotating stars, our luck runs out.

The formal structure of the Newtonian problem is, nevertheless, very regular and interesting. The operator of which ω is the eigenvalue is, of course, essentially the time-evolution operator. This turns out (Schutz 1980a) to be symmetric with respect to a non-positive-definite inner product [in fact the symplectic inner product of Eq.(44) below]. Such operators have been studied by Bognar (1974), and extensively in a series of papers by Barston (1967a,b; 1968; 1971a,b; 1972; 1974; 1977). The structure and some aspects of the stability of rotating stars in Newtonian theory have been developed in the monograph by Tassoul (1978). See also Schutz (1983, 1984) for recent reviews of this subject in more mathematical depth than we shall go into here.

Despite a considerable amount of work on the problem, the outstanding problem for Newtonian stars is still unsolved: to give a necessary and sufficient criterion for the absence of complex eigenfrequencies in the spectrum of the operator, one which can be used without actually solving for the eigenvalues. In the spherical case, this criterion is simply the positive-definiteness of C. Here, positive-definiteness of C is sufficient but not necessary for stability, and as we shall see it is never satisfied. Therefore, we are still in the position of having to compute all the normal modes of a star to see if it is stable. For this reason, although we know the way some instabilities behave in stars, it is still possible that there are others we do not suspect. Indeed, new classes of instabilities have recently been found (Balbinski 1985, Papaloizou & Pringle 1984).

In some respects, the relativistic problem is easier: the destabilizing influence of gravitational radiation makes it possible to show that positive-definiteness of the relativistic counterpart of C is in fact necessary and sufficient for stability. Here, the fact that C is not positive means that all stars are formally unstable. But because the growth rate of the instability depends on the efficiency with which gravitational waves can carry energy away, the importance of this instability depends strongly on how relativistic the star is. Moreover, the instabilities of Newtonian stars should still be present as well in the relativistic case. Again, therefore, to answer the physical questions we have to compute normal modes.

Many of these difficulties have been addressed outside the context of astrophysics, especially in meteorology and oceanography. Useful references are Drazin & Reid (1981), Greenspan (1968), and Holm, *et al* (1985).

THE MACLAURIN SPHEROIDS

I shall begin our study of rotating stability by describing the modes and instabilities of the simplest self-gravitating rotating system in Newtonian theory, the Maclaurin spheroid. This is not just because it is well to begin simply; it is also because these models are still the only sequence of models that has received extensive study. What intuition we have about rotational instabilities in astrophysics, we have to a large extent developed in the Maclaurin spheroids.

The Maclaurin spheroids are axisymmetric models of uniform density ρ and uniform angular velocity Ω. When the equilibrium equations are solved in Newtonian gravity, one finds that the surface is a perfect ellipsoid. These models are obviously a crude approximation to realistic stars, but the instabilities we see in them also seem to be present in compressible models. For reviews of their structure and some aspects of their stability, see Lyttleton (1953) and Chandrasekhar (1969). Their perturbations were first studied by Bryan (1889), who obtained a complete analytic solution to the problem: the mode equation separates in ellipsoidal harmonics, and the eigenfrequency equation is a polynomial. Unfortunately, the problem was by and large too difficult to compute by hand, and so many features of the solutions of the eigenfrequency equation remained undiscovered until the work of Comins (1979a,b).

The nonaxisymmetric modes

I shall restrict myself to the discussion of the nonaxisymmetric modes; axisymmetric modes and stablility are easier to treat but (apparently) less interesting. The modes which seem easiest to destabilize for a given spheroidal index l are those with $l = m$. Figure 5 (next page) shows the behavior of the pattern speed ω_p of the $m = 2$ and $m = 5$ modes, with various significant points indicated.

First note the general shapes of the curves. The vertical axis is normalized to Ω, which goes to zero as e approaches zero. (Therefore, the curves near the left margin go to infinity; this way of displaying frequencies is not the best for slow rotation.) For very slow rotation, the curve for any fixed m is fairly symmetrical about the horizontal axis. This is because the nonrotating star has a single eigenvalue ω^2, which gives equal values $\pm|\omega|$ for the forward- and backward-going modes. These two branches respond differently to increasing rotation. The backward-going mode is dragged forward by the rotation of the star, and eventually it is pulled so far that it goes forward in the inertial frame, although still backward relative to the star ($\omega_p/\Omega < 1$). Its forward-going counterpart also gets dragged faster forward, but not as much as the rotation of the star, so ω_p/Ω decreases.

At some point the two modes join. This is the onset of instability: I have not shown it in the diagram, but after this the modes have complex-conjugate eigenfrequencies. This signature, that instability in Newtonian theory sets in by the merging of two related eigenfrequencies, turns out to be very general (Schutz 1980b). This instability is called the *dynamical instability*, and it typically grows on the dynamical time scale, one rotation period.

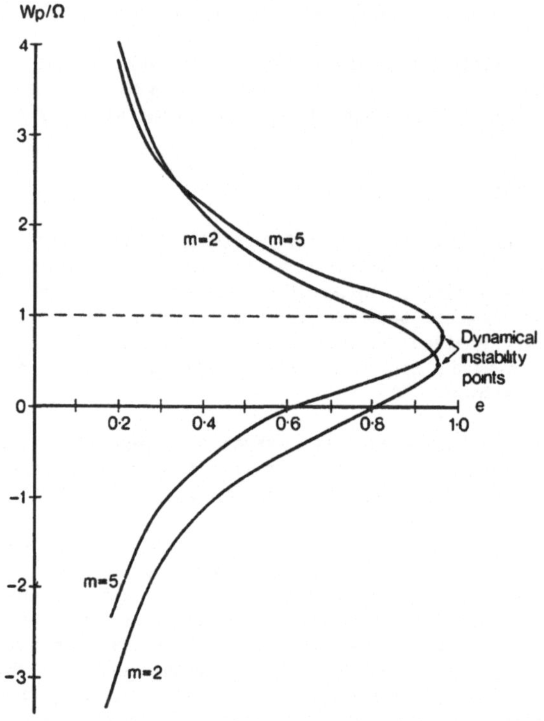

Fig. 5. The pattern speeds of the m = 2 and m = 5 modes of the Maclaurin spheroids as a function of the eccentricity e of the surface of the unperturbed star; e increases monotonically with angular momentum. Various important stability points are explained in the text. [From Schutz (1983).]

The secular instabilities

The addition of a small amount of viscosity or gravitational radiation reaction to the problem can induce other instabilities, which are called *secular instabilities* because their (long) time scale depends on the size of the added effect (the coefficient of viscosity or the ratio GM/Rc^2). The viscous instability was first studied by Roberts & Stewartson (1963), who showed that the instability sets in when a mode has zero frequency in the frame rotating with the star. This means that its pattern speed will be the same as Ω in the inertial frame, so the signal for instability is that the forward-going mode should drop below the horizontal line at $\omega_p/\Omega = 1$. Clearly, the m = 2 instability occurs before the m = 5, and this is part of a general pattern: the viscous secular instability sets in first for m = 2, and after that for each successive m in turn. Note that stars that are dynamically stable still can be secularly unstable to viscosity.

The gravitational-radiation-induced secular instability was discovered by Chandrasekhar (1970), and had been completely unexpected. In retrospect, we will see that it has an easy explanation. It sets in when the backward-going mode is dragged forward in the inertial frame, so the instability point is where the bottom branch of a curve in Fig.5 goes through $\omega_p = 0$. Chandrasekhar calculated where the m = 2 instability point was, and made the reasonable assumption that it represented the onset of this secular instability along the sequence. But Fig.5 shows a different story: the m = 5 mode goes unstable earlier. This is again part of a general pattern. The m = 2 mode is actually the *last* to go unstable, while in any Maclaurin spheroid there is some value of m such that all m larger than this are unstable. Therefore every Maclaurin spheroid is unstable. This is a special case of the generic gravitational radiation instability of all rotating stars discovered by Friedman & Schutz (1978b) and proved rigorously in general relativity by Friedman (1978a).

The Maclaurin spheroids also provide us with the "escape route" from this instability, that is, they tell us why it is not catastrophic for rotation in ordinary stars. Comins (1979b) shows that the growth time for the instability increases exponentially with m, so that even for very relativistic stars the instability in the modes for $m \geqslant 10$ or so grows too slowly to matter. Moreover, he shows that the growth rate is very sensitive to the compactness of the star, so that it is unimportant even for m = 2 in main-sequence stars. And finally, Detweiler & Lindblom (1977) and Lindblom & Detweiler (1977) showed that gravitational radiation and viscosity actually *compete*, the one acting to stabilize where the other destabilizes. Since viscosity is more effective on short length scales, it becomes relatively stronger with increasing m. It is no surprise, therefore, that Comins was also able to show that even a tiny viscosity would stabilize all m larger than some minimum, for a star of given compactness. Thus, since real stars have some viscosity, slowly rotating stars are stable even in principle.

The T/W criterion for instability

The general picture painted by the Maclaurin spheroids seems to be essentially unchanged for sequences of compressible stars. It is true that there are likely to be other instabilities as well, but at least the ones we see in the Maclaurin spheroids are also seen in calculations of compressible stars and disks, and even collisionless systems like model galaxies. A rule of thumb for calculating where some of these instabilities should set in emerged from one of the earliest extensive studies of differentially rotating, compressible stars, by Ostriker and colleagues (Tassoul & Ostriker 1968, Ostriker & Tassoul 1969, Ostriker & Bodenheimer 1973). If we let T stand for the rotational kinetic energy of the star and W for the absolute value of its gravitational potential energy, then it is found that the dynamical instability for m = 2 sets in when the ratio T/W is roughly 0.26. Since this may be the earliest significant dynamical instability on a sequence, this gives a rough idea of how far a sequence of rotating stars can reasonably be pushed. The m = 2 secular instability to gravitational radiation sets in when T/W is roughly 0.14. These criteria seem to be fairly robust, giving predictions to perhaps 10% for a wide variety of systems. But the T/W criterion does not extend as a good predictor to the more interesting gravitational radiation instabilities for m = 4 or 5. For these we either have to compute the normal modes or use the variational principle for zero-frequency modes of Newtonian stars recently discovered by Ipser & Managan (1985).

When, then, does the gravitational radiation instability matter? Only a mode calculation seems to be able to tell us. I will return to this question below, after we have understood more about how this instability operates.

A RELATIVISTIC APPROACH TO STABILITY

Having learned from the Maclaurin spheroids a number of things which we would like to understand better, we now turn to an analysis of stability in general relativity. In many ways the formalism of perturbation theory is more natural in a 4-dimensional context, and it is certainly true that it is easier to find general stability criteria in general relativity than in Newtonian gravity. We will return to the Newtonian case later. The approach I will take follows Friedman & Schutz (1975, 1978c), and has been reviewed by me in somewhat greater detail elsewhere (Schutz 1984).

Perfect fluids in general relativity

We will consider in detail only perfect fluid systems, although the main results can be extended to any dissipationless physical theory. The fluid has an equation of state of the form $p = p(\rho,S)$, a stress-energy tensor

$$T^{\alpha\beta} = (\rho+p)U^{\alpha}U^{\beta} + pg^{\alpha\beta}, \tag{20}$$

and dynamical equations

$$T^{\alpha\beta}{}_{;\beta} = 0, \tag{21}$$

$$(nU^{\alpha})_{;\alpha} = 0, \tag{22}$$

$$U^{\alpha}S_{,\alpha} = 0. \tag{23}$$

In fact, Eq.(21) and either of (22) or (23) imply the other. It is useful to define the *specific momentum* associated with a fluid element,

$$V_{\alpha} \equiv \frac{\rho+p}{n}U_{\alpha}. \tag{24}$$

In terms of the specific momentum, the vorticity conservation law has a very simple expression. If the fluid is *isentropic* (uniform entropy), then we have

$$\pounds_U(\nabla_{\alpha}V_{\beta} - \nabla_{\beta}V_{\alpha}) = 0. \tag{25a}$$

Here $\pounds_U$ is the Lie derivative with respect to the vector field U^{α}. For an introduction to the Lie derivative, see Carter's lectures in this volume or Schutz (1980c). It can be taken to be the ordinary partial derivative of a tensor's components if the coordinate system includes U^{α} as on of its coordinate basis vectors. If there is an entropy gradient in the fluid, then this law is replaced by the slightly weaker version,

$$\pounds_U(\nabla_{[\alpha}S\,\nabla_{\beta}V_{\gamma]}) = 0, \tag{25b}$$

where square brackets denote antisymmetrization. This is called *Ertel's theorem* (Ertel 1942a,b), and its relativistic form was found by Friedman (1978a).

Definition of a perturbation in terms of a sequence of solutions

Consider a smooth sequence of manifolds $M(\epsilon)$, each a sloution of Einstein's equations, the family parametrized by ϵ. Let $\epsilon = 0$ denote a *stationary* solution, which we call the unperturbed manifold. The other members of the sequence deviate more and more from the stationary state, but the limit $\epsilon \to 0$ is sufficiently differentiable in ϵ, in a sense which we won't need to define more precisely. Technically, this sequence is a trivial fiber bundle (Schutz 1980c), with base space R^1 (coordinate ϵ) and fiber R^4 (coordinates t,x,y,z). It is clear that in general there is no preferred or natural map from one spacetime in the sequence to another, no natural way to associate a point of a perturbed manifold with one of the unperturbed spacetime. Such associations are useful, however, so we imagine introducing a family of maps such that $f(\epsilon)$ maps points of the $\epsilon = 0$ manifold in a 1-1 fashion to points of $M(\epsilon)$.

We can use these maps to define a perturbation (Schutz & Sorkin 1977). Suppose there is a tensor field $Q(\epsilon)$ on each manifold, also smooth in the limit $\epsilon \to 0$ (for example, the metric tensor or the four-velocity). Then we can drag $Q(\epsilon)$ from any manifold $M(\epsilon)$ back to $\epsilon = 0$, thereby defining a tensor field $Q_f(\epsilon)^*$ on the unperturbed manifold. As ϵ varies, we therefore have a family of tensor fields on $M(0)$ which are the images under f of the family on the sequence. Given a sufficiently smooth family, then it will be approximated for small ϵ by the expansion

$$Q_f(\epsilon)^* = Q(0) + \epsilon\, \delta_f Q + O(\epsilon^2), \tag{26}$$

where we define

$$\delta_f Q \equiv \left.\frac{dQ_f(\epsilon)^*}{d\epsilon}\right|_{\epsilon=0} . \tag{27}$$

This is *defined* as the first-order perturbation in Q following f. It is a tensor field on $M(0)$. Clearly we could define second-order and higher perturbations in terms of higher derivatives with respect to ϵ along f. But without introducing f, it is impossible to define a perturbation as a tensor field on $M(0)$.

Since f is arbitrary, we could do the same with another 1-1 family of maps $h(\epsilon)$. Then there will be a different definition of a first-order perturbation, $\delta_h Q$. To see how this differs, let us consider a smooth family of *maps* $m(\epsilon;\lambda)$, such that $m(\epsilon;0) = f(\epsilon)$ and $m(\epsilon;1) = h(\epsilon)$. This family gives us a smooth transition from one map to the other. Now, if we hold ϵ fixed, say at ϵ_1, then in the manifold $M(\epsilon_1)$ the map $m(\epsilon_1;\lambda)$ traces out a one-parameter family of points as λ varies; *i.e.*, it defines a curve in $M(\epsilon_1)$. Let us denote the tangent vector to this curve as $\chi^\alpha(\epsilon_1;\lambda)$. Since all the maps reduce to the identity as $\epsilon \to 0$, χ^α goes to zero in that limit: it is a first-order quantity in ϵ. It therefore has an approximation in $M(0)$ in the spirit of Eq.(26) given by

$$\chi^\alpha(\epsilon;\lambda) = \epsilon\, \chi^\alpha{}_1(\lambda) + O(\epsilon^2). \tag{28}$$

Moreover, if the maps f and h are not too far apart for some ϵ_1, then $\chi^\alpha(\epsilon_1;\lambda)$ will be well approximated by $\epsilon_1 \chi^\alpha{}_1(0)$. This will always be true for sufficiently small ϵ_1, since the maps f and h approach each other in this limit. Therefore, keeping only the lowest order terms, it is not hard to show that the first-order perturbations are related by

$$\delta_h Q = \delta_f Q + \pounds_{\chi_1(0)} Q(0). \tag{29}$$

This change of the perturbation under a change of the map is called a *gauge transformation*. The gauge transformations of linearized theory (Misner, *et al* 1973, Schutz 1985) can be viewed in this framework.

Two preferred perturbations: Eulerian and Lagrangian

Although perturbation theory should be covariant under changes of gauge, some choices are singled out by either computational or physical considerations. Normally, tensor fields on $M(\epsilon)$ will be described in terms of some family of coordinate systems $\{x^\alpha(\epsilon)\}$ that is also smooth as $\epsilon \to 0$. I shall define the *Eulerian* map as the one which connects points that are at the same coordinate positions in the different manifolds. It is conventional to represent the perturbation with respect to this map simply by δ, with no subscript. Thus we have the component equations

$$\epsilon\ \delta g_{\alpha\beta}(x^\mu) \simeq g_{\alpha\beta}(x^\mu,\epsilon) - g_{\alpha\beta}(x^\mu,0),$$

$$\epsilon\ \delta\rho(x^\mu) \simeq \rho(x^\mu,\epsilon) - \rho(x^\mu,0),$$

etc. In the Newtonian theory, "Eulerian" has a somewhat different meaning, although in practice the difference is not usually important. Since Newtonian theory has a fixed Euclidean metric and a universal time, there exist isometries between the different manifolds. Choosing an isometry defines the Eulerian map. In practice one always chooses coordinates on the family $M(\epsilon)$ such that the isometry preserves the coordinates, in which case the relativistic and Newtonian definitions of Eulerian coincide. But in general relativity there is generally no isometry, so our definition is the only one possible.

The existence of a fluid in the manifolds allows us to define another map, called the *Lagrangian* map, which connects the "same" fluid elements in different manifolds. This is physically reasonable only in certain circumstances, namely where the sequence can be thought of as a deformation of a single system. More precisely, since the motion of the fluid preserves entropy, particle numbers, and vorticity (or Ertel's constant), then we shall require that the Lagrangian map also preserve these quantities. (That this usually defines the map almost uniquely was shown by Friedman & Schutz 1978a.) It is customary to denote the Lagrangian perturbation by Δ. The vector field χ relating δ to Δ by Eq.(29) is called the Lagrangian displacement vector field, because it can be interpreted as representing the first-order change in the position of a fluid element relative to the Eulerian map:

$$\Delta = \delta + \pounds_{\chi}. \tag{30}$$

The defining conditions for Lagrangian perturbations are therefore

$$\Delta S = 0, \qquad \Delta(nU^\alpha g^{1/2}) = 0, \qquad \text{and} \quad \Delta(\nabla_{[\alpha}S\ \nabla_\beta V_{\gamma]}) = 0, \tag{31}$$

where g is the absolute value of the determinant of the metric components.

Perturbations of Einstein's Equations

Using these definitions, Friedman & Schutz (1975, 1978c) were able to show that the first-order perturbed Einstein equations have an important symmetry property. The Eulerian perturbation of the field equations can be written

$$g^{-1/2}\delta(G^{\alpha\beta}g^{1/2}) = -\tfrac{1}{2}\,\epsilon^{\alpha\gamma\lambda\tau}\epsilon^{\beta\mu\sigma}{}_{\tau}\,\nabla_{(\gamma}\nabla_{\mu)}\delta g_{\lambda\sigma} + G^{\alpha\beta\lambda\sigma}\delta g_{\lambda\sigma}, \tag{32}$$

$$G^{\alpha\beta\lambda\sigma} \equiv \tfrac{1}{2}R^{\alpha(\lambda\sigma)\beta} - (\tfrac{1}{4}R^{\alpha(\lambda}g^{\sigma)\beta} + \tfrac{1}{4}R^{\beta(\lambda}g^{\sigma)\alpha} - \tfrac{1}{2}R^{\alpha\beta}g^{\lambda\sigma} - \tfrac{1}{2}R^{\lambda\sigma}g^{\alpha\beta})$$
$$+ \tfrac{1}{4}R(g^{\alpha\lambda}g^{\beta\sigma} + g^{\alpha\sigma}g^{\beta\lambda} - g^{\alpha\beta}g^{\lambda\sigma}). \tag{33}$$

Here round brackets denote symmetrization on a pair of indices. It is easy to see that the tensor $G^{\alpha\beta\lambda\sigma}$ has the symmetries

$$G^{\alpha\beta\lambda\sigma} = G^{(\alpha\beta)(\lambda\sigma)} = G^{\lambda\sigma\alpha\beta}, \qquad G^{\alpha\beta\lambda}{}_{\lambda} = 0. \tag{34}$$

Similarly, the Lagrangian perturbation of the stress-energy tensor density is

$$g^{-1/2}\Delta(T^{\alpha\beta}g^{1/2}) = W^{\alpha\beta\lambda\sigma}\Delta g_{\lambda\sigma}, \tag{35}$$

$$\Delta g_{\lambda\sigma} = \delta g_{\lambda\sigma} + \nabla_\lambda \chi_\sigma + \nabla_\sigma \chi_\lambda,$$

$$W^{\alpha\beta\lambda\sigma} = \tfrac{1}{2}(\rho+p)U^\alpha U^\beta U^\lambda U^\sigma + \tfrac{1}{2}p(g^{\alpha\beta}g^{\lambda\sigma} - g^{\alpha\lambda}g^{\beta\sigma} - g^{\alpha\sigma}g^{\beta\lambda})$$
$$- \tfrac{1}{2}\gamma p(g^{\alpha\beta} + U^\alpha U^\beta)(g^{\lambda\sigma} + U^\lambda U^\sigma). \tag{36}$$

This tensor has the same symmetries as $G^{\alpha\beta\lambda\sigma}$, Eq.(34), but not the traceless property.

The result of these symmetries (which reflect the fundamental fact that the combined hydrodynamical and Einstein equations can be derived from a variational principle, provided the Lagrangian conditions in Eq.(31) are respected by the variations -- see Schutz & Sorkin 1977) is that we have the following identity for any $\delta g_{\alpha\beta}$, χ^α, $\tilde{\delta g}_{\alpha\beta}$, $\tilde{\chi}^\alpha$, regardless of whether they satisfy the field equations or not:

$$16\pi\tilde{\chi}_\alpha\Delta(\nabla_\beta T^{\alpha\beta}) + \tilde{\delta g}_{\alpha\beta}\delta(G^{\alpha\beta} - 8\pi T^{\alpha\beta})$$
$$= 16\pi\chi_\alpha\tilde{\Delta}(\nabla_\beta T^{\alpha\beta}) + \delta g_{\alpha\beta}\tilde{\delta}(G^{\alpha\beta} - 8\pi T^{\alpha\beta}) + \nabla_\alpha R^\alpha, \tag{37}$$

where R^α is bilinear in $\delta g_{\alpha\beta}$ and $\tilde{\delta g}_{\alpha\beta}$.

<u>A stability criterion</u>

Now we shall see how to use Eq.(37) to derive a stability criterion. We suppose that $h_{\alpha\beta}$ and ψ^α are the eigenfunctions of a normal mode solution of the first-order perturbed field equations, with complex frequency ω, so that the time-dependent solutions are

$$\chi^\alpha(t,x^k) = \psi^\alpha(x^k)e^{i\omega t}, \qquad \delta g_{\alpha\beta}(t,x^k) = h_{\alpha\beta}(x^k)e^{i\omega t}.$$

We define the tilde-perturbation in terms of this eigenfunction:

$$\tilde{\chi}^\alpha(t,x^k) = [\psi^\alpha(x^k)]^*\, e^{-i\omega t}, \qquad \tilde{\delta g}_{\alpha\beta}(t,x^k) = [h_{\alpha\beta}(x^k)]^*\, e^{-i\omega t}.$$

Here * denotes the complex conjugate. Notice that χ^α and $\tilde{\chi}^\alpha$ are not complex conjugates of each other, since ω is not real. When substituted into Eq.(37), the time-dependences of these functions cancel and the complex-conjugate relationships of the eigenfunctions combine with the symmetry of the equation to give the following:

$$\omega^2 A - \omega(iB) - C + \nabla_k R^k = 0, \tag{38}$$

where A, iB, and C are *real* (*i.e.*, by virtue of the symmetry in Eq.(37), they are Hermitian forms).

Now suppose we integrate Eq.(38) over a spacelike hypersurface which is asymptotically null outgoing, so that it intersects future null infinity. Then

Friedman & Schutz (1975, 1978c) have shown that the integral over the sphere at infinity becomes, with an outgoing-wave boundary condition,

$$\oint R^k n_k dS = -4i\omega|\text{Bondi news function}|^2. \tag{39}$$

Because the r.h.s. of this equation is pure imaginary if ω is real, we immediately have from Eqs.(38) and (39) that *ω can be real only if it is zero.* But now imagine a sequence of unperturbed stellar models, such as a relativistic version of the Maclaurin spheroids. If a particular mode is stable somewhere along the sequence, then it can go unstable only by going through a real value of ω. By our result here, this can only be zero. *Instability in modes sets in only through zero frequency.* Moreover, if somewhere along a sequence the form C is positive-definite for all possible eigenfunctions ($h_{\alpha\beta}$, χ^{α}), then instability occurs at the first point along the sequence where C becomes semi-definite. (It is easy to see that this is necessary by substituting $\omega = 0$ into Eq.(38). To show that it is sufficient requires more work.)

This would be a very powerful stability criterion for perfect fluid configurations, except for one unfortunate fact: it turns out that C is not positive definite for any star, because the term $-W^{\alpha\beta\lambda\sigma}[\Delta g_{\alpha\beta}]^*\Delta g_{\lambda\sigma}$ in C contains derivatives with respect to ϕ quadratically, and these are negative-definite. One can always choose the trial function and the azimuthal eigenvalue m in such a way that this term dominates all others. Therefore the simple idea of a sequence going unstable at some point does not hold. One can still show, however, that the indefiniteness of C does imply the existence of unstable modes (Friedman 1978a). These are the gravitational-wave-excited modes we encountered in the Maclaurin spheroids. The generality of our treatment shows that they are a feature of all compressible stars as well.

There are two remarks on the general problem which I wish to make. One is that the proof of instability offerred here is mode-based. It would be more satisfying and perhaps illuminating if we had one which did not rely so heavily on modes. The second is that the only thing we needed to make the proof work was the symmetry property of $W^{\alpha\beta\lambda\sigma}$, and that followed from the fact that the theory has an action principle. Many other systems therefore have the same basic stability theorem, that instablility sets in through a zero frequency mode: see for example Carter & Quintana (1972) for a discussion of elastic media in general relativity.

A SIMPLE APPROACH TO THE RADIATION INSTABILITY

Now that we have seen that the radiation-excited instability of the Maclaurin spheroids is in fact a general feature of relativistic systems, we should expect to be able to find some very general and fundamental way of understanding it even in a Newtonian context. The conserved quantities of the Newtonian problem prove to be the key to such an understanding.

The equations for a perturbation of a rotating Newtonian perfect fluid, when written in their Lagrangian form (Lynden-Bell & Ostriker 1967), follow from a variational principle. From the action we can derive, via Noether's theorem, two interesting conserved quantities: the canonical energy E_c (the value of the Hamiltonian) and the canonical angular momentum J_c. Both are conserved if the unperturbed star is stationary and axisymmetric. Both are quadratic in the Lagrangian displacement vector χ^i. (The fact that they are quadratic might suggest that they are the second-order changes in the energy and angular momentum of the system. In fact, however, they are only *part* of these changes. We will examine this in more detail in the next section.)

Conserved quantities for wave fields

These conserved quantities have such useful properties that it is worthwhile studying them in some detail. Let us consider *any* dynamical system for a field χ^i whose equations follow from the unconstrained variations of a Lagrangian of the form

$$L = \tfrac{1}{2}\,\chi^i_{\ ,t}\,A_{ij}(\chi^j_{\ ,t}) + \tfrac{1}{2}\,\chi^i_{\ ,t}\,B_{ij}(\chi^j) - \tfrac{1}{2}\,\chi^i C_{ij}(\chi^j), \tag{40}$$

where A_{ij} and C_{ij} are selfadjoint operators, B_{ij} is antiselfadjoint, and all are independent of time t and azimuthal angle ϕ. These properties will hold for the perturbations of essentially any conservative, stationary, and axisymmetric system. If the field is a tensor rather than a vector, similar properties will still obtain if we interpret i and j as multi-indices.

The field equations that follow are

$$0 = A_{ij}(\chi^j_{\ ,tt}) + B_{ij}(\chi^j_{\ ,t}) + C_{ij}(\chi^j). \tag{41}$$

The conserved energy is (for possibly complex solutions, such as normal modes)

$$E_c = \tfrac{1}{2}\int [\chi^*_{i,t} A^i_{\ j}(\chi^j_{\ ,t}) + \chi^*_i C^i_{\ j}(\chi^j)]\,d^3x, \tag{42}$$

while the conserved angular momentum is

$$J_c = -\mathrm{Re}\int \chi^*_{i,\phi}[A^i_{\ j}(\chi^j) + \tfrac{1}{2} B^i_{\ j}(\chi^j)]\,d^3x. \tag{43}$$

Both of these are closely related to and derivable from a simpler conserved quadratic form, which I shall call the *symplectic form*: given any two independent fields χ^i and ψ^i, we define

$$W(\chi,\psi) = \int \{\psi^*_i[A^i_{\ j}(\chi^j_t) + \tfrac{1}{2} B^i_{\ j}(\chi^j)] - \chi^*_i[A^i_{\ j}(\psi^j_t) + \tfrac{1}{2} B^i_{\ j}(\psi^j)]\}\,d^3x. \tag{44}$$

Notice that the terms in square brackets in this equation are the canonical momenta conjugate to χ^i and ψ^i, respectively, so that this is just the antisymmetric product of the canonical coordinates and momenta. It obviously has a close relation to the Poisson bracket, and so we should not be surprised if it is also related to the conservation laws. (This relationship is discussed in more detail in Schutz 1980c.) In fact, if both χ^i and ψ^i satisfy the dynamical equation, then their symplectic product is conserved. Moreover, if χ^i satisfies the dynamical equation, then so do its derivatives $\chi^i_{\ ,t}$ and $\chi^i_{\ ,\phi}$, and the associated conserved quantities are none other than E_c and J_c:

$$E_c = \tfrac{1}{2}\,W(\chi_{,t},\chi), \qquad J_c = -\tfrac{1}{2}\,W(\chi_{,\phi},\chi). \tag{45}$$

Now we come to the fundamental result we have been building up to. Since a normal mode solution is proportional to $\exp(im\phi+i\omega t)$, the derivatives in Eq.(45) are trivial, and we immediately find that for any normal modes,

$$mE_c = -\omega J_c. \tag{46}$$

This has a number of consequences: (i) If ω is complex, then the fact that m, E_c and J_c are real means that $E_c = J_c = 0$. (ii) If ω is real, then consequence (i) and a continuity assumption imply that a mode is marginally stable if and only if $W(\chi^*,\chi) = 0$. This is an *intrinsic* characterization of marginal stability, and is as close as we have come to a stability criterion for rotating Newtonian stars. But since it is mode-based, it is not very useful. This criterion for marginal stability is arrived at by very different methods in Schutz (1980b). (iii) For nonaxisymmetric modes, where $m \neq 0$, we have a relation that illustrates the importance of the pattern speed:

$$E_c = \omega_p J_c. \tag{47}$$

This is a general property of linear waves, be they stellar perturbations or gravitational waves.

Mechanism for the gravitational wave instability

This equation will be the basis of our understanding of the way the gravitational wave instability develops along a sequence of stellar models. If we remind ourselves that J_c is at least part of the second-order change in the angular momentum of the star, then it will perhaps not be hard to believe that the sign of J_c will be determined by the relative rotation rates of the star and the wave's pattern: if the mode goes faster than the star then it has positive angular momentum, and conversely if it is slower then it has negative angular momentum. (This is not exactly true, but becomes more and more accurate as m gets larger. Rigorous bounds on the range of pattern speeds in which J_c can change sign are given in Friedman & Schutz 1978b.) On the other hand, the relative sign of E_c and J_c is determined by the sign of ω_p, the mode's pattern speed in the inertial frame. If we denote the star's angular velocity by $\Omega > 0$ (in differentially rotating stars, this should be taken to be the *mean* of the angular velocity over the mode's eigenfunction), we have the following table of signs:

	J_c	E_c
$\omega_p > \Omega > 0$	+	+
$\Omega > \omega_p > 0$	-	-
$\Omega > 0 > \omega_p$	-	+

The crucial entry is the one where E_c is negative: *if a mode rotates forwards in the inertial frame and backwards relative to the star, then its canonical energy will be negative.* This is exactly the region of instability of the gravitational wave excited modes of the Maclaurin spheroids, as in Fig.5. Why does this signal an instability of a mode when it is coupled to radiation, but not necessarily when it obeys simply the Newtonian equations? The answer is that since the Newtonian equations preserve E_c, its sign is not automatically an indication of stability: it is necessary that the symplectic form change sign, not the energy. But if the Newtonian system is coupled to another system in such a way that E_c must decrease with time, then if E_c is already negative its absolute value will increase, and the coupling will make the mode unstable. This is excactly what happens when the mode is coupled to gravitational radiation. Gravitational waves also obey equations like (41) and (47), and far from the star their physical energy *is* their E_c. So conservation of total E_c and an outgoing-wave boundary condition ensure that the Newtonian E_c will decrease.

Gravitational wave instability as a two-stream instability

Readers familiar with hydrodynamical instabilities may recognize this instability mechanism, for this is just a version of the *two-stream instability*. Energy arguments like these were first used by Sturrock (1962) to explain the two-stream instability; the only difference was that there the linear momentum of the mode replaces the angular momentum in our argument. One finds that the unstable modes have pattern speeds intermediate between the speeds of the two fluids. In our case, the two "fluids" are the star and the nonrotating inertial vacuum outside it; when modes exist which rotate at an angular velocity intermediate between the two, the instability is present. Another related instability is the Kelvin-Helmholtz instability, which is the mechanism by which the wind raises the waves on the ocean. This is often discussed in a nonlinear context, but its initial linear development is that of a two-stream instability, with the unstable modes again being intermediate in speed between the two fluids. The ocean waves travel at a fixed speed. As long as the wind moves faster than this, the ocean waves will be unstable and grow.

This analogy can help us to understand at least one reason why all rotating stars are unstable. The waves in a rotating star will, in the short-wavelength limit, move at the speed of sound relative to the fluid of the star. If the star rotates "supersonically," so that in some region even the backward-going sound waves will go forward in the inertial frame, then it will be unstable. But all compressible perfect fluid stellar models have a speed of sound that goes to zero at the surface, so any star will be unstable if the surface has any finite angular velocity. Another way of seeing that we should expect this instability in all rotating stars is the observation we made in our discussion of the p-modes of nonrotating stars, that there are p-modes of arbitrarily small pattern speeds for sufficiently large m. If a star rotates slowly, then the p-modes will have the same pattern speeds relative to the star, and so these very slowly moving ones will go unstable.

Other ways of exciting the instability

Our explanation for this instability used only one property of gravitational radiation, that it causes E_c to decrease. Therefore it can be excited by any mechanism that does the same, and in particular by any coherent form of radiation. *Magnetic fields* in stars will emit electromagnetic radiation when the star is perturbed, but one can show that the energy radiated in electromagnetic waves is always smaller then that in gravitational waves, essentially because the magnetic field energy is always less than (usually much less than) the gravitational potential energy. Therefore this does not seem to be an important mechanism except where gravitational radiation is absent, as in the l=1 (dipole) modes of the star. This has not received any attention, to my knowledge. But a mechanism that might indeed be important is *acoustic radiation*. Imagine a collapsing stellar core surrounded by an envelope that is collapsing on a much longer time-scale. Then nonaxisymmetric perturbations of the core will "stir" the envelope and this will extract energy from the modes of the core. As long as the core rotates rapidly compared to the envelope (a condition likely to be easy to satisfy in a rotating collapse), then the energy argument will parallel the one for gravitational radiation. I have made a crude estimate of the size of this effect, and it may dominate the one due to gravitational radiation near the onset of instability (Schutz 1983).

The instability due to viscosity

Finally, we may ask about viscosity. In the Maclaurin spheroids, it does not have at all the same effect as gravitational radiation. Do our energy arguments help us here? Friedman & Schutz (1978b) have discussed this at some length, but the essence of the result is that viscosity dissipates energy measured in the rest frame of the star rather than in the inertial frame. It has no systematic effect on E_c, but rather on its analogue in the rotating frame. We shall see why in the next section. This fact can be used to devise as systematic a description of the viscosity secular instability as we have given of the one due to gravitational radiation.

THE PERTURBED ENERGY OF A ROTATING SYSTEM

In the previous section we used the conserved canonical energy to build stability criteria for Newtonian stars, while cautioning that it is not the whole of the second-order change in the energy of a rotating star. Indeed, our remark at the end about the effect of viscosity would not make sense if the canonical energy in the inertial frame were the total energy, since surely viscosity dissipates energy. But if E_c is not the total energy, what is the missing piece? And if E_c and the total energy are both conserved, then so must be their difference, so is there yet another conserved quantity in the problem that we have overlooked until now? In this section I shall address these questions directly. The best way to start is with the simplest system that exhibits these features: the energy of a particle in a perturbed, nearly circular orbit in a central force.

Orbiting particle: an elementary example

Consider a particle in a central potential $V(r)$ in an orbit with angular momentum J. Its energy is

$$E = \tfrac{1}{2}\, m\dot{r}^2 + J^2/2mr^2 + V(r), \tag{48}$$

which is a nonlinear function of r. The solution which is stationary in r is the circular orbit, which we will take to be the unperturbed state. For this we have

$$\dot{r} = 0 \quad \text{and} \quad \left.\frac{\partial E}{\partial r}\right|_{J=\text{const}} = 0. \tag{49}$$

In other words, a stationary solution is an extremum of the energy provided the angular momentum is held constant. Now suppose that the orbit is perturbed, but remains in the same plane for simplicity. The first-order change in the energy is

$$\delta E = \frac{\partial E}{\partial \dot{r}}\,\delta\dot{r} + \frac{\partial E}{\partial r}\,\delta r + \frac{\partial E}{\partial J}\,\delta J. \tag{50}$$

But the first two terms vanish because of Eq.(49), and it is easy to see that the third term is just

$$\delta E = \Omega\,\delta J, \tag{51}$$

where Ω is the angular velocity of the unperturbed orbit. Physically, it is clear why there *has* to be a term of this type: part of the unperturbed energy is kinetic, and it is possible to change the kinetic energy to *first* order by changing the component of the velocity *along* the unperturbed velocity. Notice that this change is conserved: we expect energy to be conserved at all orders, and at this order this follows from angular-momentum conservation.

The second-order change in the energy is simplest if we set the first-order change δJ to zero:

$$\delta^2 E = \Omega\,\delta^2 J + \tfrac{1}{2}\, m(\delta\dot{r})^2 + \tfrac{1}{2}\,[V''(r) + 3J^2/mr^4]\,(\delta r)^2. \tag{52}$$

The first term is the second-order analogue of Eq.(51), allowing for a second-order change in J. The remaining terms are the Hamiltonian of the radial oscillation that the particle's first-order perturbed motion undergoes. When $V(r)$ is the Kepler potential, then evaluating the coefficient of $(\delta r)^2$ gives an oscillation frequency exactly equal to Ω: the perturbed orbit is (of course) closed. In general this frequency is called the epicyclic frequency.

So here we see that even in the simplest example, the total energy at second order is not the Hamiltonian of the second-order motion. There is another piece, related to the second-order change in the angular momentum. Both pieces are separately conserved if the motion obeys the original dynamical equations.

The second-order energy of a rotating fluid

The energy of a fluid, perturbed about a stationary, differentially rotating state, has a strong analogy with Eqs.(51) and (52), but there are important differences that arise because the system is a continuum. To first order the change is (Schutz & Sorkin 1977)

$$\delta E = \int [nT\Delta S + \rho\Omega\Delta j + g^{-1/2}\mu\Delta(ng^{1/2})] d^3x, \tag{53}$$

where j is the specific angular momentum of the fluid, μ is the injection energy per particle ($\mu = \frac{1}{2}v^2 + h + V$, where h is the enthalpy), and Δ is the usual Lagrangian change. Each of these terms has a ready interpretation: the first is the energy change if we add heat; the second is the kinetic energy change, as in the particle analogy above; and the third is the energy added if we add particles. It is significant that the kinetic energy term involves the specific angular momentum, because it is possible to show that this term is *conserved* by virtue of the vorticity-conservation law of the perfect fluid (Schutz 1984). It is obvious that the evolution equations for a perfect fluid also preserve each of the other two terms separately as well, so the analogy with Eq.(51) is very good.

If we set the first-order changes ΔS, Δj, and of the vorticity to zero, then we have what we have earlier defined to be a Lagrangian perturbation of the fluid, and δE vanishes. Then the second-order change in E turns out to be (Friedman & Schutz 1978a, Schutz 1984)

$$\delta^2 E = \int [nT\Delta^2 S + \rho\Omega\Delta^2 j + g^{-1/2}\mu\Delta^2(ng^{1/2})] d^3x + E_C. \tag{54}$$

The first group of terms is just the analogue at second order of Eq.(53), and the remaining term is the canonical energy, the Hamiltonian of the first-order perturbation equations. This is very closely analogous to Eq.(52). The difference between E_c and $\delta^2 E$ is indeed a separately conserved quantity, provided the perfect-fluid dynamical equations are satisfied.

None of this is absolutely necessary for computing the gravitational wave instability, but it does illuminate the viscosity instability. Viscosity does not conserve vorticity, so the second term in Eq.(54) is not constant. The total $\delta^2 E$ must decrease, but it can (and does) sometimes happen that the vorticity term decreases faster, allowing E_c to increase. A further investigation of the second-order changes in J reveals that the canonical energy in the rotating frame of the star monotonically decreases under the action of viscosity, and that this happens when a forward-going mode starts going backwards in the rotating frame. (It is possible to speak of a single rotating frame here, because we must exclude the possibility that the unperturbed star is differentially rotating. In the presence of viscosity, only a rigidly rotating star can be stationary.)

MAXIMUM ROTATION RATE OF NEUTRON STARS

The most immediate consequence of the gravitational radiation instability is that no star can rotate faster than whatever rate would render it unstable on a sufficiently short timescale. This limit is set by the largest value of m for which a mode grows on an interesting timescale (such as the age of the universe), since the larger values of m are unstable in the more slowly rotating stars but have the longer growth times. Unfortunately, this is a difficult calculation to make, since it involves knowing the normal modes of the star, their growth rates when coupled to gravitational radiation, and the size of the damping effect of viscosity. For main sequence stars and white dwarfs, the gravitational radiation timescales are too long to be of interest. But for neutron stars it seems that even modes as high as $m = 4$ or 5 may be important. Friedman (1983) estimated the likely effect of viscosity in neutron stars and concluded that the $m = 4$ mode would set the limit on rotation, since the $m = 5$ mode would be damped by viscosity. He used the Maclaurin spheroids to estimate the various timescales. Based on these estimates, Friedman concluded that the existence of the millisecond pulsar PSR 1937+214 with a rotation rate of some 642 Hz might already rule out the stiffer equations of state if the star has a baryon mass of 1.4 $M_\odot$.

Realistic models of rotating neutron stars, using the same equations of state as Arnett & Bowers (1977) used, have recently been calculated by Friedman, *et al* (1984). They concluded again that the stiffest equations of state were on the verge of being ruled out if the millisecond pulsar is a 1.4 $M_\odot$ star. Conversely, if the millisecond pulsar's rotation rate is limited by this instability, then the equation of state must be fairly stiff.

In a very recent paper, Lindblom (1986) has taken a new look at the relative importance of gravitational radiation and viscosity in realistic stars rather than in the Maclaurin spheroids, and concluded that Friedman (1983) may have overestimated the effect of viscosity, and that consequently the rotation limit may in fact be set by the $m = 5$ mode. This significantly lowers the critical rotation rate and rules out the two stiffest equations of state used by Arnett & Bowers (1977), again provided that the millisecond pulsar has a baryon mass of 1.4 $M_\odot$. Figure 6 shows the result of his calculations for various assumed kinematic viscosities ν, in units of $cm^2 s^{-1}$. But these results must also be treated with some caution, because the stellar models which he used were nonrotating, and rotation may make some changes in one's estimates of the critical quantities. We need full calculations of the normal modes of realistic rotating relativistic stars to answer these questions, and we don't have any yet. This is probably the most important untouched problem in this subject today. Not only would it help us to constrain equations of state, but such calculations would be a useful testbed for comparison with the results of full nonlinear three-dimensional hydrodynamics codes in general relativity, which will certainly need such comparison problems to ensure that their results are reliable.

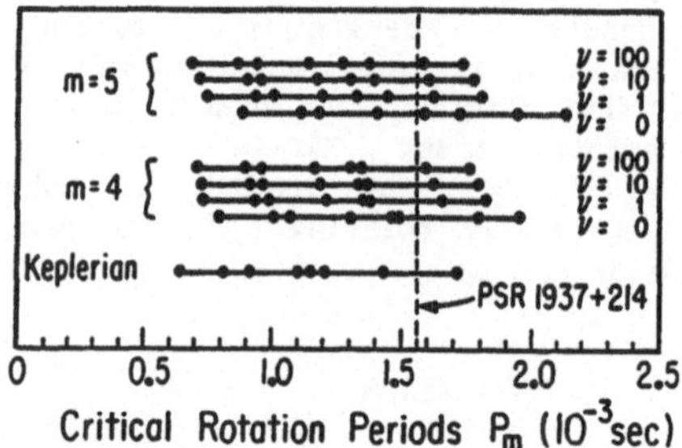

Fig. 6. Critical rotation periods of neutron stars with various kinematic viscosities (ν) and various equations of state (dots) for the $m = 4$ and 5 modes. The millisecond pulsar (dashed line) appears to rule out the stiffest equations of state. (From Lindblom 1986.)

STABILITY OF THE KERR BLACK HOLE

Just as the stability of rotating stars is much harder to analyze than that of nonrotating ones, so is the stability of Kerr more difficult than that of Schwarzschild. There is an unpublished calculation by Whiting (1985) that is reliably said to establish that all the normal modes of Kerr are stable, but I have not seen it. However, because there is some uncertainty even for Schwarzschild about whether the modes are complete (or even finite in number for any l), the stability of the modes of Kerr does not establish the stability of the metric itself. This reinforces the importance of studying the completeness problem for radiating systems.

The fullest discussion of this problem in the literature is by Chandrasekhar (1983). The analysis of Leaver (1986a,b) has made a substantial advance, and gives hope of further important progress soon. I have elsewhere given a brief introduction to the mode problem from a point of view analogous to that which I took earlier in this article for Schwarzschild (Schutz 1984). There is no space here for a full discussion of this interesting and complex problem, but I should not leave it without drawing attention to its relation to the rotating star problem. The possibility of instability in Kerr comes from the negative-energy nonaxisymmetric modes of wave fields that must exist because of the ergosphere. (The axisymmetric modes of Kerr are known to be stable: Friedman & Schutz 1973.) When a star has an ergosphere (Schutz & Comins 1978), these modes *do* result in an instability, for the same reason as in the star: as they lose energy to infinity, their already negative energy must get more negative, hence larger in absolute value (Friedman 1978b, Comins & Schutz 1978). But the boundary conditions for the Kerr problem are different: the ingoing waves at the horizon have, for these negative-energy modes, an *outward* energy flux, which in all calculated modes seems to more than compensate for the energy radiated to infinity, and allows the mode amplitude to decrease. The question is, does this happen for every wave disturbance of Kerr? Nobody yet knows, but all the evidence is that it does.

REFERENCES

Abramowicz, M. A., & Wagoner, R. V., 1987, Astrophys. J., 226:1063.

Arnett, W. D., & Bowers, R. L., 1977, Astrophys. J. Suppl., 33:415.

Balbinski, E. F. L., 1985, Mon. Not. R. astr. Soc., 216:897.

Balbinski, E., Detweiler, S., Lindblom, L., & Schutz, B. F., 1985, Mon. Not. R. astr. Soc., 213:553.

Barston, E. M., 1967a, J. Math. Phys., 8:523.

Barston, E. M., 1967b, J. Math. Phys., 8:1886.

Barston, E. M., 1968, J. Math. Phys., 9:2069.

Barston, E. M., 1971a, J. Math. Phys., 12:116.

Barston, E. M., 1971b, J. Math. Phys., 12:1867.

Barston, E. M., 1972, J. Math. Phys., 13:720.

Barston, E. M., 1974, J. Math. Phys., 15:675.

Barston, E. M., 1977,, J. Math. Phys., 18:750.

Blome, H.-J., & Mashhoon, B., 1984, Phys. Lett., 100A:231.

Bognar, J., 1974, "Indefinite inner product spaces," Springer, Berlin.

Bryan, G. H., 1889, Proc. Roy. Soc. London, A180:187.

Carter, B., & Quintana, H., 1972, Proc. Roy. Soc. London, A331:57.

Chandrasekhar, S., 1939, "Stellar structure," Dover, New York.

Chandrasekhar, S., 1964, Astrophys. J., 140:417.

Chandrasekhar, S., 1965, Astrophys. J., 142:1519.

Chandrasekhar, S., 1969, "Ellipsoidal figures of equilibrium," Yale University Press, New Haven.

Chandrasekhar, S., 1970, Phys. Rev. Lett., 24:611.

Chandrasekhar, S., 1975, Proc. Roy. Soc. London, A343:289.

Chandrasekhar, S., 1983, "The mathematical theory of black holes," Oxford University Press, Oxford.
Chandrasekhar, S., & Detweiler, S. L., 1975, Proc. Roy. Soc. London, A344:441.
Chandrasekhar, S. & Friedman, J. L., 1972, Astrophys. J., 176:745.
Comins, N. 1979a, Mon. Not. R. astr. Soc., 189:233.
Comins, N. 1979b, Mon. Not. R. astr. Soc., 189:255.
Comins, N., & Schutz, B. F., 1978, Proc. Roy. Soc. (London), A364:211.
Cox, J. P., 1980, "Theory of Stellar Pulsation," Princeton University Press, Princeton.
Detweiler, S. L., 1979, in: "Sources of Gravitational Radiation," L. Smarr, ed., Cambridge University Press, Cambridge.
Detweiler, S. L., & Lindblom, L., 1977, Astrophys. J., 213:193.
Drazin, P. G., & Reid, W. H., 1981, "Hydrodynamic stability," Cambridge University Press, Cambridge.
Dyson, J. F., 1980, Ph.D. thesis, University College Cardiff, Cardiff, Wales.
Eisenfeld, J., 1969, J. math. anal. appl., 26:357.
Ertel, H., 1942a, Phys. Zs. Leipzig, p.526.
Ertel, H., 1942b, Meteorolog. Zs. Braunschweig, p.277.
Ferrari, V., & Mashhoon, B., 1984, Phys. Rev. D, 28:2929.
Finn, L. S., 1986, Mon. Not. R. astr. Soc., 222:393.
Fowler, W. A., 1964, Rev. Mod. Phys., 36:545.
Friedman, J. L., 1978a, Commun. Math. Phys., 62:247.
Friedman, J. L., 1978b, Commun. Math. Phys., 63:243.
Friedman, J. L., 1983, Phys. Rev. Lett., 51:11.
Friedman, J. L., Ipser, J. R., & Parker, L., 1984, Nature, 312:255.
Friedman, J. L., & Schutz, B. F., 1973, Phys. Rev. Lett., 32:243.
Friedman, J. L., & Schutz, B. F., 1975, Astrophys. J., 200:204.
Friedman, J. L., & Schutz, B. F., 1978a, Astrophys. J., 221:937.
Friedman, J. L., & Schutz, B. F., 1978b, Astrophys. J., 222:881.
Friedman, J. L., & Schutz, B. F., 1978c, Astrophys. J., 222:1119.
Goebel, C. J., 1972, Astrophys. J., 172:L95.
Greenspan, H. P., 1968, "The theory of rotating fluids," Cambridge University Press, Cambridge.
Harrison, B. K., Thorne, K. S., Wakano, M., & Wheeler, J. A., 1965, "Gravitation Theory and Gravitational Collapse," University of Chicago Press, Chicago.
Hartle, J. B., 1967, Astrophys. J., 150:1005.
Hartle, J. B., & Munn, M. W., 1975, Astrophys. J., 198:467.
Hartle, J. B., & Thorne, K. S., 1968, Astrophys. J., 153:807.
Hartle, J. B., & Thorne, K. S., 1969, Astrophys. J., 158:719.
Hartle, J. B., Thorne, K. S., & Chitre, S. M., 1972, Astrophys. J., 176:177.
Holm, D. D., Marsden, J. E., Raitu, T., & Weinstein, A., 1985, Physics Reports, 123:1.
Ipser, J. R., 1969, Astrophys. J., 158:17.
Ipser, J. R., & Horowitz, G., 1979, Astrophys. J.,232:863.
Ipser, J. R., & Managan, R. A., 1985, Astrophys. J., 292:517.
Islam, J. N., 1970, Mon. Not. R. astr. Soc., 150:237.
Iyer, S., 1986, in preparation.
Iyer, S., & Will, C. M., 1986, submitted for publication.
Kokkotas, K., & Schutz, B. F., 1986, Gen. rel. gravit., to appear.
Laval, G., Mercier, C., & Pellat, R., 1965, Nucl. Fusion, 5:156.
Leaver, E. W., 1985, Proc. Roy. Soc. London, A402:285.
Leaver, E. W., 1986a, J. Math. Phys., 27:1238.
Leaver, E. W., 1986b, Phys. Rev.D, 34:384.
Lebovitz, N. R., 1966, Astrophys. J., 146:946.
Ledoux, P., 1958, Handb. der Physik, 51:605.
Ledoux, P., & Walraven, Th., 1958, Handb. der Physik, 51:353.
Lindblom, L., 1986, to be published.
Lindblom, L., & Detweiler, S. L., 1977, Astrophys. J., 211:565.
Lindblom, L., & Detweiler, S. L., 1983, Astrophys. J. Suppl., 53:73.
Lynden-Bell, D., & Ostriker, J. P., 1967, Mon. Not. R. astr. Soc., 136:293.

Lyttleton, R. A., 1953, "The stability of rotating liquid masses," Cambridge University Press, Cambridge.
Misner, C. W., Thorne, K. S., & Wheeler, J. L., 1973, "Gravitation," Freeman & Co., San Francisco.
Oppenheimer, J. R., & Serber, R., 1938, Phys. Rev., 54:530.
Ostriker, J. P., & Bodenheimer, P., 1973, Astrophys. J., 180:171.
Ostriker, J. P., & Tassoul, J.-L., 1969, Astrophys. J., 155:987.
Papaloizou, J. C. B., & Pringle, J., 1984, Mon. Not. R. astr. Soc., 208:721.
Price, R. H., 1972, Phys. Rev. D, 5:2419.
Regge, T., & Wheeler, J. A., 1957, Phys. Rev., 108:1063.
Roberts, P. H., & Stewartson, K., 1963, Astrophys. J., 137:777.
Schutz, B. F., 1980a, Mon. Not. R. astr. Soc., 190:7.
Schutz, B. F., 1980b, Mon. Not. R. astr. Soc., 190:21.
Schutz, B. F., 1980c, "Geometrical methods of mathematical physics," Cambridge University Press, Cambridge.
Schutz, B. F., 1983, Lect. in Appl. Math., 20:99.
Schutz, B. F., 1984, in "Relativistic astrophysics and cosmology," X. Fustero & E. Verdaguer, eds., World Scientific, Singapore.
Schutz, B. F., 1985, "A First Course in General Relativity," Cambridge University Press, Cambridge.
Schutz, B. F., & Comins, N., 1978, Mon. Not. R. astr. Soc., 182:69.
Schutz, B. F., & Sorkin, R., 1977, Ann. Phys. (N.Y.), 107:1.
Schutz, B. F., & Will, C. M., 1985, Astrophys. J., 291:L33.
Shapiro, S. L., & Teukolsky, S. A., 1983, "Black holes, white dwarfs, and neutron stars," Wiley, New York.
Shapiro, S. L., & Teukolsky, S. A., 1985a, Astrophys. J., 292:L41.
Shapiro, S. L., & Teukolsky, S. A., 1985b, Astrophys. J., 298:34.
Shapiro, S. L., & Teukolsky, S. A., 1985c, Astrophys. J., 298:58.
Shapiro, S. L., & Teukolsky, S. A., 1986, in "Dynamical Spacetimes and Numerical Relativity," J. Centrella, ed., Cambridge University Press, Cambridge.
Sorkin, R. D., 1982, Astrophys. J., 257:847.
Sturrock, P. A., 1962, in: "Plasma hydromagnetics," D. Berhader, ed., Stanford University Press, Stanford.
Tassoul, J.-L., 1978, "Theory of rotating stars," Princeton University Press, Princeton.
Tassoul, J.-L., & Ostriker, J. P., 1968, Astrophys. J., 154:613.
Thorne, K. S., 1966, Astrophys. J., 144:201.
Thorne, K. S., 1968, Phys. Rev. Lett., 21:320.
Thorne, K. S., 1969, Astrophys. J., 158:1.
Thorne, K. S., & Campolattaro, A., 1967, Astrophys. J., 149:591.
Unno, W., Osaki, Y., Ando, H., & Shibahashi, H., 1979, "Nonradial oscillations of stars," University of Tokyo Press, Tokyo.
Vishveshwara, C. V., 1970, Phys. Rev. D, 1:2870.
Whiting, B., 1985, unpublished.
Zel'dovich, Ya. B., & Novikov, I. D., 1971, "Stars and relativity," (eds. Thorne, K. S. & Arnett, W. D.), University of Chicago Press, Chicago.
Zerilli, F. J., 1970, Phys. Rev. D, 2:2141.

ACCRETION AND COLLAPSE

D. Lynden-Bell

Clare College and Institute of Astronomy
The Observatories
Cambridge, CB3 OHA

I. THE GRAVOTHERMAL CATASTROPHE

1. Specific Heats

There is a three-line proof that all specific heats are positive. In a canonical ensemble of systems at equilibrium, the average energy of a system is given by the sum over all energy levels i of the system

$$\langle E\rangle = \sum_i E_i\, e^{-\beta E} / \Sigma e^{-\beta E_i}$$

where $\beta = (kT)^{-1}$. Evidently $C_V = d\langle E\rangle/dT = -k\beta^2 \frac{d}{d\beta}\langle E\rangle$ and so

$$\beta^{-2} C_V = k[\sum_i E_i{}^2 e^{-\beta E_i}/\sum_i e^{\beta E_i} - (\Sigma E_i e^{-\beta E_i}/\Sigma e^{-\beta E_i})^2] = k(\langle E^2\rangle - \langle E\rangle^2) = k\langle(E-\langle E\rangle)^2\rangle$$

which is clearly positive.

However, all astronomers know that stars, star clusters and other self-gravitating systems such as black holes, behave as systems of negative specific heat. Take energy out, they shrink and get hotter; put energy in, they expand and get cooler. This is perfectly clear from the Virial theorem which for an isolated system in dynamical equilibrium reads

$$2\tau + V = 0$$

Hence $\quad E = \tau + V = -\tau$

Here τ and V are the kinetic and potential energies. But if the system consists of N particles in motion and the temperature is T, the kinetic energy is (3/2)NkT.

Hence $\quad dE/dT = -3/2\ Nk.$

There are several important quibbles about the above result that one should clear up.

1. An isolated system cannot be at thermal equilibrium because its fastest particles will exceed the velocity of escape and evaporate.

2. Particles with attractive forces between them need some hard core repulsion to stop a pair coming very close and liberating energy to all the others. Without some restriction on how close the particles can get it is not clear that a thermodynamic equilibrium will exist.

Antonov in 1962 dealt with the first of these problems by

considering N particles sharing energy E inside a spherical container. Indeed, the Virial theorem is then modified to read

$$2\tau + V = 3pV \qquad 1.1$$

where p is the pressure on the containing sphere and V is its volume. The problem comes down to maximising the number of ways W in which n_j particles can be put in the jth cell of phase space subject to the constraints that there are N particles and the total energy is E. For cells of equal volume in phasespace

$$W = \frac{N!}{\pi n_j!}$$

If the cells are of unit volume, then the numbers n_j are the values of the distribution function f within the different cells. Using Stirling's approximation and $\Sigma n_j = N$

$$\ln W = N(\ln N-1) - \Sigma n_j(\ln n_j-1) = N(\ln N) - \int f\ln f d^6\tau$$

Hence, maximising lnW is equivalent to maximising the entropy $S=-k\int f\ln f d^6\tau$ Here k is Boltzmann's constant.

The constraints are $\int f\, d^6\tau = N$

and $$\int\frac{1}{2} mv^2 f\, d^6\tau - \frac{G}{2}\int\frac{mfd^3v\int mf'd^3v'}{|\mathbf{r}-\mathbf{r}'|}d^3\mathbf{r}d^3\mathbf{r}' = E$$

We perform the maximisation using Lagranges undetermined multipliers

$$\delta S = -k\int\delta f(\ln f + \alpha + \beta\,\frac{\delta E}{\delta f})\, d^6\tau$$

where $$\delta E = \int\frac{1}{2}\, mv^2\delta f d^6\tau - G\int\int\frac{m\delta f\; mf'}{|\mathbf{r}-\mathbf{r}'|}\, d^6\tau d^6\tau'.$$

In the above we have incorporated the $\delta f'$ term into the second term by using the change of variables $rv \leftrightarrow r'v'$.

This if $$\psi = -G\int\frac{mf'}{|\mathbf{r}-\mathbf{r}'|}\, d^6\tau' \qquad 1.2$$

we may write $$\frac{\delta E}{\delta f} = \frac{1}{2}\, mv^2 - m\psi \quad .$$

$$\delta S = 0 \text{ implies } f = A\exp - \beta(\tfrac{1}{2}mv^2 - m\psi) \qquad 1.3$$

where $A = e^{-\alpha}$, hence f is maxwellian in the potential ψ. However, this does not solve the problem since ψ itself is defined in terms of f. However, the velocity structure of f is now known and we may integrate to get ρ in terms of ψ. Thus

$$\rho = m\int fd^3v = mA\left(\frac{2\pi}{m\beta}\right)^{3/2} e^{\beta\psi} = \rho_o e^{\beta(\psi-\psi_o)}$$

since equation 1.2 may be thought of as Poisson's integral, we can use Poisson's equation and so find

$$\nabla^2\psi = -4\pi G\rho = -4\pi G\rho_o e^{\beta(\psi-\psi_o)}$$

The spherical solutions of this equation obey

$$\frac{1}{r^2}\frac{d}{dr}\left(r^2\frac{d}{dr}\Psi\right) = -B^2 e^{\Psi}$$

where $$\Psi = \beta(\psi-\psi_o) \text{ and } B^2 = 4\pi G\rho_o\beta$$

The factor B^2 may be absorbed by writing $r_* = Br$.
We need the solutions for which $\Psi = 0 = \Psi'$ at $r_* = 0$.
So this equation may be integrated (numerically) to give ψ as a function of r. When we reach a point at which

$$\frac{d\Psi}{dr_*} = -\frac{\beta GNmB}{r_*^2} = -\frac{\beta GM}{r^2 B}$$

then we have used all the mass, so we have reached the confining sphere.

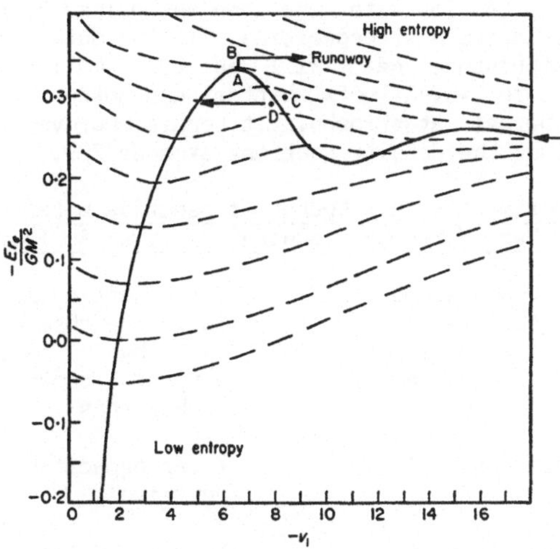

FIGURE 1.

The potential there is ψ = GM/r and this enables us to find ψ_o as a function of ρ_o and β. One may then calculate all the properties of such confined spherical systems. Figure 1 shows the graph of $-Er_e/GM^2$ as a function of the log density contrast $\ln(\rho_o/\rho_o) = \Psi_e$. Where the suffix e denotes the values at the edge where the system reaches the confining sphere. Notice that there are no solutions with $-Er_e/GM^2 > 0.335$ which corresponds to the critical density contrast of 709. Thus, if we release particles with a (negative) total energy E into a container larger than $0.335\ GM^2/(-E)$ in radius, there is no thermodynamic equilibrium for the system to go to. Some understanding of this strange result can be obtained by realising that the specific heat dE/dt has become first infinite and then negative even when the container is at a significantly smaller radius corresponding to a density contrast of only 32.2. The central density is perfectly finite and the gas behaves locally just like any non-gravitating gas. Giving the particles hard cores will not alter this equilibrium state. Thus, there must be something wrong with the proof that specific heats are positive. We shall see what it is presently.

2. A Thought Experiment

Imagine a sphere confining a system at equilibrium with a radius just less than $0.335\ GM^2/(-E)$. Now slowly expand the sphere to a larger radius. As the system does work during this expansion it loses energy so E will become more negative and the critical radius will shrink while the radius of the container increases. Thus, the system will find itself with no equilibrium to go to. Now imagine that the gas is adiabatic during the expansion. As the inner parts are held in by gravity while the outer parts are held in by the container, it is the outer parts that expand the most. Thus their adiabatic expansion leads them to a lower temperature than the central parts, see figure. Now let thermal conduction occur. Heat flows down the temperature gradient so the outer parts gain energy and the inner parts lose it. However, the inner parts have a negative specific heat so as they lose energy they shrink and get hotter. The outer parts are like a gas held in a container, so as they gain energy they also get hotter. It is now a race. Will the inner parts gain temperature by losing energy slower than the cooler outer parts gain temperature by gaining energy? Since it is the same energy that is lost to the one and gained by the other, what matters is the relative heat capacities. If the outer parts are too extensive their temperature gain will be too small and they will not catch up with the core which gets hotter and hotter as it loses heat! This is the gravothermal catastrophe. It occurs wherever the outer parts are so extensive that the density contrast exceeds 709.

A real experiment with a bucket of negative water capacity
The Heat-Water and Temperature-level Analogy

The amount of heat required to raise the absolute temperature of a substance by 1 K is its heat capacity or for one gram its specific heat.	The amount of water required to raise the absolute level of the water by 1cm is the water capacity of a bucket (for a normal bucket this is the area).
Specific heats are normally positive.	Water capacities are normally positive
However, for some self gravitating systems they can be negative.	However, water in a bucket suspended by a piece of elastic can have its absolute level fall when more water is poured in. Thus, water capacity can be negative.
If a heat bath is connected to a system of negative specific heat which is initially colder, it gets colder still. If hotter, it gets hotter still.	If a large water reservoir is connected to a negative water capacity bucket which initially has a lower level, its level gets still lower. If higher it gets still higher.
Similarly, if two negative specific heat systems are connected, they are unstable.	Similarly if two negative water capacity buckets are connected they are unstable.
However if a system of negative heat capacity is connected to a system of small positive heat capacity so that the total heat is negative, they will reach a stable equilibrium both moving up or down in temperature.	However if a bucket of negative water capacity is connected to a small positive water capacity bucket so that the total water capacity is negative, they will reach a stable equilibrium both levels moving up or down.

The error in the proof that all heat capacities are positive occurs at line zero which said "In a canonical ensemble of systems at equilibrium ." Negative heat capacity systems are unstable in canonical ensembles so ensembles of them are never in equilibrium! Nevertheless they occur and cause phase transitions.

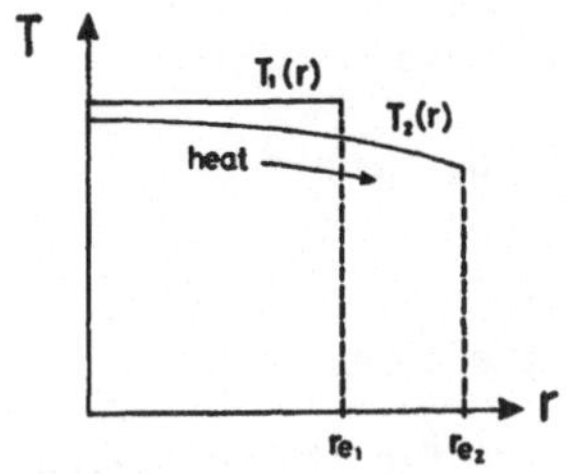

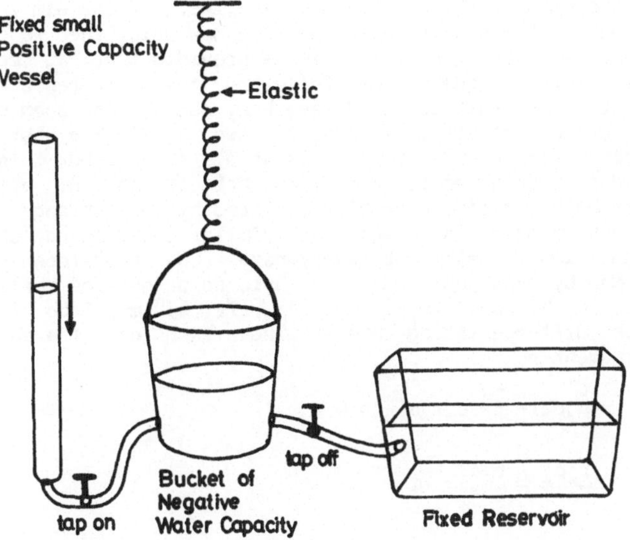

FIGURE 2.

Temperature distribution before and after expansion of the confining sphere.

Water flowing from the tubular vessel into the bucket causes the bucket to get heavier and the absolute level to fall (although the bucket gets fuller). Equilibrium is achieved as the level in the tubular vessel falls faster.

During the gravothermal catastrophe the central parts shrink to higher density, thus the point at which the density contrast reaches 709 moves inwards through the mass. Thus successively smaller masses around the centre leave behind their outer parts while concentrating their centres. The same process occurs not once but again and again and again. As a consequence we expect that the precise nature of the initial conditions will be forgotten and precisely the same process will occur on different scales. This constant rescaling of the problem shows that the solution for the density must take the form

$$\rho(r,t) = \rho_c(t)\ \rho_*(r_*) \tag{2.1}$$

where

$$r_* = r/r_c(t). \tag{2.2}$$

here $r_c(t)$ is the radius of the core.

Mathematically it is difficult to calculate precisely with an approximate similarity solution that does not hold for the whole cluster but luckily we may escape this difficulty by a ruse. Since the outer parts of the halo play no part in the evolution of the core it does not matter what the outer part of the halo is like. If our aim is the study of the core and what was once core and is now halo, then we can replace the outer halo by anything we like, always provided that it continues to do nothing! This suggests the following ruse. We replace the outer halo by the structure that it would have had, had it once been part of the core of an enourmous cluster. In practice we shall find that we have to add an infinite halo in order to do this at all scales, but this imaginary halo indeed does nothing an is so diffuse that it has no effect on the core or inner halo. With the problem doctored by this strange ruse we can now demand that we have an exact self-similar solution not only in the core and the inner halo but everywhere. We therefore seek such solutions. Now by hypothesis the outer halo does nothing while the core shrinks hence in the halo $\rho(r,t) = \rho_c(t)\rho_*[r/r_c(t)]$ must be independent of t. Thus differentiating and writing a dash for differentiation of ρ_*wrt its argument r_*

$$\dot{\rho}_c\rho_* - \frac{\dot{r}_c}{r_c}\rho_c r_*\rho'_* = 0$$

and so

$$\frac{r_*\rho'_*}{\rho_*} = \frac{r_c\dot{\rho}_c}{\dot{r}_c\rho_c} = -\alpha \tag{2.3}$$

Since the lhs is a function of r_* alone and the rhs is a function of t, must be a constant. We deduce that

$$\rho_* = Ar_*^{-\alpha} \text{ in the halo} \tag{2.4}$$

and that

$$\rho_c \propto r_c^{-\alpha} \tag{2.5}$$

Since ρ_c and r_c are function of t alone this last relationship holds everywhere. Now the core mass is some definite multiple of $\rho_c r^3_c$ (dependent on whose precise definition we use) and so $M_c \propto r_c^{3-\alpha}$. Similarly the core energy is some multiple of $-GM^2_c/r_c \propto -r_c^{5-2\alpha}$. Hence the core energy is related to the core mass by

$$E_c \propto -M_c^{\zeta} \quad \text{where} \quad \zeta = \frac{5-2\alpha}{3-\alpha} \qquad 2.6$$

Similarly calling v_c^2 the central velocity dispersion we have

$$v_c^2 \propto GM_c/r_c \propto r_c^{2-\alpha}. \qquad 2.7$$

Now the standard formula for the relaxation time in a stellar system is

$$T_r = v^3 (8\pi G m\rho \log N)^{-1} \qquad 2.8$$

For the evolution of our system we must have

$$\frac{1}{\rho_c}\frac{d\rho_c}{dt} \propto \frac{1}{T_{rc}} \propto \frac{\rho_c}{v_c^3}(8\pi G m \log N). \qquad 2.9$$

In any similarity solution it is the scales that vary so the dimensionless quantities appearing in the solution are constant (since they are scale independent). $(T_{rc}/\rho_c)(d\rho_c/dt)$ is dimensionless, and is therefore constant in the similarity solution. (In order to use such arguments strictly we should remark that we allow m the mass of a star to enter only in the relaxation time and nowhere in the momentary equilibrium of the star cluster. Encounterless stellar dynamics has equilibria independent of the masses that make them up).

In the standard formula 2.8 N is the number of stars in the system. In our infinite system N is the number of stars that interact with the core. It is natural to take N to be a multiple of the number of stars in the core itself but although there are still similarity solutions when log N is dependent on time - indeed they have the same separation constant and identical spatial form - nevertheless the extra complication of the formulae for the time dependence outweighs the gain in accuracy (Lynden-Bell 1975). We shall hereafter neglect the time variation of log N. Its value is then absorbed into the constant of proportionality in equation 2.9.

From equations 2.9, 2.7 and 2.5

$$-\alpha \frac{dr_c}{dt} \propto \frac{r_c^{1-\alpha}}{r_c^{3/2(2-\alpha)}} = r_c^{\alpha/2-2}$$

Hence

$$r_c^{3-\alpha/2} \propto (t_o - t) \qquad 2.10$$

where t_o is a constant of integration and we take $\alpha \neq 6$. From equations 2.7 and 2.9 we obtain

$$\begin{aligned} v_c^2 &\propto (t_o - t)^{(4-2\alpha)/(6-\alpha)} \\ \rho_c &\propto (t_o - t)^{-2\alpha/(6-\alpha)} \\ M_c &\propto (t_o - t)^{(6-2\alpha)/(6-\alpha)} \\ E_c &\propto (t_o - t)^{2(5-2\alpha)/(6-\alpha)} \end{aligned} \qquad 2.11$$

Notice that $r_c \to 0$ at the finite time $t = t_o$ (assuming $\alpha < 6$).
If $\alpha \geq 5/2$ then the gravitational potential energy of the central cusp becomes infinite at $t = t_o$. This is because when $\rho = A\, r^{-\alpha}$ $M \propto r^{3-\alpha}$

hence $$V = -\int \frac{GMdM}{r} = -A_1 \int \frac{GMdM}{M^{1/(3-\alpha)}}$$

and this integral diverges at small M if $\alpha \geq 5/2$. Thus the central parts would have to lose an infinite binding energy in finite time by heat conduction through their surrounding which have only finite conductivity. This cannot be so $\alpha < 5/2$. However, for the gravothermal catastrophe to occur at all, the temperature must decrease outwards. The isothermal gas sphere has $\alpha = 2$ for large r. Thus we must have $2 < \alpha < 5/2$. Detailed computations which depend on the precise nature of the energy exchange mechanism or heat conductivity give $\alpha = 2.22$ for stellar dynamics. This leads to $-E_c \propto r_c^{5-2\alpha} \propto r_c^{0.56}$ and $M_c \propto r_c^{3-\alpha} \propto r_c^{0.78}$ so both of these tend to zero as $r_c \to 0$. However, relativists are more interested in $M_{c/r} \propto r_c^{2-\alpha} = r_c^{-0.22}$. "In principle" black hole conditions will be reached $v_c \propto r_c^{-0.11}$ so to go from 300 km/s to 300,000 r_c would have to shrink by a factor $(1000)^{1/0.11} = 2 \times 10^{27}$ and M by a factor 2×10^{21}. Big Black holes are not made this way!

3. Why Self-Similar Solutions Occur in Science

The role that similarity solutions play in science is as general as the role of dimensional arguments. To be specific, consider the diffusion equation for temperature in a fixed conducting medium

$$\partial T/\partial t = Q\nabla^2 T$$

where Q is the diffusion constant. Now consider the situation in which initially $T = Br^{-\beta}$. The problem is to determine the temperature at later times.

First look at the dimensions of all the constants that occur in the specification of the problem. Calling the dimension of temperature $[\vartheta]$ and that of time [t] we find

$$[Q] = [t^{-1}L^2] \quad \text{and} \quad [B] = [L^{\beta}\vartheta].$$

The only constant independent of $[\vartheta]$ is thus Q. The problem is completely specified and no constant with the dimensions of length or of time can be made from that specification. Thus no such constant can occur in the solution. Notice however that if we give ourselves the time t after the beginning then we can make a time-dependent natural length unit.

$$L(t) = (Qt)^{1/2}$$

similarly we can make a natural time-dependent temperature unit $BL^{-\beta}$. The solution must express T as a function of r and t. This can be re-written in dimensionless form $T/(BL^{-\beta})$ must be a function of $r/L(t)$ and t/t. The latter is of course 1 and is irrelevant because t itself is the only quantity of the dimensions of [t] that can be made from Q, B and t. Thus we deduce that the solution must take the form

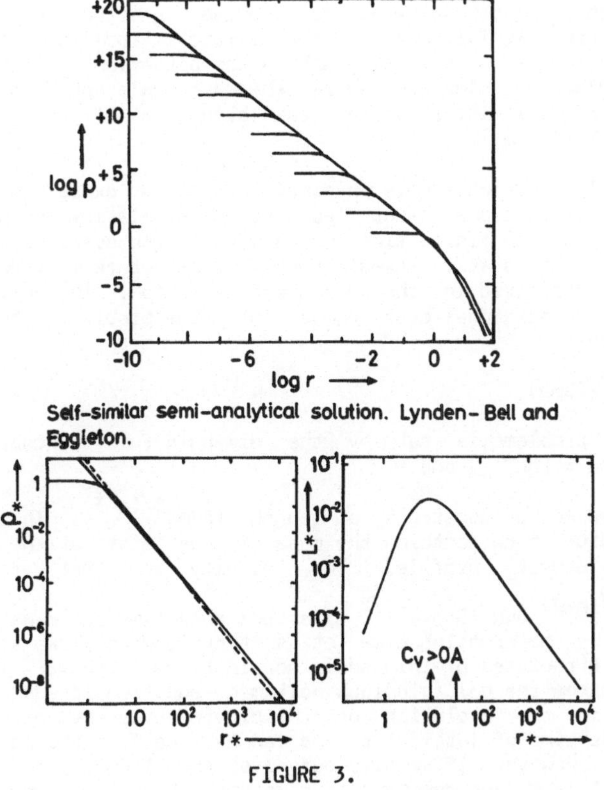

Self-similar semi-analytical solution. Lynden-Bell and Eggleton.

FIGURE 3.

Collapse - Numerical Solution of F.P. Equation, Halden Cohn

$$T = BL^{-\beta}\, T_*(r/L)$$

where $L(t)$ is defined above and T_* is a dimensionless function of its dimensionless argument. Notice this is a similarity solution because using time-dependent scales, the scales temperature is always the same function of the scaled radius. Thus the occurrence of too few dimensionful constants in the problem posed (including boundary conditions, etc.) lead inevitably to the solution being a similarity solution.

A case which occurs yet more commonly is that in which one of the pure numbers that can be made from the problem is either so large or so small that we are only interested in the asymptotic form of solution as the parameter tend to ∞ or zero. Such is the situation in the stellar dynamics of globular clusters where the mass scale set by one stellar mass m is not of real interest for the structure of the cluster as a whole because $M(r) > m$.

Lynden-Bell & Eggleton gave physical arguments explaining why the gravothermal catastrophe must lead to a similarity solution and demonstrated that in finite time the natural length scale becomes zero. At that moment the system takes a power-law structure with density $\rho = Ar^{-\alpha}$. We are interested in the subsequent evolution. The only way that the mass m of an individual star comes into the problem is through the relaxation time

$$T_{rel} \propto v^3/(Gm\rho).$$

The physical problem is set by the dimensionful constants $[G] = [M^{-1}L^3t^{-2}]$, $[A] = [ML^{\alpha-3}]$ and m.

From these we can construct a length $(A/m)^{1/(\alpha-3)}$. This is the very small radius which contains the mass of one star at the centre of the cluster's density profile. The density at that radius is $\rho_m = A(A/m)^{-\alpha/(\alpha-3)}$ and $(G\,\rho_m)^{-1/2}$ gives the only time-scale we can make. However, that is the crossing time at a tiny radius - we are neither interested in structures of this scale nor in times so short - indeed for such short times the collisionless Boltzmann equation is adequate. Our interest lies in the evolution of the cluster over several relaxation times. Thus no time of interest to us can be made from the constants of the problem. However, if we give ourselves the time since the core collapse $\tau = t - t_o$ and remember that we are interested in behaviour on the relaxation time-scale appropriate to the radius at which we look, then it must be the ratio τ/T_{rel} that arises. Notice that this ratio always involves m and τ in the combination $m\tau$. Thus we now look again at dimensional arguments using G, A and the combination $m\tau$ of dimensions [Mt] as our basis. With these assumptions we find a natural unit of length

$$r_c = [Gm^2\tau^2/A]^q \quad \text{where } q = (6-\alpha)^{-1} \qquad 3.1$$

and similarly a natural unit of mass

$$M_c = [(Gm^2\tau^2)^{3-\alpha}A^3]^q \quad \text{and of density } M_c rc^{-3} = \rho_c.$$

If we now ask again what is the form of the density evolution of the globular cluster it can only involve the constants that specify the problem and so

$$\frac{\rho}{\rho_0} = \rho_*(r/r_c). \qquad 3.2$$

Notice that ρ_c and r_c are both t-dependent and that this density is of similarity form. Using the method of dimensions we find that the characteristic radius, r, mass M_c, density ρ_c, velocity v_c, energy E_c and relaxation time T_c have the following time dependencies:

$$r_c \propto |\tau|^{2q}, \quad M_c \propto |\tau|^{2q(3-\alpha)}, \quad \rho_c \propto |\tau|^{-2\,q},$$

$$v_c^{\,2} \propto |\tau|^{-2q(\alpha-2)}, \quad E_c \propto GM_c^{\,2}r_c^{\,-1} \propto |\tau|^{2q(5-2\alpha)},$$

$$T_c \propto v_c^{\,3}/\rho_c \propto |\tau|^{(6-\alpha)q} \propto |\tau|. \qquad 3.3$$

These are identical to those found by Lynden-Bell & Eggleton, but we are now discussing τ positive instead of negative.

4. Evolution After Core Collapse

Once the central core has collapsed the re-expansion predicted by relations 3.3 requires a new energy source. This is actually provided by the formation of binary stars at the high densities formed by core collapse. Just as in Eddington's stellar structure problem, the precise nature of the energy source does not matter to us. The innermost density distribution adjusts itself so as to supply the energy needed for the self-similar expansion of the main solution. In fact, within r_c the relaxation time is shorter than the evolution time and the system becomes almost isothermal but with just sufficient radial temperature gradient to drive the heat flux needed to expand the system and increase the isothermal core. Inagaki & Lynden-Bell have discussed these re-expanding self-similar solutions. The core become almost an r^{-2} isothermal sphere and this joins the $r^{-2.22}$ halo at a point that moves out more and more slowly as time proceeds.

Bettwieser & Sugimoto have shown that the start of this process may be unstable with giant relaxation oscillations. They have the beautiful idea that if initially the binaries produce too much energy near the centre, then the central core will expand and cool to a temperature lower than their surroundings. While this persists, the gravothermal catastrophe may go into reverse with the conduction of heat inwards leading to still further expansion and cooling. This may continue until the lower temperatures of the halo are encountered when the system may return to its more isothermal state.

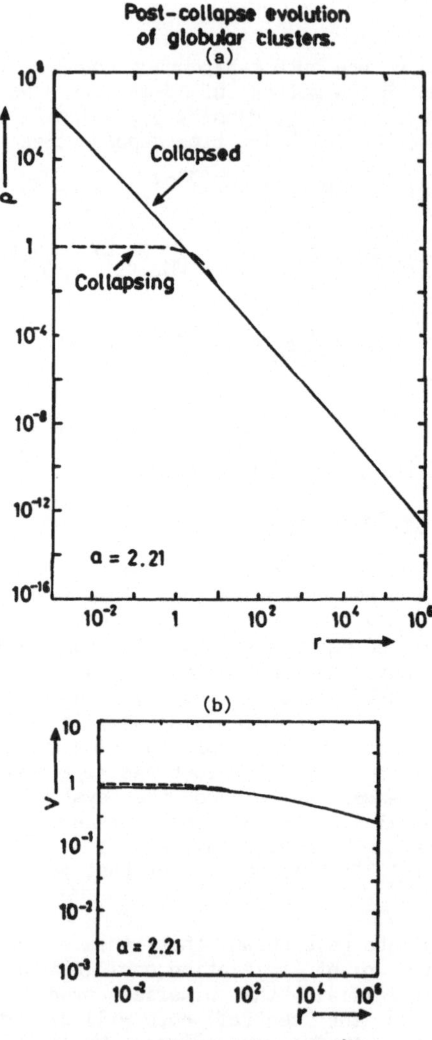

The profiles of the velocity dispersions for the same situation.

FIGURE 4.

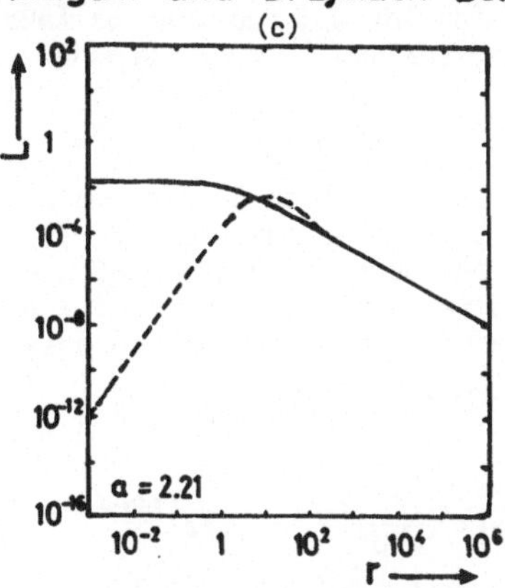

The profiles of the luminosities for the same situation.

Figure 4. Continued

II. SPHERICAL ACCRETION

5. Bondi Accretion - steady spherically symmetrical flow onto a star

Three equations govern this motion

Continuity $$4\pi\rho u r^2 = F \qquad 4.1$$

Motion $$u\frac{du}{dr} = -\frac{GM}{r^2} - \frac{1}{\rho}\frac{dp}{dr} \qquad 4.2$$

State $$p = p(\rho) \qquad \text{(adiabatic, say)} \qquad 4.3$$

Writing $c_s^2 = dp/d\rho$ for the velocity of sound the last term of 4.2 can be re-written $c_s^2 \frac{d\ln\rho}{dr}$ and from 4.1 $d\ln\rho/dr = -2/r - d\ln u/dr$. Hence 4.2 can be re-written

$$(u - c_s^2/u)\frac{du}{dr} = \frac{2c_s^2}{r}\left(1 - \frac{GM}{2c_s^2 r}\right) \qquad 4.4$$

Equation 4.4 has a critical point where $u^2 = c_s^2$ and $r = GM/2c_s^2$. Near there solutions exist with all slopes du/dr. Any solution that changes from subsonic to supersonic flow must do so through this critical point where the flow achieves the sound speed. Since the radius there is $GM/2c_s^2 = r_B$ and the velocity is c_s, it follows that such solutions have a flux

$$4\pi r_B^2 c_s\rho_B = 4\pi G^2M^2\rho_B/c_s^3 \qquad 4.5$$

where ρ_B is the density at the Bondi radius r_B. Note that c_s is not a constant but depends on ρ. To proceed further we must solve (1)-(3). Integrating 4.3 we have Bernoullis equation

$$\frac{1}{2}u^2 - GM/r + \int\frac{dp}{\rho} = w_\infty \qquad 4.6$$

where w_∞ is the value of the work function $\int\frac{dp}{\rho}$ at ∞.

We can now use 4.1 to eliminate either ρ or u. In the relativistic case elimination of u is the most helpful but since the sonic points are of great interest, it is more appropriate to eliminate ρ. To do this we need an explicit expression for $w = \int dp/\rho$ so we take the polytropic one $p = \kappa\rho^{1+1/n}$, $w = (n+1)p/\rho = (n+1)\kappa\rho^{1/n} = n$. 4.6 now reads

$$\frac{1}{2} u^2 + (n+1) \left(\frac{F}{4\pi r^2 u}\right)^{1/n} = \frac{GM}{r} + w_\infty \qquad 4.7$$

Now the right hand side is a function of r alone and will remain so if we multiply by a power of r. If that power of r is chosen correctly the right hand side can be made a function of one variable also. If we multiply by $r^{2\alpha}$ then the variable on the left will be $r^\alpha u$ provided we chose α so that the second term also involves only that variable. For this to be so we need $r^{-2n\alpha+2} u = r^\alpha u$ and so $\alpha = \frac{2}{2n+1}$ 4.8

Writing $U = r^\alpha u$ we have

$$\frac{1}{2}U^2 + (n+1) \left(\frac{F}{4 U}\right)^{1/n} = GM/r^{((2n-3)/(2n+1))} + w_\infty r^{2\alpha}$$

which is an equation of the form

$$f(U) = g(r)$$

For U large f is large and for U small f is large also. It has a minimum value of $(n+1/2)U_m^2$ when $U = U_m$, where

$$U_m^{2+1/n} = \frac{n+1}{n} \kappa\left(\frac{F}{4\pi}\right)^{1/n} = c_\infty^2\left(\frac{F}{4\pi\rho_\infty}\right)^{1/n} .$$

Provided $n > 3/2$ the function g(r) behaves rather similarly, being large at large r and at small r and having a minimum where

$$\frac{2n-3}{2n+1}\frac{GM}{r^{2-2\alpha}} = \frac{4w_\infty}{2n+1} r^{2\alpha-1}$$

that is where $r = r_m = \frac{2n-3}{4n}\frac{GM}{c_\infty^2}$

where we have used the fact that $w_\infty = nc_\infty^2$. The value of g(r) at the minimum is $\frac{1}{2\alpha}(GM/r_m^{1-2\alpha})$.

Now consider a solution starting at large r. There g(r) is large so f(U) must be large since they are equal. To get large f we must have U either large or small, but the large U choice leads to supersonic u whereas u is small at ∞ so we need the small U choice. As r decreases so U increases until we reach the point at which either g(r) or f(U) has its minimum (see figure). If g has its minimum above that of f, then further decrease in r leads to U retracing its steps back towards small U. If f has its minimum above that of g then we cannot proceed to smaller r because the corresponding g values are smaller than the minimum value of f. This problem has occurred because the chosen values of the flux F is too large. There is only one solution curve for which U continues to increase as r decreases, this is the one for which f and g have the same minimum. The condition for this is

$$(n+\tfrac{1}{2})U_m^2 = \frac{1}{2\alpha}\frac{GM}{r_m^{1-2\alpha}}$$

which reduces to

$$(n+\tfrac{1}{2})\left(c_\infty^2\left(\frac{F}{4\pi\rho_\infty}\right)^{1/n}\right)^{2n/(2n+1)} = \frac{2n+1}{4} GM\left(\frac{2n-3}{4n}\frac{GM}{c_\infty^2}\right)^{-1+4n/(2n+1)}$$

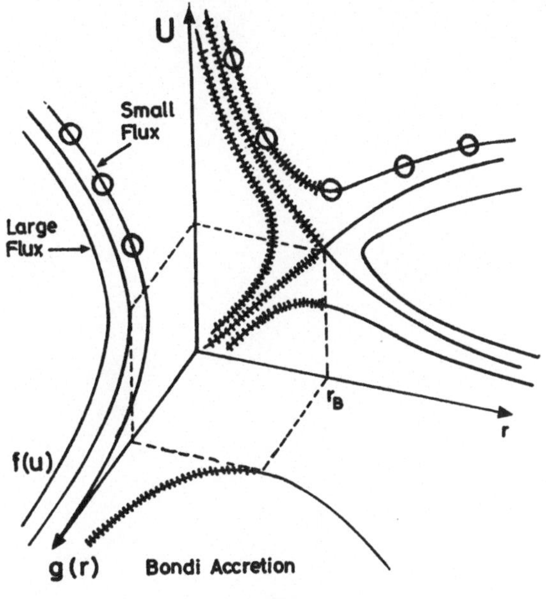

FIGURE 5.

a condition which we can use to determine F

$$F = 4\pi\rho_\infty\left(\frac{GM}{2c_\infty^2}\right)c_\infty\left(\frac{2n}{2n-3}\right)^{n-3/2}$$

It is easily checked that this solution is the one that passes through the critical point and that the factor $(2n/(2n-3))^{n-3/2}$ is merely the correction because **c** is now evaluated at ∞ and ρ is likewise, whereas in formula 3.5 both were evaluated at the critical point (which was then unknown). Thus for this special solution $r_0 = r_B$.

The other solutions remain subsonic and resist gravity by building up large pressures and densities. There are also the solutions that come in supersonically from ∞ and remain supersonic but for them the flow is not basically caused by the gravity of the star rather than the supersonic flow pattern at ∞.

It should be remarked that $n \leq 3/2$, $r \geq 5/3$ have no critical points because the pressure naturally builds up fast enough to resist gravity without supersonic motion. For such hard equations of state we need relativistic effects to bring in critical points.

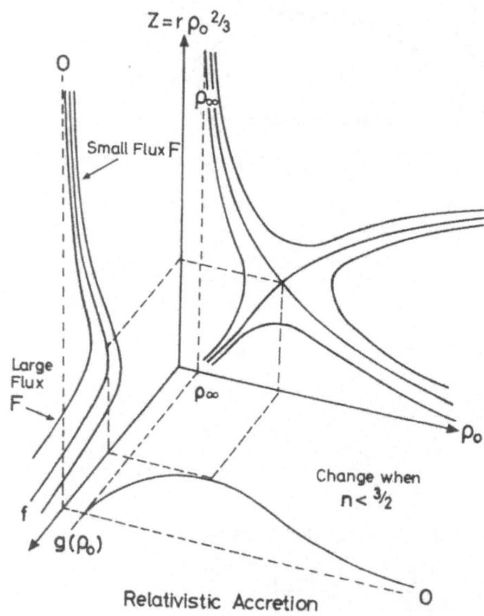

Z = r ρ₀ ^2/3
O
ρ∞
Small Flux F
Large Flux F
ρ∞
ρ₀
Change when n < 3/2
f
g(ρ₀)
Relativistic Accretion
O

FIGURE 6.

6. Relativistic Accretion

Bernoulli's Theorem

For a steady flow in a stationary space-time with timelike killing vector $\underline{k}$ we have

$$\Box.(\underline{\underline{T}}.\underline{k}) = 0 = \Box.(w\rho_o\underline{u}\ \underline{u}.\underline{k} + p\underline{k}). \qquad 6.1$$

Using continuity and $\underline{k}.\Box p = \frac{1}{c}\frac{\partial p}{\partial t} = 0$
we deduce

$$\rho_o(\underline{u}.\Box)(w\underline{u}.\underline{k}) = 0$$

hence $w\underline{u}.\underline{k}$ is constant along a streamline. Now $\underline{k}.\Box = \frac{1}{c}\frac{\partial}{\partial t}$ so k can be taken to be $k^\alpha = (0,0,0,1)$. The Bernoulli's equation reads

$$wu_o = \text{constant} \qquad 6.2$$

In Schwarschild's metric with G = c = 1

$$d\tau^2 = (1-\frac{2m}{r})dt^2 - [(1-\frac{2m}{r})^{-1}dr^2 + r^2(d\vartheta^2+\sin^2\vartheta d\varphi^2)]$$

If we write $U = -dr/d\tau$ we have for radial flow

$$u_o = g_{oo}u^o = g_{oo}\frac{dt}{d\tau} = (1-\frac{2m}{r})(1-\frac{2m}{r})^{-1/2}(1+\frac{U^2}{1-\frac{2m}{r}})^{1/2} = (1-\frac{2m}{r}+U^2)^{1/2}$$

and so
$$w(1-\frac{2m}{r}+U^2)^{1/2} = w_\infty \qquad 6.3$$

assuming $U\to 0$ at ∞.
Our other equations are rest mass conservation

$$4\pi r^2\rho_o U = F \qquad 6.4$$

and the equation of state gives $p(\rho_o)$ from which we get $w(\rho_o)$.
For polytropes

$$w = \varepsilon + p/\rho_o = c^2 + (n+1)\,\varkappa\rho_o^{1/n}. \qquad 6.5$$

Using continuity to eliminate U from Bernoulli we have

$$\left(\frac{F}{4\pi r^2\rho_o}\right)^2 - \frac{2m}{r} = \left(\frac{w_\infty}{w}\right)^2 - 1 \qquad 6.6$$

now multiply by $\rho_o^{2\beta}$ and choose β so that each term on the LHS is a function of the same variable. Evidently we need

$$(\beta-1)/2 = 2\beta \qquad \text{and hence } \beta = -1/3$$

equation 6.6 now becomes writing $z = r\rho_o^{2/3}$

$$\left(\frac{F}{4\pi z^2}\right)^2 - \frac{2m}{z} = \rho_o^{-2/3}\left[\left(\frac{w_\infty}{w}\right)^2 - 1\right] \qquad 6.7$$

$$f(z) = g(\rho_o)$$

both the functions f and g have negative minima. If we factorize to $(\frac{w_\infty}{w} - 1)(\frac{w_\infty}{w} + 1)$ then for small $\varkappa\,\rho_o^{1/n}/c^2$ the first factor behaves as $\rho_o^{1/n}$ and gives rise to the classical Bondi condition $n < 3/2$.

7. Cold Self-Similar Gravitational Collapse

A cold self-gravitating gas flows from rest at infinity to form a condensation centre that collapses spherically find all the self-similar solutions assuming that no shocks occur.

The radius of the sphere that contains mass M within obeys the equation

$$\ddot{r} = -\frac{GM}{r^2}$$

and hence $\quad \frac{\dot{r}^2}{2} = \frac{GM}{r} + \text{const}$

The constant has to be zero since the material at infinity was at rest. As the sphere collapses $\dot{r}$ is negative and so

$$\dot{r} = -(2GM/r)^{1/2} \qquad 7.1$$

hence $\quad \frac{2}{3} r^{3/2} = (2GM)^{1/2}(t_c(M) - t) \qquad 7.2$

where $t_c(M)$ is the time at which the sphere containing mass M collapses to the origin. For this non-crossing solution to be valid t_c must be an increasing or rather a non-decreasing function of M. In one sense equations 7.1 and 7.2 contain all the solutions to our problem because once $t_c(M)$ is specified, equation 7.2 gives the radius containing any mass at any time. It therefore gives the density profile and equation 7.1 can then be used to get the velocity profile. However, this solution is too general; greater interest is found in the special cases which evolve self similarly; that is when scaling functions $r_o(t)$, $M_o(t)$ exist such that the scaled radius $r_* = r/r_o$ is a function of the scaled mass $M_* = M/M_o$ only. From 7.2

$$\frac{2}{3} r_*^{3/2} = \frac{M_o^{\frac{1}{2}}}{r_o^{3/2}}(2GM_*)^{1/2}[t_c(M_*M_o)-t]$$

r_* is only a function of M_* only, if $(M_o(t)^{1/2}/r_o(t)^{3/2})[t_c(M_*M_o) - t]$ is independent of t at fixed M_*. We show in the Appendix that this is only possible if t_c is a power law plus a constant. If we write

$$t_c(M) = B\,M^{\gamma} + t_o \qquad 7.3$$

then $\quad \frac{2}{3} r^{3/2} = (2GM)^{1/2}(BM^{\gamma} + t_o - t) \qquad 7.4$

We may cast this in similarity form by writing

$$M_o(t) = [|t_o - t|/B]^{1/\gamma} \qquad 7.5$$

and $\quad \frac{2}{3} r_o^{3/2} = (2GM_o)^{\frac{1}{2}}|t_o-t| = \left(\frac{2G}{B^{\frac{1}{\gamma}}}\right)^{1/2}|t_o-t|^{1+\frac{1}{2\gamma}} \qquad 7.6$

from 7.4 we then find

$$r_*^{3/2} = M_*^{1/2}(M_*^{\gamma}\pm 1) \qquad 7.7$$

where the + is to be taken for $t<t_o$ and the minus for $t>t_o$. For large r_* we find $M_* \propto r_*\ 3/(2\gamma+1)$ so at large r_* the density $\rho_* \propto r_*^{-6\gamma/(2\gamma+1)} = r^{-\alpha}$. Evidently for a density profile defined by α one needs $\gamma = \alpha/(6-2\alpha)$. For general r we can find the density by squaring 7.4 and differentiating.

Evidently $dr^3/dM = 3r^2 dr/dM = 3/(4\pi\rho)$ where ρ is the density on the

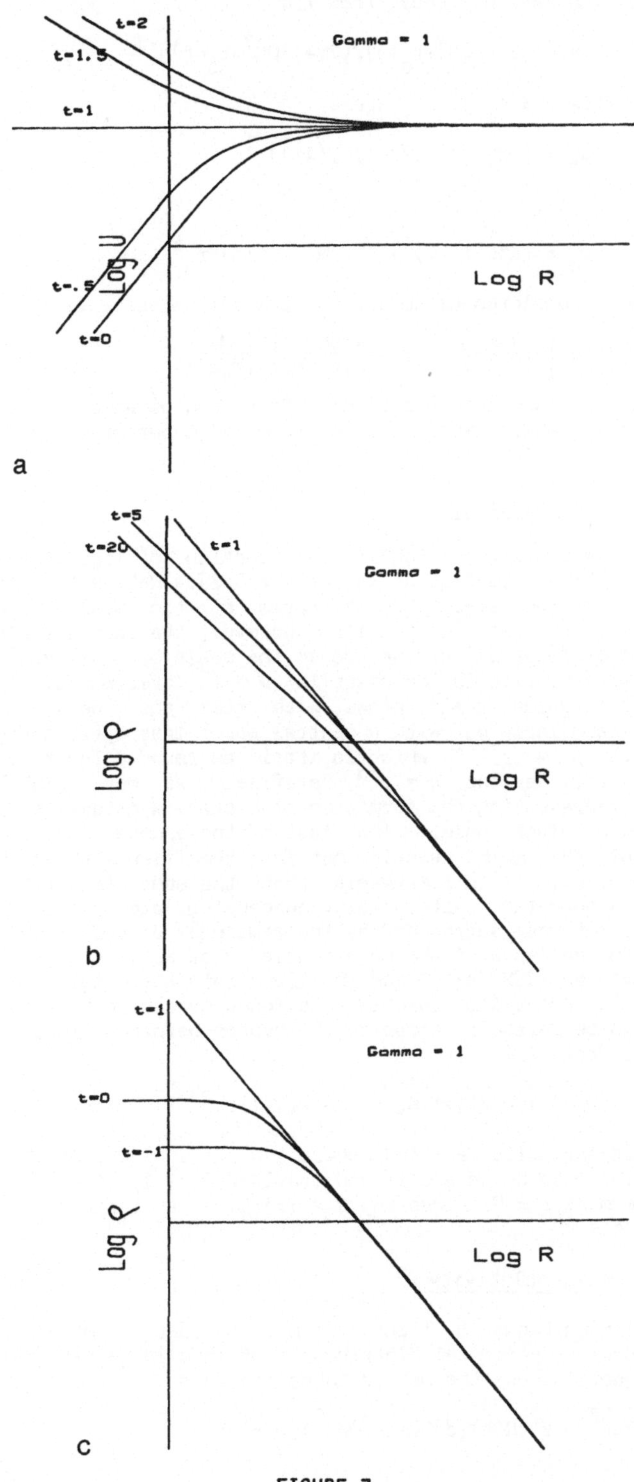

t=2
t=1.5
t=1
Gamma = 1
t=-.5
t=0
Log U
Log R
a
t=5
t=20
t=1
Gamma = 1
Log ρ
Log R
b
t=1
t=0
t=-1
Gamma = 1
Log ρ
Log R
c

FIGURE 7.

sphere containing mass M. Thus, from 7.4

$$6\pi G\rho = [(BM^{\gamma}+t_0-t)/(2\gamma+1)BM^{\gamma}+t_0-t)]^{-1} \quad 7.8$$

We may write this $\rho = \rho_0(t)\,\rho_*$ where

$$\rho_* = [(M_*{}^{\gamma}\pm1)((2\gamma+1)M_*{}^{\gamma}\pm1)]^{-1} \quad 7.9$$

and

$$\rho_0 = [6\pi G(t_0-t)^2]^{-1} = M_0/((4/3)\pi r_0{}^3) \quad 7.10$$

The solution is completed using 7.1 to give the velocity field

$$u = \dot{r} = -\left(\frac{2GM}{r}\right)^{\frac{1}{2}} = -\left[\frac{4GM}{3(t_c-t)}\right]^{1/3} = -\left(\frac{4}{3}\frac{GM_0}{|t_0-t|}\right)^{1/3}\left(\frac{M_*}{M_*\pm1}\right) \quad 7.11$$

Notice that the functions $\rho(r,t)$ and $u(r,t)$ are given parametrically in terms of M or M_* which must be eliminated in favour of r_* (or r) by use of 7.7.

Properties of the solutions

Figure 7a illustrates $\log\rho$ as a function of $\log r$ at a series of times $t \leq t_0$. [The illustration is for $\gamma = 3/2$]. Notice that the shapes of the curves are the same. Thus in terms of the scaled radius the scaled density has the same profile. However, the radius scale shrinks faster than the free fall of the gas so the scale becomes zero at t_0 and then the density profile is the power law $\rho \propto r^{-\alpha}$ everywhere. Thereafter (Figure 7b) the radius scale grows with time and the density again evolves self-similarly but with a central point mass $m_0(t)$ surrounded by a region with $\rho \propto r^{-3/2}$ which is itself surrounded for $r \gg r_0(t)$ by a region which still has the $\rho \propto r^{-\alpha}$ profile. We may think of these solutions as representing the formation of a star, a galaxy, a cluster of galaxies or a black hole. The last of these needs the relativistic discussion of the next section but for the first three Newtonian mechanics suffices. It is remarkable that the same similarity solution 7.4 represents both the finite density condensing cloud for $t < t_0$ and the growing point mass surrounded by the remainder of the cloud for $t > t_0$. After the collapse of the core centre is at $M_* = 1$, the velocity profile behaves as $(2GM_0/r)^{1/2}$ and the flux into the origin is $\dot{M}_0 = B(|t-t_0|/B)^{\gamma-1}$. Penston's solution is found by asking that the density profile should be parabolic close to the centre before collapse. With M_* small we find from 7.8

$$\rho_* \simeq 1 - 2(\gamma+1)M_*{}^{\gamma} \simeq 1 - 2(\gamma+1)r_*{}^{3\gamma}$$

hence if ρ_* is parabolic near the centre $\gamma = 2/3$. However, there is no real need for ρ to be parabolic and provided $3\gamma > 1$, ρ has a smooth density maximum at $r = 0$. Thus any $\gamma > 1/3$ gives an interesting uncusped solution for $t < t_0$.

Metrics in General Relativity

Now following Landau & Lifshitz we may consider the metric generated by a pressureless spherical distribution of mass in general relativity. In comoving coordinatess the metric takes the form

$$ds^2 = c^2dt^2 - r^2(R,t)(d\vartheta^2+\sin^2\vartheta d\varphi^2) - e^{\omega}dR^2 \quad 7.12$$

where $r^2 = e^\mu$. Writing ' for $\partial/\partial R$ and $\cdot$ for $\frac{1}{c}\partial/\partial t$ Einstein's equation reads

$$T_1^{\ 1} = 0 \Rightarrow \frac{1}{4} e^{-\omega} \mu'^2 - (\dot{\mu} + \frac{3}{4}\dot{\mu}^2) - e^{-\mu} = 0 \qquad 7.13$$

$$8\pi G c^{-4} T_0^o = -8\pi G\, c^{-2}\rho = e^{-\omega}(\mu'' + \frac{3}{4}\mu'^2 - \frac{1}{2}\mu'\omega') - \frac{1}{2}(\dot{\omega}\dot{\mu} + \frac{1}{2}\dot{\mu}^2) - e^{-\mu} \qquad 7.14$$

and $T_0^{\ 1} = 0 \Rightarrow -2\dot{\mu}' - \dot{\mu}\mu' + \dot{\omega}\mu' = 0$ 7.15

dividing the last equation by μ' and integrating w.r.t t we have

$$\omega - \mu - 2\ln \mu' = -\ln[4(1+\epsilon(R)/c^2)] \qquad 7.16$$

where the RHS is the "constant" of integration. From this we deduce

$$e^\omega = \frac{1}{4} e^\mu \mu'^2/(1+2\epsilon(R)/c^2) = r'^2/(1+2\epsilon/c^2) \qquad 7.17$$

using this expression in equation 7.13

$$(1+2\epsilon/c^2)r^{-2} - (2\frac{1}{c}\frac{\partial}{\partial t}(\dot{r}/r) + 3\,\dot{r}^2/r^2) - r^{-2} = 0$$

hence $\frac{1}{c}\frac{\partial}{\partial t}(r\dot{r}^2) = 2\frac{\epsilon}{c^2}\dot{r}$

and so $\frac{\dot{r}^2}{2} = \frac{\epsilon}{c^2} + \frac{GM(R)}{c^2 r}$ 7.18

where GM(R) is the "constant" of integration. It may be shown from (14) that $dM/dR = 4\pi r^2 \rho r'$ so that M(R) is the total mass withins coordinate R. Notice that c^2 x equation 7.18 is precisely the equation of energy conservation in the classical case which we solved earlier. When $\epsilon = 0$ the general solution is given by equation 7.2. If we now use r(R,t) as a coordinate in place of R we have

$$dr = r'dR + \dot{r}dt \text{ and hence } (r'dR)^2 = dr^2 - 2\dot{r}drdt + \dot{r}^2dt^2$$

so using 7.17 the metric is

$$ds^2 = (1 - \frac{2GM(R)}{c^2 r})c^2dt^2 - 2\sqrt{\frac{2GM}{r}}\,drdt - dr^2 - r^2(d\vartheta^2 + \sin^2\vartheta d\varphi^2)$$

To put this metric in diagonal form we write t = t(T,r) and take

$$(1 - \frac{2GM}{c^2 r})\frac{\partial t}{\partial r} = \sqrt{\frac{2GM}{r}} \qquad 7.19$$

so that the cross terms go out. We then have

$$ds^2 = (1 - 2GM/rc^2)\, c^2\left(\frac{dt}{\partial T}\right)^2 dT^2 - \frac{dr^2}{(1-2GM/rc^2)} - r^2(d\vartheta^2 + \sin^2\vartheta d\varphi^2) \qquad 7.20$$

To make this explicit, equation 7.19 must be solved for t(T,r) with r(M,t) given by 7.2 or for our special solutions 7.4. We shall not do this here but we note that outside all the mass one may take a solution of 7.19 of the form t = T - a(r) so that 6.9 reduces to the Schwarzschild metric in that region.

Appendix - Proof that self-similar solution must have t_c a power law plus a constant.

Writing $M_o^{1/2}/r_o^{3/2} = f(t)$ our problem reduces to finding the most general function t_c such that $f(t)[t_c(M_* m_o(t))-t] = F(M_*)$ where F is independent of t at fixed M_*. We write x for M_* and y for

$M_o(t)$. Differentiating with respect to x

$$ft'_c y = F'(x)$$

dividing by fy taking logarithms and then differentiating with respect to y and x we have

$$\frac{\partial^2}{\partial x \partial y}(\log t'_c) = 0$$

hence

$$\frac{\partial}{\partial x}\left(\frac{xt_c''}{t'_c}\right) = \frac{t_c''}{t'_c} + z\frac{d}{dz}\left(\frac{t_c''}{t'_c}\right) = 0$$

where z = xy and $t_c = t_c(z)$ of course. Writing $L = \ln t'_c$ we have

$$\frac{dz}{z} = -\frac{dL'}{L'}$$

hence $$\ln L' = -\ln|z| + c_1$$

$$L' = e^{c_1}/|z|$$

$$L = \ln t'_c = \pm e^{c_1}\ln|z| + c_2$$

$$t_c = B|z|^{\gamma} + t_o \quad \text{where } \gamma = \pm e^{c_1} + 1 \text{ and } B = e^{c_2}/\gamma$$

when $$\pm e^{c_1} = -1 \quad \text{then } t_c = C\ln|z| + t_o \quad \text{where } C = e^{c_2}$$

Thus the general solution is that t_c is a power law (including log) + a constant. We now consider two special cases

Logarithmic case $$\frac{2}{3}r^{3/2} = (2GM)^{1/2}(c\ln M + t_o - t)$$

Now write $$M_o = e^{-(t_o-t)/c}$$

then $$\frac{2}{3}\frac{r^{3/2}}{c/e^{-(t_o-t)}} = (2GM/M_o)^{1/2}\ln(M/M_o)$$

hence $$r_*^{3/2} = M_*^{1/2}\ln M_*$$

where $$r_* = r/r_o(t) \quad \text{and}$$

$$\frac{2}{3}r_o^{3/2} = (2G\,e^{-(t_o-t)/c})^{\frac{1}{2}}\,c$$

$$r_o \propto e^{-1/3(t_o-t)/c}$$

$$\frac{3}{4\pi\rho_*}\frac{dr_*^3}{dM_*} = (\ln M_*)^2 + 2\ln M_*$$

$$\frac{3}{\pi\rho} + 1 = (\ln M_* + 1)^2$$

$$M_* = \exp\left[\left(\frac{3}{4\pi\rho} + 1\right)^{1/2} - 1\right]$$

$$r_* = \left[\left(\frac{3}{4\pi\rho_*} + 1\right)^{1/2} - 1\right]^{2/3}\exp\frac{1}{3}\left[\left(\frac{3}{4\pi\rho_*} + 1\right)^{1/2} - 1\right].$$

Special case $\gamma = 1/2$ $\qquad \alpha = 3/2$

$$M_*^{1/2}(M_*^{1/2} \pm 1) = r_*^{3/2}$$

$$(M_*^{1/2} \pm \tfrac{1}{2})^2 = r_*^{3/2} + \tfrac{1}{4}$$

$$M_* = \left(\pm \tfrac{1}{2} + \sqrt{r_*^{3/2} + \tfrac{1}{4}}\right)^2 = r_*^{3/2} + \tfrac{1}{2} \pm \sqrt{r_*^{3/2} + \tfrac{1}{4}}$$

$$\rho \to r_*^{-3/2} \qquad \text{small } r_*$$

$$\rho \to (2r_*)^{-3/2} \qquad \text{large } r_*$$

III. DISK ACCRETION

8. Energy, Angular Momentum and Dissipation

In many places in astronomy, material has too much angular momentum to allow its concentration onto a central source of gravity. Not only Saturn's rings, planetary satellite systems and the solar system but also spiral galaxies are obvious illustrations. Here we are more concerned with disks about stars or about black holes. Spectacular examples occur when in a binary star one member expands due to stellar evolution and overflows its Roche lobe. This produces a mass of gas that swirls down onto the other star. If the other star is compact, much energy is released and many of the bright X-ray binaries rely on this mechanism.

In all such disk systems the evolution of the disk is driven by some dissipative process but normal molecular viscosity is quite inadequate so some other dissipative mechanism is necessary. There are fluid dynamical instabilities in swirling disks and it seems likely that these will give rise to turbulence since the Reynolds numbers involved are very large. Thus a widely invoked source of dissipation is the turbulent viscosity - a cover-all concept introduced to allow for the fact that turbulence dissipates energy and this energy must arise from the shear of the mean flow. However, there are the phenomena such as the generation of magnetic fields by dynamo action which may be even more effective as a source of dissipation. Here we shall try to formulate the theory in a way that is almost indepedent of the mechanism of energy dissipation. However, when a specific model is needed we shall treat the system as though it had a constant kinematic viscosity ν large enough to make significant changes in the angular momenta of a fluid element but only over a number of rotation periods of the disk.

We first consider three simple examples from the mechanics of a particle moving in a fixed gravitational field $\underline{g} = \underline{\nabla}\,\psi$ described by a potential $\psi(R,z)$.

a) If the angular momentum of the particles' motion is fixed, what is the orbit of least energy?

$$E_1 = m_1[\tfrac{1}{2}(u_R^2 + u_z^2 + h^2/R^2) - \psi(R,z)] + m_1c^2 \qquad 8.1$$

This minimises when $u_R = u_z = 0$ and also at $z = 0$ (where we assume that at given R the potential ψ maximises on the plane z = 0).

Minimising E_1 over R we have the minimum at R_h where

$$h^2/R_h^{\ 3} = -\frac{\partial\psi}{\partial R}(R_h,0) \tag{8.2}$$

8.2 is the condition for centrifugal force gravity balance. Thus minimum energy orbits are planar circular orbits. If we define the specific energy of such orbits as $\varepsilon(h)$ we have

$$\varepsilon(h) = \frac{h^2}{2R_h^{\ 2}} - \psi(R_h,0) + c^2 \tag{8.3}$$

Since we have already minimised over R, we notice that

$$d\varepsilon/dh = \frac{h}{R_h^{\ 2}} = \Omega \tag{8.4}$$

where Ω is the angular velocity of the circular orbit.

b) Now we consider minimising the energy of two particles each in a circular orbit when their total angular momentum is fixed.

$$E = m_1\ \varepsilon(h_1) + m_2\ \varepsilon(h_2)$$

$H = m_1h_1 + m_2h_2$ is fixed. Evidently

$$dE = m_1dh_1\ \varepsilon'(h_1) + m_2dh_2\ \varepsilon'(h_2) = m_1dh_1\ (\Omega_1-\Omega_2) \tag{8.5}$$

Hence E is decreased if the particle in the orbit of least Ω gains angular momentum. Thus, since Ω decreases outwards, the energy is decreased if the angular momentum is transfered outwards.

c) Now consider minimising the energy of two particles when the total mass is kept constant but not the individual masses. One may then write

$$dE = d(m_1h_1)(\Omega_1-\Omega_2) + dm_1(J_1-J_2) \tag{8.6}$$

where $J_1 = \varepsilon(h_1) - \Omega_1h_1$ 8.7

Notice that $\frac{dJ}{dR} = RV(-\frac{d\Omega}{dR}) > 0$

so J increases outwards.

The energy is decreased by passing mass inwards as well as by passing angular momentum outwards.

From these examples we learn that dissipation will lead towards planar circular orbits (disks) in which the mass moves inwards in the central parts and the angular momentum moves outwards. These general deductions are independent of the mechanism that provides the dissipation.

9. Viscous Newtonian Accretion Disks

It is assumed that the viscosity is sufficiently small that it can be ignored in the force balance between centrifugal force and gravity but that its secular effects of passing angular momentum from ring to ring and in dissipating energy are our chief interests.

Let C(R) be the couple by which the material outside R brakes the rotation of the material further in inside the disk. Let σ be the surface density of the disk ν the kinematic viscosity and $u = -u_R$. Then

$$C = (2\pi R\sigma)\ \nu\ R^2\ (-\frac{d\Omega}{dR}) \qquad 9.1$$

angular momentum

$$2\pi R\sigma u\ \frac{dh}{dR} = \frac{\partial C}{\partial R} \quad \therefore \quad 2\pi R\sigma u = \frac{\partial C}{\partial h} \qquad 9.2$$

Continuity

$$\frac{\partial}{\partial t}(2\pi R\sigma) - \frac{\partial}{\partial R}(2\pi R\sigma u) = 0$$

Substituting for $2\pi R\sigma$ and $2\pi R\sigma u$ from 6.1 and 6.2 we have

$$\frac{1}{\nu R^2(-\Omega')}\frac{\partial C}{\partial t} = h'\ \frac{\partial^2 C}{\partial h^2} \qquad 9.3$$

This is an equation of diffusion type with R or h dependent diffusion coefficient $[\nu R^2(-\Omega')h'] = \frac{3}{4}\ \frac{\nu GM}{R}$. Hence

1) as time progresses C smoothes to a linear gradient in h.

2) it does so move rapidly at small R where the diffusion coefficient is large.

Steady Accretion Disks

Integrating 9.2 with a constant flux F we have

$$C = F(h-h_o) \qquad 9.4$$

where h_o is the value of the angular momentum at which the couple vanishes, i.e. where $\Omega' = 0$. For accretion onto a star rotating with angular velocity Ω_* the accretion disk remains Keplerian until it meets the stellar atmosphere. There the rapidly rotating disk is braked by the more slowly rotating atmosphere, so Ω falls below its Keplerian value then maximises and falls catastrophically to the rotation rate of the star. This behaviour is caused by the viscosity of the material so it occurs in a boundary layer whose narrow width ℓ is determined by the small viscosity $\ell \sim \nu^{2/3}\ \frac{\Omega_*}{R_*^{1/3}(\Delta\Omega)}$ where Ω_K is the Keplerian angular velocity at the star's surface which is at radius R_*. The orbital velocity of the material in this disk just above the star has an energy equal to that lost by the material in settling to that orbit. Thus for a slowly rotating star the luminosity of the boundary layer in which the Kepler motion is dissipated should be as great as the total luminosity from the rest of the disk. The thinness of the boundary layer will give its radiation a high temperature and it may have been detected in the weak X-rays from some accreting binaries containing white dwarfs. However, in visible light there is little evidence so far for emission from the boundary layers.

The surface density and inward velocity in a steady disk follow from 9.1 and from flux conservation

$$\sigma = C/[2\pi R^3\nu(-\Omega')] \qquad 9.5$$

and $u = F/(2\pi R\sigma) = \nu R^2(-\Omega')/(h-h_o) \to \frac{3}{2}\nu/R$ for $R \gg R_*$ 9.6

Notice that if ν is a function of R formulae 9.4, 9.5 and 9.6 are still correct.

The energy budget of an infinitesimal ring of the disk is illustrated in Figure 8.

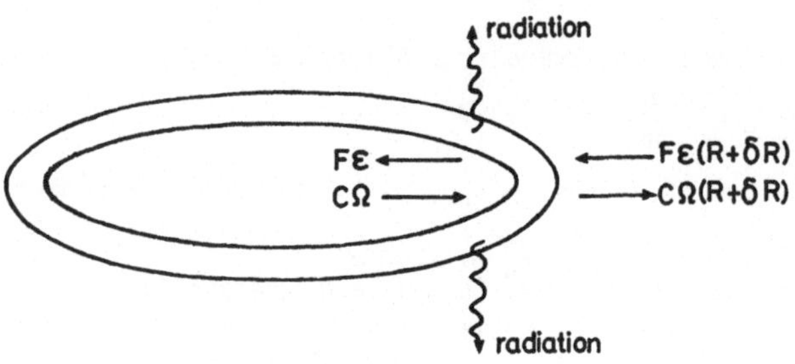

FIGURE 8.

here $\varepsilon(R)$ is the specific energy (per unit rest mass) of an orbit of radius R and $C\Omega$ is the rate of working of the viscous couple. The energy dissipated by viscosity in R, R + δR is

$$\delta R\ D(R) = \delta R \frac{d}{dR}(F\varepsilon - C\Omega) \qquad 9.7$$

If it is assumed that this dissipated energy is radiated locally a black body radiation then

$$D(R) = 2 \times 2\pi R\, \sigma_s\, T(R^4) \qquad 9.8$$

where σ_s is Stefan's constant.

For Keplerian orbits

$$\varepsilon(R) = c^2 + \tfrac{1}{2} v^2 - GM/R = c^2 - GM/(2R)$$

$$h(R) = \sqrt{GMR} \quad , \qquad \Omega(R) = \sqrt{GM/R^3}$$

Thus $T^4 = \dfrac{F}{4\pi\sigma_s R}\dfrac{d\varepsilon}{dR}[\varepsilon-(h-h_o)\Omega]$

$$T^4 = \frac{3}{8\pi}\frac{FGM}{\sigma_s R^3}[1 - (\frac{R_*}{R})^{\frac{1}{2}}] \qquad 9.9$$

hence for most of the disk T falls like $R^{-3/4}$. These formulae still hold whatever the R behaviour of the viscosity ν.

10. Relativistic Accretion Disks

The above theory needs a few interesting amendments when relativistic effects are important.

1. The formulae for ε, h and Ω have to be derived for Schwarzschild space or Kerr Space and these used in place of the Keplerian values.

2. The equation for angular momentum transport needs to be modified since in highly relativistic situations the energy radiated away can carry with it a significant angular momentum flux.

3. In accretion disks about black holes there is a last stable circular orbit where the angular momentum of circular orbits is a minimum; any braking of such an orbit by the disk exterior to it causes it to spiral downwards into the hole on the orbital rather than the viscous timescale. This rapid motion leaves little surface density within the last stable circular orbit so the disk effectively ends there. Actually there is a new type of boundary layer near the last stable circular orbit. The boundary layer thickness is of order $\nu^{2/7}$ and the radial velocity of order $\nu^{3/7}$ while the surface density becomes of order $\nu^{4/7}$ times its values further out. There is no strong shear or strong dissipation in this layer whose only function is to match the stable disk outside onto the dynamically falling rarified disk within.

The formulae for ε, h, Ω in circular orbits are easily derived for the Schwarzschild metric but need some labour in Kerr (see Appendix).

Schwarzschild circular orbits

$$\varepsilon c^{-2} = (r-2m)/[r(r-3m)]^{\frac{1}{2}} \qquad 10.1$$

$$h/(mc) = r/[m(r-3m)]^{\frac{1}{2}}$$

$$\Omega m/c = (m/r)^{3/2}$$

$$(\varepsilon-\Omega h)/c^2 = (1-3\tfrac{m}{r})^{\frac{1}{2}}$$

The last stable circular orbits are given by

Schwarzschild

$$\varepsilon_o c^{-2} = 2\sqrt{2}/3 \qquad 10.2$$

$$h_o/(mc) = \sqrt{12}$$

$$\Omega_o m/c = 6^{-3/2}$$

at $r = 6m$

Limiting Kerr a = m direct orbits

$$\varepsilon_o c^{-2} = 3^{-\frac{1}{2}} \qquad 10.3$$

$$h_o/(mc) = 2/\sqrt{3}$$

$$\Omega_o m/c = 1/4$$

$r = m$

As each element of mass passes through the disk, the binding energy is radiated. The efficiency of an accretion disk as a process for converting rest mass into radiated energy is therefore $(1-\varepsilon_o/c^2)$. This is 5.7% for Schwarzschild but 42% for limiting Kerr. We shall see presently that Kerr posesses inner energy reserves that allows it to attain even greater overall efficiencies.

We now consider the angular momentum balance of a ring when the radiated angular momentum is accounted for.

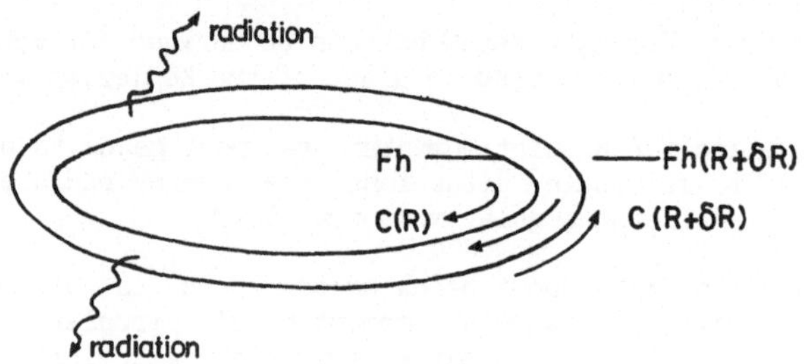

FIGURE 9.

The angular momentum put into the ring by the couples in unit time is $C(R) - C(R+\delta R)$. The angular momentum transported into the ring by the mass flux is $F(h(R+\delta R) - h(R))$. Now a rotating ring radiates angular momentum and energy in proportion to those it possesses due to the aberration of the directions of emission. (This proportionality follows because the 4 momentum of all the radiation emitted from a small square element of the disk in a small time interval is parallel to the 4 momentum of the disk element itself). Hence the angular momentum emitted is h/ε times the energy emitted. Thus the momentum balance for the ring of the disk reads

$$\frac{d}{dR}(Fh-C) = \frac{h}{\varepsilon} D(R) = \frac{h}{\varepsilon}\frac{d}{dR}(F\varepsilon = c\Omega) \qquad 10.4$$

Hence

$$(\varepsilon-\Omega h)\frac{dC}{dR} - \Omega' hC = -F\left(h\frac{d\varepsilon}{dR} - \varepsilon\frac{dh}{dR}\right)$$

This equation simplifies significantly when one remembers that for circular orbits $d\varepsilon/dh = \Omega$ so that $d\varepsilon/dR = \Omega\, dh/dR$. We then find

$$\frac{d}{dR}[(\varepsilon-\Omega h)C] = F(\varepsilon-\Omega h)\frac{dh}{dR} \qquad 10.5$$

integrating from the last stable circular orbit, where the couple effectively vanishes, to R we have

$$C = \frac{F}{(\varepsilon-\Omega h)}\int_{R_0}^{R}(\varepsilon-\Omega h)\frac{dh}{dR}\,dR. \qquad 10.6$$

which should be compared with the old formula $C = F(h-h_0)$; in fact, integration by parts gives

$$C = F[(h-h_0) + \frac{1}{\varepsilon-\Omega h}\int_{R_0}^{R}(h-h_0)h\,\Omega'\,dR]. \qquad 10.7$$

which shows the correction term explicitly. Once C is evaluated from this formula, D(R) and T(R) follow from equations 8.7 and 8.8. It is therefore possible to calculate the emission spectrum from a black-body disk. Over most of such a disk the temperature falls like $R^{-3/4}$. It is therefore of interest to calculate the integrated spectrum of a black-body disk with a power law temperature distribution. If

$$T = A\,R^{-2/a} \qquad 10.8$$

then the integrated spectrum is given by

$$S_\nu = \int_0^\infty \frac{c}{u} u_\nu(T)\ 4\pi R\ dR = \frac{8\pi^2 h}{c^2} \int_0^\infty \frac{\nu^3 R dR}{\exp(h\nu/uT)-1}$$

where we have added the emission from both sides of the disk. Writing $x = h\nu/uT(r) = h\nu\, r^{2/a}/(kA)$ we find

$$S_\nu = \frac{4\pi^2 h}{c^2} \left(\frac{kA}{h}\right)^a \nu^{3-a} \int_0^\infty \frac{a\, x^{a-1}}{e^x - 1}\, dx \qquad 10.9$$

the final integral is just a $\Gamma(a)\ \zeta(a)$ where Γ is the gamma function and ζ is the Zeta function. For $a = 8/3$ we see that $S_\nu \propto \nu^{1/3}$. Before using such a formula it is important to check to find out where it applies. We need the $T - A R^{-3/4}$ to hold down to the radii at which the integral has its major contribution at the chosen frequency ν. This occurs around $h\nu \sim kT$. To get $T \sim AR^{-3/4}$ we need $R \sim 15\ GM/c^2$ for a Schwarzschild hole. Thus the $S_\nu \propto \nu^{1/3}$ formula will hold for

$$\nu < \frac{kT}{h} = \frac{kA}{h} \left(\frac{15GM}{c^2}\right)^{-3/4} \qquad 10.10$$

where A is given by equating the power radiated at R to $4\pi R \sigma_s T^4$. Hence

$$A^4 = \frac{3FGM}{8\pi\sigma_s} \ . \qquad 10.11$$

Evolution from Schwarzschild to Kerr holes

Fed by its accretion disk a Schwarzschild hole will grow in mass not by the full rest mass flux F but by the energy flux $F\ \varepsilon_0/c^2$. However, as it grows by accreting mass-energy it also accretes angular momentum. Thus the hole does not remain spherical but it begins to spin. There are a whole family of spinning Kerr holes characterised by their masses M and angular momenta J. Kerr holes with the same value of J/M^2 are just scale models of one another. We write $J = aMc = \alpha\, m^2 c^3/G$. The dimensionless quantity α characterises the ratio of the angular momentum to the square of the mass. Notice that $a = \alpha m$.

In the Appendix we show that the energy and angular momentum of the last stable circular orbit in a Kerr metric are given by

$$\frac{\varepsilon_0}{c^2} = \left(1 - \frac{2}{3}\zeta\right)^{\frac{1}{2}} \ ; \quad h_0 = \frac{2}{3\sqrt{3}}\, mc[1+2(3\zeta^{-1}-2)^{\frac{1}{2}}] \qquad 10.12$$

where $\zeta = m/r_0$

r_0, the radius of the last stable circular orbit is given implicitly in terms of the α of the Kerr metric by

$$\alpha = 1/3\ \zeta^{-\frac{1}{2}}[4-(3\zeta^{-1}-2)^{\frac{1}{2}}] \qquad 10.13$$

For later use we note that

$$d\alpha/d\zeta = -1/6\zeta^{-3/2}[4-(3\zeta^{-1}-2) - 3\zeta^{-1}(3\zeta^{-1}-2)^{-\frac{1}{2}}] \qquad 10.14$$

On accretion, the growth rate of the hole's mass is

$$\dot{M} = F\ \varepsilon_0/c^2 = \left(1 - \frac{2}{3}\zeta\right)^{\frac{1}{2}} F \qquad 10.15$$

while the hole's angular momentum growth rate is

$$\dot{J} = F\ h_0 \qquad 10.16$$

$$\text{Hence } \frac{dJ}{dM} = \frac{h_0 c^2}{\varepsilon_0} = \frac{2}{3}\frac{mc}{\zeta^{\frac{1}{2}}}[(3\zeta^{-1}-2)^{-\frac{1}{2}} + 2] \qquad 10.17$$

$$\frac{d(\alpha m^2)}{dm^2} = \alpha + m^2 \frac{d\zeta}{dm}\frac{d\alpha}{d\zeta} = \frac{1}{3\zeta^{\frac{1}{2}}}[(3\zeta^{-1}-2^{-\frac{1}{2}}+2] \qquad 10.18$$

so using expressions 10.13 and 10.14 for α and $d\alpha/d\zeta$ we find

$$m^2 \frac{d\zeta}{dm^2}\frac{d\alpha}{d\zeta} = \zeta\frac{d\alpha}{d\zeta} \qquad 10.19$$

hence either $\frac{d\alpha}{d\zeta} = 0$ (i.e. $\zeta = 1$) or else $\frac{m^2}{\zeta}\frac{d\zeta}{dm^2} = 1$ which gives $\zeta \propto m^2$.

Now for Schwarzschild holes $\zeta = 1/6$ so if an originally Schwarzschild hole grows by disk accretion so that the mass increases by a factor $\sqrt{6}$, then ζ will tend to 1 and the hole will become an extreme Kerr hole. Thereafter it evolves self similarly with $\alpha = 1$. Such holes are much better than Schwarzschild holes at converting mass into energy since their basic efficiencies are $1-(1/\sqrt{3})$=42% in place of the 5.7%. But this is not all! Kerr holes have inner reserves of accessible energy stored in the gravimagnetic fields (dragging of inertial frames) that they produce. By Hawking's theorem the area of the hole must increase, but the same area as an extreme Kerr hole could be achieved by a Schwarzschild hole of mass $m/\sqrt{2}$. Thus $(1-1/\sqrt{2})$ = 29% of an extreme Kerr hole's mass is extractable. If we imagine this done by electromagnetic means as Professor Thorne will describe, then the efficiency of an extreme Kerr hole in converting mass into energy can be as high as 0.42 + (1-0.42) x .29 = .42 + .17 = 57%. In reality these figures are too high because accretion disks lose some of their radiation down the hole. In place of $\alpha = 1$ the self-similar growing hole is nearer to $\alpha = .998$ which gives an efficiency down from 42% to near 33%.

So far we have treated all our disks as thin and have imagined that the radiation can escape from them without difficulty. At the same time we have assumed them to be optically thick. Zel'dovich and Salpeter followed by Shakura and Sunyaev explored the breakdown of these assumptions when the mass flux was near that at which the Luminosity of the source achieved the Eddington limit. At the Eddington limit the momentum absorbed per second from the radiation field by an electron + proton exactly balances the gravity on them. That is

$$c\left(\frac{L}{4\pi r^2}\right) \sigma_{T/c^2} = \frac{GMm_H}{r^2}$$

where m_H is the mass of the hydrogen atom and σ_T is the Thompson scattering cross section.

Since beyond the limit the outer parts will be blown away, we have

$$\frac{L}{M} \leq \frac{4\pi G m_H c}{\sigma_T} = 3.10^4 \frac{L_\odot}{M_\odot} \sim 6.10^4 \text{ ergs}(\text{gm})^{-1}\text{s}^{-1}$$

But $L \propto \dot{M}$ for a black hole so Eddington's limit ensures long growth times. If $L = 1/3\ \dot{M}c^2$ then

$$\frac{\dot{M}}{M} << 6.10^{-9}/\text{yr}$$

Hence black hole growth by such a process needs periods of the order of 10^8 years.

Shakura and Sunyaev pointed out that the situation was worse than this in the inner parts of accretion disks since only a fraction z/R of the gravity holds the material down onto the disk. Thus the vertical radiation pressure could blow the disk apart and will certainly thicken it. This leads us to the study of thick disks.

IV. OPTICALLY THICK ACCRETION

11. Self-Similar Solutions

So far we have considered the case when the energy readily escapes from the sides of a disk. We now consider the opposite extreme in which the material is so optically thick that the radiation is trapped within the fluid. If we begin with a flat disk, the energy dissipated by the time the material has moved in by a factor 2 in radius is sufficient to make its thickness comparable to its radius. Thereafter pressure, gravity and centrifugal force are all important for the configuration. It is interesting to ask whether pressure does not eventually become dominant so that the angular momentum becomes unimportant at the smaller radii. In an accreting system this cannot be the case for if we neglect the rotation, the accretion will become a Bondi flow. Any initially unimportant rotation will be advected by such a flow and becomes more and more important as the centrifugal force due to it behaves like h^2/R^3 which is a more rapid dependence on R than the gravitational attraction. Thus centrifugal force cannot consistently be left out of the problem. Since we have already pointed out that pressure becomes important even if it is not initially, it is clear that pressure, gravity and centrifugal force are all important at all radii. Now we are interested in configurations that occupy many decades of radius so if all three forces are to be important at all radii they must have similar radial dependencies. We shall see presently that the assumption that pressure forces and centrifugal forces have the same ratio at all radii leads us to consider self-similar solutions. In the Newtonian case this problem was first studied by S. Gilham. In spherical coordinates the balance between pressure gravity and centrifugal forces is given by the equations

$$r\,\Omega^2\sin^2\vartheta = \frac{1}{\rho}\frac{\partial p}{\partial r} + \frac{GM}{r^2}$$

$$r\,\Omega^2\sin\vartheta\cos\vartheta = \frac{1}{r\rho}\frac{\partial p}{\partial \vartheta}$$

The viscosity causes the fluid to flow through the configuration. The balance between the advection of angular momentum into any region and the viscous couples that remove it from the surface of that region is given by

$$\rho\underline{u}\cdot\underline{\nabla}\,(r^2\sin^2\vartheta\Omega) = \underline{\nabla}\cdot(\rho\nu r^2\sin^2\vartheta\nabla\Omega)$$

and the equation that gives the equation of state we take to be $p = \varkappa\rho^{1+1/n}$

where $\varkappa$ is not constant but depends on the specific entropy. The equation for the entropy generation tells us how changes from point to point. Now $T.\nabla s = p/\rho \nabla \varkappa^n$ and so

$$\rho \underline{u} . \left[\frac{p}{\rho} \underline{\nabla}/n(p^n/\rho^{n+1}) \right] = \rho \nu \sin^2\vartheta \; (\underline{\nabla}\,\Omega)^2$$

To these we must add the continuity equation

$$\underline{\nabla}\cdot(\rho\underline{u}) = 0$$

Now since by our earlier argument the ratio of $1/\rho\ \partial p/\partial r$ to $r\Omega^2\sin^2\vartheta$ must be independent of r, we find that $r^2/\rho\ \partial p/\partial r$ and $r^3\Omega^2$ must be independent of r. We therefore introduce rescales variables by

$$\Omega = \left(\frac{GM}{r^3}\right)\ \omega\,(\vartheta)$$

$$u = \nu/r\ u^*$$

$$\rho = 1/r\nu\ \rho^*$$

$$p/\rho = \nu\rho^{1/n} = p^*/\rho^*\ GM/r.$$

To satisfy our conditions we need all the starred variables to be independent of r. The solution is then self-similar. An interesting property of the solutions that carry a new flux is that the velocity has no component. If the flux into the origin is F and the flux crossing a portion of a sphere at radius r and between ϑ and $\vartheta + d\vartheta$ is dF, then the self-similarity of the solution demands that dF/F is constant independent of r. Since the radial flux going down any curve is constant, the continuity equation tells us that $1/r^2\sin\vartheta\,\partial/\sigma\vartheta(\rho r^2\sin\vartheta u_\vartheta) = 0$. From which we deduce $\rho r^2 \sin\vartheta\, u_\vartheta$ = constant. As this is zero on the equator by symmetry it will be zero everywhere and $u_\vartheta = 0$. Eliminating everything except ω from the above equations and taking $\nu \propto r^{\frac{1}{2}}$ (as is appropriate for a turbulent viscosity with length scales αr and velocity scaling as $(GM/r)^{\frac{1}{2}}$) we find

$$\omega'' - \frac{\omega}{1-\omega^2\sin^2\vartheta}\left[\frac{3\omega'\omega t\vartheta}{\omega} + \frac{9}{2}\,\omega\,\omega'\sin\vartheta\cos\vartheta + \frac{9\alpha}{4}\,\omega^2\sin^2\vartheta + \alpha + 2\omega'^2\sin^2\vartheta - \frac{3}{4}\,(1 - \omega^2\sin^2\vartheta)\right] = 0$$

where

$$\alpha = \frac{5}{4(n-3/2)}$$

Gilham computed the solutions to this ω equation and showed that there was only one regular one. His solution is illustrated in Figure 10 along with the lines of constant pressure, density and entropy. Although these solutions are of great interest, they have serious drawbacks.

1. The entropy by construction increases inwards because entropy is generated and kept by the fluid elements as they flow. Thus there is a stratification that is unstable to convection. Actually angular momentum gradients stabilise this enar the equator but not for certain other directions.
2. The solution does not have an empty vortex region near the pole and more seriously the fluid is motionless at the pole.

Contours of h and kappa (n=1.501)

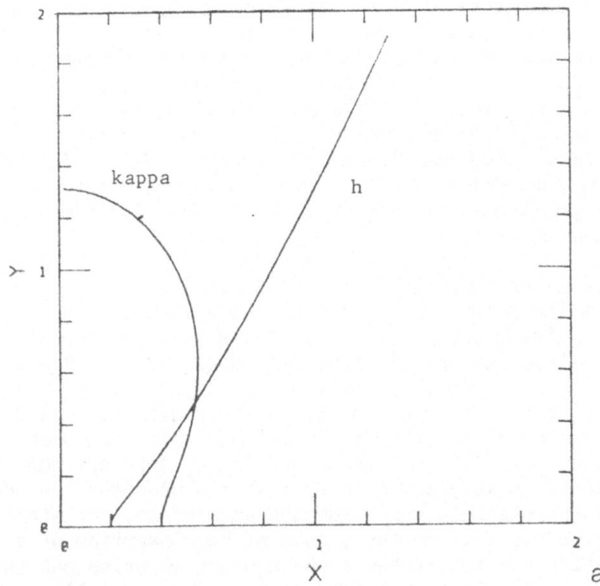

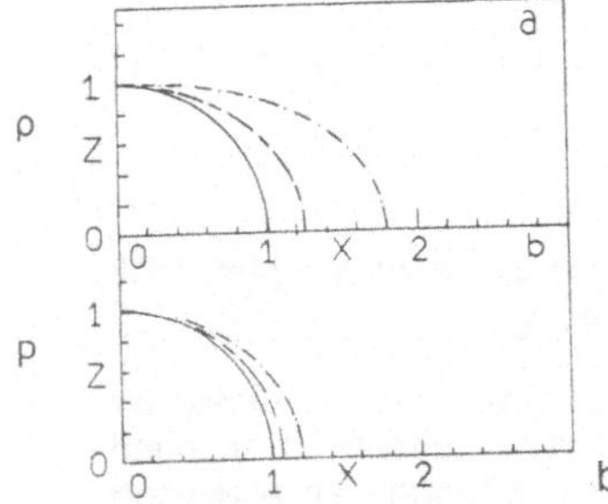

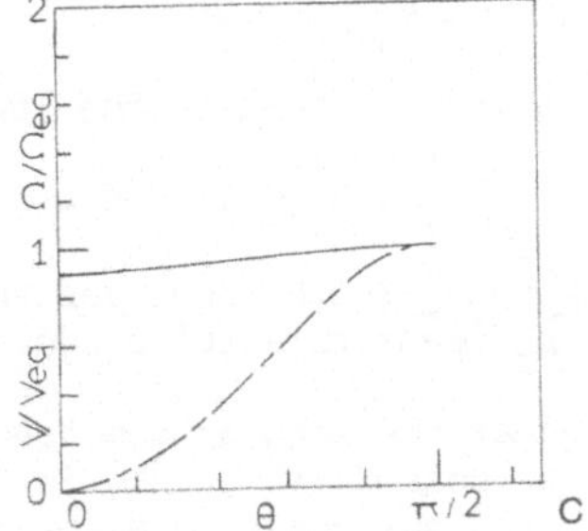

n = 1.501 ——
n = 2 ——- ——-
n = 3 ——.——.

FIGURE 10.

3. Anderson has found why any fluid at the pole must be static in the solution and why the radial motion does not reverse at any latitude. If u reversed then it would have to be zero somewhere. At such a point $r \sin \vartheta\Omega$ and $r \sin \vartheta d\Omega/d\vartheta$ must be zero so that entropy is not generated there because there is no way to it away when u is zero. But Ω is not zero so u can only be zero when ϑ is zero. On the polar axis no entropy is generated since the lateral motion is zero. However the dependence of $\varkappa$ is the same as it is at other latitudes. If u were non zero, the advection of entropy would lead to changes of its distribution in contradiction to our steady flow assumption. Thus u must be zero on axis unless $\varkappa$ is zero there. Anderson has generalised Gilham's solution to allow for some thermal conductivity, this removes the necessity for the solution to be static on axis as expected and it is possible to find solutions with outflows on axis.

In a relativistic situation with a central black hole it is hard to see what can hold up the polar column of fluid in Gilham's solution; however there are no steady self-similar solutions to this problem with the axis evacuated and this is still true if large radial motions are allowed. It is perhaps worth remarking that the argument for self-similarity given above does not apply to the region very close to the axis where angular momentum is not important. It could be that self-similarity breaks down there. Astrophysically jets are observed along rotation axes so it is very inviting to insert a Blandford and Rees jet on the axis and then integrate away from the edge of the jet starting with pressure balance across it. However, such ad hoc insertion of a jet gives no true explanation of its existence which ought to arise naturally from the analysis. An indication that jets might arise naturally from a full analysis comes from the enthalpy theorem to which we now turn.

12. Enthalpy Theorem and Jet Production

For steady state axially symmetrical flows in general relativity there are two Killing vectors $\underline{k}$ and $\underline{\ell}$ corresponding to time displacements and rotational displacements. These obey Killing's equation

$$K_{\mu\nu} = k_{\mu;\nu} + k_{\nu;\mu} = 0$$

and lead to vectorial conservation laws since

$$\underline{\Box}\cdot(\underline{\underline{T}}\cdot\underline{k}) = \tfrac{1}{2}\underline{\underline{T}}:\underline{\underline{K}} + \underline{k}\cdot(\underline{\Box}\cdot\underline{\underline{T}}) = 0 + 0$$

Thus both $\underline{\underline{T}}\cdot\underline{k}$ and $\underline{\underline{T}}\cdot\underline{\ell}$ are conserved vectors. Conservation of proper rest mass density ρ_0 leads to a third conserved vector $\rho_0\,\underline{u}$.

Now consider the angular momentum advected into a region bounded by a surface S per unit time. This is $\int h\rho_0\,\underline{u}.d\underline{S}$. If the region is bounded internally by the surface of a black hole then a flux of angular momentum h_0F will leave the region. Thus the net gain of advected angular momentum is $\int(h-h_0)\,\rho_0\,\underline{u}.d\underline{S}$. Since the fluid is in a steady flow this must be balanced by the viscous couple. The latter is zero at the hole and so

$$C = \int (h-h_o)\, \rho_o\, \underline{u}.d\underline{S} \qquad 12.1$$

This is reminiscent of our flat disk formula $C = F(h-h_o)$. If we now choose the surface S to be one of constant Ω, then the rate of working of the viscous couple is just $C\Omega$. Now consider the volume bounded by the sphere at and on the inside by the surface S. The rate of working of the couple at ∞ goes to zero because $C \propto r^{\frac{1}{2}}$ but $\Omega \propto r^{-3/2}$. The energy advected in at ∞ is $w_\infty F = \int w_\infty \rho_o \underline{u}.dS$ and the energy advected down into S is $\int_S w(u.k/c)\, \rho_o\, \underline{u}.dS$.

In a steady state these must balance the rate of working of the viscous couple. Hence

$$C\Omega = \int \left(w \frac{\underline{u}.\underline{k}}{c} - w_\infty\right) \rho_o\, \underline{u}.d\underline{S} \qquad 12.2$$

Subtracting $C\Omega$ evaluated from the angular momentum equation we obtain the enthalpy theorem.

$$\int_S \left[\left(w\frac{\underline{u}.\underline{k}}{c} - c^2\right) - \Omega(h-h_o)\right] \rho_o\, \underline{u}.dS = 0 \qquad 12.3$$

where we have assumed the fluid to be cold at ∞ to $w_\infty = c_2$. The different parts of this equation all have interesting physical meaning.

$$w\frac{\underline{u}.\underline{k}}{c^2} - c^2$$

is the relativistic Bernoulli constant for a motion in the absence of viscosity.

$$\Omega(h-h_o)$$

is of the order of the square of the circular velocity of rotation while

$$\rho_o\, \underline{u}.dS$$

is the flux of rest mass. Thus 12.3 tells us that the flux weighted average of the Bernoulli constant over a surface of constant Ω is the flux weighted average of $\Omega(h-h_o)$ which is roughly V_c^2. If we were to ask an oil man to drill a hole and insert a pipe then fluid would flow out and in the absence of friction the Bernoulli constant will give us its γ-factor at ∞. In fact, the Bernoulli constant at will be $(\gamma-1)c^2$. Hence by determining $\Omega(h-h_o)$ for models of accreting systems we may estimate the velocity with which such flows could in principle drive jets. For accretion onto a Schwarzschild hold, Anderson obtains a maximum of $\Omega(h-h_o)$ of 0.015 c_2 and a maximum γ of 1.015 corresponding to $v/c = 0.17$. This may be compared with SS 433s jet which has $v/c = 0.26$. However, it is likely that real black holes will spin and Kerr metrics almost certainly give greater values of $\Omega(h-h_o)$. Thus it is not impossible that jets arise naturally from the fluid mechanics of accreting black holes alhough no proper model exists as yet.

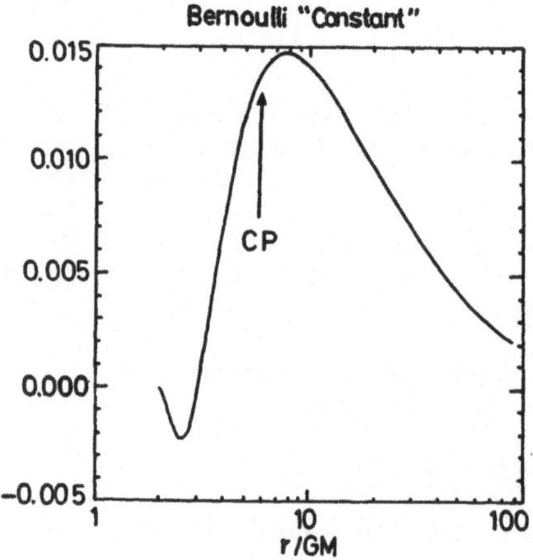

FIGURE 11.

One further remark on jets is that even after expansion the jets seen in the superluminal sources must be relativistic. This means that if the jets are followed backwards to their stagnation points the material must have very highly relativistic pressures and temperatures. These are so hot that the matter content must be totally dominated by electron-positron pairs rather than by baryons. Current calculations make it unlikely that there is time for these pairs to annihilate during the expansion of the jet into the outside world. One is therefore led to the conclusion that the jets from astrophysical sources should be made primarily of electron-positron pairs. It is important to think of ways to discover whether this is true!

Appendix

$$ds^2 = \rho^2 \Delta^{-1} dr^2 + \rho^2 d\vartheta^2 + \rho^{-2}\sin^2\vartheta[(r^2+a^2)d\varphi - adt]^2$$
$$- \Delta\rho^{-2}(dt - a \sin^2\vartheta \, d\varphi)^2 \quad (1$$

where

$$\Delta = r^2 - 2mt + a^2 \quad (2)$$

$$\rho^2 = r^2 ; a^2 \cos^2 \vartheta \quad (3)$$

and $m = GM/c^2$. M is the mass of the body and ac its angular momentum per unit mass. When m is set to zero ds^2 reduces to the familiar form for oblate spheroidal coordinates based on a disk of radius a, r is constant on confocal ellipsoids, ϑ on confocal hyperboloids and r,ϑ,φ tend to spherical polar coordinates at infinity $(r \gg a)$. In the general Kerr metric the circumference of a circle with coordinates r, ϑ is $2\pi R$ where

$$R^2 = (r^2 + a^2 + 2ma^2 \rho^{-2} r \sin^2\vartheta) \sin^2\vartheta \tag{4}$$

which becomes

$$R^2 = r^2 + a^2 + 2ma^2/r \text{ on } \vartheta = \pi/2 \tag{5}$$

The equations of geodesic motion are $\delta(\int L\, d\tau) = 0$ where L the Lagrangian is $[-g_{\mu\nu}\dot{x}^\mu \dot{x}^\nu]^{\frac{1}{2}}$ and dots denote derivation with respect to the proper time τ along the orbit so $d\tau^2 = -ds^2$. φ is ignorable so the specific angular momentum $h = \partial L/\partial\dot{\varphi}$ is conserved

$$\begin{aligned} h &= \rho^{-2}\sin^2\vartheta[(r^2 + a^2)\dot{\varphi} - a\dot{t}](r^2 + a^2) \\ &\quad + \Delta\rho^{-2}(\dot{t} - a \sin^2\vartheta\dot{\varphi})a \sin^2\vartheta \\ &= R^2\dot{\varphi} - 2mra\rho^{-2} \sin^2\vartheta\, \dot{t}. \end{aligned} \tag{6}$$

In the second form we see the angular momentum of the particle split from the field angular momentum that the particle carries along with it just like a charged particle in a magnetic field. t is ignorable and so the specific energy $\varepsilon = \partial L/\partial\dot{t}$ is conserved.

$$\begin{aligned} \varepsilon &= a\rho^{-2}\sin^2\vartheta[(r^2 + a^2)\dot{\varphi} - a\dot{t}] + \Delta\rho^{-2}(\dot{t} - a \sin^2\vartheta\dot{\varphi}) \\ &= \dot{t}(1 - 2mr\rho^{-2}) + 2ma\rho^{-2}r \sin^2\vartheta\dot{\varphi} \end{aligned} \tag{7}$$

from eqs. (6) and (7)

$$\Delta(\dot{t} - a \sin^2\vartheta\dot{\varphi}) = -ah + (r^2 + a^2)\varepsilon \tag{8}$$

and $\sin^2\vartheta[(r^2 + a^2)\dot{\varphi} - a\dot{t}] = h - a \sin^2\vartheta\varepsilon$. (9)

In the equatorial plane $\rho^2 = r^2$ so the metric equation $(ds/d\tau)^2 = -1$ reads $-\Delta = r^2\dot{r}^2 + \Delta r^{-2}(h-a\varepsilon)^2 - r^{-2}[(r^2+a^2)\varepsilon-ah]^2$ which may be written $r^2\dot{r}^2 = \Phi$ where Φ is given by

$$\Phi = (\varepsilon^2-1)(r^2+a^2+2ma^2/r) - 4\varepsilon hma/r - h^2 + 2m[r+(h^2+a^2)/r] \tag{10}$$

At the turning points $\dot{r} = 0$ and so $\Phi = 0$.

Circular orbits are distinguished by having coincident inner and outer turning points. Hence $\Phi = 0$ must have double roots for circular orbits, and so for them $d\Phi/dr = 0$ also. This gives us

$$(\varepsilon^2-1)(r^2-ma^2/r) + 2\varepsilon hma/r+m[r-(h^2+a^2)/r] = 0 \tag{11}$$

Adding twice eq. (11) to eq. (10) with $\Phi = 0$ yields

$$Q(3r^2 + a^2) - h^2 + 4mr = 0 \tag{12}$$

where

$$Q = \varepsilon^2 - 1. \tag{13}$$

Squaring eq. (11) we have

$$Q^2(r^2-ma^2r^{-1})^2 + 2Qm[r-(h^2+a^2)/r] + m^2[r-(h^2+a^2)/r]^2$$
$$= 4(Q+1)h^2m^2a^2/r^2.$$

Hence substituting for h^2 from eq. (12) and collecting powers of Q we find

$$Q^2[r^2(r-3m)^2 - 4ma^2/r] + 2mQ[r(r-3m)(r-4m) - a^2(r+5m)]$$
$$+ m^2[(r-4m-a^2/r)^2 - 16ma^2/r] = 0.$$

Solving this equation for Q we find

$$Q = \frac{m}{r} \times \left[\frac{r(r-3m)(r-4m)-a^2(r+5m)\pm 2a(m/r)^{\frac{1}{2}}(r^2-2mr+a^2)}{r(r-3m)^2 - 4ma}\right].$$

By looking at the zero of the denominator as a function of "a" we find a common factor for numerator and denominator of $r-3m\pm 2(m/r)^{\frac{1}{2}}$ a, hence

$$Q = -\zeta\left[\frac{1-4\zeta-\alpha^2\zeta^2\pm 4\alpha\zeta^{3/2}}{1-3\zeta\pm 2\alpha\zeta^{3/2}}\right] \text{where } \alpha = a/m \text{ and } \zeta = m/r. \tag{14}$$

Substitution of this value for Q into eqs. (13) and (12) yields for the energy and angular momentum of circular orbits

$$\epsilon = (1+Q)^{\frac{1}{2}} = \frac{1-2\zeta\pm \alpha\zeta^{3/2}}{(1-3\zeta\pm 2\alpha\zeta^{3/2})^{\frac{1}{2}}} \tag{15}$$

$$h = \pm \frac{m}{\zeta^{\frac{1}{2}}}\frac{1 +\alpha^2\zeta^2 \pm \alpha\zeta^{3/2}}{(1-3\zeta \pm 2\alpha\zeta^{3/2})^{\frac{1}{2}}} \tag{16}$$

The angular velocity of circular orbits as seen from infinity is

$$\Omega = \dot{\varphi}/\dot{t} = \pm m^{-1} \zeta^{3/2}/(1\pm\alpha\zeta^{3/2}) = \pm \frac{m^{\frac{1}{2}}}{r^{3/2}\pm am^{\frac{1}{2}}} \tag{17}$$

We now wish to find the last stable circular orbits. For such orbits Φ must change from a having a cup shaped maximum to having an inflection; thus not only $\Phi = \Phi' = 0$ but also $\Phi'' = 0$. In particular therefore $(\Phi+r\Phi')' = 0$ and thus the derivative of eq. (12) is also zero. We thus deduce

$$Q3r + 2m = 0 \quad \text{and so} \quad Q = -\tfrac{2}{3}\zeta\,. \tag{18}$$

Using this in eq. (14) and solving the resultant equation for α we find

$$\alpha = \tfrac{1}{3}\zeta^{-\frac{1}{2}}\,[4-(3\zeta^{-1}-2)^{\frac{1}{2}}]. \tag{19}$$

Substituting this value into eq. (12) together with $Q = -\frac{2}{3}\zeta$ we obtain

$$h_o = \frac{2}{3\sqrt{3}}m[1+2(3\zeta^{-1}-2)^{\frac{1}{2}}]$$
$$\epsilon-\Omega h = \frac{(1-3\zeta\pm 2\alpha\zeta^{3/2})^{\frac{1}{2}}}{1 \pm \alpha\zeta^{3/2}}\,. \tag{20}$$

REFERENCES

Antonov, V.A., 1962, Vest Leningrad. gos Univ. 7, 135.
Bardeen, J., 1970, Nature 226, 64.
Begelman, M.C. & Meyer, D.L., 1982, Astrophys. J. 253, 873.
Bettwieser, E. & Sugimoto, D., 1984, Mon. Not. R. astr. Soc., 208, 493.
Blandford, R.D. & Rees, M.J., 1974, Mon. Not. R. astr. Soc., 169, 395.
Bondi, H., 1952, Mon. Not. R. astr. Soc., 112, 195.
Cohn, H., 1980, Astrophys. J., 242, 765; also IAU Symp. 113, Reidel.
Fillmore, J.A. & Goldreich, P., 1984, Ap. J. 281, 1, 9.
Gilham, S., 1981, Mon. Not. R. astr. Soc., 195, 755.
Hachisu, I., Nakada, Y., Nomoto, K. & Sugimoto, D., 1978, Prog. Theor. Phys. 60, 393.
Heggie, D.C., 1985, Goodman, J. & Hut, P., eds IAU Symp. 113, Reidel.
Henon, M., 1961, Ann. Astrophys. 24, 369.
Hoyle, F. & Lyttleton, R.A., 1939, Proc. Camb. Phil. Soc. 35, 405.
Hunt, R., 1971, Mon. Not. R. astr. Soc. 154, 152.
Inagaki, S. & Lynden-Bell, D., 1983, Mon. Not. R. astr. Soc. 205, 913.
Katz, J. & Lynden-Bell, D., 1978, Mon. Not. R. astr. Soc. 184, 708.
Landau, C.D. & Lifshitz, E.M., 1966, Classical Theory of Fields, MIR, Moscow.
Lynden-Bell, D. & Wood, R., 1968, Mon. Not. R. astr. Soc. 138, 495.
Lynden-Bell, D., 1969, Nature 223, 690.
Lynden-Bell, D. & Pringle, J.E., 1974, Mon. Not. R. astr. Soc. 168, 603.
Lynden-Bell, D. & Lynden-Bell, R.M., 1977, Mon. Not. R. astr. Soc. 181, 405
Lynden-Bell, D., 1978, Physica Scripta 17, 185.
Lynden-Bell, D. & Eggleton, P.P., 1980, Mon. Not. R. astr. Soc. 191, 483.
Michel, F.C., 1972, Astrophys. & Sp. Sc. 15, 153.
Penston, M.V., 1969, Mon. Not. R. astr. Soc. 144, 425.
Tolman, R.C., 1934, Relativity Thermodynamics & Cosmology.

ACCRETION DISK ELECTRODYNAMICS

M. Kuperus

University of Utrecht, Observatory "Sonnenborgh"

Utrecht, Netherlands

1. THE STANDARD THIN DISK

A plasma cloud orbiting around a compact object of mass M with the Kepler velocity $v_\phi = (GM/r)^{\frac{1}{2}}$ at a distance r from the center of the object has an angular momentum $I = m\Omega r^2$, where m is the mass of the cloud and $\Omega = v_\phi/r$ is the angular velocity. The inner parts of the cloud have a larger angular velocity then the outer parts do thus creating a shear. The result of this differential rotation is twofold. First the cloud is stretched in the azimuthal direction until a ring is formed. Secondly because of the shear the cloud diffuses in the radial direction transporting mass inward as well as outward. The innermost parts of the ring are slowed down due to the viscous drag caused by the slower moving outer parts while the outer parts are accelerated due to the viscous drag of the faster moving inner parts. Consequently the inner parts start drifting inwards and the outer parts start drifting outwards. Since the angular momentum is proportional to r^2 the outflowing parts of the plasma cloud transport more angular momentum outward then the inward moving parts transport inward. Due to the viscous processes inside the disk the outward transport of angular momentum is achieved thus facilitating the process of accretion.

Let us assume that all the accreting matter is assembled in a thin disk where the matter is slowly drifting inwards with velocity v_r compared to the orbital Kepler velocity and where the vertical structure is determined by hydrostatic equilibrium. Let us also assume that the matter accreted by the compact object is supplied sufficiently fast from the outside world so that a steady disk may be formed. The structure of such a thin steady accretion disk has been analysed by Shakura and Sunyaev [1] and is called the standard disk model. The standard disk model, a point of departure for all analyses and considerations of thin accretion disks, is based on the conservation of mass, angular momentum and energy, where the energy source is the viscous dissipation of the differentially rotating disk represented by the viscous stress tensor $t_{\phi r}$ and the energy sink is the black body radiation of an optically thick medium.

Introducing the vertically integrated density $\Sigma(r,t) = 2\,h\rho$ and the integrated stress tensor $W(r,t) = 2\,h\,t_{\phi r}$ where h is the half thickness of the disk the vertically averaged equations are:

$$\frac{1}{r}\frac{\partial}{\partial r}(r\Sigma v_r) = 0, \tag{1.1}$$

$$\Sigma v_r(GMr)^{\frac{1}{2}} = -2\frac{\partial}{\partial r}(r^2W), \tag{1.2}$$

$$F = \frac{3}{4}\left(\frac{GM}{r^3}\right)^{\frac{1}{2}} W = \frac{\frac{2}{3} cb\ T^4}{\Sigma\bar{\kappa}}, \tag{1.3}$$

$$\bar{\kappa} = 0.4 + 0.32 * 10^{23}\left(\frac{\Sigma}{h}\right) T^{-7/2}, \tag{1.4}$$

$$p = \frac{\Sigma}{h}\frac{k}{m}\frac{T}{} + \frac{1}{3} bT^4, \tag{1.5}$$

$$p = \frac{1}{2}\Sigma h \frac{GM}{r^3}, \tag{1.6}$$

where equation (1) states that the mass accretion rate $\dot{M} = r\Sigma v_r$ is constant. Equation (2) equates the radially transported angular momentum Iv_r to the viscous stress torque. Equation (3) describes that the radiation flux F due to the energy generation by viscous dissipation is given by the energy loss of a black body where the opacity $\bar{\kappa}$ is given in (4) assuming free-free and Compton scattering. Eq. (5) gives the total pressure which consists of gas pressure and radiation pressure and eq.(6) results form the assumption of hydrostatic equilibrium after integrating over the thickness of the disk. The major assumptions in the standard model are: $h/r \ll 1$; $v/c \ll 1$; $v_r \ll v_\phi$; $v_s \ll v_\phi$ together with $\tau \gg 1$ and energy transfer by radiation only.

The key assumption in the standard model is that the viscous stress tensor is assumed to be proportional to the pressure $t_{\phi r} = \alpha p$, where α is a constant independent of the radius. The integrated stress is then given by $W = 2\alpha hp$. A solution of the equations (1) to (6) is possible in terms of the three parameters M, $\dot{M}$ and α .

It has been noticed from the beginning that ordinary molecular viscosity is orders of magnitude too small to give a stress sufficiently large to transport the angular momentum. Hence turbulent viscosity and magnetic viscosity have been proposed as the most likely transport processes in the disk. The proportionality between $t_{\phi r}$ and p can be easily argued since $t_{\phi r} \sim \eta\ (\partial v_\phi/\partial r)$ as long as $v_r \ll v_\phi$. For turbulence with v_t being the turbulent turnover time of an eddy of size ℓ the viscosity is $\eta \sim \rho v_t \ell$. From eq. (6) we find $p/\rho = (\Omega^2 h^2)/2$ which results with $v_s{}^2 = \gamma p/\rho$ into $t_{\phi r} = \alpha p$ with $\alpha \sim v_t \ell / v_s h$ Since $\ell \lesssim h$ it follows that $0 < \alpha < 1$ for subsonic turbulence. In the case of a turbulent magnetic field the magnetic stresses should be added to the turbulent stresses. Shakura and Sunyaev argue that the equations can be solved for any value of α thus giving the quantities Σ, T, v_r and h as a function of r starting from the surface of the compact object or in the case of a black hole from the innermost stable orbit which is located at r = 3 Rs, where Rs is the Schwarzschild radius. For a detailed discussion of the standard model the reader is referred to Frank, King and Raine [2] and Pringle [34] . The temperature decreases from high values inside to much lower values outside. All quantities are mildly dependend on α which assures the construction of a disk is not critically dependend on α but on the other hand makes it difficult to determine the value of α from the observations.

It is important for further discussions to mention here that the physical state inside the disk may change from radiation pressure dominated in the inner parts to gaspressure dominated in the outer parts. The ratio of radiation pressure p_r to gas pressure p_g is given by

$$\frac{p_r}{p_g} = 2.8 \times 10^{-3} \alpha^{1/10} \dot{M}_{16}^{7/10} r_{10}^{-3/8} f^{7/5} . \qquad (1.7)$$

Where $\dot{M}_{16}$ is the mass accretion rate expressed in units of 10^{16} g/sec and r_{10} in units of 10^{10} cm. Only for large accretion rates close to the Eddington limit $\dot{M} \sim 10^{18}$g/sec. a radiation pressure dominated inner region may be formed in a disk.

Since it is the diffusion of mass and angular momentum by viscous stresses that determine the formation of a disk let us consider the equations of mass and angular momentum conservation allowing for time variations:

$$\frac{\partial \Sigma}{\partial t} = - \frac{1}{r} \frac{\partial}{\partial r} (r\Sigma v_r) , \qquad (1.8)$$

$$\frac{\partial}{\partial t} (\Sigma\Omega r^2) = - \frac{1}{r} \frac{\partial}{\partial r} (r\Sigma v_r \Omega r^2) - \frac{1}{r} \frac{\partial}{\partial r} (r^2 W) . \qquad (1.9)$$

After elimination of $(r\Sigma v_r)$ we obtain the mass diffusion equation:

$$\frac{\partial \Sigma}{\partial t} = \frac{2}{r} (GM)^{-\frac{1}{2}} \frac{\partial}{\partial r} (r^{\frac{1}{2}} \frac{\partial}{\partial r} (r^2 W)) . \qquad (1.10)$$

Since $t_{\phi r} = \rho\nu \frac{\partial v_\phi}{\partial r}$ the integrated stress $W \sim \nu\ \Sigma\Omega$ so that from (1.9) and (1.10) it follows that $v_r \sim \nu/r$ and the diffusion time $\tau_{visc} = r/v_r$. The dynamical time scales in the disk are the Kepler time $\tau_k \sim 1/\Omega$ and the time to establish hydrostatic equilibrium in the vertical direction $\tau_\ell = h/v_s$.
From (1.6) we obtain $v_s/v_\phi = h/r$, so that $\tau_h \sim \tau_k$.
The fourth time scale is the thermal time scale determined by the viscous heating in the disk $\varepsilon \sim \nu\Sigma v_\phi{}^2/r^2$ (see Ref. 34).
The thermal time scale is $\tau_t = \Sigma v_s^2/\varepsilon \sim v_s^2 \tau_{visc}/v_\phi^2$.
Since $v_s << v_\phi$ the thermal time scale is much smaller than the viscous time scale. Thermal fluctuations are therefore wiped out much faster than density fluctuations.
The diffusion equation (1.10) demonstrates that density fluctuations are smeared out provided W is proportional to Σ^γ with $\gamma > 0$.
However this situation does not necessarily occur in the inner region where radiation pressure is dominant.
As a result the effective diffusion coëfficient in equation (1.10) is negative which leads to a clumping of density enhancements where the denser parts instead of smearing out are becoming more concentrated thus leading to rings of matter with little matter in between. Actually matter is pushed in regions with small W. Eventually the low density regions become optically thin and thermally unstable [3, 4]. The temperature increases and the unstable inner parts of the disk will expand. This so called "Lightman-Eardley instability" is the basis for the two temperature model of the inner disk as advocated by Shapiro, Lightman and Eardley [5] . This model has been particularly favorite for the explanation of the hard X-ray emission of Cygnus X-1. In section 3 we will return to the problem of the hard X-ray emission in terms of a thin disk sandwiched by a hot corona.

2. TURBULENT DYNAMO IN ACCRETION DISKS

A small magnetic field in the accreting matter will be amplified in a Kepler disk by the differential rotation. At the same time the turbulence in the disk will generate a turbulent magnetic field. The magnetic field in the disk therefore consists of an organized spiral type field $\vec{B}_o = (B_{ro}, B_{\phi o}, 0)$, and a turbulent magnetic field $\vec{B}_1$. Introducing the turbulent helicity

$$\alpha = -\frac{1}{3}\tau \langle \vec{v} \cdot (\nabla \times \vec{v}) \rangle , \tag{2.1}$$

where τ is the characteristic correlation time of the turbulent motions [6], it can be shown applying so called "mean field magnetohydrodynamics" that the turbulent motions may generate a non zero mean field or organized field apart of the effect of differential rotation. The field generation is thus described by the equation

$$\frac{\partial \vec{B}_o}{\partial t} = \nabla \times (\vec{v}_o \times \vec{B}_o + \alpha \vec{B}_o) + \beta \nabla^2 \vec{B}_o, \tag{2.2}$$

where β is the total diffusion coefficient (turbulent + magnetic). Since $\vec{v}_o$ is the differential Kepler rotation the field generating equations written in cylindrical symmetry are

$$\frac{\partial B_{or}}{\partial t} = -\frac{\partial}{\partial z}(\alpha B_{o\phi}) + \beta(\nabla^2 B_o)_r \tag{2.3}$$

$$\frac{\partial B_{o\phi}}{\partial t} = r\frac{d\Omega}{dr} B_{or} + \frac{\partial}{\partial z}(\alpha B_{o\phi}) + \beta(\nabla^2 B_o)_\phi \tag{2.4}$$

If we assume the $B_{o\phi}$ field primarily generated by the differential rotation, $B_{o\phi}$ may be eliminated from eq. (2.3). We then find for B_{or}

$$\frac{\partial^2 B_{or}}{\partial t^2} = -\frac{\partial}{\partial z}\left(\alpha \left[r\frac{d\Omega}{dr}\right] B_{or}\right) + \beta \frac{\partial^2 B_{or}}{\partial z^2 \partial t} \tag{2.5}$$

describing the principal equation for the magnetic dynamo in a thin accretion disk. The last term results from a thin disk approximation $(\nabla^2 B_o)_r = \partial^2 B_{or}/\partial z^2$ since $h/r \ll 1$.
The generation of B_{or} is described by the first term and the decay by the second term. Since the generation depends on α as well as on Ω this type of dynamo is called an $(\alpha\Omega)$ dynamo.

The efficiency of magnetic field generation in the disk is described by the so called dynamo number N comparing the growth time τ_g to the decay time τ_d which can be estimated from equation (2.5): $N = (\tau_g/\tau_d)^2 = \alpha\Omega h^3/\beta^2$.
This quantity is large in accretion disks so that we expect an efficient dynamo action in disks leading to strong turbulent field fluctuations. As a matter of fact the non uniform rotation together with the density stratification leads to a strong net helicity α . Following Pudritz [7, 8], we can make a rough estimate of the dynamo number $\alpha \sim \tau v^2/3\ell$ under the assumption that $\tau \sim \tau_k$ and the turbulent correlation length $\ell \sim h$. Then with $v = M_t v_s$, where M_t is the turbulent Mach number and $h/r = v_s/v_k$ for thin disks it follows that $\alpha \sim h M_t^2/3\tau_k$.
If we furthermore assume that the dissipation is mainly due to turbulent

dissipation it follows that $\beta \sim v^2 \tau \sim M^2_t h_o^2/\tau_k$, so that $N \sim (3M_t^2)^{-1}$. A value $N \sim 10 - 100$ seems natural for subsonic turbulence. Pudritz [8] derives analytically steady state solutions for the magnetic field solving the dynamo equations (2.3) and (2.4) He concludes that the first mode to be excited is the even mode which means that B_ϕ and B_r are even functions of z and B_z is an odd function of z.

It should be noted here that in the dynamo equations the Coriolis force and the gravity force are neglected. These forces will pose a severe constraint on the interaction of the magnetic field inside the disk with the turbulence. The Maxwell stresses become comparable to the Coriolis force when

$$\frac{B_\phi B_r}{4\pi} \sim \rho\, v_{turb}\, \Omega\, h \sim \rho v_s^2 M_t \sim M_t p_{tot} \,. \tag{2.6}$$

We see that in this case the magnetic stress scales with the total pressure (gas + radiation). Another mechanism that limits the growth of magnetic fields is buoyancy caused by gravity acting on magnetic flux tubes which have a lower density than the ambient medium provided they are in pressure equilibrium [9]. Burm [10] estimated the Maxwell stress limited by buoyancy:

$$\frac{B_r B_\phi}{4\pi} \sim \left(\frac{3}{2}\frac{\alpha^3}{h^3\Omega^3}\right)^{\frac{1}{2}} p_{gas} \sim \frac{1}{6}\sqrt{2}\, M^3_t\, p_{gas} \,. \tag{2.7}$$

As pointed out by Burm [10] the stress scales in this case with the gas pressure instead of the total pressure. We thus observe that buoyancy puts severe limits to the field strength generated in the disk. Because of the third power of M_t, only near sonic turbulence may generate equipartition magnetic fields.

As a consequence large scale field of any appreciable strength will emerge from the disk thus forming magnetic layers on both sides of the disk presumably consisting of magnetic loops and streamer type structures similar as observed in the solar corona. For small scale magnetic field the buoyancy argument given before should be revised taking into account the magnetic tension. This will result in a much longer rise time or even no buoyancy at all if gravity and Lorentz force balances.

An important conclusion from this is that equipartition type fields or even much stronger fields can be expected particularly on small scales which are stronger "line tied" while the large scale fields must be weak, a result that has been found also by Pudritz and Fahlmann [11]. It is therefore probably not a bad approximation to consider the disk field to exist primarily of relatively small turbulent magnetic cells and then consider the growth and decay of such a cell. Such a heuristic analysis has been made by Eardley and Lightmann [12] and Coroniti [13, 14] focussing on another field limiting mechanism namely magnetic reconnection. Consider a magnetic flux cell of dimension ℓ_o and field strength B_o. The differential rotation shears the field and thus stretches the flux cells until the magnetic stresses oppose the shear. After the cell stretching is stopped by the magnetic stress the magnetic pressure gradients pinch the plasma in the center of the cell. The magnetic field lines are thus forced to reconnect which ultimately results in the fission of the original elongated flux cell into two smaller flux cells. This fission process occurs on a typical reconnection time scale. Hence the distortion stores shear motion into magnetic energy which is released in a reconnection process primarily into radial motion. The two remaining cells do not follow the Kepler rotation since they are slowed down. As a consequence the two cells will move radially in order to adjust their angular momentum. Hence the reconnection in a disk acts as an intermediary to convert the azimuthal (differential) motion into radial as well as azimuthal motions.

It has been pointed out by Burm [10] that the scaling of the turbulent stress tensor is with total pressure (i.e. including magnetic pressure) if field amplification is limited by reconnection. In most cases the stress tensor scales with the total pressure except for buoyancy limited fieldamplification where scaling with the gas pressure is appropriate.

3. ELECTRODYNAMIC COUPLING OF ACCRETION-DISK CORONAE

The hard X-ray emission of sources such as Cygnus X-1 can also be explained if the inner disk is surrounded by a hot optically thin corona (see Ref. 15). Such a corona is likely to be present since magnetic fields emerge from the disk by magnetic buoyancy as has been argued in section 2. If one can stabilize the inner parts of the accretion disk, a magnetically dominated corona can be formed which is driven by the turbulent motions in the accretion disk [16, 17, 18].

Let us now illustrate how an optically thick accretion disk is modified when an amount of energy L_{cor} is transferred from the disk into an overlying optically thin corona taking into account the coronal stressing on the angular momentum transport. The total energy flux, at a particular radius r is given by the work done by the total stresses W given by (1.3): $L_{tot} = 3\Omega W/4 = 3\alpha ph\Omega/4$.

We now separate the internally acting stresses due to (magneto-hydrodynamic) turbulent processes inside the disk from the externally applied stresses caused by a corona, which is magnetically tied to the disk. The presence of a corona coupled to the disk exerts an extra torque on the disk whose effect is lumped into the parameter α describing the total stresses and hence the total energy input. The distinction between internal and external stresses is made by setting $\alpha \equiv \alpha_i + \alpha_e$, where α_i parameterizes the internal stress and α_e the external stresses. Let us now define L_{disk} as that part of the disk luminosity maintained by internal viscous dissipation

$$L_{disk} = \frac{3}{4} \alpha_i \, ph\Omega \, . \tag{3.1}$$

It then follows that

$$\alpha = \alpha_i \left(1 + \frac{L_{cor}}{L_{disk}}\right) . \tag{3.2}$$

Since α parameterizes the total stresses in the disk-corona system and noting that most of the angular momentum transport is taking place inside the disk because the corona has a small amount of mass as compared to the disk, the angular momentum conservation equation is given by

$$\dot{M}\Omega r^2[1 - (r_*/r)^{\frac{1}{2}}] = \alpha_i \; p \; r^2 h\left(1 + \frac{L_{cor}}{L_{disk}}\right) . \qquad (3.3)$$

With regard to energy conservation one has to take into account the fact that part of the disk luminosity is maintained by radiative and conductive feedback from the corona to the disk. Therefore the disk energy conservation equation is

$$\frac{4\sigma_B T_c^{\;4}}{\tau} = \frac{3}{4}\alpha_i \; p \; h \; \Omega + \frac{\delta \; L_{cor}}{\pi r^2} , \qquad (3.4)$$

where $\tau = (\sigma_T \sigma_{ff})^{\frac{1}{2}} \rho H$ is the optical depth assuming that the Thomson opacity, $\sigma_T = 0.4 \; cm^2/g$ exceeds the free-free opacity, $\sigma_{ff} = 6.6 \times 10^{22} \; T_c^{-7/2} \rho \; cm^2/g$ and where δ is the fraction of the coronal heating rate L_{cor}, reabsorbed by the disk due to thermal conduction and hard x-radiation. Note that L_{disk} and L_{cor} are theoretically determined luminosities that are related but not necessarily equal to the observed X-ray luminosities L_{soft} and L_{hard}.

Equations (3.3) and (3.4) together with (1,5) and (1,6) describe the structure of the disk using the disk internal parameter α_i , modified by the coronal stressing through the factor $(1 + L_{cor}/L_{disk})$, taking into account the resulting feedback of energy from the corona. Note that the additional torque on the disk due to coronal stressing is phenomenologically represented by the factor $(1 + L_{cor}/L_{disk})$ in equation (3.3).
Only a fraction $(1 + L_{cor}/L_{disk})^{-1}$ of the accreting gravitational energy is radiated by the disk; the remainder is radiatively vented by the overlying corona. Coronal dissipation tends to raise the instability threshold since the ratio P_r/P_g is decreased in proportion to the rate at wich the disk couples nonthermal energy into the corona.

Kuperus and Ionson [18] estimated the upper limit of L_{cor} following Ionson [19, 20] who demonstrated that a fruitfull approach to this problem is to consider a coronal magnetic loop the coronal "load" in an equivalent (LRC) global current circuit with resistance R, capacity C and self inductance L. For a loop with length ℓ the equivalent self inductance $L = 4\ell/\pi c^2$, the equivalent capacity $C = c^2\ell/4\pi v_A^{\;2}$, with v_A the Alfvén velocity, and the resistance $R = L/\tau_{diss}$ where τ_{diss} is the characteristic dissipation time due to Joule, viscous and compressional damping. The efficiency of the coupling of the disk and the corona is determined by the quality factor of the equivalent electric circuit

$$Q = \frac{1}{R}\left(\frac{L}{C}\right)^{\frac{1}{2}} = \frac{2\pi\tau_{diss}}{\tau_A} , \qquad (3.5)$$

while the heating flux is given by

$$F_H = 4\,\frac{B_{cor}}{B_{disk}}\,\frac{4v_A^{\;d}}{v_A^{\;c}}\;v_A^{\;d}\;\langle\frac{1}{2}\,\rho v_{turb}^2\rangle_d\;\varepsilon , \qquad (3.6)$$

where B_{cor} and B_{disk} are the coronal and disk magnetic field respectively, $v_A^{\;d}$ is the disk Alfvén velocity $v_A^{\;c}$ the coronal velocity, $\tau_A = \ell/v_A$ is the loop Alfvén time and ε is an electrodynamic coupling coefficient describing the partitioning between kinetic energy and associated magnetic stresses. For a turbulent spectrum with peak frequency

$\nu = \tau_p^{-1}$ and half width τ_c around the peak Ionson [20] gives for a weakly damped system (i.e. $Q > 1$) $\varepsilon \sim \tau_c/\tau_A$ so that

$$F_H = 4 \frac{B_{cor}}{B_{disk}} \frac{4v_A^d}{v_A^c} v_A^d < \frac{1}{2} \rho v^2_{turb} > \frac{\tau_c}{\tau_A} . \tag{3.7}$$

We will assume that all the energy that is delivered to the corona through electrodynamical coupling is used to heat the corona. Part of this energy is transfered via the inverse Compton scattering of the disk soft X-ray photons into the hard X-ray photons, while part is conducted back to the disk. If a fraction f of the disk surface is covered with coronal loops the amount of energy dissipated in the corona which we call the coronal luminosity is given by

$$L_{cor} = 4 \frac{B_{cor}}{B_{disk}} \frac{4v_a^d}{v_A^c} v_A^d < \frac{1}{2} \rho v^2 >_d \frac{\tau_c}{\tau_A} f. \tag{3.8}$$

For the correlation time a reasonable estimate is the turnover time of the disk turbulence in a disk with thickness h: $\tau_c \sim h/v_{turb}$.
The expression (3.8) for the coronal luminosity then reduces to

$$L_{cor} = 8 f \frac{B_{cor}}{B_{disk}} (v_A^d)^2 \frac{h}{\ell} . \langle \rho v_{turb} \rangle_d . \tag{3.9}$$

It has been demonstrated in sections 2 that loops generated in the turbulent disk and emerging because of magnetic buoyancy must have a length of the order of the disk thickness (see Ref. 13). Hence we adopt $\ell \approx h$.
For subsonic turbulence $\alpha < 1$ and $v_{turb} \sim \alpha v_s$ is a good approximation so that

$$L_{cor} \approx 8f \frac{B_{cor}}{B_{disk}} (v_A^d)^2 \alpha \rho v_s . \tag{3.10}$$

Comparing the coronal luminosity with the disk luminosity given by eq. (3.1) we obtain using the hydrostatic equilibrium condition $h/r = v_s/v_\phi$:

$$\frac{L_{cor}}{L_{disk}} \sim 10 f \left(\frac{B_{cor}}{B_d}\right)\left(\frac{v_A^d}{v_s^d}\right)^2 . \tag{3.11}$$

The expression is remarkable in the sense that it neither depends on the disk thickness nor on the parameter α. The last statement however is only correct for subsonic turbulence i.e. $\alpha < 1$.

Kuperus and Ionson [18] demonstrate a mild dependence on α if $\alpha \lesssim 1$. If we further assume that the disk field fan out into the corona in the same way as the photosperic fields fan out into the solar corona we may adopt $f = 1$. This means that we assume the whole corona to be filled with magnetic field. The ratio of coronal to disk emission is now directly proportional with the ratio of magnetic pressure and gas pressure in the disk at those positions where the magnetic loops emerge into the corona. Since $B_{cor} \lesssim B_{disk}$ and $\beta_{disk} = (v_s^d/v_A^d)^2 \gtrsim 1$ on the average one is inclined at first sight to consider the factor 10 as an upper limit for the energy flux ratio. However, the turbulent magnetic field fluctuations may outgrow the mean magnetic field as has been argued in section 2. In that case the ratio $(v_a/v_S)_{disk}$ may be locally much larger than unity. It is in these regions of local field concentrations that the electrodynamic coupling of the disk to the corona takes place. The more concentrated the magnetic field is the larger is the local disk magnetic field and hence the larger becomes the ratio L_{cor}/L_{disk}. As a first approximation we may thus adopt $L_{cor}/L_{disk} \sim 10$ realizing that this factor may be enhanced if the field is strongly concentrated. L_{cor} and L_{disk} are theoretically determined quantities which are related to the observed quantities L_{hard} and L_{soft} [17].

L_{hard}/L_{soft} for Cyg X-1 has been measured as large as 7 (see Ref. 21). It therefore seems that the observed ratio of hard to soft X-ray emission of black hole accretion disks can be understood by the emission of an accretion disk corona which is magnetically structured and electrodynamically coupled to the accretion disk. The driving occurs through the disk turbulent motions. The energy dissipated into the corona is lost by inverse Compton emission and converted into hard X-ray emission.

4. THE INTERACTION OF A NEUTRON STAR WITH AN ACCRETION DISK

4.1 Spin up of neutron stars

Neutron stars have a dipole magnetic field. Matter accreting on a neutron star will feel the effect of the neutron star magnetic field at the so called Alfvén radius where the fluid stresses balance the magnetic stresses: $\rho v_r v_\phi \sim B_\theta B_\phi / 4\pi$ where B_θ is the poloidal magnetic field and B_ϕ the toroidal or azimuthal field. If μ is the magnetic moment of the neutron star the field decreases with r as $B \sim \mu r^{-3}$. In the case of spherical accretion a simple estimate of the Alfvén radius r_A can be made in terms of the mass accretion rate $\dot{M} = 4\pi r^2 \rho v_r$ from which it follows that $r_A = \text{const}\ \mu^{4/7} (GM\dot{M}^2)^{-1/7}$.

The problem of the interaction of a disk with the neutron star magnetic field around the Alfvén radius has been extensively discussed [22, 23, 24, 25]. For disk accretion Gosh and Lamb [24] give for the Alfvén radius

$$r_A = 5.1 \times 10^8\ M_o^{-1/7}\ \dot{M}_{16}^{-2/7}\ \mu_{30}^{4/7}\ . \qquad (4.1)$$

where M_o is expressed in units of one solar mass, $\dot{M}_{16}$ in units of 10^{16} g/sec and μ_{30} in units of 10^{30} gauss cm^3.

The neutron star field is threading the disk and therefore exerts a torque on the disk. Matter will leave the disk and follow the fieldlines so that it falls on the neutron star magnetic poles, where dissipation takes place in a shock wave. As a result angular momentum is transported from the disk towards the neutron star. Depending on the sign of the angular momentum this may cause a spin up or a spin down of the neutron star. This problem is of great importance for the interpretation of rotation rates of X-ray pulsars which are supposed to be magnetically oblique rotators with column accretion on their magnetic poles. The recent discovery of the millisecond pulsar and the QPO's (Quasi Periodic Oscillators) throws new light on the complicated interaction of a neutron star-disk system. The region inside r_A is supposed to rotate rigidly with the stellar angular velocity Ω^*. At a certain distance r_{co} the so called corotation radius the disk has the same angular velocity as the magnetosphere.

If $r_A < r_{co}$ the disk touches upon a magnetosphere which is rotating slower than the disk. In this case the transfer of angular momentum to the neutron star in the region $r_A < r_{co}$ may cause the neutron star to spin up. The regions $r > r_{co}$ exert a negative torque on the neutron star. For a slow rotator $r_A \ll r_{co}$ so that the neutron star is spun up. If $r_A > r_{co}$ the neutron star may be spun down.
Gosh and Lamb [24] calculate the torque N of the disk on the neutron star in terms of the so called fastness parameter

$$\omega_s = \frac{\Omega^*}{\Omega_k(r_A)} = 1.19\ P^{-1}\ \dot{M}_{17}^{-3/7}\ \mu_{30}^{6/7}\ (M_o)^{-5/7} \tag{4.2}$$

They find

$$N = \dot{M}\ (GMr_o)^{1/2}\ n(\omega_s) \tag{4.3}$$

where

$$n(\omega_s) = 1.39(1 - \omega_s)^{-1}\{1 - \omega_s[4.03(1 - \omega_s)^{0.173} - 0,878]\}\ . \tag{4.4}$$

and $P = 2\pi/\Omega$ is the rotation period of the neutron star. If I is the angular momentum of the neutron star it follows from $I\,\dot{\Omega} = N_o n(\omega_s)$, that

$$-\ \dot{P} = 5.9 \times 10^{-5}\ \mu_{30}^{2/7}\ M_o^{-3/7}\ R_6^{6/7} I_{45}^{-1}\ P^2\ L_{37}^{6/7}\ n(\omega_s)\ , \tag{4.5}$$

where $L \sim \dot{M}$ has been used.

(4.2) Quasiperiodic oscillations

The discovery of millisecond radio pulsars in Low-mass X-ray binaries (LMXRB) has lead to the conclusion that during the later evolution of some binaries consisting of a neutron star and a normal companion star mass transfer from the companion to the neutron star takes place so that due to the angular momentum transfer the neutron star is spinning up. Therefore, when the mass transfer stops one will observe the neutron star as a radio pulsar with a short period in a wide binary. Assuming that the radio pulsar loses its rotational energy due to magnetic dipole radiation a magnetic field strength of the neutron star ranging from $10^{8.5}$ to $10^{11.0}$ G is found [26]. The corresponding magnetospheric radius for sperical accretion is given by $1.6\ 10^6$ cm $< r_a < 4.2\ 10^7$ cm . This means that the inner parts of the accretion disk do exist around these neutron stars and thus magnetic activity resulting in coronal structures may exist.

That some of the neutron stars in LMXRB may have magnetic fields in the order of 10^8 to 10^{10} G and periods in the millisecond range has been suggested by Alpar and Shaham [27], by interpreting the quasi-periodic oscillations (QPO), seen in GX5-1, in terms of a beat frequency model. The QPO in GX5-1 were discovered by Van der Klis et al. [28] . Power spectra derived from EXOSAT observations show a broad peak from 20-40 Hertz superimposed on a red noise component. The energy observed in the broad peak consists of 5 to 10 percent of the total luminosity of the star.

In GX5-1 the centroid frequency and the width of the peak as well as the slope of the red noise are correlated with the X-ray intensity of the source. Alpar and Shaham [27] have suggested that the observed frequency of the peak in the power spectrum is due to a beat effect between the Kepler frequency at the inner edge of the accretion disk (taken to be the magnetospheric radius, r_a) and the rotation frequency of the neutron star. This results in a relation between the luminosity of the source and the beat frequency, that fits well to the data of GX5-1 and which predicts a rotation period of the neutron star of the order of 10 ms and a magnetic field of the order of $6\ 10^9$ G, leading to a magnetospheric radius of $r_a \sim 60$ km.

Models so far have assumed that in order to produce a beat frequency, an asymmetry in the stellar magnetic field must interact with structures in the inner region of the disk. It has been suggested that the magnetic field acts as a gating mechanism for the accretion flow from these structures [29, 30]. In the model put forward by Lamb et al. [29] these structures are thought of as physical blobs created by some form of instability and are rotating in the inner region of the disk with the Kepler frequency. Berman and Stollman [30] have argued that the interaction of the field-asymmetry with these blobs probably could not provide the degree of modulation observed, but rather that the structures must be regions of reduced conductivity.

After the discovery of QPO in GX5-1 more sources have been discovered, which show quasi-periodic oscillations like SCO X-1 and CYG X-2. Some of these sources show quite a different behaviour from GX5-1, which seems difficult to explain within the context of a beat frequency model [31].

Stollman and Kuperus [32] assume that the neutron stars in some of the LMXRB have magnetic fields, which are capable of holding the accretion disk at some distance from the stellar surface, and than develope a model for the interaction of the neutron star magnetic field with magnetic structures in an accretiondisk corona. They show that energy can be transferred from the neutron star to the disk-coronasystem. We will follow briefly their analysis.

Consider the interaction of a magnetic loop with an obliquely rotating neutronstar magnetic field. The loop is anchored in the disk and hence is rotating with the Kepler angular velocity $\Omega(r)$, while the neutronstar magnetic field is, within the magnetosphere $r < r_A$ supposed to rotate with the neutronstar angular velocity Ω^*. Further assume that the magnetic field of the neutron star is not axially symmetric, due to an inclination of the magnetic axis with respect to the spin axis. In a coordinate system fixed to the loop, the loop experiences an oscillating stellar field with a frequency equal to the beat frequency Ω_B given by $\Omega_B = \Omega - \Omega^*$. The magnetic periodicity at radius r is thus given by

$$B_s(r) = B_0(r) + B_1(r) \cos \Omega_B t. \qquad (4.6)$$

In order to estimate this external source we make a number of symplifying assumptions. Firstly we assume the loop to be a streched filament with length ℓ and diameter a, located at distance r from the center of the neutron star. Secondly we assume that every part of the loop experiences the same external magnetic field, which is of course not strictly true for an oblique rotator, but not too serious an approximation as long as the aperture of the loop is small. Thirdly we assume that the external field is perpendicular to the axis of the stretched loop.

The total Lorentz force on the filament with length ℓ is given by $F = I B_s \ell/c$, where I is the total induced current intensity which can be estimated from the induction equation $L\dot{I} = c^{-1} \dot{\Phi}$, where the magnetic flux $\Phi \sim B_1\ell^2 \cos(\Omega_B t)$ is derived from the oscillating part of the magnetic field. Taking as an order of magnitude for the selfinductance of the loop $L \sim c^{-2}\ell$ we find $I/c \sim B_1 \ell \cos \Omega_B t$, so that assuming $B_1 \ll B_o$ it follows that $F = B_1 B_o \ell^2 \cos \Omega_B t$.
In order to estimate the power consumption consider the whole loop as a forced oscillator with mass m, repulsive force $-kx$, where x is the deviation from equilibrium, and damping force $- \lambda \frac{dx}{dt}$. The equation of motion of the filament is then given by

$$m \frac{d^2x}{dt^2} + \lambda \frac{dx}{dt} + kx = F_o \cos \Omega_B t, \tag{4.7}$$

where $F_o = B_o B_1 \ell^2$. The solution of the forced oscillator (4.7) is

$$x = \frac{F_o / m}{\left((\Omega_B^2 - \omega_o^2) + (\frac{\lambda}{m})^2 \Omega_B^2\right)^{1/2}} \sin(\Omega_B t - \beta) \tag{4.8}$$

where β is the phase angle, given by

$$\tan \beta = \frac{m}{\lambda} \left(\frac{\Omega_B^2 - \omega_o^2}{\Omega_B}\right) . \tag{4.9}$$

The oscillation is damped and the amplitude grows until the damping force balances the driving force. From eq. (5) we can derive the velocity

$$v = \frac{dx}{dt} = \frac{F_o}{\left((m\Omega_B - \frac{k}{\Omega_B})^2 + \lambda^2\right)^{1/2}} \cos(\Omega_B t - \beta) . \tag{4.10}$$

The maximum amplitude is reached when $\Omega_B = \omega_o = (k/m)^{\frac{1}{2}}$ which is the characteristic frequency. Then the phase angle $\beta = 0$ and the velocity is in phase with the force. This is called energy resonance because the power delivered by the force is a maximum. The denominator in (4.10) is the impedence $Z = (X^2 + R^2)^{\frac{1}{2}}$, where the reactance $X = m\Omega_B - k\Omega_B^{-1}$ and the resistance $R = \lambda$. From this it follows that $\tan\beta = X/R$ and the $v_o = F_o/Z$. The power produced by the force is given by

$$P = F \cdot v = \frac{F_o^2}{Z} \cos \Omega_B t (\cos \Omega_B t - \beta). \tag{4.11}$$

Averaged over a period the absorbed power is

$$\langle P \rangle = \frac{F_o^2}{2Z} \cos \beta = \frac{1}{2} F_o v_o \cos \beta = \frac{F_o^2 R}{2Z^2} = \frac{1}{2} R(v_o)^2, \tag{4.12}$$

where v_o is the velocity amplitude.

The damping should be known in order to estimate the velocity amplitude and the power. The damping of an oscillating filament in the solar corona has been studied by Kleczek and Kuperus [33]. Under the assumption that the main damping mechanism is magneto acoustic radiation of the vibrating filament with an effective surface area S they found

$\tau_D \sim m/\rho_c v_A^c S$, where ρ_c is the coronal density and $v_A{}^c$ is the coronal Alfvén-speed. From (4.7) we observe that $\tau_D = 2\pi m/\lambda$ so that with $S=2a\ell$ one finds $\lambda \sim 4\pi a\ell\rho_c\, v_A$. The Alfvén speed in the corona is of the order of $2\ 10^7$ G, assuming a dipole field from the star with a value of $\sim 5\ 10^9$ G at the pole and a magnetospheric radius of ~ 60 km. In order to get an estimate for the damping constant, $R = \lambda$, we express ℓ in units of 10^6 cm, a in units of 10^5 cm and ρ_c in units of 10^{-5} g/cm^3. We then find

$$R = \lambda \sim 2.2\ 10^{16}\ a_5 \ell_6 (\rho_c)_{-5}^{1/2} \ \text{g/sec} . \qquad (4.13)$$

An estimate of v_o can be made in the following way. Notice that $v_o = F_o/Z = B_1 B_o \ell^2/Z$. The value of Z is given by $Z = (X^2 + R^2)^{1/2} = (\lambda^2 + m^2\Omega_B^2 - 2mk + k^2/\Omega_B^2)^{1/2}$. The mass of the filament is $m = \pi a^2 \ell \rho_c \sim 3.10^{11}$ g. The beat frequency Ω_B is determined by the rotation period of the neutron star, $\Omega^* \sim 10$ ms, and by the radius of the magnetosphere, $r_A \sim 60$ km. One then finds $\Omega_B \sim 300\ \text{sec}^{-1}$. The value of k is determined by the eigenfrequency of the loop, $\omega_o = \pi v_a/\ell = (k/m)^{1/2} \sim 10^3\ s^{-1}$. Using the value found for m, this gives $k \sim 7.5\ 10^{18}$ g/sec^2. These quantities yield $Z \sim 3.3\ 10^{16}$ g/sec
Under the assumption $B_1 \sim 0.1\ B_o \sim 2\ 10^6$ G we then find $v_o \sim 1.2\ 10^9$ cm/sec from which the power in the oscillations can be estimated using eq. (4.12):

$$\langle P \rangle = \tfrac{1}{2} R\, (v_o)^2 \sim 2.3\ 10^{34}\ \text{erg/sec}. \qquad (4.14)$$

If the loop is in resonance $\omega_o \sim \Omega_B \sim 300\ ^{\ddagger}$, which means $\ell \sim 10^7$ cm. In this case according to eq. (4.13) $Z = R = \lambda \sim 2.2\ 10^{18}$ cm, if $a \sim 0.1\ \ell$. We then find $v_o \sim 2.8\ 10^9$ cm/sec and $\langle P \rangle \sim 4\ 10^{36}$ erg/sec. Compared with the Eddington limit $L_{Edd} \sim 10^{38}$ erg/sec one isolated coronal disk loop, powered by electrodynamic coupling of the neutronstar magnetic field, may produce a few percent of the X-ray flux.
Since it seems plausible to assume that more than one loop is present, the luminosity coming from the disk may be increased by a substantial fraction.

The model presented by Stollman and Kuperus [32] is a beat frequency model but is not based on the modulation of the accretion flow. It does, therefore, not necessarily assume that the field is able to guide the matter onto the star. This mechanism does also work when the magnetospheric radius is larger than the corotation radius, leading to an anti-correlation between the frequency and the observed intensity.

The mechanism described here, in which energy is transferred from the rotating, magnetized neutron star to the disk, seems suited to described the properties of the power spectra of QPO's in which a broad peak and a red noise component is present.

REFERENCES

1 Shakura, N.I., and Sunyaev, R.A., 1973, Astron. Astrophys. 24, 337
2 Frank, J., King A.R., and Raine, D.J., 1985, "Accretion power in Astrophysics", Cambridge Univ. Press, Cambridge
3 Lightman, A.P., 1974, Astrophys. J. 194, 419
4 Lightman, A.P., and Eardley, D.M., 1974, Astrophys. J. 187, L1

5 Shapiro, S.L., Lightman, A.P. and Eardley, D.M., 1976, Astrophys. J. 204, 187
6 Zeld'dovich, Ya.B. Ruzmaikin A.A., and Sokoloff,D.D., 1983, Magnetic Fields in Astrophysics", Gordon & Breach Publ.
7 Pudritz, R.E., 1981a, Mon. Not. Roy. Astron. Soc. 195, 881
8 Pudritz, R.E., 1981b, Mon. Not. Roy. Astron. Soc. 195, 897
9 Meyer, F., and Meyer-Hofmeister, E., 1982, Astron. & Astrophys. 106, 34
10 Burm, H.M.G., 1985, Astron. & Astrophys. 132, 143
11 Pudritz, R.E., and Fahlmann, G.G., 1982, Mon. Not. Roy. Astron. Soc. 198, 689
12 Eardley, D.M., and Lightman, A.P., 1975, Astrophys. J. 200, 187
13 Coroniti, F.V., 1981, Astrophys. J. 244, 587
14 Coroniti, F.V., 1983, IAU symp. 107, 453, eds. M.R. Kundu and G.D. Holman
15 Liang, E.P.T., and Price, R.H., 1977, Astrophys. J. 218, 247
16 Galeev, A.A., Rosner, R., and Vaiana, G.S., 1979, Astrophys. J. 229, 318
17 Ionson, J.A., and Kuperus, M., 1984, Astrophys. J. 284, 389
18 Kuperus, M. and Ionson, J.A., 1985, Astron. & Astrophys. 148, 309
19 Ionson, J.A., 1982, Astrophys. J. 254, 318
20 Ionson, J.A., 1984, Astrophys. J. 276, 357
21 Liang, E.P.T., 1980, Nature 283, 642
22 Gosh, P., and Lamb, F.K., 1978, Astrophys. J. 223, L83
23 Gosh, P. and Lamb, F.K., 1979a, Astrophys. J. 232, 259
24 Gosh, P. and Lamb, F.K., 1979b, Astrophys. J. 234, 296
25 Kaburaki, 1986, Mon. Not. Roy. Astron. Soc. 220, 321
26 Dewey, R.J., Maguire, C.M., Rawley, L.A., Stokes, G.H., and Taylor, J.H., 1986, preprint
27 Alpar, M.A., and Shaham J., 1985, Nature 316, 239
28 Van der Klis, M., Jansen, F., van Paradijs, J., Lewin, W.H.G., van den Heuvel, E.P.J., Trumper, J.E. and Sztajano, M., 1985, Nature 316, 225
29 Lamb, F.K., Shibashaki, N., Alpar, M.A. and Shaman, J., 1985, Nature 317, 681
30 Berman, N.M., and Stollman, G.M., 1986a, Astron. & Astrophys. 154, L23
31 Lewin, W.H.G., 1986, talk presented at workshop on "The Physics of Accretion onto Compact Objects", Tenerife, Spain
32 Stollman, G., and Kuperus, M., 1986, Astron. & Astrophys. (submitted)
33 Klezcek, J., and Kuperus, M., 1969, Solar Phys. 6, 72
34 Pringle, J.E., 1981, Ann. Rev. Astron. Astrophys. 19, 137

THE MEMBRANE PARADIGM FOR BLACK-HOLE ASTROPHYSICS

Kip S. Thorne

Theoretical Astrophysics
California Institute of Technology
Pasadena, California 91125 USA

In the 1960s and early 1970s most theoretical research on black holes was devoted to understanding their fundamental properties: Does the gravitational collapse of a star with weak, generic perturbations from spherical symmetry produce a black hole? [Yes.] Once a black hole forms and settles down into a stationary, equilibrium state, what is its spacetime metric? [The Kerr metric.] Are all such Kerr black holes stable against small perturbations? [Probably yes.] Are there necessarily spacetime singularities inside black holes? [Yes.] What are the general laws that govern the changes in a black hole as it evolves, due to interaction with the external universe, from one Kerr equilibrium state to another? [The four laws of black-hole dynamics.] ... .

By the mid-1970s, when the fundamental properties of black holes were fairly securely established, there had accumulated considerable observational evidence for the existence and importance of black holes in the astrophysical universe; and, consequently, black-hole theorists turned their attention to studies of how a black hole should interact with a complex astrophysical environment: What is the influence of a hole's rotation on the orientation of an accretion disk? [It drives the disk into the hole's equatorial plane, the "Bardeen-Petterson effect".] If an accretion disk deposits a magnetic field on a black hole, how does the field distribute itself over the horizon? [In that manner which minimizes the dissipation (the rate of entropy increase) due to the rotation of the horizon relative to the field.] Can the magnetic field threading the hole be a vacuum field? [No; the interaction of the hole's rotation with the magnetic field will generate electric fields that trigger a breakdown of the vacuum, creating a rich electron-positron pair plasma at a few horizon radii.] What will be the result of the interaction of this electron-positron-ladened field with the hole? [It will drive currents in the hole's magnetosphere, which extract rotational energy from the hole and deposit it in outpouring plasma and Poynting flux, the "Blandford-Znajek effect". This energy flux may well be the power source for jets that are seen emerging from the centers of quasars and active galactic nuclei.]

The research tools used to discover the fundamental laws of black holes in the late 60's and early 70's were not well suited to the astrophysical studies of the late 70's and the 80's. The fundamental-law tools, sometimes called the *Black-Hole Paradigm*, were couched in the language of curved, 4-dimensional spacetime; they relied heavily on spacetime diagrams and on topological studies of the causal structure (light-cone structure) of spacetime. By contrast, plasma physics, which had to be merged with general relativity in order to study holes in realistic astrophysical environments, is generally formulated in the 3-dimensional language of a specific reference frame. Thus, astrophysical studies of black holes required either rewriting plasma physics in the 4-dimensional language of black holes, or rewriting the fundamental laws of black holes in the 3-dimensional language of plasmas.

Plasma physics is a much more complicated subject than general relativity. Those who have studied both find it far more difficult to deduce, intuitively, how a plasma will behave in a given situation than how a black hole will behave. (Recall the travails of the designers of machines for controlled fusion!) Correspondingly, it has proved far more fruitful to rewrite black-hole physics in 3-dimensional language than to rewrite plasma physics in 4-dimensional language.

The 3-dimensional rewrite of black-hole physics, and its merger with plasma physics and other topics required for astrophysical studies, is called the *Membrane Paradigm* for black holes. The reason for this name will become clear below.

The Membrane Paradigm is based on a 3+1 *split* of 4-dimensional spacetime into 3-dimensional space plus one time dimension. Such a split relies on measurements of space and time as made by some specific family of fiducial observers (*FIDOs*). In a generic situation, where spacetime is strongly curved and highly dynamical and lacks symmetries, the choice of FIDOs is quite arbitrary. However, for a Kerr black hole there is a unique, preferred family of FIDOs: the observers whose world lines (i) orbit the hole at fixed latitude and radius so the FIDOs see unchanging gravitational fields in their vicinities; (ii) are hypersurface-orthogonal so the local 3-spaces of simultaneity of all the FIDOs merge to form 3-dimensional hypersurfaces (hypersurfaces of constant *universal time* t) in 4-dimensional spacetime; and (iii) cover the entire spacetime region outside the hole's horizon. These unique FIDOs are sometimes called Zero-Angular-Momentum observers. Far from the hole they are at rest in the hole's asymptotic rest frame; near the hole they are dragged into orbital motion by the frame-dragging force of the hole's rotation; at the horizon they precisely co-rotate with the horizon.

The Membrane Paradigm chooses its FIDOs in this preferred way in the case of a Kerr black hole; and for a weakly perturbed, slowly evolving Kerr hole it chooses FIDOs as close to these preferred ones as the perturbations allow. For a highly dynamical hole, the Paradigm uses a choice that is somewhat arbitrary, but the FIDOs always remain outside the horizon and their world lines always cover the entire spacetime exterior to it.

Although this choice of FIDOs has some beautiful properties, it also produces a serious pathology: Because the FIDOs remain always outside the horizon, as one approaches the horizon one sees their world lines asymptotically approach the horizon's null generators. Correspondingly, since any null curve is orthogonal to itself, the FIDOs' orthogonal hypersurfaces (slices of constant universal time t) asymptotically approach the horizon in the manner of Figure 1: They dip deep down into the past, never quite reaching the horizon.

In the Membrane Paradigm one mentally collapses all the hypersurfaces of constant universal time t into a single 3-dimensional space in which physics occurs as time t passes. This 3-space is sometimes called *absolute*

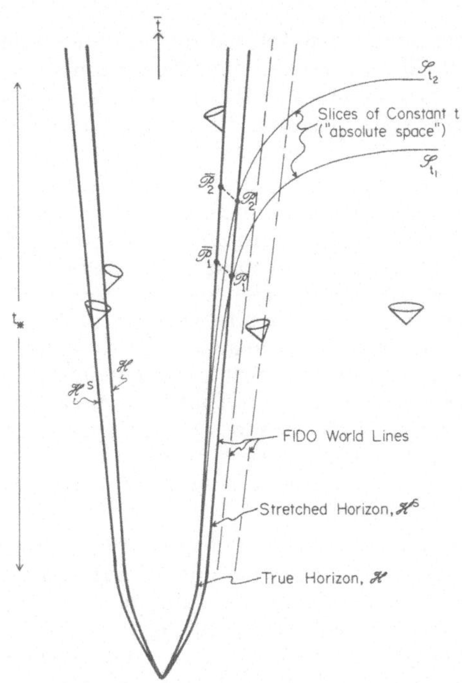

Fig. 1. Spacetime diagram exhibiting the world lines of the fiducial observers (FIDOs) near the horizon of a slowly rotating black hole, the hypersurfaces of constant universal time t which are orthogonal to those FIDO world lines, and the stretched horizon which acts as a surrogate for the true horizon in the Membrane Paradigm.

space to emphasize the fact that the Paradigm chooses it once and for all by a preferred 3+1 slicing; the slicing, once made, is not supposed to be changed.

The Membrane Paradigm describes physics in terms of quantities that are measured by the chosen FIDOs; for example, the electromagnetic field is described not by a frame-independent electromagnetic field tensor, but rather by an electric field $\vec{E}$ and a magnetic field $\vec{B}$ that are measured by the FIDOs. Every such FIDO-measured quantity is characterized mathematically by a scalar, 3-dimensional vector, or 3-dimensional tensor that resides in the absolute, 3-dimensional space; for example, $\vec{E}$ and $\vec{B}$ are vector fields in absolute space. The laws of physics then become 3-dimensional scalar, vector, and tensor relationships between the FIDO-measured quantities. For example, two of Maxwell's equations take on the familiar-looking forms $\vec{\nabla}\cdot\vec{B} = 0$ and $\vec{\nabla}\cdot\vec{E} = 4\pi\rho_e$, where $\vec{\nabla}\cdot$ denotes the 3-dimensional divergence, defined covariantly in the 3-dimensional absolute space, and ρ_e is the FIDO-measured charge density.

This use of a 3-dimensional viewpoint permits us to carry over into the vicinity of a black hole many of the intuitively useful pictures and concepts of ordinary 3-dimensional physics. For example, the FIDO-measured electric and magnetic fields can be characterized by field lines whose directions are along $\vec{E}$ and $\vec{B}$, whose FIDO-measured densities (number of lines per unit of orthogonal area) are proportional to the magnitudes of $\vec{E}$ and $\vec{B}$, and which – according to the Maxwell equations $\vec{\nabla}\cdot\vec{E} = 4\pi\rho_e$ and $\vec{\nabla}\cdot\vec{B} = 0$ – end only on electric charge in the case of $\vec{E}$-lines, and never end in the case of $\vec{B}$-lines.

The Membrane Paradigm's combined, 3-dimensional laws of black-hole physics, plasma physics, hydrodynamics, radiative transfer, ... govern the evolution of the black hole and its complex environment. In computing that evolution one must face up to a pathology introduced by the peculiar "dip-into-the-past" of absolute space (Figure 1): Because of that dip and its preventing absolute space from ever intersecting the hole's horizon, all particles and fields that fall into the hole appear to slow their motion exponentially as they approach the horizon, ultimately becoming "frozen" into a thin boundary layer just above the horizon. That boundary layer is much like the sediments on the bottom of the ocean: It is a layered-down record of the past history of the horizon's vicinity – a record of historical interest, but a record that has no influence on the present and future evolution of particles and fields above the boundary layer.

In computations of astrophysical processes this boundary layer is a nuisance. A computer code would have trouble following it as it becomes more and more compacted; the boundary layer might even force the calculation to grind to a halt – and with no compensating reward, since the interesting evolution at greater heights above the horizon is unaffected by the boundary layer. To avoid this nuisance, the Membrane Paradigm discards the boundary layer by cutting off all calculations at a fixed, 2-dimensional surface just above it. That surface-of-cutoff then becomes a surrogate for the hole's horizon; and, accordingly, it is called the *stretched horizon*.

When one takes a 4-dimensional, spacetime viewpoint, one sees the stretched horizon as a timelike (but almost null) world sheet that resides just outside the true horizon; see Figure 1. Each event on the stretched horizon is a surrogate for some event on the true horizon; for example, in Figure 1 $\mathcal{P}_1$ is a surrogate for $\overline{\mathcal{P}}_1$ and $\mathcal{P}_2$ is a surrogate for $\overline{\mathcal{P}}_2$. The properties of the true horizon and the laws that govern its evolution are thereby mirrored in properties and evolution laws of the stretched horizon. Those properties and evolution laws become, in the Membrane Paradigm, boundary conditions on the physics of the external universe; and they permit one to compute the external evolution without following the details of the boundary layer beneath the stretched horizon.

The greatest power of the Membrane Paradigm lies in the 3-dimensional mental pictures and physical intuition that go along with its equations (e.g. never-ending magnetic fields with tension along them and pressure orthogonal to them). To facilitate this intuition, the boundary conditions at the stretched horizon are written in a form, due to Thibaut Damour and Roman Znajek, that makes close contact with elementary physical intuition: The stretched horizon is regarded as a 2-dimensional membrane made of a 2-dimensional viscous fluid that is electrically charged and conducting, with finite entropy and temperature, but no ability to conduct heat; and the boundary conditions are embodied in a meshing of these membrane properties with the physics of the surrounding universe.

As an example, the electromagnetic boundary conditions are embodied in (i) Gauss's law, which equates the normal component of the FIDO-measured electric field at the stretched horizon to 4π times the stretched horizon's surface density of charge; (ii) Ampere's law, which equates the tangential component of the magnetic field to 4π times the surface current density flowing orthogonally to the tangential $\vec{B}$; (iii) Ohm's law, which equates the tangential electric electric field to the surface resistivity of the stretched horizon (377 ohms) times the surface current; and (iv) charge conservation, which says that whenever charge flows into the stretched horizon from the external universe, it does not pass on through into the hole's interior; rather, it stays on the stretched horizon in the form of surface charge density and current density and is conserved until such a time as negative charge flows in from the external universe and annihilates it (i.e. until current flows out of the stretched horizon into the external universe).

These stretched-horizon boundary conditions, together with the 3-dimensional laws of physics outside the stretched horizon, constitute the foundation for the Membrane Paradigm's powerful set of mental pictures, diagrams, and problem-solving techniques.

My lectures at Cargese were a pedagogical introduction to the Membrane Paradigm, following somewhat closely the pedagogical treatment given in a book that my colleagues and I have recently published.[1] Because that book is now widely available, the reader is referred to it for a written version of my lectures.

REFERENCES

1. K. S. Thorne, R. H. Price, and D. M. Macdonald, eds., "Black Holes: The Membrane Paradigm", Yale University Press, New Haven, Connecticut (1986).

TIDAL DISRUPTION

J.P. Luminet

Groupe d'Astrophysique Relativiste
CNRS - Observatoire de Paris
92195 Meudon Principal Cedex, France

1. INTRODUCTION

The behaviour of an extended body moving in an external gravitational field is an old and difficult problem, even in the framework of Newtonian gravitational theory. The aim of this lecture is to provide an account of recent progress in the field, and I will discuss in particular the more dramatic and interesting case where the external tidal field varies so abruptly that it is able to destroy the self-equilibrium of an astronomical body such as a natural satellite or a star.

The process of disruption of stars by tidal forces is probably relevant in the neighbourhood of giant black holes, as expected to occur in many galactic nuclei.[1] The basic model for explaining the high luminosity of active galactic nuclei involves a massive black hole (10^5 to $10^9 M_\odot$) dragging and swallowing the surrounding gas by the accretion process, which converts a large fraction of the gravitational potential energy into radiation. The reservoir of accreting gas (for instance a disk or a torus) is probably fed by debris of disrupted stars. A black hole may indeed break up stars either by its tidal field,[2] or by accelerating stars to such high velocities that, if two stars collide, they lose most of their material.[3]

The disruption of a star deserves special attention in the most violent events, corresponding to high penetration in the tidal radius, or equivalently, high velocity interstellar collisions. As was shown recently,[4] such violent disruption events are preceded by strong squeezing of the stars, the main effect being to trigger explosive nucleosynthesis in the cores of the squeezed stars.

2. TIDAL TENSOR

In this section I recall some elementary definitions of tidal forces in Newton's theory of gravitation. Consider a body moving into the gravitational field of a faraway source. The position of any infinitesimal mass element of the body in an inertial coordinate system may be specified by a position vector, say $\mathbf{x}$, which may be decomposed as $\mathbf{x} = \mathbf{X} + \mathbf{r}$, where $\mathbf{X}$ is the position vector of the centre-of-mass of the body, and $\mathbf{r}$ is the "internal" position vector of the mass element relative to the centre-of-mass. The acceleration of the mass element under consideration, $d^2\mathbf{x}/dt^2$, is just the sum of the internal force per unit mass, $\mathbf{f}$ (including for instance self-gravitation and internal pressure of the body), and of the external gravitational acceleration, $\mathbf{g}$. It is well known that the latter derives from the Newtonian potential φ_N. The Taylor expansion of φ_N relative to the centre-of-mass leads to:

$$\varphi_N(\mathbf{x}) = \varphi_N(\mathbf{X}) + r_i \frac{\partial \varphi_N(\mathbf{X})}{\partial X_i} + \tfrac{1}{2} r_i r_j \frac{\partial^2 \varphi_N(\mathbf{X})}{\partial X_i \partial X_j} + \tfrac{1}{6} r_i r_j r_k \frac{\partial^3 \varphi_N(\mathbf{X})}{\partial X_i \partial X_j \partial X_k} + \dots \tag{2.1}$$

The first term of the expansion is just the value of the Newtonian potential at the centre-of-mass, the next one involves the value of the external gravitational acceleration at the centre-of-mass. These two terms induce no tidal acceleration at all. The latter arises from the second spatial derivative of the Newtonian potential. By definition, the *tidal tensor* $\mathbf{C}$ and the *deviation tensor* $\mathbf{D}$ are given by:

$$C_{ij}(\mathbf{X}) \equiv -\frac{\partial^2 \varphi_N(\mathbf{X})}{\partial X_i \partial X_j}, \qquad D_{ijk} \equiv -\frac{\partial^3 \varphi_N(\mathbf{X})}{\partial X_i \partial X_j \partial X_k}. \tag{2.2}$$

Substituting the expansion (2.1) into the expression for the gravitational acceleration, we get:

$$g_i(\mathbf{x}) = g_i(\mathbf{X}) + C_{ij} r_j + \tfrac{1}{2} D_{ijk} r_j r_k + O(|\mathbf{r}|^3). \tag{2.3}$$

The meaning of the deviation tensor is made clear when the total acceleration of the centre-of-mass is written as:

$$\frac{d^2 X_i}{dt^2} = \frac{1}{M_*} \int g_i(\mathbf{x})\, dm = g_i(\mathbf{X}) + \frac{1}{2M_*} D_{ijk} \int r_j r_k\, dm + O(|\mathbf{r}|^3) \tag{2.4}$$

which shows that the first order correction to pure geodesic motion (that would take place if all the mass of the body were concentrated at the centre-of-mass) is proportional to the moment of inertia tensor and to the deviation tensor. Dropping the deviation corrections (known as the geodesic approximation) is thus equivalent to ignoring the internal structure of the body. In most astrophysical situations, involving, for instance, a small natural satellite orbiting a planet or a star passing near a giant black hole, the deviation term is indeed negligible.

By definition, the tidal force per unit mass acting on any infinitesimal mass element is the acceleration of that element relative to the centre-of-mass:

$$\frac{d^2 r_i}{dt^2} = \frac{d^2 x_i}{dt^2} - \frac{d^2 X_i}{dt^2} \simeq f_i + C_{ij} r_j. \tag{2.5}$$

Therefore the action of an external gravitational field on the internal motion of the body is completely specified by the tidal tensor. This tensor is symmetric and trace-free (from Poisson's equation). Taking account of the well-known value of the Newtonian potential $\varphi_N = -GM/R$, where M is the mass of the source and R the distance from the source to the centre-of-mass of the body, the tidal tensor components can be written as:

$$C_{ij}(\mathbf{X}) = \frac{GM}{R^3} \begin{pmatrix} -1 + \frac{3X^2}{R^2} & \frac{3XY}{R^2} & 0 \\ \frac{3XY}{R^2} & -1 + \frac{3Y^2}{R^2} & 0 \\ 0 & 0 & -1 \end{pmatrix} \tag{2.6}$$

where the (X, Y, Z)-inertial coordinates of the centre-of-mass have been chosen such that the orbital motion lies in the plane $Z \equiv 0$.

The instantaneous effect of the tidal field may be seen directly by putting the tidal matrix into its diagonalized form, namely

$$\mathbf{C} = \frac{GM}{R^3}\,\mathrm{diag}(-2,+1,+1). \tag{2.7}$$

The negative eigenvalue is associated with the *radial* principal direction (i.e. pointing towards the source), and the double positive eigenvalues are associated with the "orthoradial" principal plane. Thus the instantaneous effect is to stretch the body along the radial direction and to squeeze it along the two orthoradial directions. As a direct application of this formalism, Jeans derived the equilibrium configurations of a homogeneous, non-rotating body in a fixed tidal field;[5] the solutions turn out to be prolate ("cigarlike") spheroids elongated towards the source. However, such configurations are not physical since in the real situation the body will not be fixed in a constant tidal field but will have at least some rotation due to orbital motion.

3. ELLIPSOIDAL DEFORMATIONS OF HOMOGENEOUS BODIES

The problem of equilibrium of an astronomical body in a tidal field was first considered by Edouard Roche[6] in the context of liquid satellites orbiting planets of the Solar System. Thus Roche and most subsequent workers addressed their attention to *homogeneous, incompressible* bodies. For such a case it is well known that the self-gravitational potential is a *quadratic* function of internal coordinates r_i. On the other hand, the tidal force derives from a "tidal potential" that is also a *quadratic* function of the internal coordinates: $\varphi_T \equiv -\frac{1}{2}C_{ij}r_ir_j$. Therefore, *tidally distorted bodies admit ellipsoidal equilibrium configurations* (of course, more complicated shapes such as pears, bars, donuts are also exact solutions).

3.1 The stationary rotational problem

In this respect, the problem of the equilibrium of a homogeneous body in a tidal field is quite analogous to that of an isolated *rotating* homogeneous body, since the centrifugal potential is also a quadratic function of internal coordinates. A great amount of work has been devoted to the question (see e.g. Chandrasekhar[5] for a review). Let me recall just that in the case of *rigid rotation*, MacLaurin derived a series of oblate spheroidal equilibrium configurations, whereas Jacobi obtained fully triaxial ellipsoids. The important point is that no stationary ellipsoidal equilibrium is possible when the angular velocity is greater than $(0.4493\pi G\rho_*)^{1/2}$, where G is Newton's gravitation constant and ρ_* is the density of the distorted body.

The more general case allowing *internal motion* with velocity given by a linear function of the internal coordinates, (known as the Dirichlet problem), was solved by Riemann in the stationary case. A particularly interesting sequence ("type S") of Riemann ellipsoids corresponds to the case where the angular velocity vector $\boldsymbol{\omega}$ and the shear vector $\mathbf{J}$ lie in the same direction as one of the principal axes of the ellipsoid.

3.2 The stationary tidal problem

Roche solved the problem of equilibrium of a homogeneous body moving along a circular orbit around an indeformable (pointlike) gravitational source. In such a case, the tidal field is stationary (i.e. its magnitude is independent of time), but eigendirections rotate. Furthermore, on the basis of observation of lunar motion, Roche assumed that the proper rotation of the satellite and its orbital motion had the same period. Equilibrium configurations are characterized by the condition that the self-gravitational potential, the centrifugal potential and the tidal potential compensate exactly over the surface boundary of the body. The Roche ellipsoidal solutions are parametrized by the mass ratio p between the primary and the source, and by the radius R of the circular orbit. In any case, no stationary equilibrium occurs when the tidal field increases above a critical value. In the case of an infinitesimal satellite ($p = 0$), Roche derived the celebrated formula giving the *tidal (or Roche) limit*, within which no stationary equilibrium is possible:

$$R_T = 1.523(M/\rho_*)^{1/3} = 2.455(M/M_*)^{1/3} R_*,$$

where quantities with an asterisk refer to the satellite.

The more general *Roche-Riemann ellipsoidal solutions*, which allow internal motion with uniform shear, have been derived much more recently by Aizenman.[7]

3.3 The dynamical tidal problem

Now the interesting question is: what happens when a satellite transgresses the Roche limit? A rough guess comes from the comparison between the two characteristic timescales of the problem. On one hand, the timescale of variation of the tidal field is the crossing time of radius R at free-fall velocity v: $t_{\rm orb} \simeq (R^3/GM)^{1/2}$. On the other hand, the internal response of the body to any perturbation is characterized by the internal dynamical timescale, $t_* \simeq (G\rho_*)^{-1/2}$ (a few hundred seconds for typical densities). It is clear that, as far as $t_{\rm orb} \ll t_*$, the body is able to adjust its internal structure to the external perturbation, and therefore admits an instantaneous equilibrium configuration very close to a Roche-Riemann ellipsoid. But as soon as $t_* < t_{\rm orb}$, the body is forced to undergo growing instablilities, leading ultimately to disruption. The limiting case occurs when the two timescales are comparable, namely at the characteristic tidal radius $(M/\rho_*)^{1/3}$.

The quantitative description of the behaviour of a body moving in a varying tidal field requires a numerical treatment. The first calculations were done by Nduka[8] but were very incomplete. More recently, Luminet and Carter[9] have performed exhaustive calculations along parabolic orbits. The figures 1-3 show in various cases the time variation of the principal axes of the body during its passage through the region of strong tidal forces. The exact tidal radius was found to be $R_T = 1.512(M/\rho_*)^{1/3}$. When the periastron distance R_p of the orbit is slightly greater than R_T (figure 1), no disruption occurs but the tidal perturbation induces a bifurcation from an initial spherical equilibrium configuration (the proper rotation of the body is neglected) to a quasi-stationary ellipsoidal equilibrium configuration, very close to a type S irrotational Roche-Riemann solution. When $0.166R_T < R_p < R_T$, "weak" disruption occurs (i.e. the central pressure falls down to zero) in a *cigarlike* shape, one of the principal axes increasing to infinity whereas the two other vanish after the crossing of the tidal radius (figure 2). When the body penetrates into the tidal radius by a factor greater than about 6.0, disruption occurs in a *disklike* shape, i.e. the two principal axes lying within the orbital plane stretch to infinity whereas the third one vanishes (figure 3).

4. TIDAL DEFORMATIONS OF A COMPRESSIBLE BODY

4.1 The tidal rolling mill effect

The idea of disruption of whole stars by big black holes was put forward by Hills,[2] who suggested that the gas release could power the central engine in active galactic nuclei. Hills made rough estimates on the energetics of a disrupted star on the basis of the incompressible model: he pointed out that, since the kinetic energy of debris is of the

Figure 1: Time evolution of the principal axes of the incompressible model along a parabolic orbit with penetration factor $\beta = 0.40$. The orbital time is in units of the stellar dynamical time t_*. The periastron is at $t = 0$, the satellite enters the tidal radius at $t \simeq -1$, and leaves it at $t \simeq +1$.

No disruption occurs.

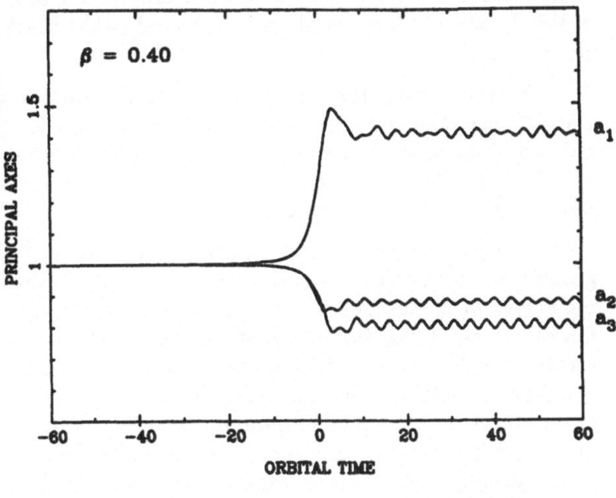

Figure 2 : Time evolution of the principal axes for the incompressible model with $\beta = 1.00$.

Cigarlike disruption occurs.

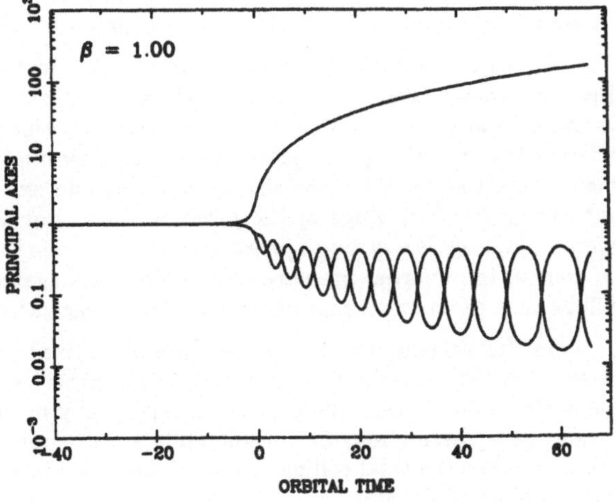

Figure 3 : Time evolution of the principal axes for the incompressible model with $\beta = 5.00$.

Disklike disruption occurs.

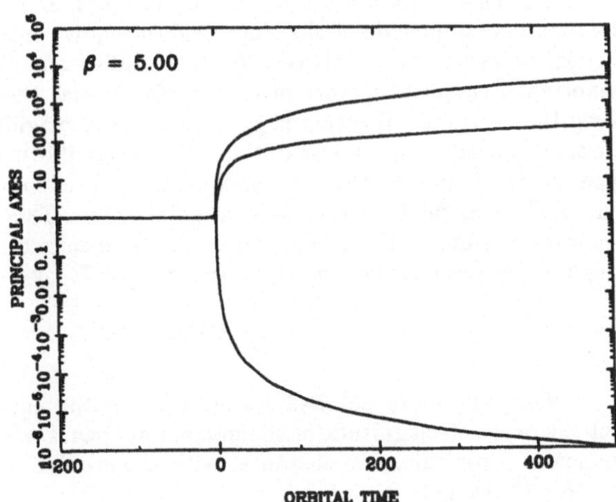

same order as its (negative) binding energy in the star before the disruption, the debris of tidally broken stars would remain bound to the black hole and might feed an accretion structure.

However, real stars are far from being homogeneous incompressible bodies. In the general case, the self-gravitational potential is no longer a quadratic function of internal coordinates, so that ellipsoidal configurations are not *exact* solutions for equilibrium of tidally distorted bodies. Full hydrodynamical calculations are needed to solve the dynamical problem as well as the stationary one.

In fact, as shown first by Carter and Luminet,[4] it is possible to draw from simple geometrical arguments a qualitative scenario exhibiting all the important features of the behaviour of a star penetrating *deeply* into a strong tidal field. The basic idea is the following: since the amplitude of the tidal force varies as $1/R^3$, as soon as a star penetrates deeply into the tidal radius of a massive body (a black hole, otherwise the star could not penetrate the tidal radius of the body without colliding with its hard surface), gravitational forces begin to dominate over all other internal forces in the star (e.g. self-gravitational and pressure forces). As a consequence, any individual particle of the star penetrating into the tidal radius undergoes initially free-fall motion (in the incompressible limit, the internal pressure was artificially contrived to adjust step by step to the tidal stresses, in order to maintain the volume constancy).

Now I recall the instantaneous effect of the tidal tensor: extension along the radial direction, compression along the two orthoradial directions. But the direction orthogonal to the orbital plane remains fixed, compressive, principal direction of the tidal tensor, whereas the two other principal directions lying within the orbital plane rotate. It follows that within the orbital plane the compressive and extensive tendencies partially cancel out during the first stage of the passage ot the star inside the tidal radius, while the overall tendency is a *strong squeezing of the star along the "vertical" direction* (figure 4). Of course, the internal pressure will finally react against the compression, and the star will bounce to an expansion phase after its passage at the periastron of the orbit.

In this scheme, the star passes through a fixed point on the orbit of its centre-of-mass, at which it looks like a "tube of toothpaste" squeezed in the middle; but since the orbital velocity near the periastron is generally much greater than the internal sound speed, the squeezing may be considered in a first approximation as simultaneous all over the star. Thus the tidal rolling-mill mechanism leads to a transitory "pancake" flattened configuration of the star.

It is clear that the amplitude of flattening at the instant of bounce depends on the maximal amplitude of the tidal field, or equivalently on the *penetration factor*, say β, defined as the ratio between the tidal radius and the periastron radius. The most important dynamical feature of the process is that the maximum compression velocity along the "vertical" direction is given in order of magnitude by $v_{\max} \simeq \beta v_*$, where v_* is the sound speed in the star at initial spherical equilibrium (a few hundred km/s for solar-type stars). It follows that the maximum compression kinetic energy of the gas in the star is β^2 times the internal energy at equilibrium. Thus, for an ordinary non-degenerate equation of state of the stellar material, the maximum "pancake" temperature will be roughly given in terms of the equilibrium value T_* by

$$T_{\max} \simeq \beta^2 T_* \tag{4.1}$$

Next, the more detailed specification of the equation of state allows us to predict the order-of-magnitude of the maximum "pancake" density and of the characteristic "pancake" timescale. For instance, with a non-relativistic polytropic equation of state $P = K\rho^{5/3}$, we get:

$$\rho_{\max} \simeq \beta^3 \rho_* \ , \qquad \Delta t_{\max} \simeq \beta^{-4} t_*. \tag{4.2}$$

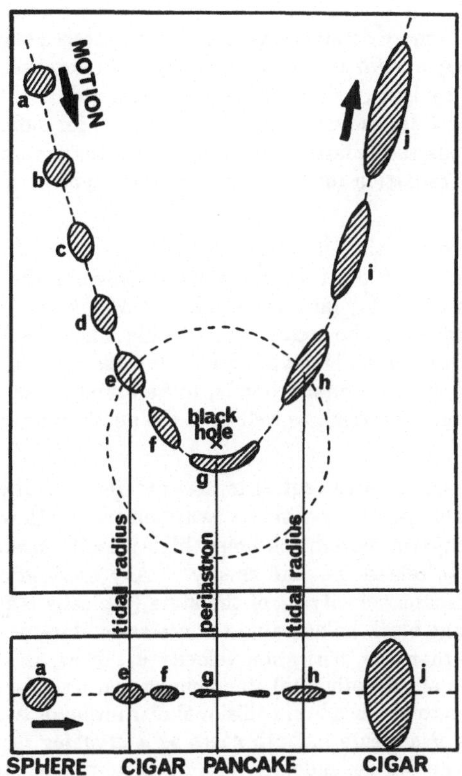

Figure 4: Deformations of a compressible star penetrating deeply within the tidal radius of a black hole (for sake of clarity the size of the star relative to the orbital distance scale has been considerably emphasized).
Upper: Configurations as projected in orbital plane. The area of the section remains practically constant between the instants of passage within the tidal radius, due to the rotation of the eigendirections of the field tensor.
Lower: Configurations in the direction orthogonal to the orbital plane.Bounce occurs slightly after the passage at periastron.

Take as an illustration a solar-type star penetrating by a factor 10 into the tidal radius of a $10^5 M_\odot$ black hole. The core of the star at spherical equilibrium is characterized by $T_* \simeq 10^7 \,^\circ K$, $\rho_* \simeq 100$g/cc, $t_* \simeq 10^3$sec. Thus the core of the star will pass through a pancake phase with $T_{\max} \simeq 10^9 \,^\circ K$, $\rho_{\max} \simeq 10^6$g/cc, during $\Delta t_{\max} \simeq 0.1$sec. Such dramatic conditions are highly favourable for *explosive nucleosynthesis,* and this is the most interesting consequence of the tidal squeezing process.

4.2 Tidal versus Collisional Disruption of Stars

Before discussing in more detail the dynamics of a squeezed star and nucleosynthetic processes, it is time to say a word about the astrophysical relevance of such a mechanism. As already pointed out by Hills,[2] since the gravitational radius of a black hole varies as its mass M whereas the tidal radius varies only as $M^{1/3}$, for sufficiently massive black holes the former encompasses the latter; then, any disruption occurs *inside* the black hole itself, and no energy of radiation may escape. For solar-type stars, the critical mass is about $10^8 M_\odot$.

Now the pancake disruption of stars requires penetration into the tidal radius by a factor greater than about 5, so that the allowed range of the black hole's mass is restricted to be less than $10^7 M_\odot$, (and more than $10^4 M_\odot$, since for smaller black holes the radius of a typical star would be greater than the black hole itself, and the geometrical considerations discussed above would not apply). This mass range is just what is expected, on both theoretical and observational grounds, to be found in the nuclei of moderately active galaxies (e.g. Seyfert galaxies), as well as in the nuclei of ordinary ones such as the Milky Way.

Modelling of quasars requires central masses greater than the Hills limit, thus excluding the astrophysical importance of tidal disruption events. However, it is well known that in the vicinity of supermassive black holes, high-velocity *interstellar collisions* may be a very efficient way to release gas and energy.[10] Any head-on collision with relative velocity exceeding the escape velocity v_* of the stars (typically 500 km/s) will partially disrupt the stars. A giant black hole is able to accelerate stars in its neighbourhood to velocities much greater than the disruption velocity v_*. Now, the interesting point[4] is that the tidal disruption of an individual star penetrating by a factor β inside the tidal radius is quite analogous to the head-on collisional disruption of two identical stars with velocity $\beta \cdot v_*$. In fact, β appears in both cases as a *crushing factor*, upon which all the dynamics of the disruption depends; for high β, disruption occurs only after the passage of the star(s) through a transitory, flattened "pancake" configuration. In conclusion, the pancake disruption of stars is probably a quite general phenomenon occurring near massive black holes, in ordinary galactic nuclei as well as in the most powerful ones.

4.3 The affine star model

The most interesting consequence of the short compression phase undergone by a star plunging into the tidal radius of a big black hole is the detonation of a fraction of its nuclear fuel. Since nuclear reaction rates depend strongly on the temperature, it is necessary to develop a more accurate description of the dynamics of a squeezed star. Without entering into heavy hydrodynamical simulations that may be sometimes misleading due to the lack of spatial resolution,[11] [12] the *affine star model* [13] [14] provides a powerful tool to handle the problem. This simplified scheme allows compressibility and inhomogeneity of the stellar material as well as entropy generation by nuclear reactions, but assumes that the layers of constant density keep an ellipsoidal shape; it is thus the most natural generalization of the oversimplified incompressible model. The affine star model is likely to provide a good approximation to the behaviour of a star's core (in which

Figure 5: Ellipsoidal deformations of the affine star model with adiabatic, non-relativistic polytropic equation of state. Same time units as in figures 1–3.

Non disruptive encounter: the star is not broken but its configuration evolves from the undistorted sphere towards a Roche-Riemann stationary solution.

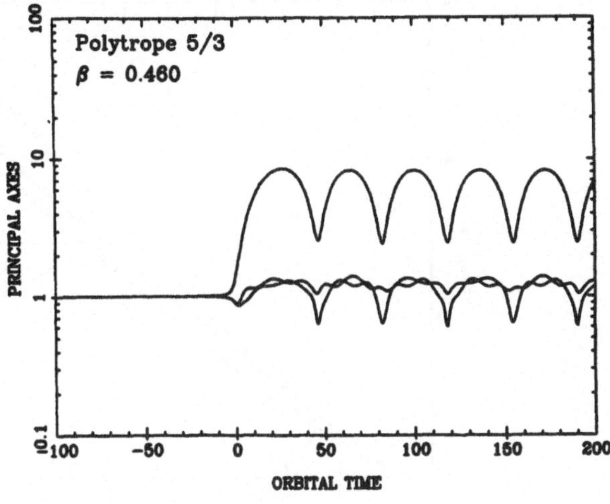

Figure 6: Weakly disruptive encounter. The compression effects are very mild, the configuration evolves quickly towards an infinitely stretched cigar.

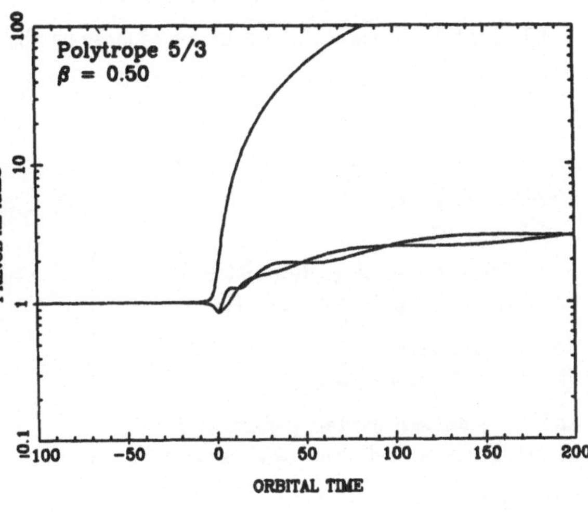

Figure 7: Strongly disruptive encounter. The transitory flattening in a pancake configuration forms soon after the passage of periastron. The final configuration is more isotropic.

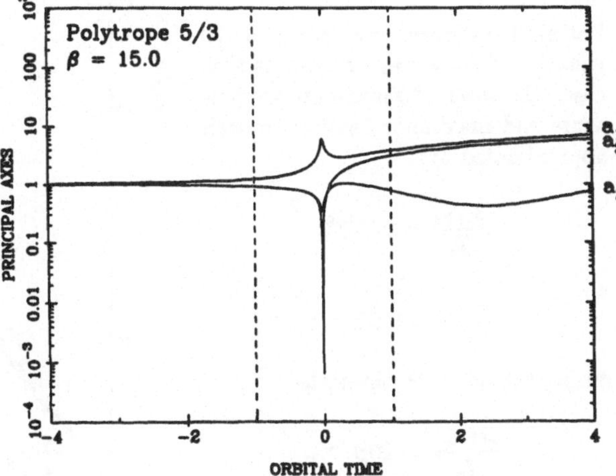

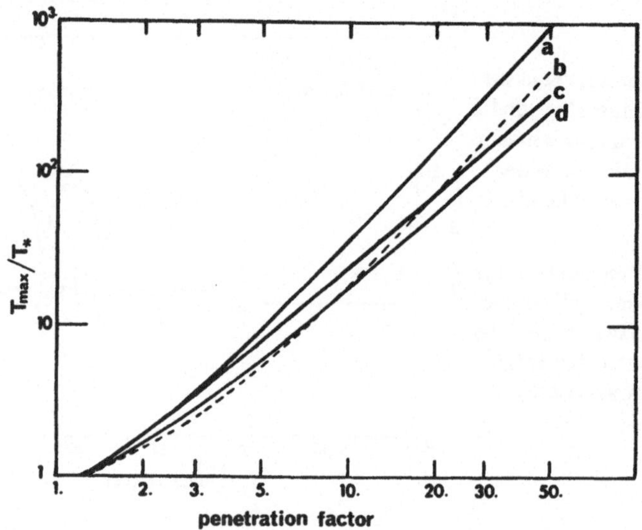

Figure 8: Influence of the equation of state on the extremal pancake quantities.

Upper: temperature
Right: density

The different curves correspond respectively to the polytropic models

(a) $\gamma = 5/3$ and
(b) $\gamma = 4/3$ (dashed line)

and the standard models (mixture of perfect gas with radiation) with mass

(c) $M_* = 3M_\odot$,
(d) $M_* = 15M_\odot$.

For most ordinary stars, the $\gamma = 5/3$ equation of state may be used. In that case, the laws of maximum temperature and maximum heating are well approximated by:

$$\frac{T_{\max}}{T_*} = 0.37\beta^2$$

$$\frac{\rho_{\max}}{\rho_*} = 0.22\beta^3.$$

during the pancake timescale

$$\frac{\Delta t_{\max}}{t_*} = 10\beta^{-4}.$$

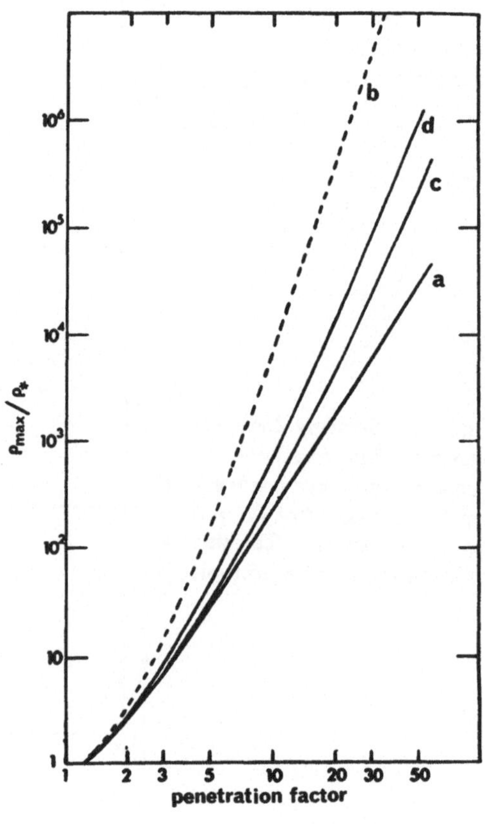

the tidally induced nuclear detonation will eventually take place) at least until the phase of the bounce, after which shock waves and significant nonlinearities may develop.

The specification of the ellipsoidal configurations and of the fluid internal motion requires a 3×3 "deformation" matrix (q_{ij}), or equivalently the 3 magnitudes of the principal axes $a_i(t)$, the 3 components of the angular velocity vector $\omega_i(t)$ and the 3 components of the shear vector $J_i(t)$. Finally the complete set of equations of motion reduce to 12, three of which describing the motion of the centre-of-mass $X_i(t)$, and the nine other corresponding to the internal degrees of freedom of the affine star model:

$$M_* \frac{d^2 X_i}{dt^2} = - M_* \frac{\partial \varphi_N}{\partial X_i} + \tfrac{1}{2} \mathcal{M}_* q_{ja} q_{ka} D_{jki} \tag{4.3}$$

$$\mathcal{M}_* \frac{d^2 q_{ia}}{dt^2} = \mathcal{M}_* C_{ij} q_{ja} + \Pi q_{ai}^{-1} + \Omega_{ji} q_{aj}^{-1} \tag{4.4}$$

where M_* is the mass of the affine star model, $\mathcal{M}_*$ is its quadrupolar mass, Π is the pressure integral depending on the equation of state and the Ω_{ij} are the components of the self-gravitational energy tensor.

Exhaustive numerical calculations of the motion of an affine star model along various parabolic orbits have been performed.[9] Figures 5 to 7 illustrate some of the characteristic results. The first diagram shows a case where the periastron distance is slightly greater than the effective disruption radius; the affine star model, initially in spherical equilibrium, is excited by the tidal perturbation and bifurcates to a nearly stationary Roche-Riemann ellipsoid with oscillations of its principal axes, angular velocity and internal motion. Now, if the star penetrates into the disruption radius but not enough to suffer strong compression effects (the overall tendency being always decompression), its configuration evolves to a more and more stretched cigar (figure 6), leading ultimately to "quiescent" disruption (for instance by fragmentation). The pancake effect develops clearly for penetration factor greater than about 5; the affine star model is strongly squeezed in the direction orthogonal to the orbital plane, then bounces and expands in all directions (figure 7).

Several equations of state were used, including the important case of a mixture of a perfect gas with radiation. The figures 8a-8b show the influence of the equation of state on the maximum temperature and density, that are the crucial quantities for "pancake" nucleosynthesis calculations.

4.4 Motion in relativistic tidal field

This school is devoted to gravitation, and so far gravitation is much better described by Einstein's General Relativity theory than by the Newtonian theory, especially when black holes are involved! In the mechanism of tidal disruption of a star, it is clear that general relativistic effects will be important if the periastron of the orbit is close to the horizon of the black hole. In General Relativity, the analog of the Newtonian tidal tensor is merely the Riemann tensor. The tidal acceleration between two nearby geodesics is given by the geodesic deviation equation

$$u^\mu u^\nu \nabla_\mu \nabla_\nu k^\sigma + R^\sigma{}_{\mu\nu\rho} u^\mu k^\nu u^\rho = 0 \tag{4.5}$$

where $\mathbf{u}$ is the quadrivector tangent to the fiducial geodesic and $\mathbf{k}$ is the separation quadrivector between geodesics. Introducing a parallel-propagated tetrad

$$\{\lambda_{(0)}, \lambda_{(1)}, \lambda_{(2)}, \lambda_{(3)}\}$$

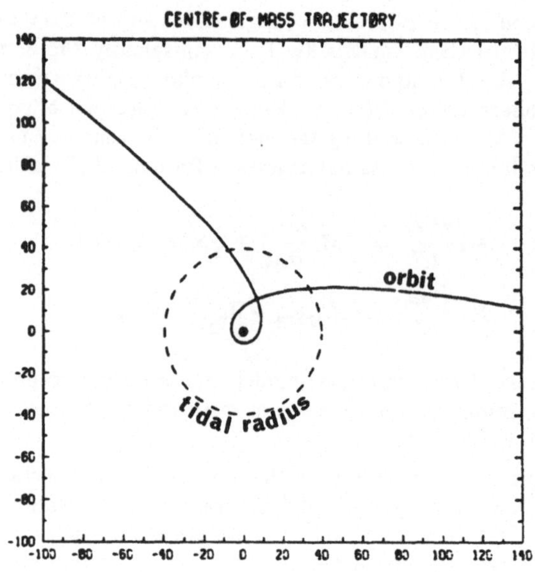

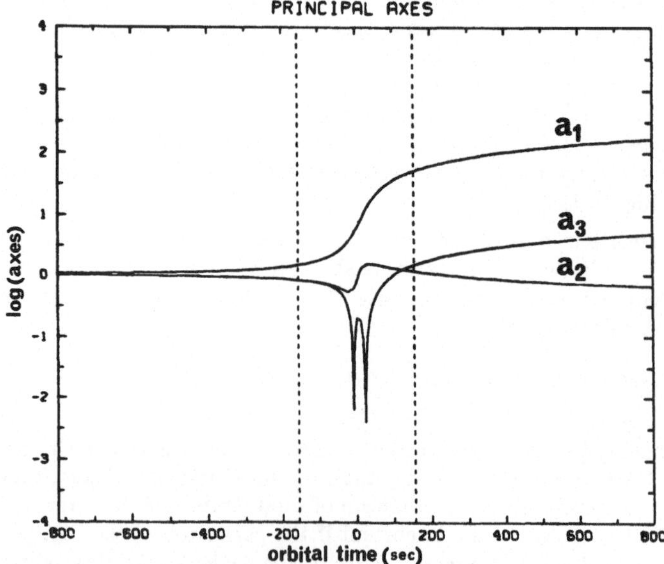

Figure 9: Deformations of the affine star model in Schwarzschild geometry for tidal radius $39.5M$ and periastron distance $56.2M$ $(G = c = 1)$, i.e. penetration factor $\beta = 7$.

Upper: Trajectory of the centre-of-mass. The position of the black hole and its Schwarzschild radius $2M$ are indicated at the origin of the coordinates. The dashed circle represents the tidal radius.

Lower: Principal axes. Proper time is in seconds. The dashed vertical lines signal the instants of passage at the tidal radius.

such that $\lambda_{(0)} \equiv \mathbf{u}$, Pirani[15] defined the "position" 3-vector $\mathbf{r}$ and the general relativistic tidal tensor $\mathbf{C}$ by:

$$r_i \equiv \lambda_{(i)\,\mu} k^{\mu} \quad (4.6)$$

$$C_{ij} \equiv - R_{\mu\nu\rho\sigma} \lambda^{\mu}_{(0)} \lambda^{\nu}_{(i)} \lambda^{\rho}_{(0)} \lambda^{\sigma}_{(j)}. \quad (4.7)$$

Then he showed that the geodesic deviation equation (4.5) may be set in exactly the same form as in Newtonian theory:

$$\frac{d^2 r_i}{d\tau^2} = C^i{}_j r^j. \quad (4.8)$$

In the context of tidal effects occurring in gravitational fields generated by black holes, it is appropriate to use the corresponding Schwarzschild or Kerr geometries. Marck[16] proved recently that the separability properties of such type D-spacetimes allow one to find analytical expressions for tetrad vectors $\{\lambda_{(i)}\}$ and tidal tensor components C_{ij}.

Motion of homogeneous bodies in black hole spacetimes has been studied by various authors,[17 18] but, like in the Newtonian approximation, the assumption of compressibility adds much more interesting features to the behaviour of the body. Luminet and Marck[19] performed the calculation of the deformation of an affine star model moving in Schwarzschild spacetime along nearly parabolic orbits. Tidal gravitational radiation can be neglected, due to the negative interference between waves emitted from different parts of the distorted body.[20] The most significant result is that several squeezing points and pancake flattenings may occur (figure 9). This is due to the property that highly parabolic orbits in Schwarzschild spacetime have a double point at finite distance; multi-pancake compressions require then that the double point lies inside the tidal radius. It turns out that the double compression effect is not at all exotic but will occur for several percent of all the solar-type stars penetrating into the tidal radius of a $10^6 M_\odot$ black hole. This purely General Relativistic effect has of course important consequences for the nuclear processes that may occur in the core of the squeezed stars.

4.5 Pancake nucleosynthesis and the fate of debris

Traditional sites of explosive nucleosynthesis, leading to the formation of heavy elements, are the supernova explosions associated with the gravitational collapse of degenerate iron cores in massive stars, and the explosions occurring on the surface of accreting compact stars (white dwarfs and neutron stars).

It is now clear that the strong gravitational fields generated by black holes may also act as detonators of explosive nucleosynthesis. In a typical "pancake" situation, the core of a squeezed star has a temperature exceeding $10^8\ {}^\circ K$ and density exceeding 10^5g/cc during less than 1 second. Under such conditions, nuclear detonation may take place but will differ from that occurring in supernovae because in pancake stars the fuel has generally the standard chemical abundance whereas in supernovae, hydrogen is already exhausted. For the most violent tidal disruption events, helium burning by the triple-alpha process occurs in a proton-rich medium.[4, 21] Such an "α-p" nuclear process leads to the production of special isotopes,[22] whose detection in the vicinity of galactic nuclei could constitute in the near future an observational clue to the existence of big black holes.

The building-up of "anomalous" isotopes by pancake nucleosynthesis would be of little astrophysical interest if the enriched stellar debris were captured by the hole. But the formation of heavy elements goes with a large nuclear energy release, whose effect on the dynamics of the stellar debris is drastic. It turns out that the (positive) nuclear energy injected into the pancake star may well exceed the initial (negative) binding energy; as a consequence, a significant fraction of the stellar gas will be free to leave the accretion radius of the black hole, while the complementary fraction will remain tightly bound to the hole. Thus, enrichment of the interstellar medium in heavy isotopes is possible. In the Galactic Center, there is indeed observational evidence for gas ejection from a very compact source.[23]

REFERENCES

1. D. Lynden-Bell, Nature **223**, 690 (1969).
2. J.G. Hills, Nature **254**, 295 (1975).
3. L. Spitzer, W.C. Saslaw, Astrophys. J. **143**, 400 (1966).
4. B. Carter, J.P. Luminet, Nature **296**, 211 (1982).
5. S. Chandrasekhar, "Ellipsoidal Figures of Equilibrium," Yale University Press, New Haven (1969).
6. E. Roche, Mem. Acad. Sci. Montpellier **1**, 243 (1847-50).
7. M.L. Aizenman, Astrophys. J. **153**, 511 (1968).
8. A. Nduka, Astrophys. J. **170**, 131 (1971).
9. J.P. Luminet, B. Carter, Astrophys. J. Suppl. **61**, 219 (1986).
10. P.J. Young, G.A. Shields, J.C. Wheeler, Astrophys. J. **212**, 367 (1977).
11. R.A. Nolthenius, J. Katz, Astrophys. J. **269**, 297 (1983).
12. G.V. Bicknell, R.A. Gingold, Astrophys. J. **273**, 749 (1983).
13. B. Carter, J.P. Luminet, Astron. Astrophys. **121**, 97 (1983).
14. B. Carter, J.P. Luminet, Monthly Notices Roy. Astron. Soc. **212**, 23 (1985).
15. F.A.E. Pirani, Acta Physica Polonica **15**, 389 (1956).
16. J.A. Marck, Proc. Roy. Soc. London **A385**, 431 (1983).
17. L.G. Fishbone, Astrophys. J. **185**, 43 (1973).
18. B. Mashhoon, Astrophys. J. **197**, 705 (1975).
19. J.P. Luminet, J.A. Marck, Montly Notices Roy. Astron. Soc. **212**, 56 (1985).
20. M.P. Haugan, S.L. Shapiro, I. Wasserman, Astrophys. J. **257**, 283 (1982).
21. B. Pichon, *Nucleosynthesis in Pancake Stars*, in : "Nucleosynthesis and its Implications on Nuclear and Particle Physics," J. Audouze & N. Mathieu, eds., pp. 223-231. D. Reidel Pub. Co, Amsterdam (1986).
22. B. Pichon, J.P. Luminet, in preparation, Observatoire de Paris.
23. M.J. Rees, in: "The Milky Way Galaxy," H. van Woerden ed., p. 379, D. Reidel Pub. Co., Amsterdam (1985).

NAKED SINGULARITIES IN SPHERICAL GRAVITATIONAL COLLAPSE

D. M. Eardley

Institute for Theoretical Physics
University of California
Santa Barbara, California 93106 USA

1. INTRODUCTION

The Cosmic Censorship Conjecture of Penrose is perhaps the most important open question in nonquantum general relativity theory.[1,2] According to the weak form of this conjecture,[3] if one starts with regular initial conditions for gravitational collapse for a physical system in an asymptotically flat setting, then any spacetime singularities which appear later must lie within black holes, so that they will be hidden from external observers. The strong form of the conjecture[1] states that no observer can see a singularity, even if the observer has fallen into a black hole. This strong form is equivalent to the statement that, generically, spacetimes are globally hyperbolic.

In particular, naked singuarities can occur at the center of a gravitationally collapsing configuration of dust or[4,5,6] or null fluid[7] in spherical symmetry. We will call these *shell-focusing* singularities[4,7] since they are reminiscent of the shell-*crossing* singularities which are well known to occur for pressureless collapse.[8] The main difference is that a shell-crossing singularity is in essence a 2-dimensional caustic surface of dust trajectories in space, whereas shell-focusing singularities occur when dust trajectories focus down to a point locus at the center of spherical symmetry. In Section 2, I will review what is known about these singularities.

Are shell-focusing singularities counterexamples to cosmic censorship? An important objection is that dust are null fluid are not fundamental forms of matter, even on the classical level. They should be regarded as approximations to more fundamental entities: For instance, null fluid approximates a massless scalar field in the eikonal approximation, while dust approximates a massive scalar field. Perhaps shell-focusing singularities are merely artifacts of this approximation. The intriguing question is then: Do shell-focusing singularities occur for a scalar field coupled to gravity? In Section 3 I will mention the results of Maithreyan[9,10] on a particular class of solutions, and of Christodoulou[11] on general classes, of spherically symmetric scalar-field collapses.

2. DUST COLLAPSE AND SHELL FOCUSING SINGULARITIES

Pressureless matter or "dust" is described by the the timelike 4-velocity field U^μ of the matter world lines and by the density ρ. The flow lines are timelike geodesics, $U^\alpha D_\alpha U^\mu = 0$, and the dust density is conserved, $D_\alpha(\rho U^\alpha) = 0$. The stress energy tensor of dust is $T_{\mu\nu} = \rho U_\mu U_\nu$. The hard question is what do do when the flow lines begin to cross, as they often do.

2.1 Shell-Crossing

Generically the crossing can be thought of as 2-dimensional shells of dust piling up at a 2-dimensional caustic locus. One option for handling such "shell-crossing" is to assume that the dust is collisionless — this is not very realistic but it is quite consistent mathematically. Such a locus where the flow lines cross collisionlessly is often called a "shell-crossing singularity" but it really should not be considered a singularity, since the spacetime causal structure is extendible through the locus. Though there is a breakdown of differentiability, the spacetime metric can be defined as a distribution in a neighborhood of the "shell-crossing singularity."[8] The dust density ρ, and some components of the curvature tensor, generically blow up like $s^{-1/2}$, where s is proper spatial distance from the caustic.

To describe the dust distribution in this situation, it is necessary to keep track of many shells of dust at the same point — most generally would be necessary to go over to a kinetic theory description, where one has a 3-dimensional velocity space $\{U^\nu \mid U_\alpha U^\alpha = -1\}$ at each point x^μ in spacetime, and a dust distribution function $\Phi(x^\mu, U^\nu)$ in phase space. Then one would have the collisionless Boltzmann equation for Φ coupled to the Einstein equation. This is just what is studied under the name of relativistic stellar dynamics — see, *e.g.*, Ref. 12.

Another option is to assume that shells of dust which try to cross at a caustic will rebound elastically; a third is to assume an inelastic collision, producing either a shell mass or some pressure; these latter two options do not seem to have been studied in the literature despite the fact that they are more "realistic" than the the collisionless assumption — perhaps because they are technically even harder.

For simplicity I will assume that shell-crossing does not occur, except perhaps right at the center.

2.2 Tolman-Bondi Solutions

The general solution for spherically symmetric dust in general relativity was given by Tolman[13] and, later, Bondi.[14] It is simple to state. The metric is (in comoving geodetic coordinates so that $g_{tt} = -1$)

$$ds^2 = -\,dt^2 + X^2(r,t)\,dr + Y^2(r,t)(\,d\theta^2 + \sin^2\theta\,d\phi^2). \tag{2.1}$$

The coordinate r is a label on concentric spherical shells of dust, and t is proper time along the world line of a dust particle. We will assume for simplicity that $r = 0$ is the center of spherical symmetry and that $\rho > 0$ at $r = 0$. The Einstein equations imply

$$X(r,t) = \frac{1}{W(r)}\frac{\partial Y}{\partial r} \tag{2.2}$$

$$\left(\frac{\partial Y}{\partial t}\right)^2 = W(r) - 1 + \frac{2M(r)}{Y} \tag{2.3}$$

where $W(r)$ is a function of integration which has the significance of *relative binding energy of shell* r, and $M(r)$ is a function of integration which has the significance of *the active gravitational mass within shell* r.

Equation (2.3) is an ordinary differential equation in t, at each r. It is essentially the Friedmann equation replicated for each shell r. The case $W(r) < 1$ (bound) corresponds to $k = +1$ (closed Friedmann model); the case $W(r) = 1$ (marginally bound) corresponds to ($k = 0$) (flat Friedmann model); and the case $W(r) > 1$ corresponds to ($k = -1$) (open Friedmann model). One solves (2.3) easily.

Upon solving the first order equation (2.3), a third function of integration $t_0(r)$ appears — the *crunch* time — defined by $Y(r, t_0(r)) = 0$. This is the time when a shell of dust is crushed to zero area. In a model of gravitational collapse, the crunch time represents the occurrence of the spacetime singularity. In the bound case, $W(r) < 1$, another singularity also occurs at the earlier *bang time*

$$t_1(r) = t_0(r) - \frac{2\pi M(r)}{(1 - W^2)^{3/2}}. \tag{2.4}$$

The marginally bound limit $W(r) \to 1$ corresponds to $t_1(r) \to -\infty$ (bang time infinitely far in the past). For bound collapse, one often considers the *time symmetric* subcase $t_1(r) \equiv -t_0(r)$. The particular case of a bound, time symmetric collapse wherein the dust density is homogenous ($\rho =$ const at given t) out to some surface boundary $r = R$, and vanishing beyond, is called *Oppenheimer-Snyder* collapse.

Thus to give a solution, one gives three functions $W(r)$, $M(r)$, $t_0(r)$ of r. Only two of these functions are physically significant because one can still make arbitrary, monotonic coordinate transformations $r \to r' = r'(r)$. Henceforth I will consider only the bound and marginally bound cases; for these, one can equally well give three other functions $W(r) \leq 1$, $t_1(r) \geq -\infty$, $t_0(r) \leq \infty$ of r to give a solution.

Shell crossing is signalled by $X \to \infty$ at finite $Y \neq 0$. In order to forbid shell crossing, it is sufficient to require $t_0'(r) \geq 0$, $t_1'(r) \leq 0$, $W'(r) \leq 0$.

2.3 Causal Structure of Tolman-Bondi Solutions

Spheres of a spherically symmetric spacetime (2.1) are trapped surfaces if $D_\mu Y$ is timelike (with Y decreasing to the future), and not otherwise. The condition that spheres be trapped is that

$$1 - \frac{2M}{Y} < 0. \tag{2.5}$$

This condition will be satisfied everywhere in a neighborhood of the spacetime singularity $Y = 0$, except where $M = 0$, *i.e.*, at the center r=0. Therefore Y will be strictly decreasing along any future-directed causal curve in that neighborhood. It follows that the singularity is spacelike everywhere except perhaps at $r = 0$, and therefore that strong cosmic censorship holds in these spacetimes unless it breaks down at $r = 0$.

What about the center $r = 0$ on the singularity? The situation is delicate. A careful estimation of the behavior of light rays near the center reveals that, if

$$\lim_{r \to 0^+} \frac{t_0'(r)}{M'(r)} > \frac{26}{3} + 5\sqrt{3} \approx 17.327 \tag{2.6}$$

then light rays emitted from arbitrarily near center of the singularity at $r = 0, t = t_0(0)$ can escape into the spacetime. Certain radial causal geodesics are incomplete. Therefore there is a naked singularity at the center that can be regarded as consisting of the past endpoints of the incomplete geodesics. The naked singularity lives for a finite time as seen by nearby observers, and then merges into the spacelike singularity which forms inside the matter. The latter, spacelike singularity is similar to the singularity at $r = 0$ in the Schwarzschild metric, and in fact extends into it if the dust density drops to zero outside some radius $r = R$. The naked singularity cannot be eliminated by extending the spacetime.[4]

Thus, in the case (2.6), global hyperbolicity breaks down at the center, and one begins to worry about strong cosmic censorship. This behavior is actually generic among spherically symmetric dust collapses, although it is unknown (and to this author doubtful) whether the naked singularity is stable under small deviations away from spherical symmetry.

On the other hand, if inequality (2.6) is reversed,

$$\lim_{r\to 0^+} \frac{t_0'(r)}{M'(r)} < \frac{26}{3} + 5\sqrt{3} \approx 17.327 \tag{2.7}$$

no naked singularity occurs at the center and the spacetime singularity is totally spacelike.

When one of these naked shell-focusing singularities is formed, it may or may not be globally naked, depending on the nature of the solution well away from the center. It is easy to show that a wide class of such examples exists for which the naked singularity is indeed globally naked, so that a violation of weak cosmic censorship does occur.

For further information and results about these naked singularities in Tolman-Bondi solutions, see Ref. 4, where the marginally bound case is discussed and many examples are given, and Ref. 6, where a rigorous discussion is given for bound, time symmetric collapses. The time-reversed versions of these spacetimes can also be regarded as models of white holes.[15]

2.4 Interpretation of Shell-Focusing Singularities

There is no doubt that these spacetimes contain naked singularities, but how are we to interpret the spacetimes? According to the condition (2.6), a naked singularity occurs for collapses which are sufficiently *slow* near the center. I say *slow* because $t_0'(0)$ is large, which means that successive dust shells are crushing to zero area at markedly later and later times — if the collapse is fast, *i.e.*, nearly homogeneous, then $t_0'(0)$ is *small*. This means that naked singularities occur for highly nonrelativistic collapse, which seems paradoxical.

The resolution to this paradox is in the assumptions which underlie the Tolman-Bondi solutions (2.1,2.2,2.3). The metric (2.1) has built into it that only one dust stream exists at any one point of spacetime. In nonrelativistic gravitational collapse, one would expect that collapsing dust shells would pass right though the center and come out again, to give a region containing at least two (and perhaps many) dust streams — but (2.1) is inadequate to describe this. Thus the Tolman-Bondi solutions have built into them the assumption that *dust shells cannot pass through the center — a shell which collapses to the center stays there as a point mass.* In other words, dust shells are assumed to behave completely inelastically at the center. It is no wonder that singularities occur there!

What if we assume different boundary conditions at the center? If we assume that shells are collisionless, then we should in particular allow them to pass collisionlessly through the center. This is technically hard to implement — in general we would need a Boltzman-equation approach to the problem, as outlined in Section 2.1 above. Nobody has done this to my knowledge. Alternately, we might assume that shells collide totally elastically with each other, and recoil elastically upon reaching the center — as regards behavior at the center, this is no different from simply assuming that shells can pass though the center. Again, this is technically hard to implement. With either modification of the boundary condition, the following conjecture seems attractive:

Conjecture 1. *With collisionless or elastically colliding dust shells, there are no naked singularities of the shell-focusing type for spherically symmetric dust collapse. Given a collapse (2.6) which has a naked singularity according to the inelastic (Tolman-Bondi) boundary condition at the center, its correct evolution according to the collisionless or elastic boundary condition will show no breakdown of cosmic censorship at the center, but rather will have a region of spacetime containing two or more dust streams.*

I believe the best way to approach this problem is to go to a Boltzmann equation, or stellar dynamic, description of the matter, and to study distribution functions $\Phi(x^\mu, U^\nu)$ which are smooth in phase space. In fact this problem has been studied a good deal for the case of Newtonian gravity.[16,17] For Newtonian gravity, some powerful no-singularity (or global existence) theorems have been proven.[16] It is reasonable to conjecture that these results extend gracefully to general relativity in the spherically symmetric case, where one has a radial and time coordinates r and t, and velocity space coordinates consisting of the radial velocity $U^{\hat{r}}$ and the tangential velocity $U^{tan} \equiv [(U^{\hat{\theta}})^2 + (U^{\hat{\phi}})^2]^{1/2}$ in an orthonormal basis. The dust distribution function is then $\Phi(r, t, U^{\hat{r}}, U^{tan})$.

Conjecture 2. *For spherically symmetric distributions $\Phi(r, t, U^{\hat{r}}, U^{tan})$ of dust ("stellar dynamics") in general relativity, gravitational collapse from regular initial conditions never produces naked singularities. Either no singularity at all occurs, or the singularity is totally spacelike and hidden inside a black hole. (Here Φ is required to be smooth and to fall off fast enough at large velocity in velocity space; cf. Ref. 16).*

2.5 Collapse of Null Fluid

One can also study the collapse of null fluid, which is similar to dust except that the matter streams along null geodesics rather than timelike ones. The matter is described by the 4-velocity field l^μ of the matter world lines and by the density ρ. The flow lines are null geodesics, $l^\alpha D_\alpha l^\mu = 0$, and the dust density is conserved, $D_\alpha(\rho l^\alpha) = 0$. The stress energy tensor is $T_{\mu\nu} = \rho l_\mu l_\nu$.

Spherically symmetric collapse of inward radially streaming null fluid is described by the Vaidya metric, wherein $l_\mu = \partial_\mu v$,

$$ds^2 = -\left(1 - \frac{2m(v)}{r}\right) dv^2 + 2\, dv\, dr + r^2(\, d\theta^2 + \sin^2\theta\, d\phi^2) \tag{2.8}$$

where v is an advanced time and $m(v)$ is an arbitrary, nondecreasing proper mass as a function of v. An interesting family of collapses can be obtained by setting $m(v) \equiv 0$ for $v < 0$, so that spacetime is flat to the past of the null surface $v = 0$, which marks the front of the inward streaming matter. In this family, a naked shell-focusing singularity occurs at the center $r = 0$ if[7]

$$\lim_{v \to 0^+} \frac{1}{m'(v)} > 16, \tag{2.9}$$

i.e., if the collapse is slow enough. The situation is parallel to that for the Tolman-Bondi solutions, with (2.9) replacing (2.6), and similar comments about the boundary condition at the center apply. For null fluid the collisionless and the elastically colliding cases are identical.

Conjecture 3. *With collisionless radially streaming null fluid, there are no naked singularities of the shell-focusing type for spherically symmetric collapse. Given a collapse (2.9) which has a naked singularity according to the inelastic (Vaidya) boundary condition at the center, its correct evolution according to the collisionless boundary condition will show no breakdown of cosmic censorship at the center, but rather will have a region of spacetime containing two radial streams, one inward going and the other outward-going.*

Conjecture 3 is easier to study than Conjecture 1 because, in spherical symmetry, only two streams of radially moving null matter exist; for dust a one parameter family exists. Maithreyan and I have found some evidence in favor of Conjecture 3 by studing a particular class of Vaidya metrics, namely the self-similar ones given by $m(v) = \varsigma v$ for $v > 0$, where $1/16 > \varsigma > 0$ is a constant. For these metrics we are able to construct explicitly the two-stream evolution which has no naked singularity.[18]

For a calculation of quantum mechanical particle creation by these naked singularities in Vaidya metrics, see Ref. 7.

3. COLLAPSE OF SCALAR FIELD CONFIGURATIONS

A scalar field ϕ is more fundamental than dust or null matter. Do such shell-focusing singularities occur for spherically symmetric gravitational collapse of scalar field configurations? The answer seems to be *NO*, at least for the massless scalar field obeying the (minimally coupled) wave equation coupled to the Einstein equation:

$$\Box\phi = 0 \tag{3.1a}$$

$$G_{\mu\nu} = 8\pi(\partial_\mu\phi\partial_\nu\phi - \frac{1}{2}g_{\mu\nu}\partial^\alpha\phi\partial_\alpha\phi). \tag{3.1b}$$

3.1 Self-Similar Collapse

T. Maithreyan has studied this problem under the restriction of self-similarity (or homothetic symmetry).[9,10] The initial data at past null infinity are $\phi(v,r) = 0$ for $v \leq 0$, $\phi(v,r) \to cv/r$ for $v > 0$, where c is a constant. Here v is, as above, and advanced time, and r is the radial coordinate, with $r \to \infty$ at past null infinity.

Maithreyan is able to solve this problem in closed form. In summary, the main result is that there is never a naked singularity, as long as the proper boundary condition is used on ϕ (regularity of ϕ) at the center of spherical symmetry. For sufficiently weak initial data, $c < \sqrt{8\pi}$, there is no graviational collapse; the ingoing scalar radiation rebounds from the center to become outgoing radiation, and the solution is time-symmetric. For strong enough initial data, $c \geq \sqrt{8\pi}$, a apparent horizon and a spacetime singularity form. The singularity is always spacelike or null, *i.e.*, never naked, so that no counterexample to cosmic censorshop occurs.

3.2 Generic Spherical Collapse

In an important series of papers, Christodoulou[11] studies the generic spherical gravitational collapse of a massless scalar field using global existence theory of partial differential equations. The main result of the first paper is that global, singularity-free solutions exist for weak enough initial data. The main result of the second paper is that generalized (roughly speaking, distributional) solutions exist globally for arbitrarily strong initial data. Such methods promise to provide the ultimate solution to the problem of cosmic censorship.[19]

Acknowledgements

This research was supported in part by the National Science Foundation under Grant Nos. PHY85-06686 and PHY82-17853, supplemented by funds from the National Aeronautics and Space Administration, at the University of California at Santa Barbara. Many of these results are from joint work with Tara Maithreyan. I am grateful to Tom Sideris for pointing out Refs. 16,17.

REFERENCES

1. R. Penrose *in:* Gravitational Radiation and Gravitational Collapse, C. DeWitt–Morette, ed. Reidel, Dordrecht, (1974).
2. G. Horowitz and R. Geroch *in:* Einstein Centennary Volume, S.W. Hawking and W. Israel, eds., Cambridge University Press, Cambridge (1979).
3. R. Penrose, *in:* Confrontation of Cosmological Theories with Observational Data, M.S. Longair, ed. Reidel, Dordrecht (1974).
4. D.M. Eardley and L. Smarr, *Phys. Rev. D* **19**:2239 (1979).
5. D.M. Eardley, *Gen. Rel. Gravitat.* **10**:1033 (1979).
6. D. Christodoulou, *Comm. Math. Phys.* **93**:171 (1984).
7. W.A. Hiscock, L.G. Williams and D.M. Eardley, *Phys. Rev. D* **26**:751 (1982).
8. A. Papapetrou & A. Hamoui, *Ann. Inst. H. Poincaré A* **9**:179 (1968).
9. T. Maithreyan, Ph.D. Thesis, Boston University, USA, 1985.
10. T. Maithreyan and D. Eardley, ITP preprint (1986).
11. D. Christodoulou, *Comm. Math. Phys.* **105**:337 (1986); *ibid.* **106**:587 (1986).
12. S.L. Shapiro & Saul A. Teukolsky, *in:* Dynamical Spacetimes and Numerical Relativity, J.M. Centrella, ed. Cambridge University Press, Cambridge (1986).
13. R.C. Tolman, *Proc. Nat. Acad. Sci. (U.S.)* **20**:169 (1934).
14. H. Bondi, *Mon. Not. Roy. Astron. Soc.* **107**:410 (1947).
15. D.M. Eardley, *Phys. Rev. Lett.* **33**:442 (1974).
16. J. Batt, *J. Diff. Eq.* **25**:342 (1977).
17. E. Horst, *Math. Meth. in the Appl. Sci.* **3**:229 (1981); *ibid.* **4**:19 (1982).
18. D. Eardley & T. Maithreyan, in preparation, 1987.
19. V.Moncrief and D.Eardley, *Gen. Rel. Gravitat.* **13**:887 (1981).

PART II :

GRAVITATION IN COSMOLOGY

SOME TOPICS IN RELATIVISTIC COSMOLOGY

John D. Barrow

Astronomy Centre
University of Sussex
Brighton BN1 9QH, U.K.

INTRODUCTION

These lectures do not aim to provide a complete survey of relativistic cosmology. Rather, we shall confine attention to a number of specific aspects of relativistic cosmology which might be of interest to current investigations of high energy physics in the early universe and to the wider audience of relativists and astrophysicists at this School. In particular, we shall focus upon features of general relativistic cosmology that are of relevance to the inflationary universe theory and the cosmological questions that it confronts[1,2]. A number of new results will be described in the second half of the notes. The detailed particle physics motivation for the existence of an inflationary phase during the early history of the expanding universe will be covered in other lectures.

A number of research problems arising from the subject matter covered in the lectures have been added to supplement these notes. These are signalled in the text by the symbol □ and are, as far as the author is aware, unsolved; relevant associated references which might be helpful in their solution have also been added where possible.

ORIENTATION

The aim of modern cosmology is to determine the structure of the Universe -- its age, shape, composition, history and so forth -- by observation and to understand those observations within the framework of a unified mathematical theory of gravitation and matter, locally and independently tested by experiment. This grandiose programme must come to terms with a number of unusual problems that are characteristic of cosmology:

The Universe is unique

This tautological observation is relevant to any deductions regarding the stability or generality of particular properties of the Universe with respect, say, to all the possibilities admitted in the initial data space of the Einstein equations. The fact that a particular property is not open dense[3] (*generic*) in the solution space of Einstein's equations need not be an argument against it being a property of the observed Universe.

There may also exist unique selection effects in operation if the Universe is infinite. For example, there is an old argument that exhaustively random infinite initial data sets possess the awkward property that any property that *can* arise, *must* arise somewhere infinitely often with probability unity[4-6].

There may also prove to exist fundamental problems[7,91] of principle when dealing with any quantum cosmological theory unless the many worlds interpretation of quantum mechanics is adopted, in which case what is meant by the uniqueness of the Universe must be more carefully defined. Other lecturers will discuss what is meant by "prediction" in a quantum cosmological model.

Non-local influences

The Einstein equations do not fix the topology of space yet this global feature of space-time can have very strong local effects upon current observables as well as upon quantum processes in the early universe. Although it is conventional to assume that the Universe possesses the natural spatial topology (S^3 or R^3) it may be far more probable for any universe "created out of nothing"[8] to possess one of the myriad of non-standard topologies. If extra spatial dimensions exist they are not expected to possess a natural topology[14].

Some cosmological boundary conditions may be necessary either at an initial singularity or at past infinity (the alternative -- that all timelike and null geodesics are closed, perhaps with periods >> 10^{10} yrs is not appealing). The influence of initial conditions is occluded by the possibility that some portions of the initial singularity may be timelike (that is they lie in the causal future of other parts of it) so there is a breakdown of our usual picture of determinism.

The Universe may be significantly inhomogeneous over very large length scales. This is actually predicted by most inflationary universe scenarios[1].

Mach's Principle[9], if it true and can be suitably formulated, may reveal some global constraints upon the local structure of space-time. Even more speculatively, quantum non-locality may have some unexpected cosmological manifestation within a future theory of quantum gravitation.

Horizons

The causal structure of space-time ensures[10] that astronomical observations are limited to that part of space-time which lies in or on our past null cone. Our direct observations are limited at present to redshifts of order[11] 3.8 (if quasar redshifts are entirely cosmological) and our earliest indirect information comes from primordial nucleosynthesis at redshifts ~ 10^{10}. We have no direct or indirect observational evidence of any sort regarding the structure of the Universe at earlier times than ~ 1s. The most exciting observational development in this respect would be the discovery of a primordial mini black hole or super-massive monopole (again?!).

How many spatial dimensions are there?

Recent developments in particle physics have resurrected the old idea that gauge invariances in the observed three spatial dimensions may be interpreted as coordinate invariances of a gravitation theory with additional spatial dimensions. All Kaluza-Klein[12,13] and superstring[14] theories predict the existence of additional spatial dimensions. This considerably expands the possible evolutionary histories for the very early history of the Universe in ways that are as yet not fully understood. The additional dimensions are always assumed to be spatial although there seems to be no fundmental reason for this restriction.

Variation of fundamental "constants"

The mathematical formulations of physical laws that we have found most expedient necessarily contain certain proportionality factors whose precise values are in general not constrained by the solution of those equations. These proportionality constants we call the "constants of Nature". It is possible that those quantities we assume to be fundamentally constant do vary on cosmological timescales[6]. Such a variation of the "constants" we observe in three dimensional space is predicted to occur necessarily in cosmological theories possessing uncompactified extra spatial dimensions, but is strongly constrained by observation[15-17].

Unknown Physics

As yet, we possess no very strong-field tests of general relativity and no high-energy tests of asymptotic freedom. The structure of the very early universe will be strongly affected by any deviation from these standard theories at high energies: for example by the influence of quadratic corrections to the Einstein gravitational lagrangian in regions of high space-time curvature, or the breakdown of quantum field theory at high energies. We may soon run out of practical and affordable local tests of the theoretical ideas that are necessary to model the early history of the Universe. The widespread, and essentially teleological view that all the "right" theories must have testable consequences that *we* will be able to check by experiment is no more than a hope. In particular, if the Universe is as close to the critical density today as some inflationary theories predict[18] then we may never be able to determine whether the Universe is open or closed[19].

Selection Effects

Astronomical observations of distant objects are beset by well-known evolutionary selection effects – for instance, because of light-travel time-delays, distant galaxies are now seen when they were younger and hence intrinsically different from nearby galaxies of similar appearance.

The truth of some theories of galaxy formation requires some quantitative assessment to be made of the (as yet) purely subjective impression of "filaments" in the clustering of galaxies[21].

If there is no law of Nature that fixes initial conditions uniquely, completely and self-consistently, then our own existence may act as an irreducible selection effect[6]. This would moderate our surprise at finding the Universe to possess properties that are unlikely *a priori*. These properties (for example the great size and age of the observed Universe) may possess no explanation other than that they are necessary for the evolution of biological complexity.

Unknown Matter Fields

We have only scant knowledge of the quantity of matter in the Universe[22] (the so called "missing matter problem") and the form that it takes. We do not know whether the initial and present stages of the expansion are dominated by the effects of non-classical fluids, examples of which are those described by the cosmological constant or vacuum strings. We do not know which energy conditions it is reasonable to expect the material content of the early universe to obey. For instance, violation of the once inviolate[23] strong energy condition ($\rho + 3p \geq 0$ for an isotropic perfect fluid having density ρ and pressure p) is now accepted as a necessary ingredient of inflationary universe theories.

□ Examine the observable effects of infinite cosmic strings and inflation in a finite open universe with 3-torus topology.
□ Determine whether it is more probable for a finite universe that is created 'out of nothing' to possess an exotic topology than the natural one (S^3).
□ Find a probability measure for the space of cosmological initial data. Can you exploit the fact that chaotic cosmological models (like the Mixmaster models[33,56,132]) preserve a natural dynamical measure as the singularity is aproached?

HOW LITTLE COULD WE KNOW?

In order to create some perspective for later discussion of the popular standard Big Bang model described by an isotropic and homogeneous Friedman model we shall begin by summarizing what might be called 'the state of maximum ignorance'. Let us suppose that we are allowed to assume only that *gravitation is described by a metric theory*. That is, that there exists some space-time geometry with metric interval

$$ds^2 = g_{ab}\, dx^a dx^b \quad ; \quad 0 \leq a,b \leq 3 \qquad (1)$$

and so there exist geodesic equations determined by paths of stationary action between any two points in the space-time. However, lacking a gravitation theory we do not yet know how the geometry is coupled to the mass-energy content of space-time (or even if it is coupled to it).

Now, suppose that we are also allowed to assume that *all the familiar local (non-gravitational) physics holds*, that *we can ignore selection effects*, and that *our observations are perfectly accurate*. How much can we deduce about the structure of the Universe[24,38]?

In Figure 1 we show the observational situation that confronts us in our attempts to map out the structure of space-time. We are able to make direct observations by collecting photons, neutrinos, gravitational waves and any other relativistic or massless particles travelling down *our* past null cone. These observations will be occluded by various known or unknown sources of opacity in the past. For example, in Figure 1 we have indicated how far back in redshift we can observe directly using photons, neutrinos and gravitational waves using the numbers appropriate in the standard Friedman model as a guide.

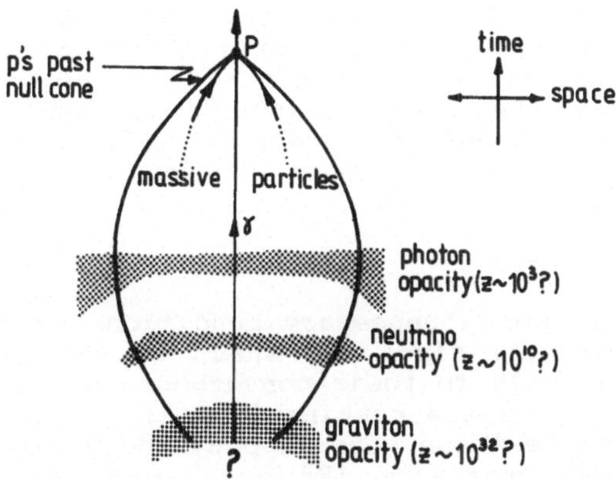

Fig. 1. The region of space-time to the causal past of an astronomer moving along the world-line γ and presently at p. The regions directly accessible to observation are the past null cone along which we receive massless and relativistic particles and gravitational waves and its interior, from which we receive massive particles like cosmic rays or primordial nuclei moving at subluminal velocities.

We can also make observations of the structure of the local region inside our past null cone by observing the motions of non-relativistic massive objects like cosmic rays, planets, stars, galaxies, nuclei, massive neutrinos and so forth. Notice that we can say absolutely nothing *in principle* about the structure of space-time outside our past null cone or about the past null cones of other observers or about our future null cone, and without a theory of gravitation we cannot predict the structure on the inside of our past null cone from data observed on it.

The direct observations available to us[24,38] are *number counts of celestial objects with solid angle, proper motions; shapes, sizes and intensities of images; the spectra and intensity distributions of radiation fields*. These do not allow us, in fact, to deduce either the complete metric of space-time or the full energy-momentum tensor of its material content even down our own past null cone without the introduction of unverifiable assumptions about its structure[24,38]. No observations can demonstrate the *Cosmological Principle* that the Universe is globally homogeneous and isotropic.

This lack of progress is not really surprising. But it is interesting to note that certain very strong conclusions are possible under the minimal assumptions we have allowed ourselves. In particular, it is possible to prove a variety of singularity theorems[23]. A typical statement of the simplest of which is that: if S is a Cauchy surface with extrinsic curvature $\chi_a{}^a \geqslant C \geqslant 0$, with C constant, and the Ricci tensor of space-time obeys the reconvergence condition $R_{ab}V^aV^b \geqslant 0$, where V^a is a timelike vector field, then all inextendible timelike geodesics terminate in the past at a physical singularity. (It is straightforward to modify these assumptions so that they refer to conditions in and on our past null cone only). If we assume a theory of gravity, (for example general relativity), then the Einstein equations allow us to re-express the reconvergence condition on the Ricci tensor as the *strong energy condition* on the energy-momentum tensor, T_{ab}, of the matter; that is, for any timelike vector field V^a,

$$[T_{ab} - \tfrac{1}{2}g_{ab}\, T_a{}^a]V^aV^b \geqslant 0 \qquad (2)$$

(Note that (2) is not a necessary condition for the formation of a past singularity since the shear of the geodesics also contributes positively to their convergence and can compensate for the negativity of the combination (2).) Other theories of gravity, for example those generated by adding terms quadratic in the curvature invariants (R^2 or $R_{ab}R^{ab}$) to the Einstein gravitational action can produce space-time singularities[39] but the conditions on T_{ab} that are sufficient for them to do so will generally be considerably more complicated than (2).

Suppose that we now widen our assumptions to include[24-26,38]: *general relativity is the theory of gravitation*. It then turns out that the observations we are able to make (ignoring selection effects and practical imperfections) are *necessary and sufficient to determine the metric and the energy momentum tensor along our past null cone until any singularity is reached*. This theorem, proved by Nel and Stoeger[38], is a remarkable and unexpected result. It would be interesting to know for what other theories of gravitation it is true.

Now that we have assumed a theory of gravitation we have some hyperbolic evolution equations which enable us to determine some of the structure off the past null cone from information on it. However, because of our inability to observe outside the past null cone we could still never verify that the Universe is or was homogeneous and isotropic, even if it really is to some approximation[38]. It is interesting to recall that if current inflationary models of the early universe are correct then it is likely that the Universe is extremely inhomogeneous beyond our horizon. The only way in which we can make deductions about this region is by assuming some form of Copernican or Cosmological Principle. The only information available to us in the absence of such unverifiable assumptions is that conveyed by the elliptic constraint equations of general relativity. These allow us to "feel" the gravitational effects of these regions even though we cannot see them. The conservation equations prevent anything happening there (half the Universe outside our horizon disappearing, for example) which could give us acausal contradictions here and now.

The fact that we find we cannot base cosmology upon observations alone means that we must proceed as in other quantitative sciences by specifying some variables, in this case g_{ab} and T_{ab}, up to a set of free parameters. These parameters are then determined by fitting the model to the observations until a contradiction or a superior fit to another model is obtained. If we make the assumption of spatial homogeneity then all local information about the Universe is taken to be global information and the role of the past null cone structure upon what is directly observable is obscured.

□ Give a rigorous statistical description of approximate spatial homogeneity and a resolution of the "closure problem" that arises. Relate it to the idea of 'almost' homogeneous groups of motions[42].

□ Determine whether other theories of gravitation[43,50] (*eg* Brans-Dicke theory or general relativity plus quadratic or higher-order curvature corrections to the lagrangian[39]) allow our past null cone to be determined uniquely and completely by ideal observations.

□ Prove singularity theorems in other theories of gravity subject to energy conditions different from the usual strong energy condition (2) by using their field equations to express the geodesic convergence condition as a condition on the stress-energy tensor.

□ Prove a singularity theorem that makes use of Einstein's equations in an essential way (*i.e.* not simply to arrive at (2) as a restatement of the geodesic convergence condition $R_{ab}u_a u_b \geq 0$ or to deduce differentiability conditions).

□ Show that observables plus general relativity do not allow our past null cone to be determined if selection effects and observational limitations are allowed for.

□ Reanalyse the problem of determining our past null cone assuming that there exist $N>3$ spatial dimensions. In particular, evaluate the effect of having N even, and/or N not equal to 3 on the propagation of weak gravitational waves and other wave fields. Determine whether weak gravity waves can propagate both *inside* and on the past null cone in any of these cases[44].

NEWTONIAN GRAVITATION

Before looking in more detail at relativistic cosmological models it is instructive to compare the content of general relativity with that of Newtonian graviation, to which it reduces in the limit of slow motion and weak gravitational fields. The principal contrasts are summarized below:

Relativistic Cosmology	Newtonian Cosmology
10 field equations	1 field equation
10 potentials	1 potential
Non-linear equations	Linear equation
Intrinsically geometrical	Absolute space and time
Can cope with ∞ space	Requires finite space
All energies gravitate	Mass-density gravitates
Hyperbolic propagation	Instantaneous propagation
Singularities of space-time	Singularities in space
Horizons and black holes	No horizons or black holes
Gravitational waves	No gravitational waves

This list indicates the greater content of general relativistic cosmology compared with the Newtonian theory: Not all Newtonian cosmological models will possess relativistic analogues and not all general relativistic cosmologies possess a Newtonian analogue. We stress also the sense in which "general" relativity generalises special relativity: the latter is a theory of a space-time manifold $\mathbf{R}^4$ endowed with the Minkowski metric η_{ab}, whereas the arena of general relativity is a general four-manifold $\mathbf{M}^4$ endowed with a pseudo-Riemannian metric g_{ab}.

The field equations of general relativity are usually derived from the demand that the second-rank divergenceless energy-momentum tensor T_{ab} be determined by the most general divergenceless combination of second-rank tensors linear in the curvature. This prescription allows the presence of a cosmological constant in the Einstein equations although it had no known counterpart in Newtonian gravity. However, it is instructive to see[6] how the presence of a cosmological constant is inevitable in Newtonian theory and could even have been found by Newton.

Suppose we ask, as Newton did[58], for the most general form of the gravitational potential, $\Phi(r)$, such that the external potential of a spherical shell is identical to that of a point of equal mass to the shell located at its centre O. If the shell has surface density σ and radius a then at some point P located at a distance $r > a$ from the centre of the shell the potential due to the shell and due to a point mass M(a) at O will be equal if

$$M(a)\Phi(r) + 2\pi\sigma a\lambda(a) = 2\pi\sigma a r^{-1} \int_{r-a}^{r+a} x\Phi(x)dx \qquad (3)$$

where $\lambda(a)$ is a constant and $M(a) = 4\pi a^2\sigma$ is the mass of the shell. If we differentiate this integral equation it can be solved to give $\lambda(a) = 2Ba^2$ and the general solution of Newton's problem is

$$\Phi(r) = Ar^{-1} + \Lambda r^2 \ ; \ A, \Lambda \text{ constants} \qquad (4)$$

where $A = -GM(a)$ and Λ is what Einstein termed the *cosmological constant*. The appropriate field equation for $\Phi(r)$ is the modified Poisson equation

$$\nabla^2\Phi + \Lambda = 4\pi G\rho \tag{5}$$

□ Find the most general gravitational lagrangians describing a metric theory of gravity (assume it to be a function of the independent curvature invariants as a first hypothesis) for which the external gravitational field of a sphere can be replaced *(i)* by that of a point mass and, *(ii)* by a point of the same mass as the sphere.
□ What is the most general gravitational lagrangian that reduces to the Poisson equation in the slow-motion and weak-field limit?
□ What is the relativistic theory of gravitation in 2+1 dimensional space-time which has 2+1 dimensional Newtonian gravity as a limit? (It is *not* 2+1 dimensional general relativity[37,45]. Note: the logarithmic potential of 2+1 Newtonian gravity diverges at spatial infinity).

NEWTONIAN COSMOLOGY

The kinematical content of Newtonian cosmology[27,46] is most simply seen by considering the generalization of Hubble's law for the relative recesion velocities v of objects possessing relative separations r. If this expansion proceeds isotropically and homogeneously then we write

$$v = Hr \tag{6}$$

where the scalar function H is the usual Hubble parameter. Let us now extend (6) to a completely general expansion flow which proceeds at different rates in different directions. If the three orthogonal velocity components are v_α, $\alpha = 1,2,3$, along orthogonal directions r_α then we have, in general,

$$v_\alpha = H_{\alpha\beta}\, r_\beta \tag{7}$$

where the 3x3 matrix $H_{\alpha\beta}$ is the Hubble tensor of velocity gradients. We split $H_{\alpha\beta}$ into its symmetric ($\theta_{\alpha\beta}$) and antisymmetric ($\omega_{\alpha\beta}$) parts

$$H_{\alpha\beta} = \theta_{\alpha\beta} + \omega_{\alpha\beta} \tag{8}$$

and then decompose $\theta_{\alpha\beta}$ into its trace (H) and tracefree ($\sigma_{\alpha\beta}$) parts,

$$H_{\alpha\beta} = \sigma_{\alpha\beta} + H\delta_{\alpha\beta}/3 + \omega_{\alpha\beta} \tag{9}$$

where $\delta_{\alpha\beta}$ is the Kronecker delta symbol. Physically, the symmetric matrix $\sigma_{\alpha\beta}$ describes the shear distortion of the flow at constant volume, $\omega_{\alpha\beta}$ the vorticity and H the pure volume dilation. Observation indicates that in the present-day Universe the isotropic volume dilation of the Hubble flow exceeds any rotation or shear by at least[100,107] a factor of order 10^5.

The dynamics of Newtonian cosmology are given by the field equation (5) together with the continuity and momentum conservation equations

$$\dot{\rho} + \rho\nabla\mathbf{v} = 0 \tag{10}$$

$$\dot{\mathbf{v}} + (\mathbf{v}\cdot\nabla)\mathbf{v} = -\nabla\Phi \tag{11}$$

If we define a vorticity vector ω_λ, vorticity scalar ω, and shear scalar σ by

$$\omega_\lambda = \tfrac{1}{2}\epsilon_{\lambda\mu\nu}\,\omega_{\mu\nu} \quad ; \quad \omega^2 = 2\omega_\mu\omega^\mu \quad ; \quad \sigma^2 = 2\sigma_{\alpha\beta}\sigma^{\alpha\beta} \tag{12}$$

then equations (7)-(9) and (12) reduce (5), (10) and (11) to

$$4\pi\rho a^3/3 = M \tag{13}$$

$$(a^2\omega_\lambda)^{\cdot} = a^2\omega_\mu\sigma_{\mu\lambda} \tag{14}$$

$$\ddot{a} - a(\Lambda - \tfrac{1}{2}\sigma^2 + \omega^2)/3 + GMa^{-2} = 0 \tag{15}$$

where the expansion scale factor $a(t)$ is defined by[50]

$$a^3(t) = \exp(\textstyle\int H_{\alpha\alpha}\,dt) \tag{16}$$

If $\sigma = \omega = 0$ we obtain the zero-pressure Friedman universes of general relativity. Curiously, neither the Newtonian cosmological models[51], nor their perturbations[41,52] were analysed until after their general relativistic counterparts.

In general relativity there exists an additional non-Newtonian effect which can influence the expansion of the universe. This is the possibility of a spatial curvature anisotropy. The Newtonian analogue of the spatial curvature term in cosmological models is necessarily isotropic.

□ Give a singularity theorem for Newtonian cosmology.
□ Classify the singularities arising in irrotational Newtonian cosmological models using catastrophe theory[48,49] and find the general behaviour in the neighbourhood of the singularity.
□ Adapt the classic work of Leray and others[40,33,47] on the global and local existence of solutions to the equations (10) and (11) of Newtonian hydrodynamics to prove results about the Newtonian and general relativistic cosmological problems.
□ Determine the asymptotic behaviour of ever-expanding Newtonian cosmological models. Relate the answer to the general behaviour of general relativistic cosmological models with isotropic spatial curvature and to the asymptotic solution of the Newtonian N-body problem with potential of the form (4).
□ Which Newtonian cosmological models have no relativistic analogue?
□ Show that Newtonian black holes cannot exist.

GENERAL RELATIVISTIC COSMOLOGY

In order to appreciate the degree of generality of the particular solutions of general relativity that are known, let us consider the specification of the general solution of

Einstein's equations for cosmological evolution from a spacelike hypersurface of constant time, S. If the metric is expressed in synchronous coordinates

$$ds^2 = dt^2 - g_{\alpha\beta}(x^\alpha,t)dx^\alpha dx^\beta \quad ; \quad 1 \le \alpha,\beta \le 3 \tag{17}$$

and the stress tensor, T_{ab}, is that of a perfect fluid with pressure p, density ρ, equation of state $p(\rho)$ and normalized 4-velocity u_i, $0 \le i,j \le 3$,

$$T_{ab} = (\rho + p)u_a u_b - pg_{ab} \; ; \quad u_a u^a = 1 \tag{18}$$

then on S we can, in general, specify 6 $g_{\alpha\beta}$, 6 $\dot{g}_{\alpha\beta}$, $3u_\alpha$ and 1 $p(\rho)$ = 16 independently arbitrary functions of the three spatial variables x^α. However, 8 of these functions are not independent of the others by virtue of the four $R_o{}^a$ constraint equations and the four general coordinate covariances of the general relativity. This leaves a general solution, in the presence of perfect fluid, specified by 8 independent arbitrary functions of three spatial variables. In the vacuum case only 12-8 = 4 arbitrary functions are necessary to specify the general solution[28,29,30]. Note that the most general transformations of the time and space coordinates which leave the metric in the synchronous form (10) contain four arbitrary functions of the space coordinates.

If we have a gravitation theory that is derived from a gravitational lagrangian that contains terms *quadratic* in the curvature invariants then the resulting field equations are (except for a special case) of 4th order in g_{ab}. In vacuum, the initial data on a {t=constant} surface must now also include 6 components of $\ddot{g}_{\alpha\beta}$ and 6 of $\dddot{g}_{\alpha\beta}$ and the general vacuum solution must be prescribed by a total of 24-4-4 = 16 independently arbitrary functions of 3 variables.

□ Determine what type of global information is required to patch together local approximations to the general solution of the Einstein equations and how it affects the assessment of generality[54-56].
□ What meaning can be associated with the function-counting assessment of generality when T_{ab} is not specified by an equation of state? How is the situation altered by requiring that the Second Law of thermodynamics hold?
□ Find a method of characterising the relative generality of cosmological solutions with timelike portions.
□ Determine the number of arbitrary functions necessary to characterise the general solution of a theory of gravity in N space-time dimensions whose gravitational lagrangian contains terms linear in the curvature and quadratic in the different possible independent quadratic curvature invariants.
□ Determine what under what conditions on T_{ab} and g_{ab} cosmological models exhibit chaotic behaviour[56] as $t \to 0$ and $t \to \infty$. Hence formulate a definition of generality based upon the number of arbitrary constants necessary to describe the *asymptotes* of a solution rather than the solution itself and compare this with the standard function-counting assessments of generality[57].
□ Determine the number of arbitrary functions necessary to characterize the most general relativistic cosmological models which possess a Newtonian analogue.

THE FRIEDMAN COSMOLOGICAL MODELS

The simplest cosmological solutions are the homogeneous and isotropic Friedman models. They possess the metric

$$ds^2 = dt^2 - a^2(t)[\, dr^2/(1-kr^2) + r^2(d\theta^2 + \sin^2\theta d\phi^2)] \quad (19)$$

The spatial geometry of each {t = constant} surface is that of a space of constant curvature. The sign of this curvature is determined by that of the constant k, which, by a coordinate transformation, ($r \to |k|^{1/2} r$ and $a \to |k|^{-1/2} a$), may without loss of generality be set to the values 0, +1 or -1. The salient features of the three cases are as follows:

k = 0:

The spacelike surfaces of constant time are Euclidean. The spatial sections an the spatial volume are infinite if their topology is R^3. An identification of Cartesian coordinates (x,y,z) with (x + A, y + B, z + C) where A,B, and C are constants produces the 3-torus topology and the spatial volume is finite.

k = -1:

The spacelike surfaces of constant time possess constant negative curvature as can be seen explicitly by the transformation $r = \mathrm{sh}\chi$ which brings (19) to the form

$$ds^2 = a^2(t)[\, d\chi^2 + \mathrm{sh}^2\chi\,(d\theta^2 + \sin^2\theta d\phi^2)] \quad (20)$$

The spatial sections and volume are infinite if the natural R^3 topology is adopted. (In fact all the possible topologies have yet to be classified[62]). If we introduce a fourth pseudo-coordinate, u, along with the Cartesians x,y,z via

$$x = a\mathrm{sh}\chi\cos\theta, \quad y = a\mathrm{sh}\chi\cos\phi, \quad z = a\mathrm{sh}\chi\sin\theta\sin\phi, \quad u = a\mathrm{ch}\chi$$

then we see that

$$dx^2 + dy^2 + dz^2 - du^2 = a^2(t)[\, d\chi^2 + \mathrm{sh}^2\chi\,(d\theta^2 + \sin^2\theta d\phi^2)] \quad (21)$$

and the 3-geometry can be viewed as embedded in a 4-dimensional pseudo-Euclidean space. Remarkably, Clarke[61] has shown that *any* non-compact four-dimensional space-time can be embedded in the 89-dimensional Euclidean space, R^{89}, with metric signature (-,-,+,....+).

k = +1:

The spatial sections and volume are finite although there is no boundary to the space. The space is one of constant positive curvature as can be seen by using the transformation $r = a\sin\chi$ in (20). This displays it to be a 3-sphere of radius a(t) which may be embedded in a 4-dimensional Euclidean space by introducing an analogous pseudo-coordinate to that used to obtain (21).

In each case we see that the 3-spaces of the Friedman space-times can be considered to be curved relative to an artificial flat 4-dimensional space whose curvature radius in its fourth dimension is just a(t). The speed of expansion of the universe into this fourth dimension is described by da/dt. It is *not* an expansion into an external three-dimensional space.

The scale factor a(t) is determined by solving the two independent Einstein equations. In the presence of a *comoving* (that is $u_a = \delta_a{}^0$ so geodesic observers expanding with the universe retain constant values of their spatial coordinates (r,θ,ϕ)) perfect fluid stress (18), these equations are

$$\frac{\dot{a}^2}{a^2} = \frac{8\pi G\rho}{3} - \frac{k}{a^2} \tag{22}$$

$$\dot{\rho} + 3\dot{a}(\rho + p)/a = 0 \tag{23}$$

These equations have no content unless some restriction is placed on the behaviour of p and ρ, otherwise *any* a(t) solves (22) and (23) for some p and ρ. If we pick an equation of state

$$p = (\gamma - 1)\rho \quad ; \quad \gamma \text{ constant}, -1 < \gamma \leq 2, \tag{24}$$

then the solution of (22)-(23) will be characterised by just *two* independently arbitrary *constants* rather than the eight functions of the general solution described above.

There are two simple classes of solution to (22)-(24) which do not involve elliptic functions: when k = 0 we have the Einstein-de Sitter universes

$$a(t) \propto t^{2/3\gamma} \; ; \; \rho = 1/6\pi G\gamma^2 t^2 \tag{25}$$

In the case of blackbody radiation ($\gamma = 4/3$), the general solution has the simple form

$$a(t) \propto (At - kt^2)^{1/2} \quad ; \quad \rho = 3A^2/32\pi G(At-kt^2)^2 \tag{26}$$

where A is an arbitrary constant. When A = 0 and k = -1 we have the vacuum solution of Milne with $a(t) \propto t$.

The schematic evolution of the Friedman universes for k = 0, +1 and -1 when $2/3 < \gamma \leq 2$ is shown in Figure 2. When $\gamma = 2/3$ there are no recollapsing closed universes and $a(t) \propto t$ for all k.

By differentiating (22) and using (23) one sees that

$$\ddot{a} = -4\pi G(\rho + 3p)a \tag{27}$$

Hence, if the Friedman universe is expanding now ($\dot{a}/a > 0$) it must have had a = 0 and $\rho = \infty$ at a finite time in the past if $\rho + 3p \geq 0$.

It is instructive to consider the Friedman models in

conformal time τ defined in terms of the comoving proper time t by

$$dt = a(t)d\tau \; ; \; ' = d/d\tau \tag{28}$$

If we now make the change of variable

$$y = a^{(3\gamma-2)/2} \tag{29}$$

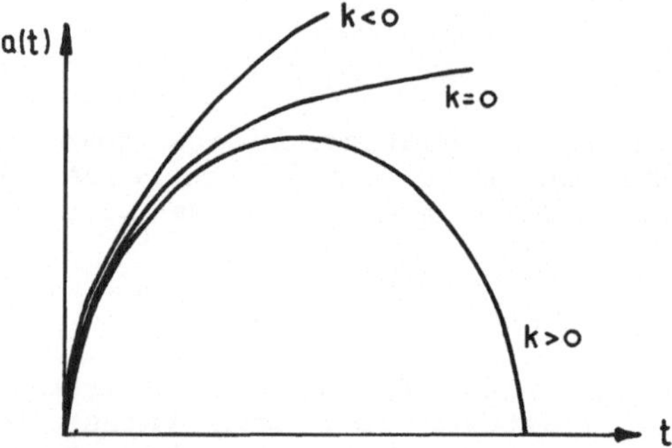

Fig. 2. The evolution of the scale-factor a(t) with respect to the comoving proper time t. The qualitative behaviour is determined by the sign of the curvature parameter k. In this time coordinate each member of the family of different closed universes with k > 0 has a different total lifetime.

then the Friedman universes are *all* described by the simple harmonic oscillator equation

$$y'' = -\tfrac{1}{4}k(3\gamma-2)^2 y \tag{30}$$

This neat result appears previously to have gone unnoticed but has recently been exploited to simplify the treatment of some quantum cosmological models[7,59,60].

Another interesting feature of the description in conformal time is that for fixed γ all closed universe have the *same* conformal lifetime. If we choose $y(\tau=0) = 0 = a(t=0)$ then this lifetime is $\Delta\tau = 2\pi/(3\gamma-2)$; see Figure 3. In t-time closed universes do not all begin and end at the same moment.

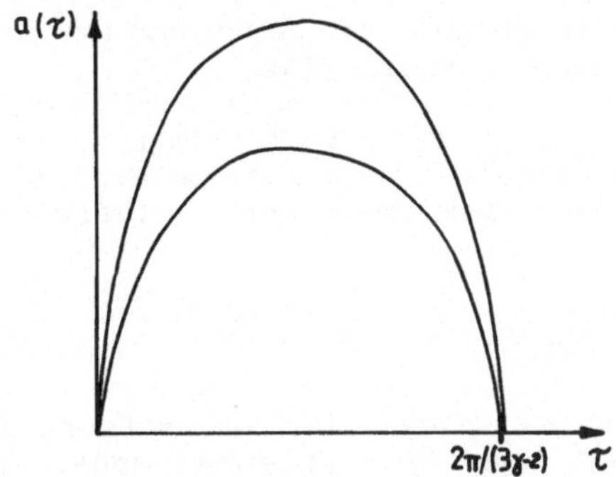

Fig. 3. The evolution of closed Friedman models in conformal time. For fixed $\gamma \neq 2/3$ they all have the same conformal lifetime and attain their expansion maxima at the same moment of conformal time.

In general the closed models are described by the simple formulae

$$a(\tau) = a_{max} \{\sin[(3\gamma-2)\tau/2]\}^{2/(3\gamma-2)} \quad (31)$$

$$t(\tau) = a_{max} \int \{\sin[(3\gamma-2)x/2]\}^{2/(3\gamma-2)} dx \quad (32)$$

where a_{max} is the value of $a(\tau)$ at the moment of maximum expansion, τ_{max} defined by $a'(\tau_{max}) = 0$, $a''(\tau_{max}) < 0$.

The conformal time coordinate also exhibits the causal structure of the Friedman models. For example, using (28) and the radial coordinate $\chi = \sin^{-1}(r/a)$ in the $k = +1$ model, the metric is

$$ds^2 = a^2(\tau)[\, d\tau^2 - d\chi^2 - \sin^2\chi(\, d\theta^2 + \sin^2\theta\, d\phi^2)] \quad (33)$$

The equations of radial ($d\theta = d\phi = 0$) light-rays ($ds^2 = 0$) are therefore $d\chi = \pm\, d\tau$. So, after a conformal time τ an observer located at $\tau = 0$ can only see everything inside the radius $\chi = \tau$. This radius is called the *particle horizon at time* τ.

□ Find a method of characterising the relative generality of cosmological solutions with timelike portions.
□ Investigate the effect of the 'wedge' structure of space-time outside a vacuum string in a Friedman space-time with 3-torus topology.
□ Examine the effects of inflation in a finite open universe with 3-torus topology.
□ Find a technique for solving equations like (22) and (23) subject to inequalities on p and ρ rather than initial conditions.

□ Investigate the behaviour of perturbations to the Friedman models with non-standard topologies.
□ Examine the evolution of a spherical overdensity to the black hole state in a closed Friedman Universe using the conformal time coordinate. Investigate ways of discriminating between the black hole and the closed universe.

OBSERVABLE PARAMETERS

It is useful to pick the two mathematical quantities needed to completely specify a Friedman model so that they are directly observable (modulo selection effects). We define the Hubble parameter

$$H = \dot{a}/a \tag{34}$$

and the deceleration parameter

$$q = -\ddot{a}\, a/\, \dot{a}^2 \tag{35}$$

Their present values (subscript '0') are found to be

$$H_0 = 75 \pm 25\ \mathrm{Kms^{-1}Mpc^{-1}}\ ;\quad q_0 < 2, \tag{36}$$

If $\Lambda = 0$ then differentiating (22) and evaluating today we have

$$H_0{}^2(2q_0 - 1) = ka_0{}^2 \tag{37}$$

Therefore $k \geq 0 \leftrightarrow q_0 \geq ½ \leftrightarrow \Omega_0 \geq 1$ where the density parameter Ω_0 is defined by

$$\Omega_0 = \rho_0/\rho_{cr}\ :\ \rho_{cr} = 3H_0{}^2/8\pi G \tag{38}$$

We note that the expansion scale factor a(t) is not observable but it is convenient to measure times and distance measures in units of the present value of a. Accordingly we define the *redshift* $z > 0$ by

$$1 + z = a_0/a(t) \tag{39}$$

For example, the angular scale, θ_h, subtended by the particle horizon, defined above, at redshift z in a $k = 0$ Friedman universe is

$$2\sin(\theta_h/2) = [(1 + z)^{½} - 1]^{-1} \tag{40}$$

So, quasars at $z = 2$ separated by more than about 84° on the sky are causally disjoint, as are regions of the microwave background radiation separated by more than about 2° on its last scattering surface at $z = 1000$.

WHEN DO CLOSED UNIVERSES RECOLLAPSE?

One of the most interesting features of the inflationary universe picture is that it predicts[18] that the present density of the Universe should be within one part in a million of the critical density of the $k = 0$ Friedman model. If there is some quantum or superstring(?) constraint upon Nature which ensures that k must be zero ($\Omega = 1$ if $\Lambda = 0$) then this prediction has no content. However, if our Universe has $k \neq 0$ ($\Omega \neq 1$) then it is very difficult to understand why the two terms on the right-hand-side of equation (22) are still of the same order of magnitude after $a(t)$ has changed by thirty-two orders of magnitude since the Planck time $t_p = G^{-1/2} \sim 10^{-43}$s. The observed state of affairs requires k/a^2 to have been initially smaller than ρ by a huge factor $> 10^{60}$. Providing an explanation for this state of affairs is equivalent to explaining why the Universe has expanded for $10^{60}t_p \sim 10^{17}$ secs.

It was first realized by Guth[18] (see also Sato[63]) that a sufficiently long period of expansion with the equation of state of the matter obeying $p = -\rho$ so that $a(t) \propto \exp(H_o t)$ with H_o constant (or even $p = (\gamma-1)\rho$ with $0 \leq \gamma < 2/3$ and $a(t) \propto t^{2/3\gamma}$ in fact) can explain why the k/a^2 term in the Friedman equation is not enormously larger than the $8\pi G\rho/3 \sim a^{-3\gamma}$ term for large a.

The inflationary picture also predicts that there should exist a spectrum of constant curvature density fluctuations out to the maximum scale of the inflation[1]. As yet there is no natural way to obtain the observed amplitude of these fluctuations ($\delta\rho/\rho \sim 10^{-4}$). However, if the theory did generate fluctuations of this magnitude with constant amplitude on every scale as it enters the horizon then the level of density fluctuations is larger than the distance of the predicted mean density from the critical density. The possibility of such a situation arising in the post-inflationary phase of the Universe led Zeldovich and Grishchuk[64] to investigate whether it is possible for local unbound sub-regions of a closed universe to avoid hitting the final singularity. They, and subsequently Bonnor[65] and Hellaby and Lake[66], examined this problem in the context of spherically symmetric space-times containing zero pressure matter. They found that under these circumstances it is not possible for negatively-curved shells of material to avoid the final singularity. Barrow and Tipler[67] examined this question in greater generality to determine the conditions under which closed universes (defined as those possessing compact Cauchy surfaces) collapse to an all-encompassing future singularity. They showed that the existence of a maximal hypersurface is a necessary and sufficient condition for the existence of both an initial and a final all-encompassing "strong" singularity in a universe with a compact Cauchy surface satisfying the strong energy condition (2) and a generic condition to the effect that all geodesics feel some tidal gravitational field.

A spacelike hypersurface, S, is said to be a *maximal hypersurface* if $z^a_{;a} = 0$ everywhere on S, where z^a is the unit normal vector to S (this corresponds to the physical notion of the moment of maximum expansion). A "strong" singularity is one which, roughly speaking, crushes out of existence objects which hit it. Examples of "strong" singularities are strong curvature singularities and crushing singularities[68]. (These restrictions on the singularities are required only for the 'necessary' part of the singularity theorem: that is, a maximal hypersurface will give all-encompassing initial and final singularities, but only "strong" initial and final singularities will have a maximal hypersurface between them).

It has been shown[6,67,71] that only closed universes with spatial topology either S^3 or $S^2 \times S^1$ or more complicated hybrids formed by connected summations and special identifications of these two basic topologies actually admit maximal hypersurfaces so long as the differentiable structure of 4-dimensional space-time does not possess one of the infinite number of "exotic" differentiable structures of Freedman[69] and Donaldson[70] that pure mathematicians are currently so excited about (that is, we want the coordinate systems covering the 4-dimensional space-time to be only those generated by pulling-up the 3-dimensional systems covering the 3-spaces). Thus closed universes with other topologies (for example the 3-torus, T^3) cannot recollapse to an all-encompassing final singularity. However, it is still an open question whether or not all closed universes with S^3 and $S^2 \times S^1$ topologies do in fact recollapse to a singularity.

By "recollapsing universe", here, we mean a non-static space-time which has a maximal compact Cauchy hypersurface, and by "all-encompassing initial singularity" we mean that all inextendible timelike curves have a length less than a constant L to the past of any Cauchy surface, where L may depend on the Cauchy surface, but not on the particular curve. "All-encompassing final singularity" is defined analogously. As stated above, a closed recollapsing universe begins and ends in an all-encompassing singularity if the strong energy and a generic condition hold. However, a recollapsing universe is distinct from the Wheeler universe defined in Marsden and Tipler[68] and in Barrow and Tipler[67]. A Wheeler universe begins and ends in all-encompassing singularities, but it could be that the singularity might not be strong, and hence a Wheeler universe might not have a maximal hypersurface. But a Wheeler universe with a maximal hypersurface is a recollapsing universe, if the strong energy and a generic condition hold.

If an equation of state is not defined then it turns out that the conditions under which even closed Friedman universes recollapse are rather subtle. a recent investigation[72] shows that old theorems of Tolman and Ward[73] concerning this question are actually incorrect. Neither the conditions[27] $\{\rho > 0$ and $p \geq 0\}$ nor $\{\rho + p > 0$ and $\rho + 3p > 0\}$ are strong enough to guarentee that a closed Friedman universe recollapses. The problem is that the pressure may diverge *before* a maximal hypersurface is reached. For example[72], if we take

$$a(t) = 1 + t - 2(1 - t)^{3/2}/3 \quad ; \; 0 < t < 1 \qquad (41)$$

then we can calculate ρ and p from equations (22) and (23) for $k = +1$ and we find $p \to \infty$ as $t \to 1$ but ρ, a and $\dot{a}$ remain finite. Sufficient conditions to obtain recollapse are found to be $\{\rho + 3p > 0$ and $|p| \leq C\rho$, C constant$\}$ together with a *matter regularity condition*: T_{ab} is regular except at space-time singularities. The conditions on T_{ab} for the recollapse of anisotropic closed universes are as yet unknown[72].

□ Prove or disprove the conjecture[72] that all globally hyperbolic closed universes with S^3 or $S^2 \times S^1$ spatial topology and with stress-energy tensors which obey (i) the strong energy condition, and (ii) the positive pressure criterion, begin in an all-encompassing initial singularity and end in an all-encompassing final singularity.
□ Find conditions on T_{ab} for the closed Kantowski-Sachs[74,76] and Taub [86] axisymmetric Bianchi type IX universes to recollapse to a final singularity.
□ Examine the consequences of the discovery[69,70] that 4-dimensional manifolds admit an infinite number of distinct differentiable structures (for all other dimensions there is one unique differentiable structure only) for cosmological models. Does the special character of 4-dimensional manifolds provide any clue as to why the observed dimension of space-time is 4 in the context of Kaluza-Klein or superstring theories exhibiting dimensional compactification[116]?

SPATIALLY HOMOGENEOUS UNIVERSES

Physically speaking, we would regard as spatially homogeneous a universe in which all comoving observers record the same picture of cosmic history. Mathematically[79], this requires that for any pair of observers A and B there is a symmetry which allows the universe model considered as centred on A to be transformed into an identical model centred on B. Spaces of this type either fall into the Bianchi classification and possess a three-dimensional group of translation symmetries or into the exceptional case of Kantowski-Sachs[74],(found also by Kompanyeets and Chernov[75]), which possesses a four-dimensional group of motions with no simply transitive three-dimensional subgroup. The Kantowski-Sachs models are not considered here because their exact solutions have been found and fully studied elsewhere[76]. Henceforth any reference to spatially homogeneous cosmologies will refer to the Bianchi[77,139] types.

The Bianchi spaces allow us to identify spacelike hypersurfaces S(t) on which we may define at least three independent Killing vector fields ε_A (where upper case Latin indices run from 1 to 3) which satisfy Killing's equations,

$$\varepsilon_{m;n} + \varepsilon_{n;m} = 0 \tag{42}$$

The commutators of the ε_A are determined by a set of structure constants, $C_{AB}{}^D$,

$$[\varepsilon_A, \varepsilon_B] \equiv \varepsilon_A \varepsilon_B - \varepsilon_B \varepsilon_A = C_{AB}{}^D \, \varepsilon_D \tag{43}$$

The components of the metric, g_{ab}, are invariant under the isometry generated by infinitesimal translations along these Killing vector fields - the functional time-dependence of the metric is the same at all points. The Einstein equations relate the energy-momentum tensor T_{mn} to the first and second derivatives of g_{mn} and so, if g_{mn} is invariant under an isometry, all physical properties of T_{mn} also remain invariant. Since the Killing vector fields span each S(t) we have a spatially homogeneous space-time.

The set of n Killing vectors possesses an n-dimensional group structure, G_n, characterised by the equivalence classes of structure constants, $C_{BC}{}^{A}$, and these can be used to classify all spatially homogeneous cosmologies. The possible group types possess a G_3, G_4 or a G_6, and in all but one exceptional (Kantowski-Sachs) case there is at least one G_3 subgroup. The classification of this subgroup leads to Bianchi's famous decomposition into types but we shall give the modified scheme developed by Estabrook, Wahlquist and Behr[78], and Ellis and MacCallum[79,80,81].

Table 1. The number of independent, arbitrary constants necessary to specify a Bianchi type universe on a spacelike hypersurface of constant time in vacuum (r) and with perfect fluid source (s). Typically, s = r + 4, except for some particular cases where geometrical constraints exclude particular degrees of freedom for the fluid motion and then s < r + 4. The Bianchi group dimension is p.

BIANCHI TYPE		p	ARBITRARY FUNCTIONS IN VACUUM, r	ARBITRARY FUNCTIONS WITH PERFECT FLUID, s
I		0	1	2
II		3	2	5
VI_0		5	3	7
VII_0	*Class A*	5	3	7
VIII		6	4*	8*
IX		6	4*	8*
IV		5	3	7
V		3	1	5
VI_h	*Class B*	6	4*	8*
VII_h		6	4*	8*
$VI_{-1/9}$		6	4*	7

On any particular spacelike hypersurface, the Killing vector basis can be chosen so that the structure constants can be decomposed as

$$C_{BC}^{A} = \epsilon_{BCE}n^{EA} + \delta^{A}_{C}a_{B} - \delta^{A}_{B}a_{C} \qquad (44)$$

where ϵ_{BCE} is the completely antisymmetric symbol ($\epsilon_{123} = 1$) and

$$n^{AB} = \mathrm{diag}(n_1, n_2, n_3) \qquad (45)$$

$$a^{B} = (a, 0, 0) \qquad (46)$$

and all the $\{n_1, n_2, n_3, a\}$ can be normalized to $\pm$ 1 or 0. If $an_2n_3 = 0$, then n_2 and n_3 can be set equal to $\pm$ 1 and a to $\sqrt{|h|}$, where h is a parameter used in the group classification.

The possible combinations of n_1 and a fix the Bianchi-Behr group types. They are divided into two classes, A and B, according to whether a is zero or non-zero, as shown in Table 1.

The isotropic Friedman models have G_6 symmetry groups with G_3 subgroups such that the zero curvature model can be thought of as a special case of Bianchi types I or VII_0, the open Friedman model as a special case of types V or VII_h and the closed model as a special case of type IX.

In the Table we have indicated the relative degree[82] of generality of each type in two ways. First, the group dimension p gives the dimension of the orbit of $C_{BC}{}^A$ as a subset of all the 9 distinct $C_{BC}{}^A$ components. Because the Jacobi identities must be satisfied by the Killing vectors, that is

$$\varepsilon^{ABC}[[\xi_A,\xi_B],\xi_C] = 0 \quad (47)$$

these imply

$$C_{[AB}{}^D C_{C]D}{}^E = 0 \quad (48)$$

and so,

$$n^{AB}a_B = 0, \quad (49)$$

and hence the orbits of any particular group type are at most *six-dimensional*.

We are interested in the relative generality of particular Bianchi universe *solutions* of the field equations rather than their intrinsic group structure. The best way[83] of chacterising the generality of particular metrics is to count the independent parameters in the initial data set giving rise to the general solution of each group type for a specified T_{ab}. The quantity r in Table 1 is the number of arbitrary constants which appear in the *general vacuum solution* of each group type (in types VI_h and VII_h this count includes the parameter h).

The most general vacuum solutions are therefore those of types VII_h, VI_h, VIII, $VI_{-1/9}$ and IX, and are parametrised by four independent arbitrary constants. (The reason for the appearance of type $VI_{-1/9}$ which appears to be more special than VI_h is[82-84] that two of the Einstein constraint equations become null identities for the choice $h = -1/9$). We recall our earlier discussion following equations (17) and (18) in which the general inhomogeneous vacuum solution was characterised locally by four free *functions* of three spatial variables rather than four *constants*. The less general Bianchi types admit fewer independent shear modes ($\dot{g}_{ij}$ components); for example, the Bianchi type I vacuum solution is the famous Kasner metric[29,85] and is determined by only one free constant.

The column labelled s in Table 1 gives the number of independent arbitrary constants necessary to specify the general solution of each Bianchi type with a *perfect fluid* energy-momentum tensor. Recall that this required the independent specification of eight functions in the general inhomogeneous case. We expect 4 additional parameters to specify the general perfect fluid solution compared with the vacuum specification. In fact, we do not have $s = r + 4$ in all cases because the constraint equations place additional restrictions upon the generality of the T_{ab} allowed in types I and II. For example, if a perfect fluid is added to the vacuum type I (Kasner) solution the Bianchi I geometry requires the time-space Ricci components to be identically zero, hence T_{mn} is constrained to have $T_{o\alpha} = 0$, $\alpha = 1, 2, 3$ and so we must have comoving velocities with $u_a = \delta_a{}^o$. Therefore only one additional piece of data is required beyond the one-parameter vacuum datum: the density ρ. The perfect fluid $VI_{-1/9}$ is also not as general as perfect fluid VI_h, VII_h, VIII and IX solutions (unlike the situation in vacuum) because the double degeneracy of the R_{oi} constraint equations only occurs in vacuum or when the velocity field is comoving, ($u_a = \delta_a{}^o$); in all other cases $s = r + 4$.

The *general* vacuum solutions[86] are known only for Bianchi types I (1 parameter), II (2 parameters) and V (1 parameter) in vacuum and only for type I (2 parameters) in the presence of perfect fluid obeying (18) and (24). Special vacuum solutions are known for type VI_0 (1 parameter), VI_h (2 parameter), $VI_{-1/9}$ (zero parameter), VII_h (2 parameter) along with various special perfect fluid solutions. The stability of these special solutions has been studied in detail recently[34,82,88].

□ Classify the homogeneous spaces and thereby the possible spatially homogeneous cosmological models in N dimensions.
□ Which Bianchi types admit the de Sitter universe?
□ Use the approximate symmetry group construction of Spero[42] to classify the known inhomogeneous cosmological solutions.
□ Produce an analogous classification of relative generality to that in Table 1 for the self-similar generalization of the Bianchi classification developed by Eardley[87].
□ Study inhomogeneous perturbations of the known exact spatially homogeneous universe solutions.
□ Extend the evaluation of the relative generality of the Bianchi universes given in Table 1 to the case where non-perfect fluid stress tensors are admitted. Does the hierarchy of generality remain the same?

THE MICROWAVE BACKGROUND AND THE DENSITY OF THE UNIVERSE

In this section we shall describe the observational consequences of small anisotropies in the Hubble expansion and curvature of the Universe. Such anisotropies *must* exist at some level because the Universe is not precisely homogeneous in space. We shall also see that this question turns out to have an unexpected connection with the observational problem of determining Ω_o, the present density of the Universe, (38).

There exist[19] a variety of theoretical prejudices regarding the likely value of Ω_o. They are worth commenting on since they all support a situation in which any observational test of whether the Universe is open or closed appears to be impossibly difficult:

Simplicity: Ω_0 *= 1 exactly*

This minimizes the number of free parameters in the isotropic Friedman model if we assume the spatial sections possess natural topology. It avoids awkward questions regarding the fine-tuning of the initial conditions. However in light of the fact that the Universe is not exactly a Friedman model and quantum gravitational fluctuations inevitably exist in the spatial curvature when the expansion emerges from the Planck era, we might want to refine this form of our belief in simplicity to something like...

Quantum Simplicity: $\Omega_0 = 1 \pm$ (quantum correction)

In some theories the sign of this quantum correction can be specified. For example, if the Universe was created spontaneously out of a quantum vacuum[64,89], then we require $\Omega_0 > 1$ if the topology is natural. However, the existence of so many more nonstandard topologies, (*e.g.*, the 3-torus), may make the quantum creation of universes with nonstandard topology out of "nothing" a more probable occurrence; however, such topologies require additional parameters to specify them. It is interesting to note that the Friedman model allows the largest number of possible topologies. One might even interpret the fact that our Universe possesses a density so close to the critical one as evidence for its original creation out of a quantum vacuum state.

Inflation

Models for the evolution of the early universe which admit either a phase transition with very special properties[63], or the existence of hypothetical scalar fields with appropriately weak nonlinear self-interactions[1,90] or quadratic lagrangians[39] will undergo a short period of de Sitter expansion. This inflationary interlude in the expansion can explain the proximity of the observed Universe to the critical density and leads us to predict that the present value of Ω_0 will satisfy[18]

$$|\Omega_0 - 1| \leqslant 10^{-6} \qquad (50)$$

This is just the condition that would have permitted sufficient inflation to resolve the horizon and monopole problems[63] (assuming, of course, that such problems exist following a quantum gravitational or superstring era). Inflation does not predict the sign of $|\Omega_0 - 1|$. In a universe obeying (50) with $\delta\rho/\rho \sim 10^{-4}$ on all scales local observations may never be a reliable guide to the global density[64].

Quantum Gravity

To deal with boundary terms in the gravitational action and associated path integrals, these models are formulated in closed universes[91]. Whether or not this restriction is fundamental is still unclear. It also appears possible[92,93] that quantum restrictions will lead to universes in which $|\Omega_0 - 1|$ is very small with high probability. One should bear in mind that mathematical cosmologists appear to have a natural prejudice for closed universes because it is that much easier to prove theorems concerning the properties of compact spaces.

Many of these theoretical prejudices lean towards universe models that will result in Ω_0 lying tantalizingly close to unity. This appears to be a depressing state of affairs for observational cosmologists aiming to determine whether or not the Universe is open, since they are already faced with a complicated and ambiguous collection of astronomical data. Direct observations[94] of luminous galaxies lead to an estimate of $\Omega_0 \sim 0.007$, while[95,96] the dynamics of individual galaxies and those in groups and clusters limit the density clustered within them to $\Omega_0 \sim 0.2$. The totality of observations of helium -4, helium -3, and deuterium place a limit on the total density of material in baryonic form of $\Omega_0 \sim 0.15$, assuming these light elements were manufactured primarily by primordial nucleosynthesis[97]. The possibility that large quantities of dark material could exist, either in the form of black holes of a size that avoids the nucleosynthesis constraints or in an unclustered sea of nonbaryonic particles, makes it impossible to draw any definite conclusions regarding the total density of the Universe that are robust enough to exclude the theoretical prejudices outlined above.

If we forget the theoretical prejudices for a very small value of $|\Omega_0 - 1|$, then the observational data still allows the dull solution that all dark matter is baryonic and resides in planetary-sized objects or faint stars. Attempts to resolve this "missing mass problem" by measuring the deceleration of the Hubble expansion are bedevilled by a number of notorious selection and evolutionary effects[31], and the lingering possibility of a nonzero cosmological constant[98] complicates the issue even further.

Despite the difficulties outlined above, we shall see there may exist[99,100] ways of using the structure of the microwave background radiation anisotropy over intermediate angular scales to determine the density of the Universe and decide whether or not a cosmological model is open or closed no matter how small the value of $|\Omega_0 - 1|$.

MICROWAVE BACKGROUND OBSERVATIONS

Temperature fluctuations of dipole signature with an amplitude $\sim 10^{-3}$ are the only ones to have been positively detected[101] in the microwave background radiation. Only upper limits exist on any possible quadrupole moment and these are summarized in Table 2

We shall be concerned with the structure of the background radiation over "intermediate" angular scales exceeding about 10° and its associated temperature profile $T(\theta,\phi)$. These scales exceed those of fine scale associated with fluctuations from embryonic protoclusters, the effects of reheating and the thickness of the last-scattering surface.

Table 2. Microwave background observations: limits on the amplitude of the quadrupole anisotropy in the microwave background at various wavelengths.

Upper Limit on $\Delta T/T_0$	Angular scale θ(deg)	Wavelength (cm)	Ref.
3.0×10^{-5}	10-90	1.2	102
4.4×10^{-5}	6	0.07	103
6.2×10^{-5}	90	0.03	104
7.0×10^{-5}	90	0.8	105
4.0×10^{-4}	6	0.8	105

CHARACTERISTIC MICROWAVE BACKGROUND PATTERNS

There have been many investigations into the form of the microwave background expected in anisotropic, homogeneous cosmological models[99,100,106-111]. We shall describe the simplest examples of each possibility for illustrative purposes.

First, we note that in general, the temperature anisotropy can contain three distinct pieces: a dipole term due to our motion relative to the universal frame in which the radiation is isotropic; a Doppler term, which need not be purely dipolar, due to any similar relative motion of the last scattering surface of the radiation; and, finally, an anisotropic distortion term due to the intrinsic expansion anisotropy of the Universe. This last distortion term can have three important forms in realistic cosmological models, depending on the density of the Universe and the form of the anisotropy:

Quadrupole

The simplest example is the axisymmetric Bianchi type I universe[109,99]. The metric is

$$ds^2 = dt^2 - X^2(t)[dx^2 + dy^2] - Y^2(t)dz^2 \tag{51}$$

The geodesic equations can be integrated exactly and when the anisotropy level is low, the temperature anisotropy has the simple quadrupolar θ-variation,

$$\frac{\Delta T(\theta,\Phi)}{T_0} \equiv \frac{T(\theta,\Phi) - T_0}{T_0} \quad \alpha\ P_2(\cos\theta) \tag{52}$$

where $P_2(\cos\theta)$ is a Legendre polynomial and T_0 is the mean radiation temperature. When the model is not axisymmetric, then the dependence with angle is given by the spherical harmonic $Y_{2m}(\theta,\Phi)$. The solutions of the geodesic equations allow no convergence or divergence of geodesic separations between last scattering (L) and observation (O) at the present epoch. This behaviour is essentially Newtonian and also occurs in some very special anisotropic models with $\Omega_0 < 1$ and also in general closed models like Bianchi IX. (In these more complicated cases it may be possible for a dipole to exist in addition to a quadrupole whereas in the simple metric (51) the Ricci tensor satisfies $R_{0\alpha} = 0$, $\alpha = 1, 2$, and 3, and hence any perfect fluid must be comoving and there can be no dipole.

Hotspot

If an anisotropic universe is open ($\Omega_0 < 1$), then there can occur a new effect that was first pointed out by Novikov[112]. Various other authors[106,108,111] have studied theoretical aspects of this problem, and Barrow *et al*[99] and Bajtlik *et al*[113] have computed observable features in detail. The simplest example is an axisymmetric Bianchi-type-V model containing pressureless matter,

$$ds^2 = dt^2 - X^2(t)e^{2z\sqrt{1 - \Omega_0}}(dx^2 + dy^2) - Y^2(t)dz^2 \qquad (53)$$

This reduces to (51) when $\Omega_0 = 1$ and is the open Friedman universe in unusual coordinates when $X(t) = Y(t)$. This metric only solves the Einstein equations if the 3-velocity possesses a nonzero component:

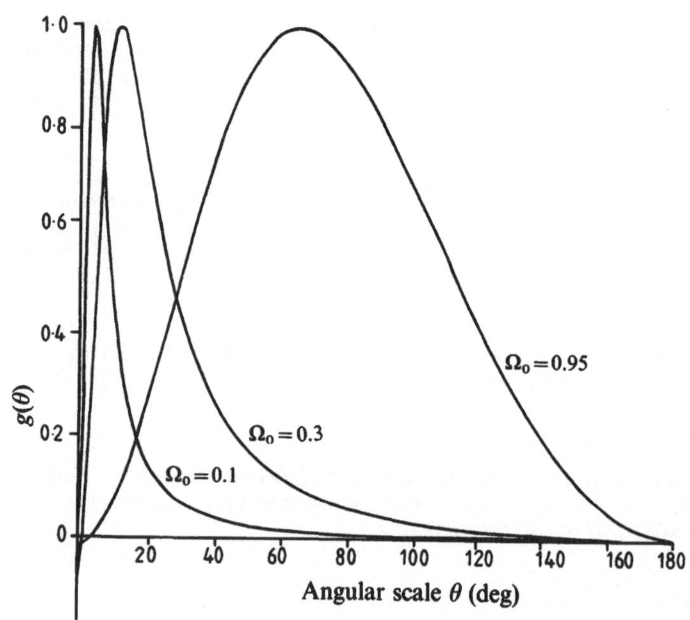

Fig.4. Quadrupole focussing. The variation of $\Delta T(\theta_0, \phi_0)/T_0 \propto g(\theta_0)$ with angular scale in type V universes for various values of Ω_0. The focussing effect increases as Ω_0 decreases.

$$u_\alpha = (0,\ 0,\ \pm\ 2\sigma_0(1 - \Omega_0)^{1/2}(1 + z)/\sqrt{3}\ H_0\Omega_0) \tag{54}$$

where the present ratio of shear to Hubble rate, σ_0/H_0, is a constant parametrizing the amplitude of the anisotropy; and z is the red shift. In both (51) and (53) the scale factors can be normalized so that X = Y at the red shift z_L = 1000 when the radiation first becomes collisionless. The resulting temperature profile observed on the sky today can be computed and has two components: a dipole, because of the velocity term (but this need not be present in a nonaxisymmetric model of this type), and a distorted quadrupole or "hotspot". This is illustrated in Figure 4 for various values of Ω_0.

As Ω_0 takes lower values, the quadrupole pattern is focussed into a region of diminishing angular scale. The diameter of the focussed quadrupole is roughly $2\Omega_0$ radians. If one follows the evolution of $T(\theta,\Phi)$ with falling red shift, from z_L to the present, one finds that the focussing starts to become particularly effective at red shifts less than $\sim \Omega_0^{-1}$ when the negative curvature dominates the effect of the matter on the dynamics. The focussing effect can be seen explicitly in the geodesic behaviour. The relation between angles θ_0 and θ_L is

$$\tan\left(\frac{\theta_0}{2}\right) = \tan\left(\frac{\theta_0}{2}\right)\ \exp\ (\tau_L - \tau_0) \tag{55}$$

where we have used the τ time coordinates defined by

$$\tau_0 = ch^{-1}(2\Omega_0^{-1} - 1),\ \tau_L = ch^{-1}\left[\frac{2(\Omega_0^{-1} - 1)}{1 + z_L} + 1\right] \tag{56}$$

Hence, if $\Omega_0 < 1$, we have $\theta_0 < \theta_L$ and there is *focussing*. This effect disappears as the spatial curvature becomes flat, $\Omega_0 \rightarrow 1$. When $\Omega_0 < 1$, then instead of (52) we have, approximately,

$$\frac{\Delta T(\theta,\Phi)}{T_0} \alpha\ P_2\ \left[\cos\left\{2\ \tan^{-1}(2\ \tan(\frac{\theta}{2})\ \Omega_0^{-1}\ )\right\}\right] \tag{57}$$

Elsewhere, we have produced detailed temperature maps of microwave skies possessing the hotspot feature[99,100] and Lukash and Novikov[114] have also discussed the structure of inhomogeneous hotspots[19]. The angular scale of the hotspot feature emphasizes the need for complete sky coverage in observational searches for anisotropies. It is also important to note that the maximum amplitude of the temperature anisotropy in the hotspot will considerably exceed that outside the hotspot region; Figure 5. We note that the hotspot effect occurs only in open (Ω_0 < 1) universes that contain anisotropies, although not in all open universes.

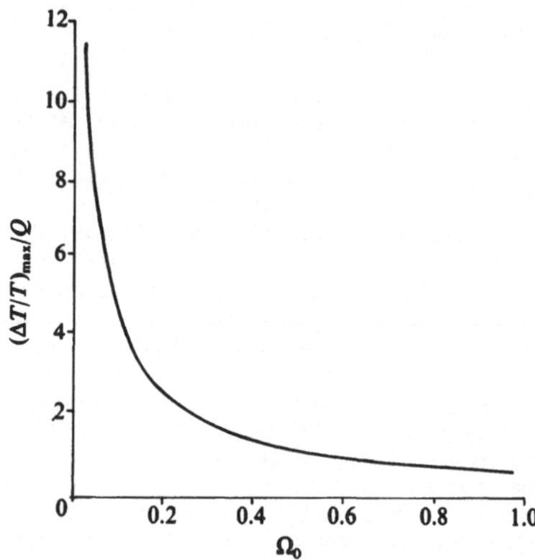

Fig.5. Hotspot contrast. The enhancement predicted at the hotspot's temperature maximum compared with the quadrupole over the rest of the sky. The enhancement disappears, along with the hotspot, as $\Omega_0 \to 1$.

The type V model we have used as a hot spot paradigm has isotropic spatial curvature and possesses a simple Newtonian analogue. The anisotropy modes in Bianchi type V models all decay as the universe expands. In order to give a detectable temperature anisotropy signal today, the anisotropy would have been so large at the epoch of primordial nucleosynthesis that the helium and deuterium abaundances produced then would have been in severe conflict with observation[115] However, this does not mean that the hotspot effect can never be physically relevant in our Universe; there exist other, more general, open homogeneous universes containing the hotspot effect in which the anisotropy level need not be so large in the distant past. Also, it transpires that the simple Bianchi type V example is closely related to the description of small-amplitude density inhomogeneities in open universes[19,114]. Notice also that the hotspot effect, although allowing one to infer the total density of the Universe directly if that density lies significantly below the critical value, is of no use if the density lies very close to the critical value, as suggested by many of the theoretical prejudices discussed above. To discriminate between open and closed universes when $|\Omega_0 - 1|$ is arbitrarily small, we have to examine general relativistic effects.

Spirals

The Bianchi I and V models are not the most general anisotropic cosmological models which contain the isotropic Friedman models as special cases as is evident from Table 1. We should therefore consider the type VII_0 and VII_h models which generalise them. In each case the generalisation is equivalent to adding circularly polarized homogeneous gravitational waves to the I and V models respectively. The VII_0 and VII_h models reduce to the I and V models in the limit when the wavelength of these gravitational waves goes to infinity. Each type VII universe can be parametrized by a measure of the wavelength of these gravitational wave modes. The most convenient is the Collins-Hawking[107] parameter, x, defined as

$$x = \frac{\text{wavelength of gravitational waves}}{\text{Hubble radius}} \qquad (58)$$

The effects of finite values of x on the propagation of the microwave background radiation from its last scattering red shift are intrinsically general relativistic because the presence of these gravitational waves produces a non-Newtonian effect: anisotropic spatial curvature. It was first noticed by Collins and Hawking[22], and recently studied in more detail by Barrow *et al*[88,100], that this curvature anisotropy produces a geodesic spiralling effect in the type VII models with $\Omega_0 \leq 1$.

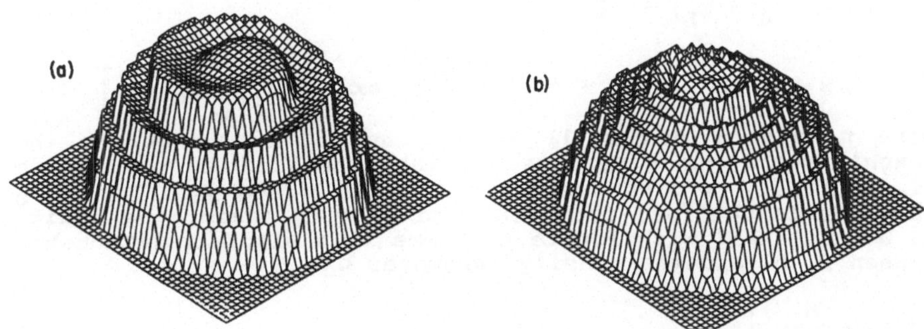

Fig.6. Spirals.*(a)* The temperature pattern, $T(\theta,\phi)$ predicted in a type VII_0 universe (with $\Omega_0 = 1$). The radial distance from the hemispheric centre gives the magnitude of $T(\theta_{obs},\phi_{obs})$; θ runs over the top of the hemisphere whilst ϕ runs from $0 \rightarrow 2\pi$ around the circular base. The value of $x = 0.067$ and the anisotropy level has been set at the largest value compatible with the observations in Table 2. *(b)* As in *(a)* but for an open universe of type VII_h (with $\Omega_0 = 0.7$). The spiral plus quadrupole seen in *(a)* has been focussed in the θ angle into a spiral hotspot.

This spiralling effect is superimposed upon a quadrupole in the flat (VII_0) case but the spiral plus quadrupole are both focussed by the hotspot effect in the open (VII_h) universe;

see Figure 6. This effect does not occur in the closed, spatially homogeneous, cosmological models of Bianchi type IX. They have an anisotropic distortion term that is essentially quadrupolar with no spiralling in the ϕ angle.

The number of twists, N, of the full spiral patterns shown in Figure 6 is determined by x as[100]

$$N \sim \frac{2}{\pi x} \tag{59}$$

and is not influenced by Ω_0 to first order. Therefore, the presence of the spiral effect allows observers to discriminate between open and closed homogeneous universes: *closed homogeneous universes cannot exhibit the spiral efect no matter how close Ω_0 is to unity*. This appears to be the only discrete cosmological observable so far discovered that has such a property.

As $x \to \infty$, the spiralling disappears. Note that the geodesic spirals are left-handed in the observer angles ($\pi - \theta_0$, $\pi + \phi_0$). The solutions[22,100] for the geodesics reveal the spiral effect and generalize the solution (55)

$$\tan\left(\frac{\theta_0}{2}\right) = \tan\left(\frac{\theta_L}{2}\right) \exp\left[-(\tau_0 - \tau_L)\sqrt{h}\right] \tag{60}$$

$$\phi_0 = \phi_L + (\tau_0 - \tau_L) - \frac{1}{\sqrt{h}} \times \ln\left[\sin^2\left(\frac{\theta_L}{2}\right) + \cos^2\left(\frac{\theta_L}{2}\right) \exp\{2(\tau_0 - \tau_L)\sqrt{h}\}\right] \tag{61}$$

where τ_0 and τ_L are still defined by (56). To obtain the Bianchi type I and V results, we take the limit $x \to \infty$ of the VII_0 and VII_h models respectively with *fixed* Ω_0. The VII_0 behaviour is obtained from (60) and (61) by taking the limits $h \to 0$ and $\Omega_0 \to 1$ with finite x. There is a simple relationship between x, h, and the density parameter Ω_0

$$x = \left[\frac{h}{1 - \Omega_0}\right]^{1/2} \tag{62}$$

The angular behaviour of the temperature profile in type-VII universes now picks up a nontrivial azimuthal variation in ϕ owing to the spiral effect, and we find in the simplest examples[100], that

$$\frac{\Delta T(\theta_0, \phi_0)}{T_0} \propto [A^2(\theta_0) + B^2(\theta_0)]^{1/2} \cos(\phi_0 + \tilde{\phi}) \tag{63}$$

where $\tilde{\phi}$ is given by

$$\cos\tilde{\phi} = \left[\left[\frac{\sigma_{12}}{\sigma}\right]_0 B(\theta_0) - \left[\frac{\sigma_{13}}{\sigma}\right]_0 A(\theta_0)\right] \times (A^2(\theta_0) + B^2(\theta_0))^{-1/2} \tag{64}$$

Here, $(\sigma_{12}/\sigma)_0$ and $(\sigma_{13}/\sigma)_0$ are constants measuring the amplitude of any anisotropy, and the functions $A(\theta)$ and $B(\theta)$

are approximately[100]

$$\left.\begin{array}{l} A(\theta) \\ B(\theta) \end{array} \sim \begin{array}{l} \sin\ [(2\cos\theta_0)x^{-1}] \\ \cos\ [(2\cos\theta_0)x^{-1}] \end{array}\right\} .f(x,\theta_0) \qquad (65)$$

for some messy function $f(x,\theta_0)$. Since there is a complete spiral turn every time $2\cos(\theta_0)x^{-1}$ is an integral multiple of 2π, we can see why (59) holds. For further details see ref. (100).

□ Determine the detailed nature of the analogue of the spiral effect in Bianchi type VI and VIII models.
□ Interpret the spiral effect and its analogues in other Bianchi types by reference to the action of the three-dimensional group of motions acting on the hypersurfaces of homogeneity.
□ Investigate the behaviour of the geodesics in inhomogeneous perturbations of homogeneous models containing hotspots and spirals.
□ Formulate Fourier analysis on a negatively-curved space of constant curvature and expand finite wavelength density perturbations of open Friedman universes in terms of the Fourier components. Relate these Fourier components to exact Bianchi V universes.
□ By recalling that the velocity-space of special relativity is a space of constant negative curvature, relate the hotspot phenomenon to that of relativistic aberration.
□ Although Bianchi type VI and VIII universes cannot completely isotropise (because they do not contain an isotropic subcase) they can become arbitrarily isotropic. Determine the form of the background anisotropy in such cases. What is the analogue of the spiral effect generated by the hyperbolic rotations in the type VI_h group of motions?
□ Discover whether the discrete difference between the microwave background sky in closed and non-closed spatially homogeneous universes still exists when inhomogeneities are admitted.
□ What spatial topologies are allowed for each Bianchi type universe and how would a non-standard topology alter the observed microwave background patterns? Is it possible for local obseravtaions of the microwave background to determine the global topology of the Universe?
□ Determine new limits on the largest spatially homogeneous magnetic and electric fields allowed in the universe by calculating the limits imposed by the microwave background isotropy in Bianchi VII and IX models. Investigate the formal analogy between magnetic field stresses and vorticity.
□ Give a geometrical argument to explain why the hotspot phenomenon occurs in Class B but not Class A of the Bianchi classification of Table 1.
□ Find an inhomogeneous cosmological model which exhibits the hotspot or spiral-like effect on geodesic propagation.

OBSERVATIONAL LIMITS

A detailed analysis of the microwave background in spatially homogeneous cosmological models containing the Friedman universes as special isotropic subcases allows us to place limits upon the large scale rotation and shear distortion of the Universe. By using the observational limits on the quadrupole listed in Table 2, and assuming that the microwave photons were last-scattered at $z = 1000$, the calculations of Barrow *et al*[100] yield the following upper limits on the rotation and shear of the Universe:

Type IX (Closed universe, $\Omega_0 \sim 1$)

$$(\sigma/H)_0 < 2.6 \cdot 10^{-5}$$
$$(\omega/H)_0 < 3.9 \cdot 10^{-13} \qquad (66)$$

Type VII (Open and Flat universes, $\Omega_0 \leq 1$)

$$\max \{(\omega/H)_0 , (\sigma/H)_0\}$$
$$< 5.10^{-9}\Omega_0^{-1}x^{-1}[1 + x^2(1 - \Omega_0)]^{1/2}[1 + 9x^2(1 - \Omega_0)]^{1/2} , \qquad (67)$$

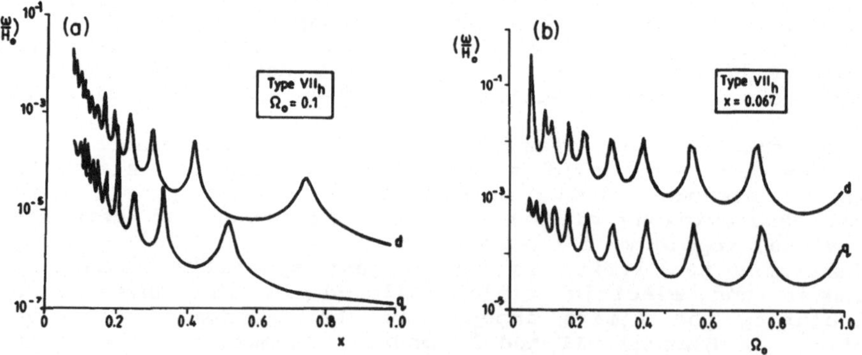

Fig.7. Upper limits[100] on the present vorticity to Hubble rate, $(\omega/H)_0$, compatible with present observations of the dipole (d) and quadrupole (q) observations of the microwave background radiation temperature isotropy: (a) in a type VII_h universe with $\Omega_0 = 0.1$ for various $x < 1$, and (b) in a type VII_h universe with $x = 0.067$ for various values of Ω_0.

For $x < 1$, this upper limit reduces to $5.10^{-9}x^{-2}\Omega_0^{-1}$; for $x > 1$ it becomes $1.5.10^{-8}\Omega_0^{-1}$.

These limits considerably improve upon the ones given earlier by Collins and Hawking[107] but they are not as strong as those obtained from primordial nucleosynthesis[115] since the latter are , in effect, imposed at $z = 10^{10}$ rather than $z = 10^3$.

We have not quoted limits for the Bianchi universes of lower generality (I and V) since these limits are obtainable from those above by specialization of parameters (see refs.(100) & (107) for full details). The limits (66) and (67) are remarkably strong and provide direct evidence for the isotropy of the Universe.

In Figure 7 we show two specific examples of numerical analyses leading to the upper limits imposed on the vorticity of the Bianchi VII universes by the dipole and quadrupole observations of the microwave background temperature isotropy[100].

The following features of Figures 7(a) & (b) are worth noting: the limits imposed by the dipole and quadrupole observations have an oscillatory character because of the spiral effect. The limits (66) and (67) result from averaging over these oscillations. The limits weaken with decreasing x and decreasing Ω in (a) and (b) because in each case the overall anisotropy pattern is becoming more poorly described by the first two *spherical* harmonics[100]. In (a) this is because of the pronounced spiral effect at small x, whereas in (b) it is because of the increasing hotspot focussing at small Ω_0.

□ Determine whether the very strong limit (66) on any vorticity in the universe if it is closed, (66), is primarily an artefact of the restriction to spatially homogeneous cosmologies, or primarily indicative of the difficulty of fitting a vortex into the restrictive geometry of a closed universe.

□ Determine limits on the largest spatially homogeneous magnetic and electric fields allowed in the Universe by calculating the accompanying microwave background anisotropy in type VII and IX models. Investigate the formal analogy between magnetic field stresses and vorticity.

□ Determine the maximum shear and vorticity allowed in models of our Universe of Bianchi VI and VIII types.

□ How do the presence of collisionless distributions of primordial neutrinos and gravitons in the Universe affect the limits derived for the rotation and shear of the Universe? Extend the calculations described to determine the relative anisotropy in photons, neutrinos and gravitons today (remember to include the effects of massive particle annihilation on the mean temperatures of the three collisionless components).

ISOTROPY AND HOMOGENEITY

Direct observations of the Hubble flow[117], radio source counts[118], the intensity distribution of the cosmic x-ray background[119] as well as the impressive temperature isotropy of the cosmic microwave background radiation discussed above, all witness to the high degree of isotropy in the expansion of the Universe back to the redshift at which the microwave photons were last scattered by electrons. Further, the excellent agreement between the predictions[97] of primordial nucleosynthesis in the standard 3 (or 4)-neutrino isotropic Friedman model of the early universe and the observed (or inferred) primordial abundances of helium-4, helium-3, deuterium and lithium-7 imply that the Universe was extremely isotropic at a redshift ~ 10^{10}. *Prior to this redshift we possess no direct or indirect observational evidence regarding the isotropy and homogeneity of the Universe.*

The isotropy of the microwave background radiation is the more surprising as regions of the microwave background separated by more than ~ 2° on its last scattering surface at z = 1000 do not appear to have been in causal communication with each other before the time when they were last scattered. How, then, did these widely separated, independent regions conspire to have the same temperature and radiation density today to better than one part in a thousand?

Traditionally this question has not been posed. Prior to the discovery of the microwave background radiation the only direct test of the isotropy of the Universe was the counting of galaxies in different solid angles around the sky, first carried out by Hubble[120]. Because the evidence gleaned by this technique is so meagre, and non-uniform, anisotropic solutions of general relativity are so hard to find cosmologists predicated their early observational and theoretical studies on the assumption of isotropy and homogeneity. Such models provided an excellent description of the present state of the Universe and hence the question of why large scale isotropy and homogeneity does exist did not arise prior to 1967. Rather, attention was focussed upon explaining the presence of the small deviations from perfect homogeneity: the heterogeneities that grew into stars and galaxies.

Soon after the microwave isotropy was first measured, Misner[121,122] stressed that it was the existence of the underlying isotropy and homogeneity that constituted the major mystery in need of explanation rather than that of the source of the small irregularities.

□ Determine whether thermodynamic aspects of classical and quantum gravitation favour the evolution of a universe from uniformity to irregularity or *vice versa*.

THE COSMOLOGICAL PRINCIPLE(S)

The power of assuming that the Universe is isotropic and spatially homogeneous to a first approximation was stressed primarily by Milne[123] and is known as the *Cosmological*

Principle. It implies that the metric of space-time is that of a space of constant curvature. It is used in one form or another to break through the Geordian knot of determining the structure of the Universe by observation which we discussed at the beginning of these notes. It is most expedient to assume isotropy and homogeneity and then look for evidence that this assumption is false. However, this is awkward in that the Universe clearly cannot be *exactly* spatially homogeneous (SH) and isotropic (I); therefore we encounter a variety of weaker versions of the Cosmological Principle; for example:

(i) the Universe is SH and I "on the average"
(ii) the Universe is SH "on the average"
(iii) we are not at the centre of the Universe
(iv) we are at a typical position in the Universe
(v) the observed portion of the Universe is a fair sample.

Usually, it is stated that because we observe the microwave background to be extremely close to isotropy and assume the truth of the Copernican Principle (either in the form *(iii)* or *(iv)*), we cannot regard this observation as unique to our location. Hence the Universe must look isotropic about any point and so must be spatially homogeneous. However, we note that although a space that is *exactly* isotropic about every point must be spatially homogeneous[31], we do not have a theorem which tells us that a space which is *almost* isotropic about every point must be *almost homogeneous*. However, it is the latter result that is being assumed in practice.

A recent attempt has been made by Stoeger *et al*[124] to formulate the Cosmological Principle in a fashion that is observationally testable. It involves comparing variations in density ρ over space and determining an upper bound on the size of these variations over each length scale L. However, it would seem to be more useful to evaluate the *metric perturbations* to space-time induced by these density inhomogeneities. This evaluation would take into account the fact that a particular level of inhomogeneity on a large scale (*eg* that of galaxy clusters) is far more significant *vis a vis* the truth of the Cosmological Principle than is an inhomogeneity of the same amplitude over the scale of the solar system. In addition, all forms of the Cosmological Principle are used solely to justify the adoption of the Friedman *metric* and so one should evaluate the distortions to the metric implied by observations in order to test this assumption. The metric perturbation ($\delta g/g$) associated with a density perturbation $\delta\rho/\rho$ over a length scale L is of order

$$\delta g/g \sim (\delta\rho/\rho)(L/ct)^2 \qquad (68)$$

where the c is the velocity of light and hence $ct \sim cH_0^{-1}$ is roughly the present size of the particle horizon. Similar expressions can be written down to give the metric perturbations associated with vortical and shear motions[126]. Finally, we note that it is the metric perturbation that is observed in any temperature anisotropy of the microwave background radiation[100,107].

□ Give as many inequivalent versions of the "Cosmological Principle" as you can and give examples of cosmological models which obey and do not obey each formulation you give.

□ Formulate the idea of an "almost symmetry" in a precise manner and determine whether a space that is almost isotropic everywhere is almost homogeneous.

□ Classify almost homogeneous spaces according to their "distance" from homogeneous ones.

□ Produce a stochastic version of the Einstein equations, treating the metric as a random variable. Determine the form of the metric under the assumptions of statistical homogeneity and isotropy employed in studies of homogeneous turbulence.

□ Give a precise version of a Cosmological Principle which can be applied to the real Universe and checked quantitatively against observation?

□ Give examples of inhomogeneous distributions of matter which obey a density but not a metric criterion for homogeneity and *vice versa*.

CAN WE PROVE A COSMOLOGICAL PRINCIPLE?

Hoyle and Narlikar[125] were the first to appreciate the idea of explaining the uniformity of the Universe without appeal to special initial conditions. In 1963, before the discovery of the microwave background radiation they pointed out that one of the advantages of the steady state theory of the expanding universe was that the de Sitter solution is stable against the growth of anisotropic and inhomogeneous perturbations..."*any finite portion of the universe gradually loses its 'memory' of an initially imposed anisotropy or inhomogeneity...the universe attains the observed regularity irrespective of initial boundary conditions*". This idea was resurrected by Misner[121,122] in the context of the Big Bang models after the discovery of the background radiation isotropy in 1967.

Misner aimed to show that large scale properties of the Universe, like its isotropy, were the inevitable consequences of physical processes occurring within the Universe and could thus be predicted to exist in any sufficiently old Universe *independent of its initial conditions*. This was termed the *chaotic cosmology programme* and focussed initially upon finding dissipative processes (*eg* shock-wave damping[127], neutrino viscosity[121-2,128-9], and quantum particle production[130-1]) which could remove arbitrarily large amounts of initial anisotropy by the present, and in finding ways of enlarging particle horizons[132] near the initial singularity so that dissipative processes could have effect over regions that would expand to extragalactic extent by today. In both these searches the "chaotic cosmologists" were unsuccessful. No solutions which allow horizon removal or enlargement with significant probability were found[133,33]; the dissipation of large amounts of anisotropy in the very early universe was shown to give rise to a very large value of the entropy per baryon, in excess of that observed today[134], if baryon number was conserved (which it is expected to be over most of the classical history of the Universe after ~ 10^{-30}s).

IS ISOTROPY A STABLE PROPERTY OF COSMOLOGICAL MODELS?

One way of evaluating the possibility of proving the inevitability of observing a Cosmological Principle to hold after 15 billion years of cosmic expansion is to evaluate whether *isotropy* is a stable property of cosmological solutions to Einstein's equations. This was attempted by Collins and Hawking[135] in 1973. Restricting themselves to spatially homogeneous cosmological models they investigated whether the isotropic flat and open Friedman universes are stable solutions of Einstein's equations as $t \to \infty$.

We recall for Table 1 that the most general class of homogeneous universe models containing the open ($k = -1$) Friedman models as a special case are those of Bianchi type VII_h. The Bianchi type VII_o is the most general class including the flat ($k = 0$) Friedman model. The Bianchi type VII equations can thus be linearized about the Friedman solutions and the stability of the latter determined as $t \to \infty$.

Collins and Hawking[135] (CH) define conditions for *isotropisation* of ever-expanding universes which include the following two requirements:

(i) the ratio of the shear to the mean Hubble expansion rate, σ/H, tends to *zero* as $t \to \infty$.

(ii) the energy-momentum tensor for the matter content of the universe obeys the *weak energy condition*:

$$T_{ab}u^a u^b \geq 0 \tag{69}$$

the *dominant energy condition*:

$$T_{oo} \geq |T_{\alpha\beta}|, \ \alpha,\beta = 1,2,3 \tag{70}$$

and the *positive pressure criterion*:

$$\sum_{\alpha=1}^{3} T_{\alpha\alpha} \geq 0 \tag{71}$$

It is then possible to prove[135] that no open set of spatially homogeneous initial data isotropises as $t \to \infty$. Only if we restrict the initial data to that which is *spatially flat*, and the equation of state of matter to be $p = 0$ to leading order, can it be proved that there exists an open neighbourhood of initial data which isotropises. Thus, it was argued that isotropic universes are a set of measure zero amongst cosmological solutions to Einstein's equations.

□ Determine whether ever-expanding Newtonian cosmological models isotropise.
□ Give a criterion to determine whether a closed universe isotropizes by defining a time coordinate which places the expansion maximum at infinity. Use your definition to determine whether the closed Friedman universe is a stable solution of Bianchi type IX.
□ Extend the Collins and Hawking stability analysis of the Friedman model to the self-similar generalization of the Bianchi classification developed by Eardley.[87]

□ Determine whether inhomogeneous perturbations alter the stability properties of isotropic universes as $t \to \infty$.
□ Determine the conditions on g_{ab} and T_{ab} for which isotropy is a stable property of singular homogeneous cosmological models as $t \to 0$.
□ How large is the class of Bianchi type VI and VIII universes which can approach arbitrarily close to isotropy as $t \to \infty$?

IS ISOTROPY REALLY UNSTABLE AND DOES IT MATTER ANYWAY?

Because the above result is widely misunderstood and because it is an important precursor to our discussion of recent attempts to prove cosmic "No Hair" theorems we shall now analyse it more carefully.[6,136,34]

First, we should appreciate that the energy conditions, (ii), ensure that as $t \to \infty$ "matter does not matter" in open universes; that is, the Friedman stability problem reduces to the *vacuum* stability of the isotropic Milne model (eqn. (26) with $\Lambda = 0$). But our Universe can only have been vacuum (=curvature) -dominated since a redshift $z \sim \Omega_0^{-1} \sim 15$ until the present. The observed low anisotropy of the real Universe is primarily a consequence of the evolution from the initial state ($z = \infty$ or $z = 10^{32}$?) until $z \sim \Omega_0^{-1} -1$ and the asymptotic behaviour in the curvature-dominated regime analysed by the CH stability theorems is essentially irrelevant to this.

In the case of the $k = 0$ Friedman model the matter is always important, but in order to obtain isotropisation it was necessary to require that the equation of state be $p = 0$. Isotropisation in the CH sense therefore only occurs in this Universe during the dust-dominated era from $z \sim 10^4$ to the present. However, the present isotropy level of the real Universe is determined primarily by the integrated effect of the anisotropy domination during the entire radiation era when $z > 10^4$. Again, the asymptotic stability result is irrelevant to the present isotropy level.

No asymptotic result regarding the anisotropy evolution as $t \to \infty$ can explain why the anisotropy level is lower than a particular value at any finite time. (For these reasons the CH stability analyses do not admit the Anthropic interpretation generally associated with them.[6,36,34])

Approach to a Family of Plane Waves

Notwithstanding the argument just presented, it is very instructive to investigate the $t \to \infty$ stability of the open Friedman universe a little further. The isotropisation criterion (i) of CH requires $\sigma/H \to 0$ as $t \to \infty$; that is, in mathematicians' language, *asymptotic stability* of the Friedman solution with respect to the type VII evolution equations. Collins and Hawking proved that this cannot occur in general when (ii) holds. However, it has since been shown[34,88] that $\sigma/H \to$ constant as $t \to \infty$; that is, the Friedman model is *stable*, although not asymptotically stable.

In physical problems asymptotic stability (perturbations decay to zero) is too strong a requirement to make — the solar system is not stable in this sense. The practical definition of stability is that perturbations be bounded and this is what we shall mean by *stability*.

We can in fact show[34,88] that spatially homogeneous perturbations of the ever-expanding Friedman model asymptote in general to a 2-parameter family of plane waves of Bianchi type VII_h described by a family of exact solutions to the Einstein equations first found by Lukash[137,138].

The Lukash solutions are spatially homogeneous vacuum solutions of Bianchi type VII_h described by the metric

$$ds^2 = dt^2 - g_{\alpha\beta}dx^\alpha dx^\beta, \qquad x^\alpha = (x,y,z) \tag{72}$$

where

$$g_{\alpha\beta} = \begin{bmatrix} a^2e^{2z}\ (\mathrm{ch}\mu+\mathrm{sh}\mu\cos\psi), & a^2e^{2z}\mathrm{sh}\mu\sin\psi, & 0 \\ a^2e^{2z}\mathrm{sh}\mu\sin\psi, & a^2e^{2z}\ (\mathrm{ch}\mu-\mathrm{sh}\mu\cos\psi), & 0 \\ 0\ , & 0 & ,c^2 \end{bmatrix} \tag{73}$$

where $a(t)$, $c(t)$, $\psi(t)$, $\mu(t)$ with k constant.

This metric describes two monochromatic circularly-polarized gravitational waves of wavelength $2\pi c(t)k^{-1}$, and amplitude $\mu(t)$ moving in the $\pm z$ direction $(=x^3)$ on a space of constant negative curvature. Physically, this is like a Bianchi type V metric of constant negative curvature plus gravitational waves which create a spatial curvature anisotropy. The Lukash solution is a *2-parameter family of plane waves*[83] (all curvature invariants vanish) with

$$\begin{aligned} a(t) &= t^{1/(1+\lambda^2)} \quad ; \quad c(t) = (1+\lambda^2)\ t \\ \psi(t) &= \frac{k\ln t}{1+\lambda^2} \quad ; \quad \lambda = \frac{k}{2}\ \mathrm{sh}\mu \end{aligned} \tag{74}$$

where λ is an arbitrary constant.

When $k = \mu = 0$ (74) reduces to the isotropic Milne universe; and when $k \to 0$ with $\mu \neq 0$ we approach the axisymmetric Bianchi type V model. It is worth noting here that $k = x^{-1} = h^{-1/2}$, where x is the 'spiral parameter' of Collins and Hawking introduced earlier[58] and h is the type VII_h label (see equation (62) with $\Omega_o = 0$ in vacuum). Each Lukash model is labelled by 2-parameters (k,μ) compared with the 4 parameters required for the general type VII_h vacuum solution. However, one can show that as $t \to \infty$ all vacuum and perfect fluid[34,88] VII_h universes approach the 2-parameter family of Lukash metrics. In particular, perturbations of the open Friedman universe (whether large or small) of VII_h type *all* approach the Lukash solutions and have $\mu \to$ constant, hence $\sigma/H \to$ constant as $t \to \infty$. This stability turns out to be non-trivial. If we start with a Lukash solution characterized by $\mu = \mu_o$ = constant then as $t \to \infty$ it will evolve towards another member of the family of Lukash plane waves with $\mu = \mu_\infty$ = constant $\neq \mu_o$. Since the $\mu = o$ case is the isotropic Milne model, we see that this means that *the isotropic solution is stable but not asymptotically stable* (for full details see ref. 100) In fact, one has $\mu_\infty < \mu_0$ and the model evolves closer to isotropy as $t \to \infty$ whenever

$$\gamma < \frac{2(4 + k^2 sh^2 \mu_0)}{12 + k^2 sh^2 \mu_0} \tag{75}$$

if the model contains perfect fluid with $p = (\gamma - 1)\rho$ equation of state when $2/3 < \gamma \leq 2$.

We can conclude that open isotropic universes are not unstable to spatially homogeneous perturbations as $t \to \infty$. However, the present observed isotropy cannot be a consequence of this asymptotic behaviour because we live at a finite time after the initial state, that is in a real sense far from the asymptotic region.

□ Numerically integrate the null geodesics of the Lukash solutions to determine the exact nature of the spiral pattern produced. Compare it with the results presented earlier for the case of small anisotropy.
□ Determine whether all Bianchi type VII_h cosmological models approach the Lukash metrics in the presence of electromagnetic fields.
□ Investigate inhomogeneous perturbations of the Lukash solutions and use them to analyse the general behaviour of cosmological solutions to Einstein's equations as $t \to \infty$.
□ Since the Lukash solutions are plane waves one might expect there to be no quantum particle production in such a space-time. Determine whether this is so.

NO HAIR THEOREMS

The attempt to prove asymptotic stability of isotropic expansion by Collins and Hawking[135] imposed very restrictive energy conditions by demanding positive pressures, (71). This ensures that open universes approach vacuum solutions as $t \to \infty$ as we have already mentioned, but it also excludes the presence of a positive cosmological constant, Λ. Since 1981 there has been growing interest[1] in the cosmological consequences of a finite period of "inflationary" early expansion history during which the dynamics were dominated by an effective cosmological constant. From the viewpoint of general relativity, inflation reduces to the possibility that some set of hypothetical matter fields in the early universe give rise to an energy momentum tensor of the form

$$T_{ab} \approx \Lambda g_{ab} \tag{76}$$

where Λ is a constant. By comparison with equation (18) we see this is equivalent to a perfect fluid with $p = -\rho = -\Lambda$, a point first made by McCrea[139]. If we examine the Friedman equation for an isotropic universe containing a perfect fluid with equation of state $p = (\gamma - 1)\rho$ and the stress T_{ab} in (76) we see that (22)-(23) give

$$\frac{\dot{a}^2}{a^2} = \frac{\rho}{3} - \frac{k}{a^2} + \frac{\Lambda}{3} \tag{77}$$

where $\rho \propto a^{-3\gamma}$. Hence as $a \to \infty$, for $\gamma > 0$, we see that

$$\frac{\dot{a}^2}{a^2} \to \frac{\Lambda}{3} \tag{78}$$

and $a \to \exp(t\sqrt{\Lambda/3})$ That is, the dynamics approach the de Sitter universe.

The *'No Hair' conjecture*[140-1] is that this asymptotic approach to de Sitter is true in general even in the presence of anisotropies, inhomogeneities and other fluid stresses. It is important to discover under what circumstances this might be true since this will delineate the set of initial conditions which can be driven towards isotropy and homogeneity by inflation.

Various 'proofs' of versions of the No Hair conjecture by Hoyle & Narlikar[125], Barrow[142], Boucher & Gibbons[143-4], Starobinskii[145] and Wald[146] have appeared in the literature for particular situations (*eg* perturbations close to isotropy and homogeneity, spatial homogeneity). For example, Starobinskii shows that if $\Lambda > 0$, then as $t \to \infty$ the Einstein equations admit a series approximation of the form

$$ds^2 = dt^2 - [e^{2t\sqrt{\Lambda/3}}A_{\alpha\beta}(\mathbf{x}) + B_{\alpha\beta}(\mathbf{x}) + e^{t\sqrt{\Lambda/3}} \; C_{\alpha\beta}(\mathbf{x}) + \\ + \ldots]dx^{\alpha}dx^{B} \tag{79}$$

where, in the presence of a perfect fluid with $2 < \gamma \leq 1$, *eight* components of the spatial functions, $A_{\alpha\beta}$, $B_{\alpha\beta}$ and $C_{\alpha\beta}$ are left arbitrary by the field equations. Deviations from the de Sitter state are seen to decay exponentially rapidly in time within the event horizon of a geodesic observer. Outside that horizon the inhomogeneity ($A_{\alpha\beta}(\mathbf{x})$) may still be large though. Inflation does not remove inhomogeneity: it simply dilutes it by expanding it to exponentially large length scales.

Other examples of No Hair "theorems", this time for ever-expanding spatially homogeneous models, but without the need for deviations from isotropy to be assumed small initially so that a series expansion is not required for an asymptotic analysis, are those of Wald[146], Jensen & Stein-Schabes[148], Turner & Widrow[149] and Moss & Sahni[147]. Exact inhomogeneous solutions which approach de Sitter in accord with (79) have also been found[150-1] They show, in particular, that if the universe contains an effective stress of the form (76) created by a self-interacting scalar field then as $t \to \infty$, (ever-expanding universes) the anisotropy falls-off exponentially rapidly and the de Sitter universe is approached so long as the remaining matter fields obey the *weak energy condition* (69) (*ie* the energy density of the scalar field is positive) and also the *strong energy condition* (2).

□ What do non-geodesic (non-hypersurface orthogonal) observers see as $t \to \infty$ in the inflationary approach to de Sitter?
□ What is the range of gravitation theories for which the No Hair conjecture holds in the form (79)?

INFLATION AND THE INITIAL VALUE PROBLEM

It is important to stress that even if the No Hair conjecture were true and all conceivable cosmological initial conditions asymptotically approached the de Sitter solution locally this would not provide an explanation of the present low isotropy of the Universe. There can exist no such explanation of the isotropy level in the Universe at a finite time[153].

There have been many claims that a finite period of de Sitter expansion has the attraction of simultaneously solving the 'flatness', 'horizon' problems as well as explaining why the universe is so old[1,18]. It is worth noting that these are in fact all the same mathematical problem under different names. It is not possible, in general, to solve one without automatically solving the others. The remaining independent cosmological problem that inflation aims to resolve -- that of the present isotropy of the universe -- has been the subject of numerous recent papers that focus upon displaying the wide range of cosmological initial conditions that inevitably evolve towards the isotropic de Sitter state.

Two points of principle need to be made about claims that inflation can explain the present isotropy of the Universe. (We shall assume that there do not arise any general relativistic or quantum gravitational processes which prevent inflation from actually occurring). Confine attention to spatially homogeneous but anisotropic cosmological models with everywhere spacelike surfaces of constant density so the Einstein equations are a set of ordinary differential equations ($\dot{\mathbf{x}} = F(\mathbf{x})$) obeying the local Lifschitz and continuity conditions (so $\|\partial F/\partial \mathbf{x}\|$ is continuous). They will contain some matter source, for example a scalar field, ϕ, with potential $V(\phi)$ able to drive inflation for a finite period. This constitutes a well-posed initial value problem which has the property that its solution at any time t is a continuous and unique function of conditions at any earlier time $T < t$. This makes it impossible *in principle* for inflation, or any other classical or quantum dissipative process operating in classical space-time, to explain the present isotropy of the Universe independently of cosmological initial conditions. We can always choose a cosmological model that today is more anisotropic than observation allows and evolve it backwards through any period of inflation to determine a set of pre-inflationary initial conditions which will therefore fail to evolve a universe consistent with observation today. At present we have no unique probability measure to apply to cosmological initial conditions and so it is not possible to say how 'large' the set of such counter-examples is.

It is clear that if one *first* chooses cosmological initial data then one can subsequently choose those constants of

Nature which fix the amount of inflation so that by the present time any initial anisotropy will be less than any pre-set level, no matter how small. This is what is often done in the literature[147-9]. It does not conflict with our first principle above, which is equivalent to the statement that if the constants of Nature which determine the duration of the inflationary epoch are chosen *first* then it is always possible to subsequently choose cosmological initial data which cannot become more isotropic than some pre-chosen level by the present day. There is a tendency for particle physicists to choose the inflationary parameters *after* the initial data to solve the isotropy problem whereas, as in all other areas of cosmology and physics, we choose the fundamental constants *before* the initial data. If that is done, inflation cannot explain the isotropy of the Universe irrespective of the initial data.

INFLATION AND THE STRONG ENERGY CONDITION

Many authors have interpreted the above-mentioned No Hair 'theorems' as a proof of the effectiveness of inflation in explaining the present high level of isotropy in the Universe independent of initial conditions. For, it is claimed, no matter how anisotropic the initial conditions, there will be an exponentially rapid approach to the isotropic de Sitter solution within the event horizon of any geodesic observer during a period of inflation.

The No Hair conjecture is, roughly speaking, also equivalent to the statement that in the presence of a positive effective cosmological constant, the de Sitter space-time is a stable asymptotic solution of the Einstein equations. A variety[142-9] of mathematical demonstrations of such a result, subject to particular assumptions, have appeared in the literature. All contain, either implicitly or explicitly , one major technical weakness: *the assumption that the energy-momentum tensor, T_{ab} of the material content other than that of the cosmological constant or inflating matter field obeys the strong energy condition (2).*

The imposition of the strong energy condition is unsatisfactory because the positive cosmological constant arising from an effective stress like (76) necessarily violates the strong energy condition (2). Indeed, violation of the strong energy condition is a necessary condition for inflation or generalised inflation[152] to occur and resolve the flatness-horizon-age problem. It is therefore quite unreasonable to expect all the other matter sources near the Planck time to obey the condition (2). We shall now show that when the unreasonable strong energy condition is dropped the cosmological No Hair conjecture fails[153-4].

The Deflationary Universe

We consider a zero-curvature Friedman model containing a perfect fluid with pressure p and energy density ρ having equation of state (15) but we shall also assume the presence of a bulk viscosity $\eta = \alpha\rho$, with α constant. The total pressure is therefore given by p' where[31]

$$p' = (\gamma - 1)\rho - 3H\alpha\rho \qquad (80)$$

The two essential field equations are, the Friedman equation

$$3H^2 = \rho \qquad (81)$$

(where $H(t) = \dot{a}/a$ is the Hubble expansion rate and we have picked units with $8\pi G = 1$)), and the conservation equation ('·' = d/dt)

$$\dot{\rho} + 3H(\rho + p') = 0 \qquad (82)$$

These yield a single differential equation for H(t):

$$2\dot{H} = 3H^2(3H\alpha - \gamma), \; \rho = 3H_0{}^2 = \gamma^2/3\alpha^2 \qquad (83)$$

There exists a special de Sitter solution of (83) with constant expansion rate

$$H = H_0 = \gamma/3\alpha \qquad (84)$$

This special solution is a stable attractor as $t \to -\infty$, but is *unstable* as $t \to +\infty$. This can be seen by solving (*83*) to obtain H(t):

$$H(t) = H_0(1 + Aa^{3\gamma/2})^{-1}; \quad A \geq 0, \text{ constant} \qquad (85)$$

Integrating (85), we obtain for the expansion scale factor, a(t), of the Friedman metric

$$\ln a + Aa^{3\gamma/2} = 3H_0\gamma t/2 \qquad (86)$$

We see that this cosmological model evolves from an initial de Sitter state at $t = -\infty$ and actually *deflates* to the zero-curvature Friedman state with $a \propto t^{2/3\gamma}$ as $t \to +\infty$. The dominant energy condition, (70), $\rho + p' \geq 0$ is always obeyed since $H \leq H_0$, but the strong energy condition (2), which reduces to $\rho + 3p' \geq 0$ is *violated* at early times.

This solution, which was first found by Murphy[155] in another context as an example of a non-singular cosmological model, reveals the restrictive role played by the strong energy condition and shows that the de Sitter state need not be stable. The stress (80) responsible for the instability can just be viewed mathematically as a form of energy-momentum tensor which de-stabilizes a de Sitter state that it has initially created after the manner of chaotic inflation. However, it is also physically motivated as a classical bulk viscosity[31]. If we pick $\gamma = 2$ then the perfect fluid stress becomes equivalent to a massless scalar field, ϕ, with $p = \rho = \dot{\phi}^2/2$ and a dissipative coupling $-3\eta\dot{\phi}$. Classically, bulk viscous stresses arise because the expansion of universe is continually trying to pull the fluid out of thermal equilibrium. They should vanish when the equation of state is pure radiation. Bulk viscosities also appear as phenomenological descriptions of quantum particle production near the Planck time[156-7]. This correspondence is of partiucular relevance when interpreting the above result in the context of early universe inflation. Since it is necessary for inflation to come to an end if the Universe is to resume the Friedman-like expansion we now observe, there must arise a violation of one of the energy conditions used to prove the existing cosmic no hair theorems. It is the particle production arising from the rapidly-varying coupling to the inflaton field to other matter fields when it oscillates about the global minimum of the potential that gives rise to the subsequent decay of the inflaton field. This decay is usually put into the equation of motion of the inflaton field by hand. The model described by (85) and (86) can be viewed as an exact description of this decay by particle production.

Models of chaotic inflation[1], say with a massive scalar field, necessarily involve fluids with a *time-varying* equation of state spanning the domain $-\rho \leq p \leq \rho$. The state (86) is an exactly soluble example of a fluid with a time-varying equation of state.

If one chooses the viscosity coefficient, η, to be constant in (80) then a different type of *inflationary* solution to (80)-(82) exists. There is again a special de Sitter solution with $H = H_0 = \eta/\gamma$, but it is now stable as $t \to \infty$ but unstable as $t \to -\infty$. The solution subject to $a(0) = 0$ is[158]

$$a^{3\gamma/2} = 2(e^{3\eta t/2} - 1)/3\eta \qquad (87)$$

This solution begins at $t = 0$ in the Friedman state and evolves to de Sitter as $t \to \infty$. If one takes $\eta = \alpha\rho^n$ then solutions with $0 < n < 1/2$ display this inflationary behaviour typified by (87) but those with $n > 1/2$ have the *deflationary* behaviour of (85)-(86). The peculiar intermediate case $n = 1/2$ exhibits power-law inflation with

$$a \propto t^{2(\gamma - \alpha\sqrt{3})/3} \quad ; \ \gamma > \alpha\sqrt{3} \qquad (88)$$

The deflationary example (86) of how the No Hair conjectures can fail when (2) is relaxed may be interesting in the context of the very first inflationary universe model which was suggested by Starobinskii[159]. This assumed an initial de Sitter state destabilised by 1-loop quantum effects and one suspects that these can therefore be modelled by a phenomenological bulk viscosity. In the model (86) one also has a natural and inevitable transition *out* of an initial de Sitter state. This deflationary evolution provides an interesting possible resolution of the "exit problem" that besets most inflationary models[1]. The fact that the model (86) is exactly soluble means also that it is possible to carry out detailed calculation of the generation of irregularities at the classical and quantum level. The initial de Sitter phase of the solution (86) should result in the generation[160] of density and graviton fluctuations with constant curvature spectra. The deviations from this state caused by the transition to the exact Friedman asymptote should leave traces in the gravitational-wave background.

We note in passing that the special de Sitter solution (84) is a gravitational equilibrium state of the field equations: it possesses an equilibrium Hawking temperature T_H

$= H_0/2\pi$ seen by[140] geodesic observers. However, this state is not in local thermal equilibrium since entropy is being generated by the bulk viscosity at a constant rate to maintain the static de Sitter state (7). It would be interesting to study the unusual relationship between the gravitational and local thermodynamics in this situation.

Finally, one might wonder whether the deflationary behaviour (86) is stable when anisotropies are added to the spatially flat Friedman universe. The case of a Bianchi type I universe containing perfect fluid and bulk viscosity was examined by Belinskii and Khalatnikov[162] who were interested in determining whether the non-existence of a singularity in the isotropic solution (86) was stable. The field equations governing the type I evolution are given by (80) and (82) but (81) is generalised to include the shear anisotropy $\sigma^2 = \Sigma^2 a^{-6}$,

$$H^2 = \rho/3 + \Sigma^2 a^{-6} \quad ; \Sigma \text{ constant} \tag{89}$$

where a is now the geometric-mean scale-factor of the three orthogonal directions of expansion.

The resulting evolution can be determined qualitatively by recourse to a phase plane portrait in H and ρ. It was claimed[162-3] that when $\Sigma \neq 0$ the initial de Sitter state is unstable to the formation of an anisotropic Kasner-like singularity with

$$a(t) \propto (\Sigma t)^{1/3}, \qquad \rho \propto t^{-\gamma}\exp(-\alpha/t^2) \tag{90}$$

as $t \to 0$ (we still have $a \propto t^{2/3\gamma}$ as $t \to \infty$). However, this asymptote is unphysical since it violates the dominant energy condition; *i.e* $\rho + p' \to (\gamma - \alpha t^{-1})\rho$ approaches zero *from below* as $t \to 0$. The phase plane is shown in Figure 8.

The dominant energy condition will be violated whenever $\dot{\rho} > 0$ and this requires the solution to satisfy

$$H \leq H_0 = \gamma/3\alpha \tag{91}$$

The only solution trajectory that does not violate the constraint (91) is the isotropic solution (85).

In Figure 8 the exact inviscid Friedman asymptote is located at the origin ($H = 0$, $\rho = 0$) of the phase-plane whilst the de Sitter solution (84) is the critical point A. Isotropic solutions lie on the parabola $3H^2 = \rho$, whilst anisotropic ($\sigma^2 \neq 0$) solutions lie outside it. In general, in such portraits one can read-off where the dominant energy

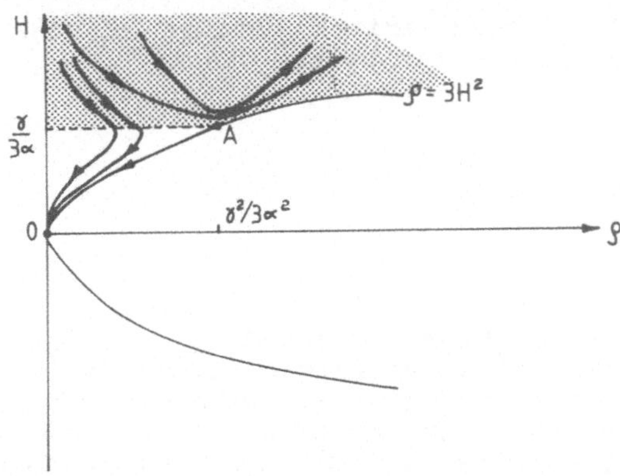

Fig. 8. The (H, ρ) phase portrait[162] for the Bianchi type I universes possessing perfect fluid $p = (\gamma - 1)\rho$ and a bulk viscous stress $\eta = \alpha\rho$. Arrows indicate the direction of increasing time. The shaded region is that in which the dominant energy condition $p' + \rho \geq 0$ is violated. The only solution trajectory that does not enter this physically forbidden region is the isotropic solution (86) which runs along the parabola $3H^2 = \rho$ from A, the de Sitter solution (84), to 0 the isotropic Friedman model with $a \propto t^{2/3\gamma}$. In accord with (89) no solution trajectories lie on the interior of this parabola.

condition will be violated by considering (82), from which we see that $\rho + p'$ will be negative in regions of the phase space where $H > 0$ and ρ is increasing or where $H < 0$ and ρ is decreasing.

It would be interesting to ascertain how general is the feature that for T_{ab} of a particular form, the dominant energy condition alone is sufficient to fix the solution uniquely. Parenthetically it is worth noting that the unphysical asymptote (90) in the type I viscous model possesses a Weyl curvature singularity (the shear $\sigma^2 = \Sigma^2 a^{-6} \to \infty$ as $t \to 0$) but $\rho \to 0$ as $t \to 0$. This illustrates a remark made in the opening pages of these notes, (following equation (2)), that the strong energy condition (2) is not necessary for the existence of a singularity. The solution (90) shows that both the strong and the dominant energy conditions can be violated and yet a singularity still arises because of the over-riding geodesic convergence created by the shear[164].

RESUMÉ

In these notes we have attempted to provide a guided tour through those areas of relativistic cosmology which should be of interest to those wanting to apply general relativity to current research areas in cosmology. We have deliberately avoided treating aspects of particle physics associated with the missing mass and galaxy formation problems since these have been given saturation coverage in other review articles and conference proceedings. We have focussed upon the following topics and issues: the relationship between observational and theoretical cosmology, the status of the Friedman universes with respect to the most general inhomogeneous and homogeneous solutions of Einstein's equations, and the use of these solutions to extract the maximum of information from observations concerning the extent to which the Universe is isotropic. The extremely high level of isotropy established by these analyses of the microwave background radiation leads us to seek an explanation for it. Hence the stability of the isotropic cosmological models under various assumptions was critically investigated. These investigations culminate naturally in an analysis of the effectiveness of 'inflation' as an explanation of the present level of isotropy. We showed that it is not possible for inflation to explain the structure of the Universe independently of initial conditions. Finally, we argued that attempts to prove the Cosmic No Hair conjecture should not use the strong energy condition and then demonstrated that when this condition is dropped the No Hair conjecture is false. A simple isotropic example involving a bulk viscious stress shows it to fail in the most dramatic way: an initial de Sitter state deflates into an Einstein-de Sitter universe.

Acknowledgements

I would like especially to thank G Galloway, R Juszkiewicz, R Matzner, A Ottewill, S Siklos, D Sonoda, J Stein-Schabes, F J Tipler and J Wainwright, with whom I have enjoyed working on some of the topics described herein. I have also benefitted from helpful discussions and correspondence with S Bhavsar, A Burd, B Carter, C B Collins, G F R Ellis, G Gibbons, J Hartle, B L Hu, W Kolb, V Lukash, M MacCallum, W McCrea, M Madsen, M Mejic, I Moss, J Narlikar, I Novikov, D Page, V Sahni, J Silk, R J Tayler and R Wald, and finally my thanks go to our long-suffering secretaries at Sussex who helped produce the final typescript.

REFERENCES

1. A. Linde, Rep. Prog. Phys., 47:925 (1984)
2. G.Gibbons, S.W. Hawking & S.T.C.Siklos, (Eds.) "The Very Early Universe", C.U.P., Cambridge (1983).
3. S.W.Hawking, Gen. Rel. & Gravn., 1: 393 (1971)
4. F.J.Tipler, Q. Jl. R. astron. Soc., 22:133 (1981).
5. G.F.R. Ellis & G.B.Brundrit, Q. Jl. R. astron. Soc., 20:37 (1979).
6. J.D.Barrow & F.J.Tipler, "The Anthropic Cosmological Principle", Oxford U.P., Oxford & N.Y. (1986).
7. F.J.Tipler, Phys. Rep., 137: 231 (1986).
8. A. Vilenkin, Phys. Lett., B117:25 (1982).
9. D.Raine & M. Heller,"The Science of Space-Time", Pachart, Arizona (1981).
10. W.Rindler, Mon. Not. R. astron. Soc., 116:662 (1956)
11. C.Hazard, R.G.McMahon & W. Sargent, Nature, 322:38 (1986).
12. T. Kaluza, Sber. preuss. Akad. Wiss. Phys. Math. Kl., 966 (1921).
13. O. Klein, Z. Physik., 37:895 (1926)
14. M. Green & J.Schwarz, Phys. Lett. B., 149: 117 (1984).
15. W. J. Marciano, Phys. Rev. Lett., 52: 489 (1984).
16. E.Kolb, M.Perry & T.P.Walker, Phys. Rev. D., 33:869 (1986).
17. J.D.Barrow, Phys. Rev. D, (in press) (1986).
18. A.Guth, Phys. Rev. D., 23:347 (1981).
19. J.D.Barrow, Can. J. Phys., 64:152 (1986).
20. E.Wampler, C.Gaskell, W.L.Burke & J.A.Baldwin, Astrophys. 276:403 (1984).
21. J.D.Barrow & S.P.Bhavsar, Q. Jl. R. astron. Soc., (in press) (1987)
22. J.D.Barrow, P.J.E.Peebles & D.W.Sciama, eds. "The Material Content of the Universe", The Royal Society, London (1986).
23. S.W. Hawking & G.F.R. Ellis, "The Large scale Structure of Space-Time", C.U.P., Cambridge, (1973).
24. J. Kristian & R.K. Sachs, Astrophys. J., 143: 379 (1966).
25. G. Daucourt, J. Phys. A., 16: 3507 (1983).
26. W.H.McCrea, Zeit. f. Ap., 18: 98 (1939).
27. G.F.R. Ellis, "General Relativity and Cosmology", Varenna Lectures in Physics, ed. R.K.Sachs, Acad. Press, London (1971).
28. A.Z.Petrov, "Einstein Spaces", Pergamon Press, Oxford (1969).
29. L.Landau & E.M. Lifshitz, "The Classical Theory of Fields" Pergamon, Oxford (1974).
30. S.T.C. Siklos, in "Relativistic Astrophysics and Cosmology", eds. E.Verdaguer & X.Fustero, World, Singapore (1984).
31. S.Weinberg, "Gravitation and Cosmology", Wiley, New York, (1972).
32. F.J.Tipler, Phys. Rev. D., 17:2521 (1978).
33. J.D.Barrow, Phys. Rep., 85:1 (1982).
34. J.D.Barrow & D.H.Sonoda, Phys. Rep., 139:1 (1986).
35. J.D.Barrow, in "Proceedings of the 1st International Conference on Phase Space", ed. A.Zachary, Plenum, NY, (1986).
36. A.Vilenkin, Phys. Rev. D., 24:2082 (1981).
37. A. Staruskiewicz, Acta Phys. Polon., 24:734 (1963).
38. G.F.R.Ellis, S.D.Nel, R.Maartens, W.R.Stoeger and

A.P.Whiteman, Phys. Rep., 124:315 (1985).
39. J.D.Barrow & A.Ottewill, J. Phys. A., 16: 2757 (1983).
40. J.Leray, Acta Math., 63:193 (1934).
41. E.M.Lifshitz & I. Khalatnikov, Adv. Phys., 12:185 (1963).
42. A.Spero & R.Baierlein, J. Math. Phys., 19:1324 (1978)
43. C.Will, "Theory and Experiment in Gravitational Physics", C.U.P., Cambridge (1983).
44. J.D.Barrow, Phil. Trans. Roy. Soc., A310:337 (1983).
45. J.D.Barrow, A.Burd & D.Lancaster, Class. & Quantum Grav., 3:551 (1986).
46. P.Landsberg & D.Evans, "Mathematical Cosmology", O.U.P., Oxford (1977).
47. J.Marsden, "Applications of Global Analysis in Mathematical Physics", Publish or Perish Inc., Boston,(1974).
48. V.Arnold, "Catastrophe Theory", Springer, NY (1985).
49. V.Arnold, S.F.Shandarin & Y.B.Zeldovich, Geophys. Astrophys. Fluid. Dyn., 20:111 (1982).
50. J.V.Narlikar & A.K.Kembhavi, Fund. Cosmic Phys., 6:1 (1980).
51. E.Milne & W.H.McCrea, Quart. J. Math. Oxford Ser., 5:73 (1934).
52. W.Bonnor, Mon. Not. R. astron. Soc., 117:104 (1957).
53. E.M.Lifshitz, J.Phys. USSR 10:116 (1946).
54. J.D.Barrow & F.J.Tipler, Phys. Reports 56:371 (1979).
55. J.D.Barrow & F.J.Tipler, Phys. Lett. A.
56. V.Belinskii, E.M.Lifshitz & I.Khalatnikov, Sov. Phys. Usp., 13:745 (1971).
57. J.D.Barrow, Sussex Preprint (1986).
58. I. Newton, "Philosophiae naturalis principia mathematica" (1713), transl. A. Motte (revised F.Cajori), Univ. California Press, Berkeley, (1946).
59. J.D.Barrow, unpublished (1984).
60. J.D.Barrow & F.J.Tipler, in preparation (1986).
61. C.J.S. Clarke, Proc. Roy. Soc. A., 314:417 (1970).
62. G.F.R.Ellis, Gen. Rel. Gravn., 2:7 (1971).
63. K.Sato, Mon. Not. R. astron. Soc., 195:467 (1981).
64. Y.B.Zeldovich & L.Grishchuk, Mon. Not. R. astr. Soc., 207:23P (1984).
65. W.Bonnor, Class & Quantum Gravity, 2:781 (1985).
66. C.Hellaby & K.Lake, Astrophys. J., 290:381 (1985), errata 300:000 (1986).
67. J.D.Barrow & F.J.Tipler, Mon. Not. R. astr. Soc., 216:395 (1985).
68. J.E.Marsden & F.J.Tipler, Phys. Reports 66:109 (1980).
69. M.Freedman, J. Diff. Geometry, 17:357 (1982).
70. S.K.Donaldson, J. Diff. Geometry 18:269 (1983).
71. R.Schoen & S.-T. Yau, Manuscript Math., 28:159 (1979).
72. J.D.Barrow, G.Galloway & F.J.Tipler, Mon. Not. R. astr. Soc., (in press) (1986).
73. R.C.Tolman & M.Ward, Phys. Rev. 32:835 (1932).
74. R.Kantowski & R.K.Sachs, J. Math. Phys., 7:443 (1966).
75. A.S.Kompanyeets & A.S. Chernov, Sov. Phys. JETP 20:1303 (1964).
76. C.B.Collins, J. Math. Phys., 18:2116 (1977).
77. L.Bianchi, Mem. Soc. It., 11:267 (1898), reprinted in Opere IX, ed. A.Maxia, Editzioni Crenonese, Rome (1952).
78. F.B.Estabrook, H.D.Wahlquist & C.G.Behr, J. Math. Phys., 9:497 (1968).
79. M.A.H. MacCallum, Cargese Lectures in Physics Vol 6,

ed. E.Schatzman, Gordon & Breach, NY (1973).
80. G.F.R. Ellis & M.A.H. MacCallum, Comm. Math. Phys., 12:108 (1969).
81. G.F.R.Ellis & M.A.H.MacCallum, Comm. Math. Phys., 19:31 (1970).
82. O.Bogoiavlenskii and S.P.Novikov, Russian Math. Surveys 31:31 (1971).
83. S.T.C.Siklos, Comm. Math. Phys., 58:255 (1978).
84. M.A.H.MacCallum, Lecture Notes in Physics 109; ed. M.Demianski, Springer, NY (1979).
85. E.Kasner, Am. J. Math., 43:217 (1921).
86. D.Kramer, E.Herlt, H.Stephani and M.A.H.MacCallum, "Exact Solutions of Einstein's Equations", C.U.P., Cambridge (1981).
87. D.Eardley, Comm. Math. Phys., 37:289 (1974).
88. J.D.Barrow & D.H.Sonoda, Gen. Rel. Gravn., 17:409 (1985).
89. A.Vilenkin, Phys. Rev. D 27:2848 (1983).
90. A.Linde, Lett. Nuovo Cim., 39:401 (1984).
91. J.Hartle & S.W.Hawking, Phys. Rev. D 28:2960 (1983).
92. J.V.Narlikar & T.Padmanabhan, Phys.Rep., 100:151 (1983).
93. S.W.Hawking, Nucl. Phys. B 239:257 (1984).
94. S.Faber & J.S.Gallagher, Ann. Rev. Astron. Astrophys., 17:135 (1979).
95. M.Davis & P.J.Peebles, Ann. Rev. Astron. Astrophys., 21:109 (1983).
96. A.Bean, G.Efstathiou, R.Ellis, B.Peterson & T.Shanks, Mon. Not. R. astron. Soc., 205:605 (1983).
97. K.Olive, D.Schramm, G.Steigman, M.Turner & J.Yang, Astrophys. J., 246:557 (1981).
98. P.J.Peebles, Astrophys. J., 284:439 (1984).
99. J.D.Barrow, R.Juszkiewicz & D.H.Sonoda, Nature, 305:397 (1983).
100. J.D.Barrow, R.Juszkiewicz & D.H.Sonoda, Mon. Not. R. astron. Soc., 213:917 (1985).
101. G.F.Smoot, M.V.Gorenstein & R.A.Muller, Phys. Rev. Lett., 39:898 (1979).
102. D.Fixsen, E.Cheng & D.Wilkinson, Phys.Rev. Lett., 50:620 (1983).
103. R.Fabbri, I.Guidi, F.Melchiorri & V.Natale, Phys. Rev. Lett., 44:1563 (1981).
104. P.M.Lubin, G.Epstein & G.F.Smoot, Phys. Rev. Lett., 50:616 (1983).
105. I.Strukov & D.P.Skulachev, Sov. Astron. Lett., 10:1 (1984).
106. S.W.Hawking, Mon. Not. R. astron. Soc., 142:129 (1969).
107. C.B.Collins & S.W.Hawking, Mon. Not. R. astron. Soc., 162:307 (1973).
108. A.Doroshkevich, V.Lukash & I.D.Novikov, Sov. Astron., 18:554 (1975).
109. K.Thorne, Astrophys. J., 148:51 (1967).
110. L.P.Grishchuk, A.Doroshkevich & I.D.Novikov, Sov. Phys., JETP 28:1210 (1969).
111. R.A.Matzner, Astrophys. J., 157:1085 (1969).
112. I.D.Novikov, Sov. Astron., 12:427 (1968).
113. S.Bajtlik, R.Juskiewicz, M.Prozynski & P.Amsterdamskii, Astrophys. J. 000:000 (1986).
114. V.Lukash & I.D.Novikov, Nature, 316:46 (1985).
115. J.D.Barrow, Mon. Not. R. astron. Soc., 175:359 (1976).
116. J.D.Barrow & J.Stein-Schabes, Phys. Lett. B 167:173 (1986)

117. P.J.Peebles, "Physical Cosmology", Princeton U.P.,NJ (1971).
118. A.Webster, Mon. Not. R. astron. Soc., 175:61 (1976)
119. R.S.Warwick, J.P.Pye & A.C.Fabian, Mon. Not. R. astron. Soc., 190:243 (1980).
120. E.Hubble, "The Realm of the Nebulae", Dover, NY (1985).
121. C.W. Misner, Nature, 214:30 (1967).
122. C.W.Misner, Astrophys. J., 151:431 (1968).
123. E.A.Milne, "Relativity, Gravitation and World Structure" O.U.P., Oxford (1935).
124. W.R.Stoeger, G.F.R.Ellis & C.Hellaby, Cape Town preprint (1986).
125. F.Hoyle & J.V.Narlikar, Proc. Roy. Soc. A., 273:1, (1963).
126. J.D.Barrow, Mon. Not. R. astron. Soc., 178:625 (1977).
127. M.J.Rees, Phys. Rev. Lett., 28:1969 (1972).
128. J.M.Stewart, Mon. Not. R. astron. Soc., 145:347 (1969).
129. A.G.Doroshkevich, Y.B.Zeldovich & I.D.Novikov, Sov. Phys. JETP, 26:408 (1968).
130. L.Parker, in "Asymptotic Structure of Space-Time", eds. F.P.Esposito & L.Witten, Plenum, NY, (1977).
131. Y.B.Zeldovich & A.Starobinskii, Sov. Phys. JETP 34: 1159 (1972).
132. C.W.Misner, Phys. Rev. Lett., 22:1071 (1969).
133. A.G.Doroshkevich & I.D. Novikov, Sov. Astron., 14:763 (1971).
134. J.D.Barrow & R.A.Matzner, Mon. Not. R.astron. Soc., 181:719 (1977).
135. C.B.Collins & S.W.Hawking, Astrophys. J., 180:317 (1973).
136. J.D.Barrow, Q. Jl. R. astron. Soc., 23:344 (1982).
137. V.N.Lukash, Sov. Phys. JETP, 40:792 (1975).
138. V.N.Lukash, Nuovo Cim. B, 35:268 (1975).
139. A.Taub, Ann. Math., 53:472 (1951).
140. G.Gibbons & S.W.Hawking, Phys. Rev. D, 15:2738 (1977).
141. S.W.Hawking & I.Moss, Phys. Lett. B, 110:35 (1982).
142. J.D.Barrow, in "The Very Early Universe", eds. G.Gibbons, S.W.Hawking & S.T.C.Siklos, C.U.P., Cambridge, (1983).
143. W.Boucher & G.Gibbons, in "The Very Early Universe" eds. G.Gibbons, S.W.Hawking & S.T.C.Siklos, C.U.P. Cambridge, (1983).
144. W.Boucher, in "Classical General Relativity", eds. W.Bonnor, J.Islam & M.A.H.MacCallum, C.U.P., Cambridge (1984).
145. A.A.Starobinskii, Sov. Phys. JETP, Lett., 37:66 (1983).
146. R.Wald, Phys. Rev. D, 28:2118 (1983).
147. I.Moss & V.Sahni, Phys. Lett. B, in press, (1986).
148. L.Jensen & J.Stein-Schabes, Fermi Lab preprint (1986).
149. M.S.Turner & L.Widrow, Fermi-Lab preprint (1986).
150. J.D.Barrow & J.Stein-Schabes, Phys. Lett.A 103:315 (1984).
151. J.D.Barrow & O.Gron, Sussex preprint (1986).
152. F.Lucchin & S.Matarrese, Phys. Lett.B 164:282 (1985).
153. J.D.Barrow, "The Deflationary Universe: an instability of the de Sitter universe", Sussex preprint (1986).
154. J.D.Barrow, "Deflationary universes with quadratic lagrangians", Sussex preprint (1986).
155. G.Murphy, Phys. Rev.D 8:4231 (1973).

156. Y.B.Zeldovich, Sov. Phys. JETP Lett. 12:307 (1980).
157. B.L.Hu, Phys. Lett.A 90:375 (1982).
158. R.Treciokas & G.F.R.Ellis, Comm.Math.Phys. 23:1 (1971).
159. A.Starobinskii, Phys. Lett. B 115:295 (1982).
160. S.W.Hawking, Phys. Lett.B 115:295 (1982).
161. L.F.Abbott & M.B.Wise, Nucl. Phys.B 244:541 (1984).
162. V. Belinskii & I.Khalatnikov, Sov.Phys. JETP Lett. 21:99(1975).
163. V.Belinskii & I.Khalatnikov, Sov.Phys. JETP 42:205 (1976).
164. J.D. Barrow & R.A. Matzner, Phys. Rev. D 21:336 (1980).

COSMIC STRINGS AND THE ORIGIN OF STRUCTURE IN THE UNIVERSE

D. M. Eardley

Institute for Theoretical Physics
University of California
Santa Barbara, California 93106 USA

1. INTRODUCTION

The observed universe is homogeneous and isotropic on the largest observable scales. The best evidence for this comes from observations of the cosmic background radiation (CBR). On smaller scales, a striking amount of structure can be seen — galaxies, clusters of galaxies, and the "large scale structure" in the form of possible filaments, bubbles, sheets or voids. The best proximate explanation for this structure is small amplitude perturbations in the early universe, which grew by gravitational instability into the observed large scale structure during the expansion of the universe. At some time in the future when we have a complete theory of the universe and its initial conditions — see James Hartle's lectures in this volume[1] for some promising ideas toward such a theory — both the overall homogeneity and the structure should be a calculable consequence of the theory. Until then, people have made partial progress toward understanding the genesis of structure on a homogeneous background, based on the laws of fundamental physics as currently known. At this time we have at least two possible fundamental mechanisms for generation of the conjectural initial perturbations, namely quantum fluctuations, or thermodynamic fluctuations of a particular sort. My purpose in these lectures is to review and outline the basic physical nature of these two mechanisms, leaving out the details. Both mechanisms are well reviewed in the literature, and the reader will be referred both to more comprehensive reviews and to the primary literature throughout these lectures.

1.1 Origin of Perturbations in the Universe

The first explanation is quantum fluctuations in the early universe. Such effects are often called "quantum particle production" and can be calculated, at least to some degree of approximation, by the methods of quantum field theory in a curved spacetime. In the standard Friedmann-Robertson-Walker (FRW) cosmological model, with an equation of state dominated by a relativistic gas, these calculations are inconclusive because the initial conditions for the quantum field are hard to pose unambiguously at the big bang singularity. This ambiguity is in turn a manifestation to the "horizon problem", the problem that the causal horizon was very small (in comoving coordinates) at early times in these models, so that we cannot understand how long range processes became synchronized. For quantum field theory, the particular manifestation is that the initial vacuum state is ambiguous in any given mode, because at early enough times the spatial wavelength of that mode is far outside the horizon. Cosmological models which contain a sufficently long inlationary era[2,3,4,5] go a long way toward resolving this ambiguity. In such models, for any given mode, the wavelength will be well inside the horizon at sufficiently early times during the inflationary era, if the inflationary era lasts long enough, so that the initial vacuum is well defined mode-by-mode. To emphasize the simplicity and generality of this point, I present in Section 2 a derivation of quantum particle creation for a free scalar field in an inflationary universe.

The second explanation is that initial fluctuations are caused by thermodynamic fluctuations at a phase transition in the matter fields in some epoch of the early universe. As long known, ordinary short wavelength fluctuations such as thermal fluctuations in energy density or particle number away from a homogeneous state cannot produce observed long wavelength perturbations. Density fluctuations are stronger at a phase transition, but still not strong enough on long wavelenths. Thus one turns to topological defects, a kind of thermodynamic fluctuation created in certain phase transitions, which can have stronger long wavelength manifestations. One finds that topological defects can work cosmologically. A case-by-case examination of kinds of topological defects shows that only "cosmic strings", defects which are 1-dimensional in space or 2-dimensional in spacetime, lead to structure which is neither so weak as to be unobservable now, nor so strong as to wreck the overall homogeneity and other overall properties of the observed universe. This explanation is reviewed in Sects. 3–5.

2. QUANTUM PARTICLE CREATION IN AN INFLATIONARY UNIVERSE

During an inflationary era[2,3,4,5], the expansion factor R of the universe increases exponentially,

$$R \propto \exp(Ht) \tag{2.1}$$

where

$$H = \sqrt{\frac{\Lambda}{3}} \sim \frac{R_{Planck}}{R_{GUT}^2} \tag{2.2}$$

is the Hubble constant during inflation, fixed by parameters of fundamental physics, namely the Planck length $R_{Planck} = (G\hbar)^{1/2} \sim 10^{-33} cm$ and the length scale R_{GUT} connected with the Grand Unified Theory (GUT Theory), the symmetry-breaking of which causes a phase transition. Typically, $R_{GUT} \sim (10^2 \text{ to } 10^4) R_{Planck}$. The phase transition gets hung up in a metastable state, which leads to vacuum stress and inflation. The main difficulty with an inflation era is in getting out of it. The causally disconnected regions must be resynchronized by whatever mechanism ends inflation, that is, the spacetime symmetry $SO(4,1)$ of the de Sitter universe must be broken back down to the lesser symmetry $SO(4)$ of the Friedman–Robertson–Walker (FRW) model. * For instance, in the "new inflationary model", the scalar field ϕ remembers the original $SO(4)$ symmetry throughout the inflationary era, during its "rolling phase", and resynchronizes the universe at the final phase transition. In this model calcuation I shall simply assume a synchronized end of inflation, without examining the mechanism for it.

2.1 A Simple Model

$SO(4)$ is the isometry of the closed ($k = +1$) FRW model. If the universe is spatially flat ($k = 0$) then the group is E_3, if open ($k = -1$), then the group is $SO(3,1)$.

Let us sketch a simple example of quantum fluctuations in an inflationary universe model. The spacetime metric is taken to be

$$ds^2 = -dt^2 + R^2(t)\,|\mathbf{x}|^2 \tag{2.3a}$$

$$= R^2(\tau)\left(-d\tau^2 + |d\mathbf{x}|^2\right) \tag{2.3b}$$

where

$$R(t) = \begin{cases} \frac{1}{\sqrt{2H}} e^{-Ht-1/2}, & t \leq \frac{1}{2H} \text{ (inflation);} \\ t^{1/2}, & t \geq \frac{1}{2H} \text{ (radiation);} \end{cases} \tag{2.4a}$$

or, equivalently in terms of conformal time τ,

$$R(\tau) = \begin{cases} -\frac{1}{H\tau}, & \tau \leq -\sqrt{\frac{2}{H}} \text{ (inflation);} \\ \frac{1}{2}\tau + \sqrt{\frac{2}{H}}, & \tau \geq -\sqrt{\frac{2}{H}} \text{ (radiation).} \end{cases} \tag{2.4b}$$

Here I have assumed a spatially flat ($k = 0$) FRW model with inflation (exponential expansion of R(t)) before some time $t = 1/2H$ and a radiation era (equation of state of a relativistic gas) after that time. The matching conditions at $t = 1/2H$ are continuity of R and dR/dt — these matching conditions are necessary avoid a spurious burst of particle creation right at $t = 1/2H$. Whether or not there was a pre-inflationary era does not matter for this model.

Now introduce a test quantum scalar field ϕ obeying the minimally coupled field equation $\Box\phi = 0$ on this background metric. (For some standard techniques of quantum field theory in curved spacetime, which will be used in the ensuing calcuation, see, *e.g.*, , Ref. 6.) Separate ϕ into modes $\phi_{\mathbf{k}}(t)$ according to the spatial translation invariance,

$$\phi(t, \mathbf{x}) = \sum_{\mathbf{k}} \phi_{\mathbf{k}}(t) e^{i\mathbf{k}\cdot\mathbf{x}}, \tag{2.5}$$

where $\mathbf{k}$ is a spatial wave vector. Now we fix the initial vacuum state on a mode-by-mode basis. Demand that, for each mode, there exist no particles at early enough times, $t \to -\infty$. This simple and appealing prescription is often implicitly assumed without further justification, but it is worth pointing out that it agrees with the "de Sitter invariant vacuum state" and with the "wavefunction of the universe[1]". This assumption is robust against changes in initial conditions in that the results would be essentially the same if, as often assumed, a radiation era preceded inflation, and the quantum field was in some sort of local thermal state in that early radiation era, as long as inflation lasts long enough.

Adopting the Heisenberg picture for the calculation (the Schroedinger picture would serve equally well), the mode $\phi_{\mathbf{k}}(t)$ obeys the equation of motion

$$\frac{d^2}{d\tau^2}\psi_{\mathbf{k}}(\tau) + \left(-V(\tau) + k^2\right)\psi_{\mathbf{k}}(\tau) = 0 \tag{2.6}$$

where

$$\psi_{\mathbf{k}} \equiv \frac{\phi_{\mathbf{k}}}{R} \tag{2.7}$$

and

$$V(\tau) \equiv \frac{1}{R}\frac{d^2R}{d\tau^2} = \frac{1}{2}\frac{dR^2}{dt^2} = \begin{cases} \frac{2}{\tau^2}, & \tau \leq -\sqrt{\frac{2}{H}}; \\ 0, & \tau \geq -\sqrt{\frac{2}{H}}. \end{cases} \tag{2.8}$$

Thus, the quantum operator $\psi_{\mathbf{k}}(\tau)$ obeys an equation of motion identical to that for a one-dimensional potential scattering problem with potential $V(\tau)$. The solution can be obtained in terms of spherical Hankel functions $h_1^{(\pm)}(k\tau)$ (during inflation — here $k \equiv |\mathbf{k}|$) and in terms of monochromatic waves $exp(\pm ik\tau)$ (during the subsequent radiation era). Matching $\psi_{\mathbf{k}}(\tau)$ and its derivative at the point $\tau = -1/2H$ gives a global solution. Matching solutions for c-number wave functions shows that a pure positive frequency solution $h_1^{(+)}$ at early times corresponds to a mixture of positive and negative frequencies at late times, which means that particle creation is going on:

$$h_1^{(+)}(k\tau) \equiv \left(1 + \frac{i}{k\tau}\right) e^{ik\tau} \Leftrightarrow \alpha e^{ik\tau} + \beta e^{-ik\tau}. \tag{2.9}$$

If the mode $\mathbf{k}$ is in the in-vacuum at early times, the number of particles in the mode at late times is given by the square of the coefficient of mixing of positive into negative frequencies. For long wavelength modes $k \ll 1$, the created particle number is approximately

$$\begin{aligned} n_{\mathbf{k}} &= \langle 0_{in} \mid N_{out} \mid 0_{in} \rangle = |\beta|^2 \\ &\approx \frac{H^2}{4k^4} \end{aligned} \tag{2.10}$$

or, in useful units,

$$n_{\mathbf{p}} \sim \frac{H^2 R_{Planck}^2}{\mathbf{p}^4 R^4} \tag{2.11}$$

giving rise to an energy density of created particles $\rho = \int p n_{\mathbf{p}}\, d^3\mathbf{p}$

$$\rho \sim \left(\frac{R_{Planck}}{R_{GUT}}\right)^4 \cdot \log \frac{t}{R_{GUT}} \cdot \frac{1}{R^4}. \tag{2.12}$$

Here $\mathbf{p}$ is physical 3-momemtum of a particle as measured by a cosmogical observer and $H \sim R_{Planck}/R_{GUT}^2$ is as in Eq. (2.2); these equations should be interpreted as measurements by a cosmological observer at late times, well within her Hubble radius.

Equation (2.11) is the celebrated Zel'dovich spectrum for the scalar field, with the same energy in each octave of total wave number k or total momentum p. The total energy density (2.12) behaves almost like a radiation gas, $\rho_{radiation} \sim 1/R^4$ – but the spectrum is very different from a thermal distribution, with much more power on longer scales. The result for a free scalar field ϕ is itself not of much cosmological interest, because there is no evidence for such a field, and no known reason for it to exist. However, the same ideas apply to two much more interesting, albeit more complicated, fields.

2.2 More Realistic Models

Rubakov, Sazhin and Veryaskin[7] calculated the spectrum of gravitational waves created during inflation. The results are similar, because the tensor (wave) modes of the perturbed gravitational field in an FRW cosmology obey the very same wave equation $\Box\phi = 0$.[8] The main consequence of this work is an upper limit $R_{Planck}/R_{GUT} \lesssim 10^{-2}$, constrained by anisotropies that would be induced in the CBR by such gravitational waves, which are not observed.

Starobinsky[9] calculated the spectrum of gravitational waves and adiabatic perturbations in a model universe with an initial DeSitter era.[10] Again, a Zel'dovich spectrum results.

Several authors (see Ref. 4 for review) calculated perturbations induced by quantum fluctuations in the Higgs potential ϕ in the "new inflationary model[3,4]" of the early universe. In this model, ϕ is coupled to itself and to the other matter fields in a special way, so that its evolution controls the "reheating" of the universe which ends the inflationary era. Quantum fluctuations in ϕ appear just as in the model above, but the dynamical effects of these fluctuations is greater due to the special coupling of ϕ in the new inflationary model. Consequently the spectrum of fluctuations is still a Zel'dovich spectrum, but the amplitude is different and typically greater; in fact for the simplest ("Coleman-Weinberg[4]" choice of the scalar self-coupling, the relative amplitudes of fluctuation would be greater than unity in the present universe, quite contradicting observation.

3. TOPOLOGICAL DEFECTS IN FIELD THEORIES

Topological defects arise in phase transitions with an order parameter that has degenerate ground states. Such topological defects are stable — they cannot decay without changing the field topology, which usually would cost an infinite energy and is forbidden both classically and quantum mechanically — and therefore they are "solitons." For review, see Refs. 11, 12. In most applications to field theory, the order parameter is a scalar field ϕ. As the early universe expands and cools, the field theory of elementary particle physics may undergo one or several phase transitions, in which ϕ takes on a vacuum expectation value in its ground state.

3.1 Cosmological Constant

In the simplest case, ϕ has only one possible vacuum expectation value in the ground state, and no topological defects exist. However, profound cosmological consequences can occur. If the energy density ρ_0 of ϕ is nonzero in the ground state, then by Lorentz invariance the whole stress tensor is a pure trace, $T_{\mu\nu} = \rho_0 g_{\mu\nu}$. This is called a vacuum stress; or, in general relativity, this situation has long been known as a nonzero value of the cosmological constant Λ. The magnitude of ρ_0 should be set by elementary particle physics $\rho_0 \sim \hbar/R^4_{GUT}$, which is huge, and violently contradicted by observation — this disaster is the famous "cosmological constant problem." Such a cosmological constant would cause the universe to expand exponentally with a huge Hubble constant. For some currently unknown reason, the cosmological constant is zero to an enormous precision ($\sim 10^{-120}$) in the present ground state of the matter fields. Despite our lack of understanding of this issue, the prospect of an era of nonzero cosmological constant during the early universe is a tempting one. Such an era has been suggested by Starobinksy[10] as due to quantum gravitational effects, and in an influential paper by Guth[2] as due to a metastable state preceding a phase transition in matter fields. Guth proposed several compelling reasons why one might want to believe in such an "inflationary cosmology". However, the inflationary era had to end because the universe became the present one, perhaps in a transition from a metastable state with a nonzero cosmological constant to a stable state with zero cosmological constant. Inflation also occurs in certain model universes selected by the Hawking-Hartle wave function of the universe.[1] As discussed above in Sect. 2, quantum fluctuations during inflation are a natural mechanism for producing the observed fluctuations of the present universe.[9,4,7]

3.2 Domain Walls

In the next simplest case, ϕ has a finite number of discrete ground states (0-dimensional manifold of ground states) available to it after the phase transition

— these points are the absolute minima of the energy as a function of ϕ. In the cosmological setting, ϕ must be uncorrelated with itself outside the horizon because of causality. Thus, the universe will break up into different domains in which ϕ takes different ground state values. The interfaces between such regions are called "domain walls." In the neighborhood of a domain wall, ϕ is forced permanently away from the ground state manifold because it must undergo a transition from one ground state to another. As the universe expands, these domain walls will remain thin, with a characteristic thickness given by some scale of elementary particle physics, say R_{GUT}.

A domain wall has energy density and stress associated with it, due to the fact that ϕ takes values outside the ground state manifold, and also due to gradients in ϕ and other fields such as gauge fields — again set by the parameters of elementary particle physics. From the macroscopic point of view, a domain wall is characterized by a mass per unit area σ, which is a constant set by elementary particle physics, and by its shape as a time-dependent $(d-1=2)$-surface in 3-space, or, equivalently, as a $(d=3)$-dimensional world hypersurface in 4-dimensional spacetime. The world hypersurface has one timelike dimension and two spacelike ones. The world sheet also has a constant surface tension, which is numerically equal to σ (in units where $c=1$). Stretching and expanding a domain wall does not change σ — the increase in total mass due to the increase in surface area is balanced by the work which must be done against the surface tension. Physically a domain wall behaves like a stretched membrane — for instance it can support waves, which travel at the speed of light. The macroscopic effective action of a domain wall, which determines its equations of motion, is just $-\sigma \int d^3V$, where V is 3-dimensional volume in the world hypersurface.

Table 1 summarizes the properties of various topological defects.

3.3 Cosmic Strings

Turn now to the case in which the ground state manifold is 1-dimensional, say a circle S^1 in the internal manifold of all possible values of ϕ. For instance, consider a complex-valued field ϕ with a self-coupling consisting of a potential $V(|\phi|)$ with a unique minimum at $|\phi| = b = \text{const} > 0$. The ground states — the absolute minima of energy — will form a circle given by $|\phi| = b$, and are parametrized by the phase of ϕ.

Since the ground state manifold is 1-dimensional instead of 0-dimensional, the dimensionality in space of topological defects will be one dimension less, or 1. The topological defects are therefore $(d-1=1)$-dimensional "cosmic strings" in space, or, equivalently, $(d=2)$-dimensional world sheets in spacetime. The situation can be visualized as follows: Trace the value of ϕ along any closed loop in space. Assume modulus b is fixed; but the phase varies. If the phase has a nonzero net rotation $2\pi n$ (n a nonzero integer) in going around the loop, then ϕ must, for reasons of continuity, vanish somewhere within any 2-surface which spans the loop. Imagine deforming such a 2-surface which spans the loop. Each 2-surface has a point where ϕ vanishes, and, under deformation, these points will trace out on a 1-dimensional locus in space where ϕ vanishes — a cosmic string — and n will be nonzero if and only if our loop links such a string. Along the string, ϕ is forced out of the ground state manifold, giving rise to an energy density in the neighborhood of the string. In mathematical language, n belongs to $\pi_1(S^1)$.[13,14,15]

A cosmic string is characterized by a mass per unit length μ and its time-dependent shape as a curve is space. It also has a tension along it, which is numerically equal to μ. A cosmic string can support waves, which move at the speed of light; in fact the motion of a cosmic string can be described as a motion transverse to itself at the speed of light. (Boosts along the direction of a string leave the string invariant,

Table 1. Matter field configurations of cosmological significance. The spacetime dimension d describes the locus of strong fields in spacetime — an instanton is effectively a point-event, a monopole a 1-dimensional world line, a cosmic string a 2-dimensional world sheet, and so on. The spatial dimension $d-1$ of the field configuration is the spacetime dimension less one — a monopole is effectively a point at any instant, a string is a 1-dimensional curve, and so on.

d	Name	Mass Parameter		Cosmological Effect
0	Instanton	(action)	$S \sim \hbar$	None (ρ_B/ρ_γ?)
1	Monopole	mass	$m \sim \frac{\hbar}{R_{GUT}}$	Disaster (if not inflated away)
2	Cosmic String	mass/length	$\mu \sim \frac{\hbar}{R_{GUT}^2}$	Fluctuations and gravity waves in a Zel'dovich spectrum
3	Domain Wall	mass/area	$\sigma \sim \frac{\hbar}{R_{GUT}{}^3}$	Disaster (if not inflated away)
4	Vacuum Stress	density	$\rho_0 \sim \frac{\hbar G}{R_{GUT}^4}$	Disaster (if not set to 0 eventually); fluctuations and gravity waves in a Zel'dovich spectrum

to the extent that it is straight). The macroscopic effective action of a cosmic string, which determines its equations of motion, is $-\sigma \int d^2S$ where S is 2-surface area in the world sheet.[11,12]

The cosmic strings have no direct relation to the "superstrings" which are currently of much interest in elementary particle physics. The action of superstring theory is related to the effective action just mentioned; however, in superstring theory the main goal is to quantize that action, whereas cosmic strings are well described by classical physics.

3.4 Monopoles

Moving up another dimension in the ground state manifold is now straightforward. For dimension 2, (*e.g.*, the ground state manifold is a 2-sphere S^2), the region of space occupied by the defect is just a point, and is measured mathematically by an element $\pi_2(S^2) = \mathbf{Z}$, *i.e.*, by an integer n, for 2-spheres S^2 surrounding that point. Such a defect is known as a "monopole," and in grand unified theories they always occur as magnetic monopoles of the electromagnetic field. In spacetime the defect is a $d = 1$-dimensional world line. Macroscopically such defects are just point particles characterized by a constant mass m, which depends only on the parameters of the elementary particle theory. The action of a monopole is $-m \int ds$ where the 1-dimension integral runs over the world line.

3.5 Instantons

Moving up still one more dimension in dimension of the ground state manifold, we arrive at a somewhat different situation. Now the defect is just a point in spacetime — it is a pointlike object which exists only for an instant, appearing from nowhere and disappearing into nowhere. One might think that such an occurrence would be forbidden by an infinite energy barrier, but it turns out in gauge theories that such transitions can and do occur as finite action (finite energy barrier) quantum fluctuations in the gauge potentials, leaving behind no sign of their occurrence in the measurable gauge fields. They are known as "instantons" (for review and references see e.g. Ref. 16). An instanton does not have a mass (a mass would have to be conserved) but it does have an action $S \sim \hbar$. The existence of instantons is of profound importance for gauge fields theories of the elementary particles — for instance, causing a breakdown of CP invariance — but their cosmological effects seem to be subtle, *e.g.*, an influence on baryon creation or destruction at relatively low temperatures.[5]

The cosmological effects of the existence of these various topological defects is summarized in Table 1. How do these effects come about?

4. COSMOLOGICAL EVOLUTION OF TOPOLOGICAL DEFECTS

In most gauge theories of the elementary particles, the full gauge symmetry is unbroken at high enough temperature. As the temperature decreases in a universe with a hot big bang, the full gauge symmetry will be lost in one or more phase transitions. At each phase transition, topological defects of the various kinds may be created. In different regions separated by more than the Hubble radius, which are causally uncorrelated, the order parameter ϕ will choose different random values in the ground state manifold. Therefore ~ 1 topological defect per Hubble volume will be created at any phase transition where they are created at all. The exact calculation of the defect density is hard, but this estimate will suffice.

4.1 Bounds on Evolution of Defect Density

The subsequent evolution of the universe will then be determined by the competition between two effects. First, the mass-energy density of the defects may come to dominate the universe, which is generally bad because the resulting universe, however interesting, does not resemble our present one. Second, the defects will to a greater or lesser extent "relax" or "anneal" — this means that defects will move around and interact with each other, and tend to annihilate as defects and anti-defects find one another. Relaxation is limited by causality — defects cannot move faster than light. If relaxation is extremely efficient, then the mass-energy of the defects will quickly change back into hot gas, and the evolution of the universe will be affected but little by the defects. Although the competition of these two effects is hard to calculate exactly, it is easy to put lower and upper bounds on the evolution of the mass-energy density of the defects as follows.

The lower limit comes from assuming that the defects relax as fast as possible, limited only by causality. Therefore, at all times after the phase transition in which they are created, there will be $\sim$ 1 defect per Hubble volume. The upper limit comes from assuming that the defects cannot relax appreciably at all — in this case the defect density in *comoving* volume is constant, *i.e.*, defects are conserved. The causality limit is constant and the no-relaxation limit on the mass-energy density of defects are shown in Table 2 for monopoles, cosmic strings, and domain walls.

Table 2. Cosmological evolution of topological defects in FriedmannRobertson-Walker universes. The spatial dimension of a defect is $d-1$. Shown also is vacuum stress, a matter field configuration which is not a topological defect. Here R is the expansion factor, a function of cosmic time, and R_H is the horizon size or Hubble radius of the universe as a function of cosmic time. These quantities are related by

$$R_H \propto t \propto R^2, \quad \text{Radiation Era;}$$
$$R_H \propto t \propto R^{\frac{3}{2}}, \quad \text{Matter Era.}$$

The causality limit is one defect per Hubble volume R_H^3; the no-relaxation limit corresponds to conserved defect density, or a constant number of defects per comoving volume R^3. Note that the horizon size R_H expands faster than the expansion factor R, so that the inequalities below become looser as the universe expands. The inequalities marked with $*$, as $\overset{*}{\lesssim}$, would be approximate equalities in nature.

d	Causality Limit	Average Density	No-Relaxation Limit
1	$\frac{Gm}{R_H^3}$	$\lesssim G\rho_{monopole}$	$\overset{*}{\lesssim} \frac{Gm}{R^3}$
2	$\frac{G\mu}{R_H^2}$	$\overset{*}{\lesssim} G\rho_{string}$	$\lesssim \frac{G\mu}{R^2}$
3	$\frac{G\sigma}{R_H}$	$\overset{*}{\lesssim} G\rho_{domain\ wall}$	$\lesssim \frac{G\sigma}{R}$
4	$G\rho_0$	$\sim G\rho_{vacuum\ stress}$	$\sim G\rho_0$
Compare to evolution of total matter density:			
Radiation Era	$\frac{1}{R_H^2}$	$\sim G\rho_{radiation}$	$\sim \frac{R_{Planck}^2}{R^4}$
Matter Era	$\frac{1}{R_H^2}$	$\sim G\rho_{matter}$	$\sim \frac{\text{const}}{R^3}$

Which limit is realized for each kind of defect? That is, will defects typically meet one another in their wanderings so that they can annihilate? This depends primarily on the spacetime dimension d of the defect. Generically, in 4-space, two objects of dimension d_1 and d_2 will intersect if

$$d_1 + d_2 \geq 4 = \text{dimension of spacetime.} \tag{4.1}$$

Therefore the condition for defects of spacetime dimension d to meet one another as they wander about in spacetime is

$$2d \geq 4 \quad \Rightarrow \quad d \geq 2$$

so that monopoles do not generically meet each other, while cosmic strings and domain walls do. This simple argument implies that *monopoles will not relax, while cosmic strings and domain walls will.* These conclusions are shown in Table 2, which denotes the actually realized limits by the sign $\overset{*}{\lesssim}$. Finally, and as already noted in

Section 3, vacuum stress eventually dominates the universe unless it somehow decays, since it does not decrease at all.

For simplicity I will henceforth confine attention to radiation dominated universes, which ours was until relatively recently. Introduction of a matter dominated era at late times does not alter the essential nature of the conclusions.

4.2 Evolution of Monopoles and Domain Walls

The main conclusion from Table 2 is that both monopoles and domain walls will eventually dominate the universe, but for different reasons. Monopoles cannot relax, and the no-relaxation limit of monopole mass density

$$\rho_{monopole} \sim \frac{m}{R^3} \tag{4.2}$$

will eventually dominate the energy density of radiation, which drops faster as the universe expands, $\rho_{radiation} \propto R^{-4}$.

On the other hand, domain walls are expected to relax as fast as possible. But even so they will eventually dominate the universe, because even their causality limit

$$\rho_{domain\ wall} \sim \frac{\sigma}{R_H} \propto \frac{1}{R^2} \tag{4.3}$$

is too big, compared to $\rho_{radiation} \propto R^{-4}$.

Both monopoles and domain walls must be banned from the present universe. One can construct models obeying this restriction either by using a model of elementary particle physics which does not generate them in the first place, or by relegating them to a pre-inflationary era. Further estimates along the same lines show that, during an inflationary era, the density of topological defects decreases rapidly. A long enough inflationary era will drive their density so low that they are not a cosmological problem[2]. In fact, it is very hard to construct models of the observed elementary particles that do not have magnetic monopole defects, so that inflation seems to be necessary to get rid of them.

4.3 Evolution of Cosmic Strings

In constrast, cosmic strings seem to be just right for cosmology.[12,17,18,19,20] Assuming they do relax at the causality limit, their mass-energy density will maintain a constant ratio with that of radiation,

$$\rho_{cosmic\ string} \sim \frac{G\mu}{R_H^2} \propto \frac{1}{R^4} \propto G\mu \cdot \rho_{radiation} \tag{4.4}$$

so that they will not dominate the universe, but they will remain of enduring importance. The dimensionless parameter $G\mu$ which appears in Eq. (4.4) is of principal importance for cosmic string theory; it is fixed by fundamental physics. As will be discussed below, a reasonable value to keep in mind is

$$G\mu \sim 10^{-6}.$$

A more detailed account of the evolution of the cosmic string complex than that just given is necessary, however.

5. MOTION AND EVOLUTION OF COSMIC STRINGS

Cosmic strings move around at speeds $\sim c$. When string happen to cross each other, it is generally thought that they can "interconnect", so that the strings are in effect broken, ends exchanged, and mended at the crossing point. Interconnection may not happen at all crossings, but it is thought to happen at a good fraction of them, which is what is important for cosmology. Interconnection is important because it allows the cosmic string complex to relax at nearly the causality limit — with one significant exception. If a string happens to cross itself, interconnection will create a pinched-off closed loop.[17,18,19,20]

5.1 Closed Loops

A closed loop may in turn cross itself repeatedly and quickly cut itself up into smaller and smaller loops. However, as first suggested by Turok, an appreciable fraction of long-lived daughter loops are created in this process, and rather quickly a closed loop will turn itself into long-lived daughters.[18,19] The daughter loops are not stationary strings, rather they represent closed loops with strong waves running around in both directions. If these daughters lived forever, their mass-energy density would come to dominate the universe, because they would represent a fraction of the total string complex which was not relaxing at the causality limit(*cf.* Table 2). In fact, each closed loop would contribute to the total mass energy as a conserved particle, so that the mean energy density in closed loops would decrease as

$$\rho_{closed\ loops} \propto \frac{\text{const}}{R^3} \tag{5.1}$$

during cosmic evolution. This would be disastrous. However, closed loops decay by emission of gravitational radiation. A simple estimate based on the quadrupole formula will suffice, even though the quadrupole formula is not very accurate for these rather relativistic systems. For a closed loop of size r, the total mass-energy is $E \sim \mu r c^2$, and the luminosity in gravitational waves is given by the familiar formula involving the quadrupole moment Q:

$$\begin{aligned} \dot{E} &\sim c^{-5} G \left| \dddot{Q} \right|^2 \\ &\sim c^{-5} G \left| \frac{\mu r \cdot r^2}{(r/c)^3} \right| \\ &\sim c G \mu^2. \end{aligned} \tag{5.2}$$

The ratio gives the lifetime to gravitational wave decay,

$$T_{GW} \sim \frac{E}{\dot{E}} \sim \left(G\mu/c^2\right)^{-1} \cdot (r/c) \tag{5.3}$$

or $T_{GW} \sim (G\mu)^{-1} r$ in units where $c = 1$. Therefore, a long lived daughter loop of size $r \sim R_H$ created at time t lives until $(G\mu)^{-1} t \sim 10^6 t$, that is, until the universe had expanded by a factor $(G\mu)^{-1/2}$ in the radiation era since its formation. Taking into account Eqs. (4.4,5.1), the net energy density in such loops is at any time during the radiation era

$$\rho_{closed\ loops} \sim (G\mu)^{\frac{1}{2}} \cdot \rho_{radiation} \tag{5.4}$$

which replaces the incomplete Eq. (4.4). Thus, in cosmic string cosmology, the main contribution to the energy density of strings at any one time comes from long-lived daughter loops created when the universe was much smaller.

5.2 Gravitational Waves from Closed Loops

The gravitational radiation from decaying closed loops is also nonnegligible.[21,22,23,24,25] The total energy density ρ_{GW} in gravitational waves is evidently about the same as that in closed loops, since the closed loops turn into gravitational waves:

$$\rho_{GW} \sim \rho_{closed\ loops} \sim (G\mu)^{\frac{1}{2}} \cdot \rho_{radiation}. \tag{5.5}$$

and more careful estimates bear this out. In fact, observational upper limits on long period gravitational waves may give the best limits for the existence of cosmic strings. Such limits have been obtained by observation of timing noise in stable radio pulsars.

5.3 Galaxy Formation by Cosmic Strings

If cosmic strings exist, the seeds for galaxy formation were present in the form of long-lived closed loops of cosmic strings. Much work has been done on this mechanism for galaxy formation, and it would be impossible to cover it adequately here. The reader can read Refs. 26, 27, 28, 29 for earlier work, and Refs. 30, 31, 32 for a more quantitative treatment. A major conclusion is that the formation of galaxies and clusters of galaxies in the present universe can be understood if $(G\mu) \sim 10^{-6}$, but values much less or much greater are ruled out.

5.4 Fluctuations in the Cosmic Background Radiation Due to Cosmic Strings

Another way to detect or bound cosmic strings is through their influence on the CBR.[33,34,35] It seems that the present observational limits on anisotropies in the CBR are compatible with a value $(G\mu) \sim 10^{-6}$, but that the effects should be observable at a level of sensitivity not much below the present observational limits.

5.5 Light Bending Due to Cosmic Strings

Finally, cosmic strings can act as gravitational lenses. In the simplest case, a straight stationary string, the bending angle is

$$\theta \approx 8\pi G\mu \sim 10^{-5}$$

or a few arc seconds. The main differences from the more familiar kind of gravitational lens is, first, that the string is a linear feature and might be expected to extend across the sky, or in a closed loop; and, second, that a string does not magnify its images, and therefore the two images of a single point source should be of equal brightness.

Acknowledgements

I am grateful for J. Traschen and R. Gregory for comments on the gravitational effects of cosmic strings. I am especially grateful to R. Brandenberger for several conversations and many references to the literature. This research was supported in part by the National Science Foundation under Grant Nos. PHY85-06686 and PHY82-17853, supplemented by funds from the National Aeronautics and Space Administration, at the University of California at Santa Barbara.

REFERENCES

1. J. Hartle, this volume (1987).
2. A. Guth, *Phys. Rev. D* **23**:347 (1981).
3. A. Linde, *Phys. Lett.* **108B**:389 (1982).
4. R. Brandenberger, *Rev. Mod. Phys.* **57**:1 (1985).
5. E. Kolb, this volume (1987).
6. N.D. Birrell & P.C.W. Davies, Quantum Fields in Curved Space, Cambridge University Press, Cambridge (1982).
7. V. Rubakov, M. Sazhin & A. Veryaskin, *Phys. Lett.* :115B (189,) 1982.
8. See, *e.g.*, E. Lifshitz and I. Khalatnikov, *Advan. Phys.* **12**:185 (1963).
9. A. Starobinsky, *Pis'ma Astron. J.* **9**:579 (1983) (English tranlation in: *Sov. Astron. Lett.* **9**:302 (1984)).
10. A. Starobinsky, *Phys. Lett.* **91B**:99 (1980).
11. A. Vilenkin, *Phys. Rep.* **121**:263 (1985).
12. T.W.B. Kibble, *Phys. Rep.* **67**:183 (1980).
13. T.W.B. Kibble, *J. Phys.* **A9**:1387 (1976).
14. D. Olive & N. Turok, *Phys. Lett.* **117B**:193 (1982).
15. G. Lazarides, Q. Shafi & T. Walsh, *Nucl. Phys.* B **195**:157 (1982).
16. R. Jackiw, *Rev. Mod. Phys.* **52**:661 (1980).
17. T.W.B. Kibble, *Nucl. Phys.* B **252**:227 (1985).
18. T. Vachaspati & A. Vilenkin, *Phys. Rev. D* **30**:2036 (1984).
19. A. Albrecht & N. Turok, *Phys. Rev. Lett.* **54**:1868 (1985).
20. N. Turok & P. Bhattacharjee, *Phys. Rev. D* **29**:1557 (1984).
21. A. Vilenkin, *Phys. Rev.* **D23**:852 (1981); *Phys. Lett.* **107B**:47 (1982).
22. T. Vachaspati & A. Vilenkin, *Phys. Rev. D* **31**:3052 (1985).
23. N. Turok, *Nucl. Phys.* B **242**:520 (1984).
24. C. Hogan & M. Rees, *Nature* **311**:109 (1984).
25. R. Brandenberger, A. Albrecht & N. Turok, *Nucl. Phys.* B **277**:605 (1986).
26. A. Vilenkin, *Phys. Rev. Lett.* **46**:1169 (1981); *ibid.* p. 1496.
27. A. Vilenkin, *Phys. Rev. D* **24**:2082 (1981).
28. Ya. B. Zel'dovich, *Monthly Notices Roy. Astron. Soc. (London)* **192**:663 (1980).
29. N. Turok, *Nucl. Phys.* B **242**:520 (1984).
30. N. Turok & R. Brandenberger, *Phys. Rev. D* **33**:2175 (1986).
31. H. Sato, Kyoto preprint (1986).
32. N. Turok, *Phys. Rev. Lett.* **55**:1801 (1985)
33. R. Brandenberger & N. Turok, *Phys. Rev. D* **33**:2182 (1986).
34. J. Traschen, N. Turok & R. Brandenberger, *Phys. Rev. D* **34**:919 (1986).
35. N. Kaiser & A. Stebbins, *Nature* **310**:391 (1984).

COSMOLOGICAL PHASE TRANSITIONS

Edward W. Kolb

Theoretical Astrophysics
Fermi National Accelerator Laboratory
Batavia, Illinois 60510 USA

1. THE EVOLUTION OF THE VACUUM

If the universe started from conditions of high temperature and density, there should have been a series of phase transitions associated with spontaneous symmetry breaking. The cosmological phase transitions could have observable consequences in the present Universe. Some of the consequences including the formation of topological defects and cosmological inflation are reviewed here.

One of the most important tools in building particle physics models is the use of spontaneous symmetry breaking (SSB). The proposal that there are underlying symmetries of nature that are not manifest in the vacuum is a crucial link in the unification of forces. Of particular interest for cosmology is the expectation that at the high temperatures of the big bang symmetries broken today will be restored, and that there are phase transitions to the broken state. The possibility that topological defects will be produced in the transition is the subject of this section. The possibility that the Universe will undergo inflation in a phase transition will be the subject of the next section.

Before discussing the creation of topological defects in the phase transition, some general aspects of high-temperature restoration of symmetry and the development of the phase transition will be reviewed.

1.1 *High Temperature Symmetry Restoration*

To study temperature effects, consider a real scalar field described by the Lagrangian

$$\mathcal{L} = \frac{1}{2}(\partial_\mu\phi)(\partial^\mu\phi) - V(\phi); \quad V(\phi) = -\frac{1}{2}M^2\phi^2 + \frac{1}{4}\lambda\phi^4. \tag{1.1}$$

The minima of the potential (determined by the condition $\partial V/\partial\phi = 0$), and the value of the potential at the minima are given by

$$\langle\phi\rangle = \pm\sqrt{\frac{M^2}{\lambda}}; \quad V(\langle\phi\rangle) = -\frac{M^4}{4\lambda}. \tag{1.2}$$

Presumably, the ground state of the system is either $+\langle\phi\rangle$ or $-\langle\phi\rangle$ and the reflection symmetry $\phi \leftrightarrow -\phi$ present in the Lagrangian is not respected by the vacuum state. When a symmetry of the Lagrangian is not respected by the vacuum, the symmetry is said to be spontaneously broken.

From the stress tensor in terms of the Lagrangian, $T_{\mu\nu} = -\partial_\mu\phi\partial_\nu\phi - \mathcal{L}g_{\mu\nu}$, the energy density of the vacuum is

$$\langle T_{00}\rangle = \rho_V = -\mathcal{L} = V(\phi) = -\frac{M^4}{4\lambda}. \tag{1.3}$$

The contribution of the vacuum energy to the total energy density today must be smaller than the critical density $\rho_C = 1.88 \times 10^{-29}h^2$ g cm$^{-3} \simeq 10^{-46}$ GeV^4. Since this number is so small, it is tempting to require $\rho_V = 0$. This can be accomplished by adding to the Lagrangian a constant factor of $+M^4/4\lambda$. This constant term will not affect the equations of motion, and the sole effect is to cancel the present vacuum energy.

There are several ways to understand the phenomena of high-temperature symmetry restoration. The most physical way is to express the effective finite-temperature mass of ϕ as the zero-temperature mass, $-M^2$, and a plasma mass, $M_{plasma} \simeq a\lambda T^2$, where a is a constant of order unity. If $M_T^2 = -M^2 + M_{plasma}^2 \leq 0$, the minimum of the potential will be at $\phi \neq 0$ (SSB), while if $M_T^2 = -M^2 + M_{plasma}^2 \geq 0$, the effective mass term will be positive and the minimum of the potential will be at $\phi = 0$ (symmetry restored). There is a critical temperature, $T_c = M/(a\lambda)^{1/2}$ above which $\langle\phi\rangle = 0$ [1].

A more rigorous approach to symmetry restoration is to account for the effect of the ambient background gas in the calculation of the higher-order quantum corrections to the classical potential. The finite temperature potential will include a temperature-dependent term that represents the free energy of ϕ particles at temperature T. To one loop, the full potential is [2]

$$V_T(\phi) = V(\phi) + \frac{T^4}{2\pi^2}\int_0^\infty dx x^2 \ln\left[1 - \exp[-(x^2 + \mu^2/T^2)^{1/2}]\right], \tag{1.4}$$

where $V(\phi)$ is the zero-temperature one-loop potential, and $\mu^2 = -M^2 + 3\lambda\phi^2$. At high temperature, Eq. 1.4 has the expansion

$$V_T(\phi) = V(\phi) - \frac{\pi^2}{90}T^4 + \frac{\lambda}{8}T^2\phi^2 + \cdots \tag{1.5}$$

The term proportional to T^4 is minus the pressure of a spinless boson, which should be the leading contribution to the free energy, and the second term is the "plasma" mass term for ϕ. Eq. 1.4 has a critical temperature, $T_c = 2M/\lambda^{1/2}$, above which the symmetry is restored.

The phase transition from the symmetric to the broken phase can be either first order or higher order. If at T_c there is a barrier between $\phi = 0$ and the SSB minimum $\phi = \sigma$, the change in ϕ will be discontinuous, signalling a first order transition. If no barrier is present at T_c, the change in ϕ will be continuous, signalling a higher order transition.

In general, at some temperature $T \leq T_c$, the $\phi = 0$ phase is a metastable phase, and will be terminated by the decay of the false vacuum by quantum or thermal tunneling. Here, quantum tunneling will refer the zero-temperature part of the tunneling rate.

The quantum tunneling occurs by the nucleation of bubbles of the new phase. The probability for bubble nucleation is calculated by solving the *Euclidean* equation of motion [3]

$$\Box_E \phi - V'(\phi) = \frac{d^2\phi}{dt^2} + \nabla^2\phi - V'(\phi) = 0 \tag{1.6}$$

(where $V' \equiv dV/d\phi$) with boundary conditions $\phi = 0$ at $\vec{x}^2 + t^2 = \infty$. The probability of bubble nucleation per unit volume per unit time is $\Gamma = A\exp(-S_E)$, where S_E is the Euclidean action for the solution of Eq. 1.6

$$S_E(\phi) = \int d^4x \left[\frac{1}{2}\left(\frac{d\phi}{dt}\right)^2 + \frac{1}{2}(\nabla\phi)^2 + V(\phi)\right]. \tag{1.7}$$

The calculation of the constant A is quite complicated, but for most applications a guess of A on dimensional grounds will suffice.

Of the many possible solutions to Eq. 1.6, the one with least action is the most important. The least action solution has $O(4)$ symmetry, and the Euclidean equation of motion becomes

$$\frac{d^2\phi}{dr^2} + \frac{3}{r}\frac{d\phi}{dr} - V'(\phi) = 0, \tag{1.8}$$

with boundary conditions $\phi = 0$ at $r^2 = \vec{x}^2 + t^2 = \infty$ and $d\phi/dr = 0$ at $r = 0$. In general solutions to this equation can not be found. However in the "thin-wall" approximation, where the difference in energy between the metastable and true vacua are small compared to the height of the barrier, the "damping" term proportional to $d\phi/dr$ can be neglected. The solution for S_E is then simply

$$S_E = \int_0^\sigma d\phi\sqrt{2V(\phi)}. \tag{1.9}$$

The tunneling rate at finite temperature [4] can be found following the above procedure, remembering that field theory at finite temperature is equivalent to Euclidean field theory with the time periodic with period T^{-1}. The finite-temperature tunneling rate is found by solving the equation of motion (only considering the least-action solution, which in this case has $O(3)$ symmetry)

$$\frac{d^2\phi}{ds^2} + \frac{2}{s}\frac{d\phi}{ds} - V_T'(\phi) = 0, \tag{1.10}$$

where $s = \vec{x}^2$. The finite-temperature tunneling rate is

$$\Gamma_T = A\frac{S_3}{T}\exp(-S_3/T), \tag{1.11}$$

where S_3 is the three-dimensional action of the solution of Eq. 1.10

$$S_3 = \int d^3x \left[\frac{1}{2}(\nabla\phi)^2 + V_T(\phi)\right]. \tag{1.12}$$

1.2 *Domain Walls* [5]

The simple model of the previous section can be used to demonstrate domain walls. The Lagrangian can be written in the form

$$\mathcal{L} = \frac{1}{2}(\partial_\mu \phi)^2 - \frac{1}{4}\lambda(\phi^2 - \langle\phi\rangle^2)^2; \qquad \langle\phi\rangle^2 \equiv \sigma^2 = \frac{M^2}{\lambda}. \tag{1.13}$$

The Z_2 symmetry of the Lagrangian is broken when ϕ obtains a vacuum expectation value $\phi = +\sigma$ or $\phi = -\sigma$. Imagine that space is divided into two regions. In one region of space $\phi = +\sigma$, and in the other region of space $\phi = -\sigma$. The transition region between the two vacua is called a domain wall. Domain walls should be produced, for instance, in the nucleation of bubbles. The bubbles of true vacuum will be either $\phi = +\sigma$ or $\phi = -\sigma$, with equal probability.

Imagine a wall in the $x-y$ plane at $z=0$. At $z=-\infty$, $\phi=-\sigma$, and at $z = +\infty$, $\phi = +\sigma$. The equation of motion for ϕ is $\Box\phi + \lambda\phi(\phi^2 - \sigma^2) = 0$. The minimum energy solution to the equation of motion, subject to the boundary conditions above, is $\phi_W(z) = \sigma \tanh(z/\Delta)$ where Δ is the "thickness" of the wall, given by $\Delta = (\lambda/2)^{1/2}\sigma^{-1}$.

The finite, but non-zero, thickness of the wall is easy to understand. The terms contributing to the energy include a gradient term and a potential energy term. The gradient term is minimized by making the wall as thick as possible, and the potential term is minimized by making the wall as thin as possible, i.e., by minimizing the distance over which ϕ is away from $\pm\sigma$. The balance between these terms results in a wall of thickness Δ.

The stress tensor with $\phi = \phi_W$ is

$$T_\mu{}^\nu = \frac{\lambda}{2}\sigma^4 \cosh^{-4}(z/\Delta)\mathrm{diag}(1,1,1,0). \tag{1.14}$$

From the stress tensor it is possible to define a surface tension for the wall, $\eta = \int T_0{}^0 dz = (4/3)(\lambda/2)^{1/2}\sigma^3$. It is also obvious from the stress tensor that since the (ii) component is equal to the (00) component, the gravitational interaction of the infinite wall will be non-Newtonian. This can lead to some strange interactions. For instance, two infinite walls in the $x-y$ plane will *repel* each other. This strange gravitational behavior only obtains for infinite and straight walls. The gravitational field at large distances from a spherical wall of radius R, would be that of a massive particle of mass $m \simeq R^2\sigma$.

The existence of domain walls can be ruled out today simply on the grounds of their contribution to the total mass of the Universe. A domain wall with $R \simeq R_{\mathrm{horizon}} \simeq H_0^{-1} \simeq 10^{28}$cm would contribute a mass of $M_{\mathrm{wall}} = \eta R^2_{\mathrm{wall}} = 10^{60}$grams. This would be about a factor of 10^5 larger than the total mass within R_{horizon}.

The simple model of this section had domain walls because of the existence of disconnected vacuum states. The general condition for the existence of domain walls in the symmetry breaking $\mathcal{G} \to \mathcal{H}$ is that $\Pi_0(\mathcal{M}) \neq I$, where $\mathcal{M}$ is the manifold of equivalent vacuum states $\mathcal{M} \equiv \mathcal{G}/\mathcal{H}$, and Π_0 is the homotopy group that counts disconnected components. In the above example, $\mathcal{G} = Z_2$, $\mathcal{H} = I$, $\mathcal{M} = Z_2$, and $\Pi_0(\mathcal{M}) = Z_2 \neq I$.

1.3 *Cosmic Strings* [6,7]

A simple model that demonstrates the existence of cosmic strings is a gauge version of the model of the previous section. The Lagrangian of the model contains a U_1 gauge field, A_μ, in addition to the complex Higgs field, ϕ,

$$\mathcal{L} = D_\mu\phi D^\mu\phi - \frac{1}{4}F_{\mu\nu}F^{\mu\nu} - \frac{1}{4}\lambda(\phi^\dagger\phi - \langle\phi\rangle^2)^2; \qquad \langle\phi\rangle^2 = \sigma\exp(i\theta) \tag{1.15}$$

Again, $\sigma^2 = M^2/\lambda$, $F_{\mu\nu} = \partial_\mu A_\nu - \partial_\nu A_\mu$, and $D_\mu\phi = \partial_\mu\phi - ieA_\mu\phi$.

Since there is a local gauge symmetry, $\theta = \theta(\vec{x})$, can be position dependent. Since ϕ is single valued, the total $\Delta\theta$ around any closed path must be an integer multiple of 2π. Imagine such a closed path with $\Delta\theta = 2\pi$. As the path is shrunk to a point (and no singularities are encountered), $\Delta\theta$ cannot change from $\Delta\theta = 2\pi$ to $\Delta\theta = 0$. There must therefore be one point contained within the path where the phase θ is undefined, i.e., $\langle\phi\rangle = 0$. The region of false vacuum within the path is part of a tube of false vacuum. These tubes of false vacuum either must be closed or infinite in length, otherwise it would be possible to deform the path around the tube, and contract it to a point without encountering the tube of false vacuum. It will turn out that these tubes of false vacuum have a characteristic transverse dimension far smaller than its length, so they appear as one-dimensional objects called "strings."

The string solution to the Lagrangian in Eq. 1.15 was first found by Nielsen and Olesen [8]. At large distances from an infinite string in the z-direction, $\phi \longrightarrow \sigma\exp(in\theta)$; $A_\mu \longrightarrow -ie^{-1}\partial_\mu\left[\ln(\phi/\sigma)\right]$, where θ is the angle in the $x-y$ plane. Note this choice of A_μ and ϕ is a finite energy solution, since at large distances from the string, $D_\mu\phi \to 0$ and $F_{\mu\nu} \to 0$.

For an infinite string in the z-direction, the stress tensor takes the form $T_\mu{}^\nu = \mu\delta(x)\delta(y)\mathrm{diag}(1,0,0,1)$, where μ is the mass per unit length of the string (string tension) given by $\mu \simeq \sigma^2$.

Far from a string loop of radius R, the gravitational field of the string is that of a particle of mass $M_{\rm string} = \mu R_{\rm string}$. For a string that stretches across the present horizon, the mass would be $M_{\rm string} = 10^{18}(\sigma/\mathrm{GeV})^2$ grams. Cosmic string networks may have very interesting astrophysical consequences, including acting as seeds for the formation of large-scale structure.

String solutions will be present in the symmetry breaking $\mathcal{G} \to \mathcal{H}$, if the manifold of degenerate vacuum states $\mathcal{M} = \mathcal{G}/\mathcal{H}$ contains unshrinkable loops, i.e., if the mapping of $\mathcal{M}$ onto the circle is non-trivial. This is formally expressed by the statement that string solutions exist if $\Pi_1(\mathcal{M}) \neq I$. In the above example $\mathcal{G} = U_1$ was broken, $\mathcal{M}$ is a circle, and $\Pi_1(\mathcal{M}) = Z$, the set of integers.

Some of the cosmological and astrophysical effects of strings are discussed elsewhere in this book [9].

1.4 *Magnetic Monopoles* [10,11]

Domain walls are topological defects in two dimensions, and strings are topological defects in one dimension. Zero-dimensional defects appear in theories with SSB as magnetic monopoles. For a simple model that illustrates the existence of magnetic monopoles, consider an SO_3 gauge theory with a Higgs triplet field ϕ^a

$$\mathcal{L} = \frac{1}{2}D_\mu\phi^a D^\mu\phi^a - \frac{1}{4}F^a_{\mu\nu}F^{a\mu\nu} - \frac{1}{4}\lambda(\phi^a\phi^a - \langle\phi\rangle^2)^2; \qquad \langle\phi\rangle^2 = \sigma\hat{\sigma}, \tag{1.16}$$

where $\sigma\hat{\sigma}$ is an isovector in the SO_3 space of magnitude σ and direction $\hat{\sigma}$ ($\hat{\sigma}$ is a unit isovector). Here

$$F^a_{\mu\nu} = \partial_\mu A^a_\nu - \partial_\nu A^a_\mu - e\varepsilon_{abc}A^b_\mu A^c_\nu; \qquad D_\mu\phi^a = \partial_\mu\phi^a - e\varepsilon_{abc}A^b_\mu\phi^c. \tag{1.17}$$

Since the theory has a local gauge symmetry, σ is a constant, but $\hat{\sigma}$ can be a function of $\vec{x}$. Imagine a configuration in which at one point $\phi^a = \sigma(0,0,1)$, at another point $\phi^a = \sigma(0,1,0)$, at another point $\phi^a = \sigma(1,0,0)$, and so forth. The lowest-energy configuration has $\phi^a =$ constant, and the x-dependence of ϕ^a can in general be gauged away. However there are configurations that cannot be deformed into a configuration of constant $\hat{\sigma}$ by a finite-energy transformation. An example of such a configuration is the "hedgehog" configuration, in which $\hat{\sigma} = \hat{r}$, where $\hat{r}$ is the unit vector in the radial direction. But for the obvious angular dependence, the solution is spherically symmetric at $r \to \infty$: $\phi^a(r,t) \to \sigma\hat{r}$; $A^a_\mu(r,t) \to \varepsilon_{\mu ab}\hat{r}_b/er$. The magnetic field at $r \to \infty$ corresponding to the hedgehog solution is

$$B^a_i = \frac{1}{2}\varepsilon_{ijk}F^a_{jk} = \frac{\hat{r}_i\hat{r}^a}{er^2}, \tag{1.18}$$

which is the magnetic field of a magnetic charge of $g = 1/e$. The mass of the field configuration is $M_{\text{monopole}} \simeq \sigma/e$.

There have been many experiments to look for magnetic monopoles. The limit on the average number density of magnetic monopoles in the Universe depends upon the properties of the monopoles (mass, charge, proton decay catalysis, etc.). If magnetic monopoles exist, they would have a multitude of astrophysical consequences.

Monopoles will be present in the symmetry breaking $\mathcal{G} \to \mathcal{H}$, if the manifold of degenerate vacuum states contains unshrinkable surfaces, i.e., if the mapping of $\mathcal{M}$ onto the two-sphere is non-trivial. This is formally expressed by the statement that monopole solutions exist if $\Pi_2(\mathcal{M}) \neq I$. In the above example $\mathcal{G} = SO_3$, $\mathcal{H} = U_1$ and $\Pi_2(\mathcal{M})$ is the set of even integers.

1.5 *The Kibble Mechanism* [6]

The existence of the above topological defects is a prediction of many gauge theories with SSB. They are inherently non-perturbative, and cannot be produced in high energy collisions. The only place they can be produced is in phase transitions in the early Universe. Although monopoles, strings, and domain walls are topologically stable, they are, of course, not the minimum energy solution. However the production of the defects in the phase transition seems unavoidable. The mechanism for the production of the defects is known as the Kibble mechanism.

The Kibble mechanism is based upon the fact that in the phase transition the correlation length is limited by the particle horizon. The particle horizon is the maximum distance over which a massless particle could propagate from the time of the bang. Imagine that a particle is emitted at coordinates ($t = 0$, $r = r_H$, $\theta = 0$, $\phi = 0$) and is detected at the origin of the coordinate system at coordinates ($t = t$, $r = 0$, $\theta = 0$, $\phi = 0$). The coordinate r_H is given by

$$\int_0^t \frac{dt'}{R(t')} = \int_0^{r_H} \frac{dr}{(1-kr^2)^{1/2}} \simeq r_H. \tag{1.19}$$

The coordinate r_H by itself is just a label. The proper distance to the horizon is given by $d_H = R(t)r_H$, so

$$d_H = R(t)\int_0^t \frac{dt'}{R(t')}. \tag{1.20}$$

If $R \propto t^n$ ($n > 1$), then $d_H = (1-n)^{-1}t$.

The correlation length in the phase transition sets the maximum distance over which the Higgs field can be correlated. In general, the calculation of the correlation length depends upon the details of the transition. However, the fact that the horizon is finite in the standard cosmology implies that at the phase transition ($t = t_c$, $T =$

T_c), the Higgs field must be uncorrelated on scales greater than the horizon, so the horizon acts as an effective upper bound to the correlation length.

Imagine that at the phase transition the Higgs field is uncorrelated on scales greater than $\xi = d_H$. The initial random nature of $\langle\phi\rangle$ is damped (remember $E_{\rm min}$ occurs for $\langle\phi\rangle =$ constant). However there are Higgs configurations that are topologically stable and will be frozen in as topological defects.

Consider monopoles as an example of the freezing in of topological defects [12]. The direction of the isovector Higgs field is random on scales greater than ξ. The probability that a random orientation of $\langle\phi\rangle$ will have a hedgehog structure is about 0.1, so there should be about one monopole (or antimonopole) per 10 horizon volumes, $n_M = 0.1d_H^3 \simeq 0.1(m_{Pl}/T_c^2)^3$, using the age of a radiation-dominated Universe $t = m_{Pl}/T^2$. The entropy density at T_c is $s \simeq T_c^3$, so the monopole-entropy ratio is $n_M/s \simeq 0.1(T_c/m_{Pl})^3$. Since monopole-antimonopole annihilation is not important, if entropy is not created after monopole production, the above monopole-entropy ratio should obtain today. For $T_c = 10^{15}$GeV, $m_M = 10^{16}$ GeV as expected in grand unified theories, $n_M/s \simeq 10^{-13}$, which gives the present energy density in magnetic monopoles $\rho_{\rm monopoles} \simeq 10^{11}\rho_C$. Obviously some mechanism must suppress monopole production, enhance monopole annihilation, or increase entropy. An increase in entropy would also dilute the abundance of strings and domain walls. It is possible that monopoles were diluted to a level accessible to observation, or that strings were produced after the dilution of monopoles. Detection of monopoles or strings would provide unique information about both particle physics and cosmology. In complicated gauge theories with several symmetry breaking steps there are often interesting hybrid creatures, such as domain walls bounded by strings, strings terminated by monopoles, monopoles with strings through them, etc. They all have unique signatures, and observation of them would provide information about the steps of symmetry breaking.

2. INFLATION

The standard FRW cosmology provides a remarkably simple and beautiful model to describe the Universe. Nevertheless, there are some aspects of the standard picture that strongly suggests that the model is not a complete one. After discussing the problems of the cosmology developed so far, a possible solution to the problems will be presented. This solution goes by the name of "inflation" [13].

2.1 *Loose Ends of the Standard Cosmology*

• *Large-Scale Smoothness*: The Robertson-Walker metric describes a space that is homogeneous and isotropic. Why is space homogeneous and isotropic? There are other possibilities, including homogeneous but anisotropic spaces, and inhomogeneous spaces. The most precise indication of the smoothness of the Universe is provided by the microwave background radiation. If the entire observable Universe was in causal contact when the radiation last scattered, it might be imagined that microphysical processes would have damped any fluctuations and a single temperature would have obtained. However in the standard cosmology the distance to the horizon increases with time. The size of the horizon is conveniently expressed in terms of the entropy within the horizon

$$S_H = s\frac{4\pi}{3}d_H^3 \simeq T^3t^3. \tag{2.1}$$

The entropy within the horizon today is $S_H(0) \simeq 10^{88}$. In a matter-dominated Universe, $S_H = S_H(0)(1+z)^{-3/2}$, while in a radiation-dominated Universe, $S_H = S_H(0)(1+z)^{-3}$. The entropy in the horizon at recombination when the radiation last scattered was $S_H(t = t_{rec}) \simeq 10^{83}$. The Universe as presently observed consisted of about 10^5 causally disconnected regions at recombination, so causal processes could not have led to smoothness. At the time of primordial nucleosynthesis, the entropy within the horizon volume was $S_H(t_{nucleo}) \simeq 10^{53}$, or about 10^{-30} of the present Universe.

The first untidy fact about the standard cosmology is that there is no physical explanation for why the Universe is smooth.

• *Density Perturbations*: Although the Universe is smooth on large scales, there is a rich structure on small scales. It is usually assumed that the structures observed today were once small perturbations on a smooth background, and have grown as the result of the gravitational instabilities in an expanding Universe. The relic photons did not take part in the gravitational collapse, and remain as fossil evidence of the once-smooth Universe.

Density inhomogeneities are usually expressed in a Fourier expansion

$$\left(\frac{\delta\rho}{\rho}\right) = (2\pi)^{-3}\int \delta_k \exp(-i\vec{k}\cdot\vec{x})d^3k. \tag{2.2}$$

Here k is a co-moving label. The *physical* wavenumber and wavelength are related to k by $k_{ph} = k/R(t)$, $\lambda_{ph} = (2\pi/k)R(t)$. It is also convenient to express the scale of the perturbation in terms of the mass in baryons contained within the perturbation. For constant B, the baryon mass on scale λ is proportional to λ^3. The baryon mass within the horizon at time t is $M_H(t) \simeq m_pBsd_H^3 \propto S_H$. The quantity usually referred to as $(\delta\rho/\rho)$ on a given scale is the r.m.s. mass fluctuations on that scale

$$\left(\frac{\delta\rho}{\rho}\right)_k^2 = (2\pi)^{-3}k^3|\delta_k|^2. \tag{2.3}$$

The exact nature of the perturbations required for galaxy formation is unknown. A promising choice for density perturbations is that as every distance scale comes within the horizon, the r.m.s. fluctuations in the density are $10^{-4}-10^{-5}$ *independent of the scale.* This is usually expressed as

$$\left(\frac{\delta\rho}{\rho}\right)_H \simeq 10^{-4}. \tag{2.4}$$

Here $(\delta\rho/\rho)_H$ is $(\delta\rho/\rho)$ on the scale $\lambda = d_H = t$ at time $t = d_H$.

The evolution of the perturbations within the horizon is determined by local physics, e.g., the Jeans criteria. The behavior of the perturbations outside of the horizon is complicated by the fact that there is a "gauge dependence" that reflects the freedom of the choice for a reference spacetime. Nevertheless, the growth of metric perturbations on scales larger than the horizon can be studied by using the uniform Hubble flow gauge (time slices chosen to give constant H). From the Friedmann equation with H constant, fluctuations in ρ are equivalent to fluctuations in the spatial curvature k/R^2

$$\delta\left(\frac{k}{R^2}\right) \Longleftrightarrow \delta\left(\frac{8\pi G}{3}\rho\right). \tag{2.5}$$

In a radiation-dominated (matter-dominated) Universe, $\rho \propto R^{-4}$ (R^{-3}), so

$$(\delta\rho/\rho) \propto \begin{cases} R^{-2}/R^{-4} \sim (1+z)^{-2} & \text{(RD)} \\ R^{-2}/R^{-3} \sim (1+z)^{-1} & \text{(MD)}. \end{cases} \tag{2.6}$$

Since $S_H \propto (1+z)^{-3}$ for (RD) and $S_H \propto (1+z)^{3/2}$ for (MD), $(\delta\rho/\rho) \propto S_H^{2/3} \propto M_H^{2/3}$ for both (RD) and (MD). So as any scale comes within the horizon, the growth that scale has experienced while outside the horizon depends upon the mass contained in the scale as it enters the horizon

$$\left(\frac{\delta\rho}{\rho}\right)_H \sim \left(\frac{\delta\rho}{\rho}\right)_0 (M_H(t))^{2/3}, \tag{2.7}$$

where t_0 is some arbitrary initial time. If $(\delta\rho/\rho)_0$ is proportional to $M^{-2/3}$, as each scale comes within the horizon, $(\delta\rho/\rho)$ will be a constant. Larger scales have smaller initial amplitudes, but they have a longer time to grow outside the horizon. If $(\delta\rho/\rho)_0$ is characterized by a steeper spectrum, the first scales that come within the horizon would have been non-linear. If $(\delta\rho/\rho)_0$ is characterized by a flatter spectrum, larger scales would have larger $(\delta\rho/\rho)$ at horizon crossing.

The standard model can shed no light on the origin of the density perturbations. It must simply assume that at $t = 0$ there are perturbations of the appropriate magnitude and spectrum impressed upon the metric.

• *Spatial Flatness - Age*: In the standard Friedmann cosmology, $\Omega - 1 = k/R^2H^2$. In the past, $H^2 \propto \rho$, which for a matter-dominated Universe gives $H^2 \propto R^{-3}$, and for a radiation-dominated Universe gives $H^2 \propto R^{-4}$. Since today $|\Omega - 1|$ is of order unity, at previous epochs

$$|\Omega - 1| \simeq \begin{cases} R/R_0 = (1+z)^{-1} & \text{(MD)} \\ (R/R_0)^2 = (1+z)^{-2} & \text{(RD)}. \end{cases} \tag{2.8}$$

At the time of primordial nucleosynthesis, $|\Omega - 1| \leq 10^{-16}$, and at the planck time $|\Omega - 1| \leq 10^{-60}$. Obviously Ω was *very* close to one at early times, i.e., the curvature term was small compared to H^2 and $8\pi G\rho/3$.

The smallness of the curvature term is necessary for the Universe to survive as long as it has without either re-collapsing (for $k = +1$) or becoming curvature dominated (for $k = -1$). The natural time scale in the Friedmann equation is the planck time $t_{Pl} = 2 \times 10^{-43}$ sec. The difference between the kinetic term (H^2) and the potential term ($8\pi G\rho/3$) is the curvature term. This must be small in order for the Universe to expand for 10^{17} sec. $\sim 10^{60} t_{Pl}$.

The standard Friedmann model has no explanation for the present spatial flatness of the Universe.

• *Cosmological Constant*: The most general form of Einstein's equations includes a cosmological constant

$$R_{\mu\nu} - \frac{1}{2}g_{\mu\nu}R = 8\pi G T_{\mu\nu} + \Lambda g_{\mu\nu}. \tag{2.9}$$

With the stress-tensor in the perfect-fluid form (U_μ is the fluid velocity vector, $U_\mu = (1,0,0,0)$ in the fluid rest frame) $T_{\mu\nu} = -pg_{\mu\nu} + (\rho + p)U_\mu U_\nu$, the effect of the cosmological constant is to add to the fluid contributions to ρ and p, terms $\rho_\Lambda = -p_\Lambda = \Lambda/8\pi G$. The generalized energy and pressure are given by $\rho^* = \rho + \rho_\Lambda$, $p^* = p + p_\Lambda$, and the Einstein equations can be written in terms of $T^*_{\mu\nu}$, which is $T_{\mu\nu}$ with $\rho \to \rho^*$, $p \to p^*$,

$$R_{\mu\nu} - \frac{1}{2}g_{\mu\nu}R = 8\pi G T^*_{\mu\nu}. \tag{2.10}$$

If ρ^* and p^* are dominated by ρ_Λ and p_Λ, the conservation and Friedmann equations become

$$\rho^* \propto R^0 = \text{constant}; \quad H^2 = \frac{8\pi G\rho^*}{3} = \frac{\Lambda}{3}, \tag{2.11}$$

which has solution $R \propto \exp(Ht)$.

Today the contribution of a cosmological constant to the energy density of the Universe must be less than ρ_C. In useful units, $\rho_C = 8.07 \times 10^{-47}h^2$ GeV^4. Among the contributions to Λ are contributions from the condensates of Higgs particles due to SSB. During cosmological phase transitions, the vacuum energy density changes by σ^4, where σ is the zero-temperature vacuum expectation value of the Higgs field. This change in the vacuum energy is 10^9GeV^4 for the electroweak transition, and 10^{60}GeV^4 for the GUT transition. A cosmological constant of this order must be present before the transition to ensure that after all transitions are complete the energy density of the vacuum is less than about 10^{-47}GeV^4.

The standard cosmology cannot explain why the present vacuum energy density is so small.

• *Unwanted Relics*: There are a variety of particles that are expected to survive annihilation and contribute to the present energy density. Particles with very large masses typically have very small annihilation cross sections and should be abundant. This is rather unfortunate, as their contribution to the mass density typically is many orders of magnitude larger than ρ_C. The magnetic monopoles produced in the GUT phase transition are an example of such an unwanted relic.

The standard cosmology has no mechanism of ridding the Universe of unwanted particles.

The problems mentioned here do not invalidate the standard cosmology. They are accommodated by the standard cosmology, but they are not explained. The goal of cosmology is to explain the present structure of the Universe on the basis of physical law, and one hopes that physical law will one day explain the above points. Inflation is a model for such an explanation.

2.2 *Inflation - The Basic Picture* [14,15]

Consider as a model for new inflation a phase transition associated with SSB with a scalar potential given by

$$V(\phi) = \frac{1}{4}\lambda(\phi^2 - \sigma^2)^2. \tag{2.12}$$

At temperatures $T \gg T_c = 2\sigma$, $\langle\phi\rangle = 0$, and $V(\langle\phi\rangle) = \lambda\sigma^4/4 \equiv \rho_V$. At temperatures $T \ll T_c$, $\langle\phi\rangle = \sigma$, and $V(\langle\phi\rangle) = \rho_V = 0$. New inflation will occur as ϕ makes the transition from the high temperature minimum of the potential to the low temperature minimum of the potential.

At some temperature $T \leq T_c$, in some region of the Universe, the Higgs field will make the transition from $\phi = 0$ to $\phi \neq 0$. Assume that in this region of the Universe ϕ is spatially uniform. The evolution of ϕ to the low-temperature ground state is not instantaneous, but requires a time determined by the dynamics of the theory. The equation of motion for ϕ can be found from $T^{\mu\nu}{}_{;\nu} = 0$, where $T_{\mu\nu} = -\partial_\mu\phi\partial_\nu\phi - \mathcal{L}g_{\mu\nu}$. With the assumption that ϕ is spatially homogeneous $\ddot{\phi} + 3H\dot{\phi} + V'(\phi) = 0$, where $V' = \partial V/\partial\phi$, and $H^2 = 8\pi G\rho/3$. The contributions to ρ include a radiation term ρ_R, a kinetic term for ϕ, and a potential term for ϕ:

$$\rho = \rho_R + \frac{1}{2}\dot{\phi}^2 + V(\phi). \tag{2.13}$$

If there is a "flat" region in $V(\phi)$, the evolution of ϕ will be "slow" and the $\ddot{\phi}$ term can be neglected in the equation of motion. In this flat region ϕ will change very slowly and $V(\phi)$ will be roughly constant. Therefore the contribution to ρ from $V(\phi)$ will be roughly constant and will rapidly come to dominate ρ_R which decreases in proportion to R^4. When ρ is dominated by potential energy the scale factor increases exponentially. If this flat region in the potential extends from ϕ_s to ϕ_e, R will increase by an amount $R(\phi_e) = R(\phi_s)\exp(H\Delta t)$, where Δt is the time it takes to make the transition from ϕ_s to ϕ_e, and $H^2 \simeq V(\phi)/m_{Pl}^2 \simeq \sigma^4/m_{Pl}^2$. For a concrete example, assume for the moment that $\Delta t = 100H^{-1}$.

Now assume that after traversing the flat region in the potential, at $\phi \geq \phi_e \simeq \sigma$ there is a "steep" region in the potential. In this steep region the oscillations in the zero momentum mode of ϕ will rapidly convert the potential energy to radiation. If this conversion is efficient, the Universe will be reheated to a temperature $T_{\rm RH}$ found by equating the potential energy density to the radiation energy density: $V(\phi) \simeq T^4$, or $T_{\rm RH} \simeq \sigma$.

This is the basis scenario for new inflation. To illustrate the scenario, take $\sigma = 10^{14}$GeV, and the initial size of the region to be the size of the horizon at T_c, $R_i = H^{-1} \simeq m_{Pl}/\sigma^2 = 10^{-23}$cm (it is reasonable to expect ϕ to be uniform on scales that are in causal contact). The initial entropy in this region is $S_i \simeq (R_iT_i)^3 \simeq 10^{14}$. The final size of the region in the example where $\Delta t = 100H^{-1}$ is $R_f = \exp(100)R_i \simeq 3\times 10^{20}$cm. With efficient reheating $T_{\rm RH} = \sigma$, and the final entropy contained in the region is $S_f = (R_fT_{\rm RH})^3 \simeq 10^{144}$.

This large creation of entropy has helped with three out of four problems. *Large-Scale Smoothness*: At $T = 10^{14}$GeV, the presently observable Universe ($S = 10^{88}$) was contained in a size of 10cm, and easily fit within the smooth region after inflation. *Density Perturbations*: To see how inflation generates density perturbations it is necessary to treat the dynamics of the scalar field in greater detail than done so far. This will be done shortly. *Spatial Flatness - Age*: After inflation R has increased by exp(100) but the final temperature is close to the initial temperature. Thus, immediately after inflation the spatial curvature term k/R^2 is a factor of $\exp(-200)$ smaller, while the energy density term is unchanged. *Cosmological Constant*: Inflation does not help the cosmological constant problem. *Unwanted Relics*: The number density of particles present before inflation is decreased by a factor of $R_i^3/R_f^3 \simeq \exp(-300)$. This is true also for the original photons. It is crucial to create entropy in the termination of inflation.

In this example it was assumed that the slow-roll period lasted for 100 e-folds. The minimum number of e-folds is the number required to fit the observed entropy of 10^{88} into a single inflation region. The final entropy in the inflation region is $S_f \simeq T_{\rm RH}^3R_f^3$. The size of the final region is related to the number of e-folds by $R_f^3 = \exp(3N)R_i^3$, assuming little or no growth during the oscillation phase. The largest possible smooth initial region is the size of the horizon at the phase transition, $R_i = H^{-1}(T_c) \simeq m_{Pl}/\sigma^2$, assuming $T_c = \sigma$. The maximum reheat temperature is

$T_{RH} \simeq \sigma$, so the final entropy is $S_f \simeq \sigma^3 \exp(3N) m_{Pl}^3/\sigma^6 \simeq \exp(3N) m_{Pl}^3/\sigma^3$. The requirement $S_f \geq 10^{88}$ gives $N \geq 58 + \ln(\sigma/10^{15}\text{GeV})$.

2.3 *Dynamics of Inflation*

The evolution of the spatially homogeneous scalar field (zero momentum mode of the scalar field) is crucial for inflation. If the coupling of the scalar field to other fields are included, the equation of motion for the zero-momentum mode of ϕ is (ϕ will denote the zero-momentum mode unless otherwise indicated)

$$\ddot{\phi} + 3H\dot{\phi} + \Gamma_\phi \dot{\phi} + V'(\phi) = 0, \tag{2.14}$$

where Γ_ϕ is the ϕ decay width. The decay width is typically $\Gamma_\phi = h^2 m_\phi$, where h is a coupling constant, and m_ϕ is the mass of ϕ [16]. The energy density and pressure of ϕ are given by

$$\rho_\phi = \frac{1}{2}\dot{\phi}^2 + V(\phi); \quad p_\phi = \frac{1}{2}\dot{\phi}^2 - V(\phi). \tag{2.15}$$

The "slow roll" regime is when the $\ddot{\phi}$ and Γ_ϕ terms in Eq. 2.14 can be neglected, and $V(\phi)$ is the dominant term in Eq. 2.13. The equation of motion during slow roll is

$$3H\dot{\phi} = -V'(\phi). \tag{2.16}$$

Neglecting $\ddot{\phi}$ is consistent if

$$|V''(\phi)| \leq 9H^2; \quad |V'(\phi) m_{Pl}/V(\phi)| \leq (48\pi)^{1/2}. \tag{2.17}$$

These conditions will determine the duration of slow roll.

The number of e-folds of inflation while ϕ rolls from ϕ_1 to ϕ_2 during slow roll is given by

$$N(\phi_1 \to \phi_2) = \int_{\phi_1}^{\phi_2} H dt = -3 \int_{\phi_1}^{\phi_2} \frac{H^2(\phi)}{V'(\phi)} d\phi, \tag{2.18}$$

where $dt = \dot{\phi}^{-1} d\phi = -3H/V' d\phi$.

With ρ_ϕ given by Eq. 2.15, $\dot{\rho}_\phi = \dot{\phi}\ddot{\phi} + \dot{\phi}V'(\phi)$, and using Eq. 2.14, $\dot{\rho}_\phi = -3H\dot{\phi}^2 - \Gamma_\phi \dot{\phi}^2$. The two terms in the equation for $\dot{\rho}_\phi$ represent the change due to the redshift of the kinetic energy in the ϕ field (proportional to H) and the change due to decay of the ϕ field (proportional to Γ_ϕ). When ϕ starts oscillating about the minimum of the potential, the energy transfers between kinetic and potential energy until ϕ decays. Over an oscillation cycle $\langle \dot{\phi}^2 \rangle = \rho_\phi$, and $\dot{\phi}^2$ can be replaced by ρ_ϕ in the equation for $\dot{\rho}_\phi$. The energy from ϕ decay is transferred into radiation, and the equation for the evolution of ρ_R becomes $\dot{\rho}_R = -4H\rho_R + \Gamma_\phi \rho_\phi$.

The equations for $\dot{\rho}_R$ and $\dot{\rho}_\phi$ can be integrated to study reheating. If oscillation about the minimum begins at $t = t_3$ and $R = R_3$ with $\rho_\phi = \sigma^4$, the ϕ energy density will decrease as

$$\rho_\phi = \sigma^4 \left(\frac{R}{R_3}\right)^{-3} \exp[-\Gamma(t - t_3)]. \tag{2.19}$$

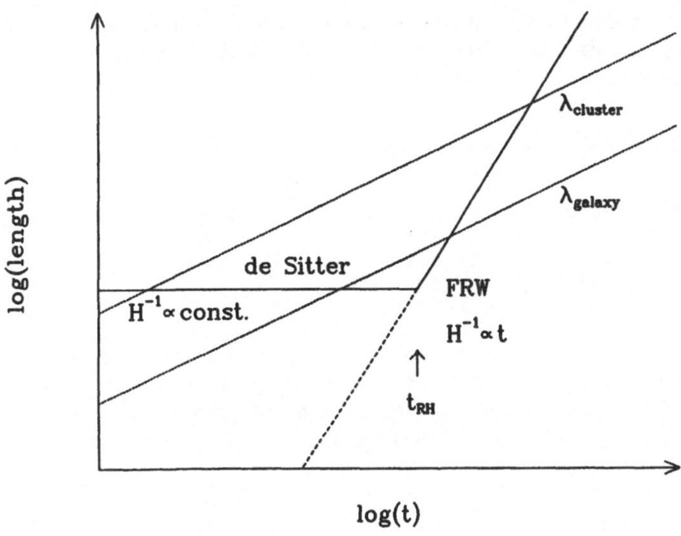

Figure 1: Physical scales cross the physics horizon twice

Until decay, the ϕ energy density decreases in expansion as the energy density for massive particles. When ϕ decays, the remaining energy is converted to radiation ($\rho_\phi \to (\pi^2/30)g_* T_{RH}^4$). Obviously, the longer ϕ oscillates before decay, the less energy will be available for conversion into radiation, and the lower will be the reheat temperature. If the decay width is large compared to the expansion rate at the start of oscillation, $H_3 \simeq \sigma^2/m_{Pl}$, reheating will occur before damping of ρ_ϕ, and $T_{RH} \simeq g_*^{-1/4}\sigma$. If the decay width is small compared to H_3, ϕ will oscillate until the age of the Universe is equal to the ϕ lifetime, i.e., until $\Gamma_\phi = H = \rho_\phi^{1/2}/m_{Pl}$. Then when $\rho_\phi \to g_* T_{RH}^4$, the reheat temperature will be $T_{RH} = g_*^{-1/4}\rho_\phi^{1/4} = g_*^{-1/4}(\Gamma_\phi m_{Pl})^{1/2}$.

Now consider the generation of fluctuations in ρ. In the FRW radiation-dominated Universe $H^{-1} \propto t$, while during the slow-roll epoch, $H^{-1} \simeq m_{Pl}/V(\phi)^2$ is constant. If H is constant, the Universe in approximately in a de Sitter phase. H^{-1} sets the scale over which microphysical processes can act. H^{-1} will be called the "physics horizon." During the slow roll phase the physics horizon is constant and physical scales increase exponentially. Eventually, physical scales once smaller than the horizon will become larger than the horizon. After termination of the slow-roll phase the Universe reheats, behaves like a FRW radiation-dominated Universe, and scales outside the horizon will eventually come (back) within the horizon. This double-cross of the physics horizon is illustrated in Fig. 1.

Notice that the *last* scales to go outside H^{-1} during the de Sitter phase are the *first* scales to come back inside H^{-1} during the FRW phase. Ignoring the σ dependence, the Hubble radius today ($\simeq$ 3000 Mpc) crossed the horizon 58 e-folds before the end of inflation. Any scale smaller than the Hubble radius today crossed the horizon $58 + \ln(\sigma/10^{15}\text{GeV}) + \ln(\lambda/3000\text{Mpc})$ e-folds before the end of inflation. Using $B = 10^{-10}$, the mass in baryons inside the horizon today is $10^{21}M_\odot$. Since $B \propto \lambda^3$, any baryon mass scale crossed the horizon $58 + \ln(\sigma/10^{15}\text{GeV}) + (1/3)\ln(M/10^{21}M_\odot)$ e-folds before the end of inflation. Scales that will eventually contain a galaxy mass ($M = 10^{11}M_\odot$) crossed the horizon 50 e-folds before the end of inflation, while scales that will eventually contain a galaxy cluster mass ($M = 10^{14}M_\odot$) crossed the horizon 53 e-folds before the end of inflation.

So far it has been assumed that the ϕ field is constant. However there are quantum fluctuations in ϕ due to the fact that during the slow-roll epoch the Universe is approximately in a de Sitter phase [18,19,20,21]. If the fluctuations $\delta\phi$ are expressed as a Fourier expansion

$$\delta\phi = (2\pi)^{-3} \int d^3k \delta\phi_k \exp(-i\vec{k}\cdot\vec{x}), \tag{2.20}$$

then the de Sitter fluctuations result in (note: $\Delta\phi \equiv \Delta\phi_k$)

$$(\Delta\phi)^2 \equiv (2\pi)^{-3}k^3|\delta\phi_k|^2 = \left(\frac{H}{2\pi}\right)^2. \tag{2.21}$$

These fluctuations obtain on scales less than the physics horizon during the de Sitter phase. As each scale goes outside the horizon during slow roll, it has fluctuations $(\Delta\phi)^2 = (H/2\pi)^2$. Since the energy density depends upon ϕ, the fluctuations in ϕ lead to fluctuations in ρ of $\delta\rho = (\partial V/\partial\phi)\Delta\phi$. Using $\rho \simeq V \simeq \sigma^4$ and $\partial V/\partial\phi = -3H\dot{\phi}$, fluctuations in ϕ lead to

$$\left(\frac{\delta\rho}{\rho}\right)_k \simeq \left(\frac{\dot{\phi}H^2}{\sigma^4}\right). \tag{2.22}$$

Once the scale is larger than H^{-1}, it can no longer be affected by microphysics. The behavior of the perturbation outside the horizon is gauge-dependent. However the behavior outside the horizon can be characterized by a parameter ς, given by

$$\varsigma \equiv \frac{\delta\rho}{\rho + p} \simeq \begin{cases} \delta\rho/\rho & \text{FRW} \\ \delta\rho/\dot{\phi}^2 & \text{de Sitter.} \end{cases} \tag{2.23}$$

When a scale comes back within the horizon during the FRW phase, ς is the same as when it first went outside the horizon during inflation. Therefore, $(\delta\rho/\rho)$ relevant for galaxy formation is given by [17]

$$\left(\frac{\delta\rho}{\rho}\right)_H \simeq \left(\frac{-3H\dot{\phi}\Delta\phi}{\dot{\phi}^2}\right)_H \simeq \left(\frac{H^2}{\dot{\phi}}\right)_H. \tag{2.24}$$

With the approximation that H and $\dot{\phi}$ are constant during the slow-roll phase, $(\delta\rho/\rho)$ as it re-enters the horizon will be scale free . In the slow-roll period, $\dot{\phi} = -V'(\phi)/3H$, and the equation for $(\delta\rho/\rho)$ becomes

$$\left(\frac{\delta\rho}{\rho}\right)_H \simeq \left(\frac{-3H^3}{V'(\phi)}\right). \tag{2.25}$$

2.4 *Specific Models*

The first example considered is the original attempt to implement new inflation. The model is based upon a SU_5 GUT with symmetry breaking via the Coleman-Weinberg mechanism [14,15]. The scalar field responsible for inflation (hereafter referred to as the *inflaton*) is in the 24-dimensional representation of SU_5 and is responsible for the symmetry breaking $SU_5 \to SU_3 \times SU_2 \times U_1$. Let ϕ denote the magnitude of the Higgs field in the $SU_3 \times SU_2 \times U_1$ direction. The one-loop, zero-temperature Coleman-Weinberg potential is

$$V(\phi) = B\sigma^4/2 + B\phi^4\left[\ln(\phi^2/\sigma^2) - 1/2\right], \tag{2.26}$$

where $B = 25\alpha_{GUT}^2/16 \simeq 10^{-3}$, and $\sigma = 2 \times 10^{15}$GeV. Because of the absence of a mass term, the potential is very flat near the origin (SSB arises due to one-loop radiative corrections). For $\phi \ll \sigma$, the potential may be approximated in the slow-roll regime by

$$V(\phi) \simeq B\sigma^4/2 - \lambda\phi^4/4; \qquad \lambda \simeq |4B\ln(\phi^2/\sigma^2)| \simeq 0.1. \tag{2.27}$$

For $\phi \ll \sigma$

$$V(\phi) \simeq B\sigma^4/2; \qquad H^2 = \frac{8\pi G\rho}{3} \simeq \frac{4\pi}{3}\frac{B\sigma^4}{m_{Pl}^2}. \tag{2.28}$$

The critical temperature for this potential is about 10^{14}GeV. The finite temperature potential has a small temperature-dependent barrier near the origin, and it is not until $T = 10^9$GeV or so that this barrier is low enough that the action for bubble nucleation drops to order unity. At this time the Universe will undergo "spinodal decomposition" and break up into irregularly shaped fluctuation regions within which ϕ is approximately constant.

Consider the evolution of ϕ in the slow-roll regime. Slow roll commences at ϕ_s and ends at ϕ_e. The end of slow roll is determined by the condition $|V''(\phi_e)| = 9H^2$, or $\phi_e^2 = 3H^2/\lambda$. For any ϕ in the region $\phi_s \leq \phi \leq \phi_e$, the number of e-folds from ϕ to ϕ_e (time t to time t_e) is given by

$$N(\phi \rightarrow \phi_e) = \int_t^{t_e} H dt = \int_\phi^{\phi_e} H\dot{\phi}^{-1} d\phi. \tag{2.29}$$

Using $3H\dot{\phi} = -dV/d\phi$ during slow roll,

$$N(\phi \rightarrow \phi_e) = \frac{3H^2}{2\lambda}\left(\frac{1}{\phi^2} - \frac{1}{\phi_e^2}\right). \tag{2.30}$$

The total number of e-folds in slow roll depends upon ϕ_s. To have enough inflation, $N(\phi_s \rightarrow \phi_e)$ must be greater than 58. Since λ is 10^{-1}, ϕ_s must be smaller than H in order to have sufficient e-folds. However de Sitter space fluctuations introduce uncertainties in ϕ of this order. The quantum fluctuations may prematurely terminate inflation. At the very least they suggest that the semiclassical equations of motion may be invalid.

More serious is the magnitude of the density fluctuations [18,19,20,21]. During slow roll for the Coleman-Weinberg potential $V'(\phi) \simeq \lambda\phi^3$, and Eq. 2.25 gives

$$\left(\frac{\delta\rho}{\rho}\right)_H \simeq \frac{3H^3}{\lambda\phi^3} \simeq \left(\frac{\lambda}{3}\right)^{1/2} [2N(\phi \rightarrow \phi_e)]^{3/2}, \tag{2.31}$$

where Eq. 2.30 has been used to express ϕ in terms of the number of e-folds before the end of inflation. Although $(\delta\rho/\rho)$ depends upon N to a power, N depends upon the *logarithm* of the length or mass scale, so the scale dependence of $(\delta\rho/\rho)$ is only logarithmic. The problem with the Coleman-Weinberg potential is not the spectrum, but the magnitude of the perturbations. Using $\lambda = 0.1$ and $N(\phi \rightarrow \phi_e) = 58 + (1/3)\ln(M/10^{21}M_\odot)$, $(\delta\rho/\rho)_H$ on the scale of galaxies is 182, and on the scale of clusters is 199. The spectrum is very flat, but about 10^6 too large. Notice that a smaller λ cures both problems.

Although the original model for new inflation was a failure, it pointed the way for the construction of somewhat more successful models. The potential of the original Coleman-Weinberg model was not flat enough, i.e., λ was too large. If ϕ couples to gauge fields, λ will be of order α_{GUT}^2, which is too large. If ϕ is a weakly-coupled gauge singlet, the effective λ can be small, and will remain small after radiative corrections. If $\lambda \leq 10^{-12}$, the density perturbations from Eq. 2.31 will be small enough. However a weakly-coupled inflaton will have a small decay width, and the reheat temperature will be low. If λ is also the magnitude for the coupling of the inflaton to other fields, the decay width at the minimum will be $\Gamma_\phi \simeq \lambda^2 m_\phi \simeq \lambda^2 \sigma$, and the reheat temperature will be $T_{\rm RH} \simeq (\Gamma_\phi m_{Pl})^{1/2} \simeq 10^5$GeV for $\lambda = 10^{-12}$ and $\sigma = 10^{15}$GeV. A more careful calculation may give one or two orders of magnitude larger value of $T_{\rm RH}$, but it is clear that a weakly-coupled field will have a low reheat temperature. This presents a problem for baryogenesis. Any baryon asymmetry present before inflation will be diluted due to the creation of the large amount of entropy, so it is necessary to create the baryon asymmetry either during or after the reheating epoch. Many inflation models are squeezed between the requirement of a weakly coupled inflaton for a flat potential and an inflaton that has a large enough decay width to give $T_{\rm RH}$ large enough for baryogenesis.

Supersymmetric models have been proposed as a mechanism to stabilize small couplings in the inflaton potential against radiative corrections. Supersymmetric models introduce several additional potential problems. The high-temperature minimum of the potential is generally not at $\phi = 0$, and $\langle\phi\rangle$ may smoothly evolve to the zero-temperature minimum. There are two possible solutions. If the high-temperature minimum is at $\phi \leq 0$, there will always be a barrier between the high-temperature and the low-temperature minimum. The other solution is to ignore the problem. Since the inflaton must be weakly coupled, it may never be in LTE, and the initial value of ϕ may be random. Another problem with supersymmetric models is the gravitino problem. Gravitinos are weakly-interacting, long-lived particles present in supersymmetric models. They will be produced in reheating in embarrassingly large numbers unless the reheat temperature is less than about 10^9GeV. Finally, in supersymmetric models where supersymmetry breaking is done with a Polonyi field, the Polonyi field can be set into oscillations that will not decay because the Polonyi field is "hidden." Since the energy density in the oscillating field behaves like non-relativistic matter, it will eventually come to dominate the Universe.

For successful new inflation, several requirements must be fulfilled. The requirements occur during different periods of inflation [22].

• *Start Inflation*: The scalar field must be smooth in a region such that the energy density and pressure associated with spatial gradients in ϕ are smaller than the potential energy. If the average value of ϕ is ϕ_0 and the region has typical spatial dimension L, this requirement implies $(\nabla\phi)^2 = O(\phi_0/L) \ll V(\phi_0) = O(\sigma^4)$. If this requirement is not met and the $(\nabla\phi)^2$ term dominates, $R(t)$ will expand as t to a power and inflation will not occur. However once $V(\phi)$ does dominate, the gradient terms rapidly become small in the exponential expansion and can be ignored.

In supersymmetric models where LTE is obtained, the high-temperature minimum of $V(\phi)$ should be at $\phi \leq 0$ to prevent ϕ from smoothly evolving to the zero-temperature minimum without inflating.

• *Start Slow Roll*: If ϕ is not a gauge singlet it may roll in the "wrong" direction. For instance for the Coleman-Weinberg SU_5 model, the steepest direction for ϕ near the origin is toward a minimum where $SU_4 \times U_1$ is the unbroken symmetry. If ϕ is a gauge singlet there is no problem with ending up in the wrong phase.

In order to have slow roll, the potential must have a flat region in which $|V''(\phi)| \leq 9H^2$ and $|V'(\phi)m_{Pl}/V(\phi)| \leq (48\pi)^{1/2}$.

• *Roll Far Enough*: The interval of slow roll, $[\phi_s, \phi_e]$, must be large enough that quantum fluctuations do not terminate slow roll. This condition will be met if $\phi_e - \phi_s \gg H$.

The number of e-folds, $N = \int H dt$ from ϕ_s to ϕ_e, must be greater than $58 + \ln(\sigma/10^{15}\text{GeV})$.

• *Small Perturbations*: The magnitude of the perturbations must be less than of order 10^{-4} on the scale of galaxies to clusters in order to avoid large fluctuations in the MBR. If the fluctuations produced in inflation are to lead to structure formation, they should be greater than of order 10^{-5}. Therefore during slow roll $H^2/\dot{\phi} \leq 10^{-4}$.

In addition to the scalar perturbations discussed so far, inflation will produce tensor perturbations. These tensor perturbations can be thought of as gravity waves. As each scale leaves the horizon during inflation the energy density of gravity waves on that scale is $\rho_{GW} \simeq H^4$. In terms of a dimensionless amplitude $h = H/m_{Pl}$ and wavelength λ, $\rho_{GW} \simeq (m_{Pl}^2 h^2/\lambda^2)_{\lambda=H^{-1}}$. These gravity waves will re-enter the horizon during the FRW phase with the same dimensionless amplitude h, and induce an anisotropy in the MBR of order h. For $\delta T/T \leq 10^{-4}$, $h = H/m_{Pl} \leq 10^{-4}$. Since $H \simeq \sigma^2/m_{Pl}$, σ must be less than about 10^{17}GeV.

• *Exit Properly*: The reheat temperature must be high enough so the Universe is radiation dominated during primordial nucleosynthesis. Using $T_{\rm RH} = (\Gamma_\phi m_{Pl})^{1/2}$, $T_{\rm RH} \geq 1$ MeV requires $\Gamma_\phi \geq 10^{-25}$GeV. If baryogenesis proceeds in the standard way, then $T_{\rm RH} \geq 10^9$GeV, which implies $\Gamma_\phi \geq 10^{-1}$GeV. In order to ameliorate the problem of low reheat temperature and baryogenesis, it has been proposed that a baryon asymmetry is created by the decay of the inflaton. The energy density in the coherent oscillations can be thought of as due to a collection of zero momentum inflatons with number density $n_\phi = \rho_\phi/m_\phi$. In reheating, $\rho_\phi \to g_* T_{\rm RH}^4$, so $n_\phi = g_* T_{\rm RH}^4/m_\phi$ at reheating. Suppose the inflaton decays into a particle, S, which, in turn, decays out of equilibrium with baryon number violation. The number density of S's that decay is the same as the number density of parent inflatons. If the CP parameter in the decay of the S is ϵ, then the asymmetry in baryons produced by the S is $n_B = \epsilon n_\phi = \epsilon g_* T_{\rm RH}^4/m_\phi$. The entropy density produced after thermalization of the inflaton decay products is $s = g_* T_{\rm RH}^3$. Therefore $B \equiv n_B/s = \epsilon T_{\rm RH}/m_\phi$. If $B \geq 10^{-10}$, then $T_{\rm RH} \geq 10^{-10} m_\phi/\epsilon$.

There is a model-dependent upper limit on $T_{\rm RH}$ to avoid making unwanted relics. For example, in supersymmetric models, $T_{\rm RH} \leq 10^9$GeV to avoid overproducing gravitinos.

The above problems and some possible solutions are given in Table 1. Although there are models that satisfy all the above requirements, none of them seem so compelling that they must be the final answer. In fact, in the past few years there has been increasing effort in the generalization of inflation as a phenomena that is decoupled from a cosmological phase transition.

2.5 *Present Status and Future Directions*

Although the general scenario of inflation presents a very attractive means to ameliorate at least some of the untidiness of the standard model, it is by no means clear that all (or even any) problems are solved or understood. It is now clear that there are models, both supersymmetric and non-supersymmetric, which can successfully implement the program of new inflation as outlined above. It is useful to normalize the more non-standard models of inflation by comparing them to these two "standard" models of inflation.

The non-supersymmetric model is a GUT model based upon SU_5. The model was first proposed by Shafi and Vilenkin [23], and refined by Pi [24]. In the latest version of the model the inflaton is the real part of a complex gauge-singlet field

Table 1: Possible problems and solutions in new inflation

EPOCH	PROBLEM	POSSIBLE SOLUTION
Start	ϕ Smooth	$(\nabla\phi)^2 \ll V(\phi)$
Inflation	Thermal Constraint	$\langle\phi\rangle \leq 0$
Execute	Roll in Right Direction	ϕ is gauge singlet
Slow Roll	Flat Region in $V(\phi)$	$\lvert V''(\phi)\rvert \leq 9H^2$, and
		$\lvert V'(\phi)m_{Pl}/V(\phi)\rvert \leq (48\pi)^{1/2}$
Roll Far	Quantum Fluctuations	$\phi_e - \phi_s \gg H$
Enough	Sufficient e-folds	$N = \int H dt \geq 58$
Small	Scalar Perturbations	$(H^2/\dot{\phi}) \leq 10^{-4}$
Perturbations	Tensor Perturbations	$H/m_{Pl} \leq 10^{-4}$
Exit Properly	Nucleosynthesis	$T_{\rm RH} \geq 1$ MeV
	Baryogenesis	$T_{\rm RH} \geq 10^{-10}\epsilon^{-1}m_\phi$
	Gravitinos	$T_{\rm RH} \leq 10^9$GeV

with a Coleman-Weinberg potential of the form in Eq. 2.26, with ϕ representing the magnitude of the complex field, and $B = O(10^{-14})$. It must be assumed that the couplings of the ϕ to all other fields in the theory are less than about 10^{-7} to prevent quantum corrections from spoiling the smallness of B. The real part of ϕ will be the inflaton, and the imaginary part of ϕ will be the axion. ϕ couples to the adjoint Higgs, and induces SU_5 breaking when it receives a VEV. This requires $\sigma = 10^{18}$GeV. Since B is so small (and will remain small after radiative corrections), the problems with the original Coleman-Weinberg SU_5 model vanish. The reheat temperature is barely high enough to produce a baryon asymmetry through the decay of the inflaton as discussed above. At the expense of introducing a small number, the model is simple and it works.

An example of a supersymmetric model that works was proposed by Holman, Ramond, and Ross [25]. The superpotential in their model has a "inflation sector" with superpotential $I = (\Delta^2/M)(\phi - M)^2$, where $M = m_{Pl}/(8\pi)^{1/2}$. The scalar potential in supersymmetric models is typically an expansion in ϕ/M, given in this case by

$$V(\phi) = \Delta^4(1 - 4\phi^3/M^3 + 6.5\phi^4/M^4 - 8\phi^5/M^5 + \cdots). \tag{2.32}$$

For $\Delta/M \simeq 10^{-4}$, ($\Delta \simeq 2 \times 10^{14}$GeV), density fluctuations are small enough and sufficient e-folds obtain. The decay width of the ϕ (which has only gravitational coupling to other fields) is $\Gamma_\phi \simeq \Delta^6/M^5$, which for Δ small enough to satisfy the perturbations constraint, leads to $T_{\rm RH} \simeq (\Gamma_\phi m_{Pl})^{1/2} \simeq 10^6$GeV. With the baryon asymmetry produced via inflaton decay, this is large enough. At the expense of the introduction of a sector whose sole purpose is inflation, the model is simple and it works.

Both the above models have two potential problems. The first problem is that to this point the calculations of the evolution of the scalar field have been semi-classical. It may be that a true quantum calculation of the evolution of ϕ, including production of density perturbations, will give a result much different than the semi-classical result. Preliminary work on this problem suggests that the semi-classical approximations are reasonable. The second potential problem has to do with the initial value of ϕ. Both fields are extremely weakly coupled and are unlikely ever to be in LTE. There is no reason to assume $\phi \simeq 0$ for an initial condition (in fact, it may not even be the high-temperature minimum for the supersymmetric example). It is tempting to say that this is not a problem, and that it is only necessary for $\phi \simeq 0$ in some region of the Universe where the kinetic contributions to ρ are small enough to start inflation.

The above two models are existence proofs that it is possible to implement new inflation. Whether new inflation is the final answer will be discussed briefly after mentioning some other approaches for inflation that do not involve SSB.

For weakly coupled scalar fields there is no reason to believe the inflaton will be in LTE at high temperature, and the value of ϕ at high temperature might be random (hence the name "chaotic inflation"). Imagine a simple scalar potential of the form $V(\phi) = \lambda\phi^4$, with minimum at $\langle\phi\rangle = 0$. Assume as initial conditions that $\phi = \phi_0 \neq 0$, and that ϕ is sufficiently smooth in a large enough region to inflate. The number of e-folds of inflation is

$$N(\phi \rightarrow 0) = \int_\phi^0 H dt \simeq \pi \left(\frac{\phi}{m_{Pl}}\right)^2. \tag{2.33}$$

In order to obtain 58 e-folds of inflation, $\phi_0 \geq 4.3 m_{Pl}$. The density perturbations are

$$\left(\frac{\delta\rho}{\rho}\right)_H \simeq \left(\frac{3H^3}{V'(\phi)}\right) \simeq \lambda^{1/2}\left(\frac{\phi}{m_{Pl}}\right)^3 \simeq \lambda^{1/2} N(\phi \rightarrow 0)^{3/2}. \tag{2.34}$$

Again, using $N = 50$, λ must be smaller than about 10^{-14} for sufficiently small density perturbations. Since Linde [26] originally proposed this model several refinements have been made. First, it has been shown that it is possible to use a $m^2\phi^2$ potential rather than a $\lambda\phi^4$ potential. Some work has been done in examining and formalizing what exactly is meant by "chaotic" initial conditions, and which regions of phase space will inflate. Linde's model is an example of how general inflation is, and that it is possible, perhaps even desirable, to separate inflation from SSB phase transitions. Chaotic inflation (at least for the $\lambda\phi^4$ case) has the possible problem of using classical gravity in the regime $\phi \geq m_{Pl}$. At present it also has the undesirable feature of involving the dynamics of a scalar field introduced for the sole purpose of inflation.

A model even further from the original idea of an SSB phase transition is a pure gravity model based upon including an ϵR^2 term in the gravity Lagrangian. Such higher-derivative terms are expected to be present in theories with extra dimensions. Mijić, Morris, and Suen [27] have examined this possibility in detail, including questions of density perturbations and reheating and find that all constraints can be met for $10^{11} \leq \epsilon^{-1/2} \leq 10^{13}$GeV.

Yet further from the original idea of inflation is the possibility that the inflaton is related to the size of extra dimensions. This will not be discussed here. A possibility discussed elsewhere in this volume is the role of quantum gravity and the program of the "wave function of the Universe."

In a Universe without inflation, the space of initial conditions that give the Universe we observe is a set of measure zero. The inflationary Universe enlarges the space of initial data that will lead to the observable Universe. However, it does not imply that every imaginable set of initial data will lead to inflation. A trivial example is a closed Universe that becomes curvature dominated, and collapses before the vacuum energy dominates and causes inflation. The question "is inflation inevitable" has not yet been completely answered. Inflation may be the final answer, part of the final answer, or none of the final answer. This is discussed further in other lectures [28].

If inflation did occur there are two general predictions. The first prediction is that Ω is very close to 1. It would be hard to imagine that *exactly* 58 e-folds of inflation occurred. With all models that give small density perturbations, the number of e-folds of inflation is enormous, and the intrinsic curvature will only appear on scales far larger than our present horizon. Of course, scale-free density perturbations would appear on the horizon today, so a $(\delta\rho/\rho) \simeq 10^{-4}$ would lead to $\Omega = 1 \pm 10^{-4}$. The second prediction is that of scale-free density perturbations. At present there is no convincing data to support either prediction. Dynamical measurements of Ω seem to give $\Omega = 0.1 \rightarrow 0.3$. This has (at least) three possible explanations. Either there are systematic uncertainties in all the measurements, there is unclustered matter (like massless particles) that give the unseen part of Ω, or there is a present vacuum energy that can account for spatial flatness (the actual prediction of inflation) and $\Omega \neq 1$. None of these explanations are compelling. If the recent determination of the velocity field on large-scales are correct, it is evidence against a scale-free spectrum. Possible ways out are the measurements are wrong, cosmic strings, and double inflation.

The last point is that some explanation must be found for the present smallness of the cosmological constant.

ACKNOWLEDGEMENTS

I would like to thank Frank Accetta, Richard Holmam, and So-Young Pi for carefully looking through these notes. The section on inflation has been strongly influenced by Michael Turner's views and writings on the subject [29]. This work has been partially supported by the Department of Energy and the National Aeronautics and Space Administration.

REFERENCES

1. D. A. Kirzhnits and A. D. Linde, Sov. Phys. JETP **40**, 628 (1974).
2. L. Dolan and R. Jackiw, Phys. Rev. D. **9**, 3320 (1974); S. Weinberg, Phys. Rev. D. **9**, 3357 (1974).
3. S. Coleman, Phys. Rev. D **15**, 2929 (1977); C. Callan and S. Coleman, Phys. Rev. D **16**, 1762 (1977); S. Coleman and F. De Luccia, Phys. Rev. D **21**, 3305 (1980). The tunneling rate is associated with a classical motion in imaginary time because the decay rate is related to the imaginary part of the energy. This is because the wave function oscillates in the classically allowed region, but is exponentially damped in the classically forbidden region.
4. A. Linde Nucl. Phys. **B216**, 421 (1983).
5. Ya. B. Zel'dovich, I. Yu. Kobzarev, and L. B. Okun, Sov. Phys. JETP **40**, 1 (1975).
6. T. W. B. Kibble, J. Phys. A **9**, 1387 (1976). This is an excellent paper that is required reading in the subject.
7. A. Vilenkin, Phys. Rep **121**, 263 (1985). This is a detailed and well-written review that contains many details not included here.
8. H. B. Nielsen and P. Olesen, Nucl. Phys. **B61**, 45 (1973).
9. See D. Eardley, this volume.

10. G. 't Hooft, Nucl. Phys. **B79**, 276 (1974); A. M. Polyakov, JETP Lett. **20**, 194 (1974).

11. J. Preskill, Ann. Rev. Nucl. Part. Sci. **34**, 461 (1984).

12. Preskill, Phys. Rev. Lett. **43**, 1365 (1979).

13. A. Guth, Phys. Rev. D **23**, 347 (1981).

14. A. Albrecht and P. Steinhardt, Phys. Rev. Lett. **48**, 1220 (1982).

15. A. Linde, Phys. Lett. **108B**, 389 (1982).

16. It is crucial to remember that $m_\phi \equiv \partial^2 V(\phi)/\partial\phi^2$ is a function of ϕ, and will be small in the flat region of the potential.

17. There should be no confusion between the sub-H which indicates the quantity is to be evaluated at the time of horizon crossing, and the expansion rate H.

18. A. Starobinsky, Phys. Lett. **117B**, 175 (1982).

19. J. M. Bardeen, P. Steinhardt, and M. S. Turner, Phys. Rev. D **28**, 679 (1983).

20. A. Guth and S.-Y. Pi, Phys. Rev. Lett. **49**, 1110 (1982).

21. S. Hawking, Phys. Lett. **115B**, 295 (1982).

22. P. Steinhardt and M. S. Turner, Phys. Rev. D **29**, 2162 (1984).

23. Q. Shafi and A. Vilenkin, Phys. Rev. Lett. **52**, 691 (1984).

24. S.-Y. Pi, Phys. Rev. Lett. **52**, 1725 (1984).

25. R. Holman, P. Ramond, and G. Ross, Phys. Lett. **137B**, 343 (1984).

26. A. D. Linde, Phys. Lett. **129B**, 177 (1983).

27. M. B. Mijić, M. S. Morris, and W.-M. Suen, "The R^2 Cosmology - Inflation Without a Phase Transition," Caltech Report CATT-68-1320, (Feb. 1986).

28. See the contributions of J. Hartle and J. Barrow in this volume.

29. M. S. Turner, in *Proceedings of the 1984 Cargese School on Fundamental Physics and Cosmology*, ed. J. Audouze and J. Tran Thanh Van (Editions Frontieres, Gif-Sur-Yvette, 1985).

PREDICTION IN QUANTUM COSMOLOGY

James B. Hartle

Department of Physics
University of California
Santa Barbara, CA 93106, USA

1. INTRODUCTION

As far as we know them, the fundamental laws of physics are quantum mechanical in nature. If these laws apply to the universe as a whole, then there must be a description of the universe in quantum mechancial terms. Even our present cosmological observations require such a description in principle, although in practice these observations are so limited and crude that the approximation of classical physics is entirely adequate. In the early universe, however, the classical approximation is unlikely to be valid. There, towards the big bang singularity, at curvatures characterized by the Planck length, $(\hbar G/c^3)^{\frac{1}{2}}$, quantum fluctuations become important and eventually dominant.

Were the aim of cosmology only to describe the present universe, expressing that description in quantum mechanical terms might be an interesting intellectual exercise but of no observational relevance. Today, however, we have a more ambitious aim: to explain the presently observed universe by a simple and compelling law of initial conditions. It is natural to expect such a law to be quantum mechanical for several reasons: A law of initial conditions must describe the early universe where quantum gravitational fluctuations are important. In quantum fluctuations we can imagine a simple origin of present complexity. Finally, if all the other fundamental laws of physics are quantum mechanical, it is only natural to expect the law of initial conditions to be so also. It is for the search for a law of initial conditions that we need a quantum mechanical description of the universe – a quantum cosmology.[1,2–5] The nature of this description is the subject of these lectures.[6]

In the application of quantum mechanics to the universe as a whole, one confronts the characteristic features of quantum theory in a striking and unavoidable manner. Some find these features uncomfortable, or unsatisfactory or even absurd. It is not the purpose of these lectures to examine whether these attitudes represent a success or failure of intuition. Rather, the purpose is to sketch how the standard quantum theory, or a suitable generalization of it, can be used to frame a law of initial conditions and to extract from it predictions for cosmological observation. We shall thus assume quantum mechanics.

We shall also assume spacetime. While it is uncertain whether Einstein's vision of spacetime as a fundamental dynamical quantity is correct, it is perhaps the most compelling viewpoint in which to frame a quantum theory of cosmology. Within this framework of quantum spacetime our discussion will, in large part, be general and not single out any particular dynamics or particular theory of initial conditions.

The aim of these lectures is to show how the single system which is our universe is described in a quantum theory of spacetime and to sketch how a prescription for the quantum state of the universe can be used to make verifiable predictions. This is discussed in general in Sections 2 and 3. However, as the observations which are accessible to us are describable in classical terms, extracting predictions in the classical limit is a particular problem of special importance. This is considered in Section 4. As an interesting by product of this discussion, we describe the connection between a quantum theory of spacetime and the approximation of quantum field theory in curved spacetime.

2. PREDICTIONS FROM THE WAVE FUNCTION OF THE UNIVERSE

2.1 The Wave Function of the Universe

In quantum mechanics we describe the state of a system by giving its wave function. The wave function enables us to made predictions about observations made on a spacelike surface; it thus captures quantum mechanically the classical notion of the "state of the system at a moment of time." The arguments of the wave function are the variables describing how the system's history intersects the spacelike surface. For example, for the quantum mechanics of a particle, the histories are particle paths $x(t)$. We write for the wave function

$$\psi = \psi(x,t). \tag{2.1}$$

The t labels the hypersurface and the x specifies the intersection of the history with it.

In the quantum mechanics of a closed cosmologies with fixed (for simplicity) spatial topology, say that of a 3-sphere S^3, the histories are the 4-geometries, ${}^4\mathcal{G}$, on $\mathbf{R} \times S^3$. The appropriate notion of a 4-geometry fixed on a spacelike surface is the 3-geometry, ${}^3\mathcal{G}$, induced on that surface. One can think of this as specified by a 3-metric $h_{ij}(\mathrm{x})$ on the fixed spatial topology. Thus for the quantum mechanics of a closed cosmology we write[7]

$$\Psi = \Psi[{}^3\mathcal{G}] = \Psi[h_{ij}(\mathrm{x})]. \tag{2.2}$$

Note that there is no additional "time" label. This is because a generic 3-geometry will fit into a generic 4-geometry at locally only one place if it fits at all. The 3-geometry itself carries the information about its location in spacetime. For example, the 4-geometry of a closed Friedman universe is described by the metric

$$ds^2 = -d\tau^2 + a^2(\tau)d\Omega_3^2, \tag{2.3}$$

where $d\Omega_3^2$ is the metric on the round 3-sphere. A 3-geometry is described by the radius of a 3-sphere. This radius locates us locally in the spacetime although in the large there are two values of τ for each value of a. This labeling of the wave function correctly counts the degrees of freedom. Of the six components of h_{ij}, three are gauge. If one of the remaining three is time, there are left two degrees of freedom — the correct number for the massless, spin-2 gravitational field.

The space of all three geometries is called superspace. Each "point" represents a different geometry on the fixed spatial topology. In the case of pure gravity that we have been describing, the wave function is a complex function on superspace. With the inclusion of matter fields the wave function depends on their configurations on the spacelike surface as well, and we write typically

$$\Psi = \Psi[h_{ij}(\mathbf{x}), \Phi(\mathbf{x})]. \tag{2.4}$$

A law for initial conditions in quantum cosmology is a law which prescribes this wave function.

2.2 Cosmological Observations and Cosmological Predictions

To make contact with observations we must specifiy the observational consequences of the state of the universe being described by this or another wave function. This is usually called an "interpretation" of Ψ. There is little doubt that what I can say here will not address every issue which can be raised on this fascinating topic and even less doubt that it will not satisfy many who have thought about the subject. I would like, however, to offer some minimal elements of an interpretation which I believe will enable an attribution of Ψ to the universe to be confronted by cosmological observations. These elements are an example of "an Everett interpretation" although the words and emphasis may be different from other interpretations in this broad catagory.[8]

There are at least three problems to be addressed (1) the special nature of cosmological predictions, (2) the quantum mechanics of single systems, and (3) the problem of time. We shall discuss them in order.

2.3 The Nature of Cosmological Predictions

The favorite paradigm of prediction in physics is evolution: If we start the system in a certain state then a time t later we predict that it will be in a certain other state. This type of problem conforms to the characteristic form of predictive statements. If "this" then "that"– a correlation between experimental conditions and observations. In placing conditions and observations in temporal order, however, it is very uncharacteristic of predictions we can make in the astrophysical and geological sciences. In geology we might predict as follows: "If we are in a certain type of strata, then we should find a certain type of dinosaur bone." "If we are in the middle of an ocean floor, then we predict an upwelling trench" and so on. Here condition and observation are at the same time. This, prediction of correlations *at the same time* is, I believe, characteristic of systems over which we have little experimental control.

Cosmology is much the same. We can, of course, imagine a 10^9 year experiment "Given the observations of the positions of the galaxies now, we predict that if we wait 10^9 years, we will see them in new positions $\cdots$." Such a prediction *is* a test of a theory of initial conditions because the longer we wait the more initial data we see. It is, however, not very practical and therefore not very interesting.

A more interesting type of prediction is a 10^9 franc experiment: Suppose you are allocated 10^9 francs to build new optical, radio, X-Ray, neutrino and gravitational wave telescopes, what do you predict you will see? A typical prediction might be the following: "Given the locally measured values of the Hubble constant and the mass density, we predict that at great distances we will see the same uniform mass density, a certain galaxy-galaxy correlation function, a certain gravitational wave background, etc." Characteristically these predictions involve conditions and observations at a single moment of time.

2.4 Quantum Mechanics of Individual Systems

The idea for dealing with the universe as a single system is to take quantum mechanics seriously. One assumes that there is one wave function Ψ defined on a preferred configuation space which contains all the predictable information about observations made on a spacelike surface. *If Ψ is sufficiently peaked about some region in the configuration space, we predict that we will observe the correlations between the observables which characterize this region. If Ψ is small in some region, we predict that observations of the correlations which characterize this region are precluded. Where Ψ is neither small nor sufficiently peaked we don't predict anything.* That's it.

The natural reaction to such a proposal for interpretation is to ask "Where is probability?" In response, two things can be said. First, probabilities for single systems have no direct observational meaning and the universe, by definition, is a single system. Second, as we shall show below, this interpretation implies the usual probability interpretation of ordinary quantum mechanics when applied to ensembles of identically prepared systems.

In cosmology, therefore, we would examine any particular proposal for Ψ to see which correlations are predicted – those on which the wave function is sharply peaked, and which are precluded – those on which it is essentially zero. We would ask, for example: "Given the value of the Hubble constant and the local mean density is the wave function sharply peaked about a form of the galaxy-galaxy correlation function?" If so we predict that correlation of variables. Note that characteristically we have conditions and observations on a single spacelike surface. This type of interpretation means that one's ability to predict in quantum cosmology is very limited. Given a value of the Hubble constant and the local mean mass density, one can ask whether the wave function sharply peaked about the number of planets in the solar system, or the architecture of this building, or the weights of the participants of this conference. I, for one, hope not. One of the central problems in quantum cosmology is therefore to find what correlations are predicted and how specific must we be in conditions to get predictions for interesting observations. (Problem 1).

Ordinary quantum mechanics can be formulated as a theory of individual systems. Indeed, a moments reflection will show that this has to be so. Quantum mechanics formulated only in terms of probabilities would make definite predictions only about infinite ensembles – an idealization we do not encounter in the real world. Any ensemble can be regarded as a single system composed of many identical parts. Quantum mechanics should be formulable as a theory of individual systems and the probability interpretation derivable from the predictions this formulation makes about single systems with many identical subsystems. In the late '60's a number of workers independently showed how to do this:[9]

Consider a single system and let it be described by a wave function ψ. Possible observations correspond to operators in the Hilbert space of states. For the physical interpretation of ψ for an individual system assume only the following: *If ψ is an eigenfunction of an observable A then an observation of A will yield the eigenvalue. For those observables of which ψ is not an eigenfunction there is no prediction for the outcome of an observation.* We can then derive the probability interpretation of ψ as follows:

Suppose the configuration space of the single system is C; the configuration space for an ensemble of N systems is C^N. An ensemble of N systems each in the identical state $\psi(q)$ is described by the wave function on C^N

$$\Psi(q_1, \cdots, q_n) = \psi(q_1)\psi(q_2)\cdots\psi(q_n). \tag{2.5}$$

On the Hilbert space of wave functions on C^N there is an operator $\hat{f}_a$ corresponding to observing q on the first system, q on the second, etc., and then computing the frequency that a given value a occurs. For an infinitely large ensemble of identical systems, each in a state ψ, is it a mathematical fact that the product wave function (2.5) is an eigenfunction of this operator

$$\hat{f}_a \Psi = |\psi(a)|^2 \Psi. \tag{2.6}$$

The predicted frequency is the square of the wave function of the single system. In this way we deduce the probability interpretation of quantum mechanics from its predictions about individual systems.

To see how this works let us consider a definite example. Consider ensembles of spin -1/2 systems. A single system has states $|S>, S=\uparrow$ or $\downarrow$, and the Hilbert space of an ensemble of N systems is spanned by the basis

$$|S_1>|S_2>\cdots|S_N>. \tag{2.7}$$

In this basis we can define the operator corresponding to a measurement of the relative frequency of say spin up, $\uparrow$

$$f_\uparrow^N = \sum_{S_1\cdots S_N} |S_1>\cdots|S_N> \left(\sum_{S_i} \frac{\delta_{S_i\uparrow}}{N}\right) < S_N|\cdots < S_1|. \tag{2.8}$$

Consider now the expectation value of $f_\uparrow^N$ in the state of an ensemble of identically prepared systems each in state $|\psi>$

$$|\psi^N> = |\psi>|\psi>\cdots|\psi>. \tag{2.9}$$

It is

$$\begin{aligned} <\psi^N|f_\uparrow^N|\psi^N> &= \sum_{S_1\cdots S_N}\left(\sum_{S_i}\frac{\delta_{\uparrow S_i}}{N}\right)|<S_1|\psi>|^2\cdots|<S_N|\psi>|^2 \\ &= |<\uparrow|\psi>|^2 \sum_{S_2\cdots S_N} |<S_2|\psi>|^2\cdots|<S_N|\psi>|^2 \\ &= |<\uparrow|\psi>|^2. \end{aligned} \tag{2.10}$$

Consider also the fluctuations about this mean value:

$$<\psi^N|(f_\uparrow^N - |<\uparrow|\psi>|^2)^2|\psi^N> = <\psi^N|(f_\uparrow^N)^2|\psi^N> - |<\uparrow|\psi>|^4. \tag{2.12}$$

The first term is

$$\begin{aligned} &\sum_{S_1\cdots S_N}\left(\sum_{i=j}\frac{\delta_{\uparrow S_i}}{N^2} + \sum_{i\neq j}\frac{\delta_{\uparrow S_i}S_{\uparrow S_j}}{N^2}\right)|<S_1|\psi>|^2\cdots|<S_N|\psi>|^2 \\ &= \frac{|<\uparrow|\psi>|^2}{N} + |<\uparrow|\psi>|^4\frac{N^2-N}{N^2}. \end{aligned} \tag{2.13}$$

Thus

$$<\psi^N|(f_\uparrow^N - |<\uparrow|\psi>|^2)^2|\psi^n> = \frac{1}{N}(|<\uparrow|\psi>|^2 - |<\uparrow|\psi>|^4) \to 0 \text{ as } N\to\infty \tag{2.14}$$

and we have indeed shown

$$||f_{\uparrow}^{N}|\psi^{N}> - |<\uparrow|\psi>|^{2}|\psi^{N}>|| \rightarrow 0, \tag{2.15}$$

which is (2.6).

The above derivation of the probability interpretation of ordinary quantum mechanics can be cast into the language used to interpret the cosmological wave function. "Superspace" is the configuration space C^N of the ensemble. Equations (2.10) and (2.15) show that for large N the wave function of an ensemble of identical systems, (2.5), is increasingly sharply peaked in the variable corresponding to a measurement of the frequency of spin - $\uparrow$ in the ensemble. The value about which it is peaked is $|<\uparrow|\psi>|^2$. Thus, we predict from the Ψ of (2.5) that a measurement of the frequency should yield this value.

This correspondance in language, however, points up an incompleteness in the interpretation of the cosmological wave function. To give a precise meaning to "sufficiently peaked" a measure is needed on configuration space. In ordinary quantum mechanics this is supplied by Hilbert space as the above derivation shows. However, there is no satisfactory Hilbert space formulation of quantum gravity. The reason is the problem of time.

2.5 The Problem of Time

Time plays a central and peculiar role in the formulation of Hamiltonian quantum mechanics. The scalar product specifying the Hilbert space of states is defined at one instant of time. States specify directly the probabilities of observations carried out at one instant of time. Time is the sole observable not represented by an operator in Hilbert space but rather enters the theory as a parameter describing evolution. In the construction of a quantum theory for a specific system, the identification of the time variable is a central issue.

In non-relativistic classical physics time plays a special role which is unambiguously transferred to quantum mechanics. In special relativistic quantum mechanics there is already an issue of the choice of time variable, but there is also a resolution. We can construct quantum mechanics using as the peculiar time variable the time associated with a particular Lorentz frame. The issue is whether the quantum theory, so constructed, is consistent with the equivalence of Lorentz frames. It is. There is a unitary relation between the quantum theories constructed in different Lorentz frames and physical amplitudes are therefore Lorentz invariant.

For the construction of quantum theories of spacetime the choice of time becomes a fundamental difficulty. A preferred foliation of spacetime by spacelike surfaces is necessary to formulate canonical quantum mechanics. The classical theory certainly singles out no such foliation, and we have no evidence that theories formulated on two different foliations are unitarily equivalent. There is thus a conflict between canonical quantum mechanics and general covariance.

To resolve this conflict we have, it seems to me, two choices (1) Modify general relativity so at the quantum mechanical level a preferred time is singled out, or (2) Modify quantum mechanics so it does not need a preferred time.

The first option has been much discussed. In these lectures I would like to offer a few thoughts about the second.

Feynman's sum over histories formulation of quantum mechanics is a natural alternative starting point for constructing quantum theories of spacetime in which the problem of time is neither as immediate nor as central as it is in Hamiltonian quantum mechanics. The basic ingredients of a sum over histories formulation are these:

(1) *The Histories*: A history $\mathcal{H}$ is the set of observables which describe all possible experiments. Examples are the particle paths of ordinary quantum mechanics or the 4-geometries of spacetime physics.

(2) *The Probability Amplitude for a History.* The joint probability amplitude for making all the observations which make up a history is

$$\Phi(\mathcal{H}) = \exp[iS(\mathcal{H})], \tag{2.16}$$

where $S(\mathcal{H})$ is the classical action for the history. For a particle this is

$$S[X(\tau)] = \int_{\tau'}^{\tau''} d\tau \left[\frac{1}{2}m\dot{X}^2 - V(X)\right], \tag{2.17}$$

or for spacetime with the dynamics of general relativity it is

$$S[{}^4\mathcal{G}] = \frac{1}{\ell^2}\int_M d^4x R\sqrt{-g} - \frac{2}{\ell^2}\int_{\partial M} d^3x h^{\frac{1}{2}} K. \tag{2.18}$$

(3) *Conditional Probability Amplitudes.* In particular experiments the observables can be divided into three classes (i) The conditions $\mathcal{C}$ – those observables fixed by the experimental arrangement. (ii) The observations $\mathcal{O}$ – the results of the experiments. (iii) The unobserved $\mathcal{U}$ – those observables neither conditioned nor observed. The conditional amplitude for $\mathcal{O}$ given $\mathcal{C}$ is (the principle of superposition)

$$\Phi(\mathcal{O}|\mathcal{C}) = \sum_{\mathcal{U}} \Phi(\mathcal{U}). \tag{2.19}$$

A measure, which is just as important as the action, is needed to define such sums.

(4) *Probability.* The relative probability that $\mathcal{O}$ occurs in a set of observations given the conditions $\mathcal{C}$ is $|\Phi(\mathcal{O}|\mathcal{C})|^2$. From this, the probabilities of one outcome of an exhaustive and exclusive set of observations can be computed by appropriate normalization according to the usual rules for probability.

Some further restrictions and caveats must be given, but this is the basic framework. The ideas will perhaps become clearer with an example illustrated in Figure 1. We consider a non-relativistic particle moving in one dimension. Suppose the particle starts at X_1 at time τ_1. It then passes through a slit of width Δ_2 at time τ_2. At τ_3 there is a coherent detector which registers whether the particle is in the interval Δ_3 disturbing the particle as little as possible. Finally at τ_4 the particle's position is detected. The conditions $\mathcal{C}$ for this experiment might be those imposed at times τ_1 and τ_2. A complete and exclusive set of observations would then be, X_4, the value of the position at τ_4 *and* whether the detector at τ_3 registered ("clicked") or did not. Given the experimental arrangement one of these possibilities must happen and no more than one can happen.

Let us compute the probability for the detector to click and the particle to be found at X_4 in a range Δ_4. The conditional probability amplitude $\Phi(X_4, \text{click}|\mathcal{C})$ is the sum over all paths which start at X_1 at τ_1 pass through the slit at τ_2, cross the detector volume at τ_3 and end at τ_4 at X_4. (Figure 1). The conditional amplitude $\Phi(X_4, \text{noclick}\,|\mathcal{C})$ would be a similar sum over paths which cross outside the detector volume at τ_3. The probability is the square of $\Phi(X_4,$ click $|C)$ normalized over the set of complete and exclusive possibilities. That is

$$P(X_4 \text{in} \Delta_4, \text{click}|C) = \Delta_4 |\Phi(X_4, \text{click}|C)|^2 \times \left[\int_R dX_4 |\Phi(X_4, \text{click}|C)|^2 + \int_R dX_4 |\Phi(X_4, \text{noclick}|C)|^2\right]^{-1}. \quad (2.20)$$

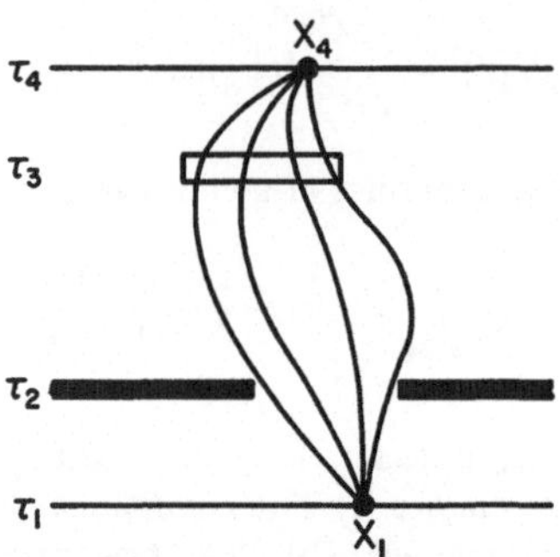

Fig. 1. An example illustrating the construction of conditional probability amplitudes. A particle is localized at τ_1, passes through a slit at τ_2, registers or does not register in a detector volume at τ_3, and its position is determined at τ_4. Given that the particle started at X_1 and passed through the slit, the conditional probability amplitude for it to register in the detector and be at X_4 is a sum over all paths which start at X_1 pass through the slit and detector volume and end at X_4.

The same result is predicted by ordinary quantum mechanics[10]. There one would say that the state at time τ_1 was $|X_1\tau_1>$, the state with the particle localized at X_1. After the localization at τ_2 the "wave packet is reduced" and the state is

$$|\psi_2\rangle \equiv N_2^{-2} \int_{\Delta_2} |x_2\tau_2\rangle\langle x_2\tau_1|X_1\tau_1\rangle, \quad (2.21)$$

with N_2 determined so the state has unit norm. At τ_3 the "wave packet is again reduced." The probability that the particle is inside the detector at τ_3 is

$$P(\text{click}) = \int_{\Delta_3} dx_3 |\langle x_3\tau_3|\psi_2\rangle|^2. \quad (2.22)$$

The state after detection is the normalized projection of $|\psi_2>$ on the interval Δ_3

$$|\psi_3\rangle = N_3^{-2} \int_{\Delta_3} dX_3 |X_3\tau_3\rangle\langle X_3\tau_3|\psi_2\rangle. \quad (2.23)$$

At τ_4 the probability that the particle is at X_4 and the detector has clicked is the product

$$\Delta_4 |\langle X_4 \tau_4 | \psi_3 \rangle|^2 P(\text{click}). \tag{2.25}$$

This is the same as (2.20) as an explicit calculation will show.

The contrast between the usual discussion of this experiment and that in the sum over histories formulation shows that the sum over histories formalism handles observations at different times democratically and efficiently, so that it is well adapted to deal with conditions and observations which are not in temporal order. In particular, the sum over histories formulation can deal efficiently and naturally with observations which lie on a general surface, $\tau = \tau(X)$, and with conditions which also lie on the surface. It is thus especially useful in cosmology.

We can now ask whether we can recover the Hilbert space formulation of quantum mechanics from its sum over histories version. This is easy to do on the surfaces of the preferred non-relativistic time. A conditional amplitude $\Phi(\mathcal{O}|\mathcal{C}) = \sum_{paths} e^{iS}$ for which the *conditions and observations are temporally ordered* can be factored into a sum over paths before an intermediate surface τ and a sum after τ (Figure 2a)

$$\Phi(\mathcal{O}|\mathcal{C}) = \int dX \psi^*_{\mathcal{O}}(X\tau) \psi_{\mathcal{C}}(X\tau), \tag{2.26}$$

where $\psi_{\mathcal{C}}(X\tau) = \sum_{paths\ in\ M_-} e^{iS}$ for paths which meet the conditions $\mathcal{C}$ and end at X. There is a similar expression for $\psi^*_{\mathcal{O}}$. This sum defines the wave function from the sum over paths. Further, it has its usual probability interpretation because the positions at τ are a set of exhaustive and exclusive observations.

How would this construction go on a more general surface? The crucial difference is that now the paths can cross and recross the surface many times. (Figure 2b). Formally, since each path is divided into parts by the surface, one could write down

$$\Phi(\mathcal{O}|\mathcal{C}) = \sum_{\left(\substack{number\ of \\ crossings,\ n}\right)} \int dY_1 \cdots dY_n \psi^*_{\mathcal{O}}(Y_1 \cdots Y_n, S) \psi_{\mathcal{C}}(Y_1 \cdots Y_n, S), \tag{2.27}$$

where

$$\psi_{\mathcal{C}} = \sum_{paths\ in\ M_-} e^{iS}. \tag{2.28}$$

Some attention is now needed to define what one means by sums over regions like M_-. This can be done by introducing a spacetime lattice and going to Euclidean time. Then there is a close connection between sums over histories and the continuum limits of stochastic processes. In particular, for a free particle the sums over histories can be defined as the continuum limit of a random walk.

When one calculates the sum for a fixed number of crossings in this manner, one finds that the amplitude vanishes! A composition law of the form (2.27) exists on the lattice but does not have a continuum limit. We do not recover a Hilbert space formulation of quantum mechanics on a general surface $\tau = \tau(X)$. The reason is that the expected number of crossings is infinite and the amplitude for any finite number of crossings is zero. Only due to the peculiar fact that the paths move forward in the preferred non-relativistic time can we recover a Hilbert space formulation of the theory on surfaces of that time because we know the paths cross them once and only once.

The absence of a Hilbert space for a general surface is not an obstacle to the computation of probabilities for observations and conditions which lie on the surface. The sum over histories formulation allows this and our specific example illustrates it. However, typically, greater care is needed to pose questions with sensible answers (unlike the amplitude of (2.28)).

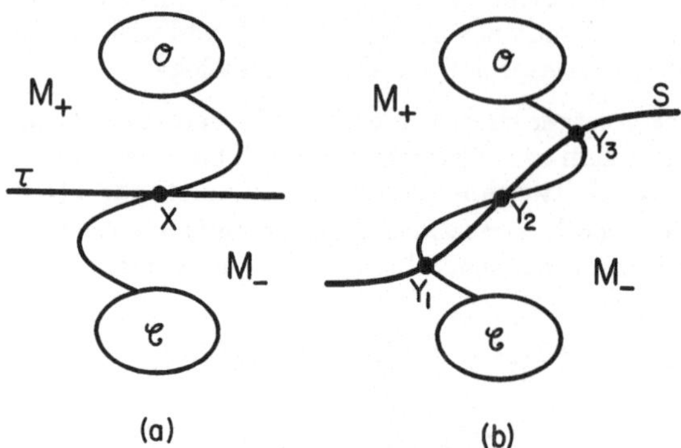

Fig. 2. (a) When conditions, $\mathcal{C}$, and observations, $\mathcal{O}$, are in temporal order, a path contributing to the conditional amplitude $\Phi(\mathcal{O}|\mathcal{C})$ intersects an intermediate surface of constant preferred non-relativistic time τ at one and only one position X. The sum defining the amplitude may thus be factored in to a sum over paths prior to τ, a sum over paths after τ and a sum over X. (b) By contrast, a path contributing to the conditional amplitude $\Phi(\mathcal{O}|\mathcal{C})$ may intersect a general surface S many times. Indeed, the expected number of intersections is infinite.

Now imagine for a moment that we had been brought up on sum over histories quantum mechanics. What would our attitude be if we encountered a theory in which there was *no* surface on which we could recover a Hamiltonian, Hilbert space formulation of the theory? Would we insist on this as a necessary requirement for a successful quantum theory? I would like to suggest that the answer should be no.

I would now like to describe two examples of theories in which it seems is not possible to recover a Hilbert space formulation, at least not by the methods we have been describing. The first is non-relativistic quantum mechanics with real clocks. The second is general relativity.

We have no direct perception of the time of non-relativistic mechanics. Psychological time is certainly poorly correlated with this variable. What we do observe are the positions of indicators in mechanical systems carefully arranged so these indicators are correlated with the time of the Schrödinger equation. Such systems are called clocks. The simplest example is a free particle moving with definite velocity in one dimension. Its position T, appropriately calibrated, gives the time.

In quantum mechanics an ideal clock would be one whose position T remained always correlated with the Schrödinger equation time τ. A possible solution to the Schrödinger equation is then

$$\psi(T,\tau) = \delta(T-\tau). \tag{2.29}$$

The corresponding Hamiltonian would be linear in the momenta

$$h_C = -i\frac{\partial}{\partial T} = P_T. \tag{2.30}$$

Such clocks do not exist. The energy of this Hamiltonian would be unbounded below.

Real clocks, such as our free particle, can exist. Their energy is positive because their Hamiltonians are quadratic in the momenta

$$h_C = \frac{P_T^2}{2M}. \tag{2.31}$$

However, such clocks are inevitably imperfect. In the case of a particle even if there is initially a sharp correlation between T and τ, eventually the wave packet will spread and the clock will lose accuracy. The spreading can be reduced arbitrarily by making the particle sufficiently massive. Such general limitations on the masses of clocks have been discussed by Salecker and Wigner.[11] These limitations are not important in non-relativistic quantum mechanics because one can imagine arbitrarily massive clocks which do not disturb the system. In the gravitational physics of closed systems, however, these limitations become fundamental.

Suppose that we try and construct a non-relativistic quantum mechanics in which the indicators of real clocks are involved directly. This is a problem which has been discussed by DeWitt,[12] Peres,[13] Page and Wooters[14] and no doubt by many others. From the sum over histories point of view one might proceed as follows: The histories are the world lines $X(\tau), T(\tau)$; moving forward in τ but both forward and backward in T. The observables and conditions will involve T and X and therefore *sums* over the unobserved parameter τ. An interesting set of correlations are the values of X for a given T. If a further part of the conditions is to put the T - component of the system into "a good clock state" we recover an approximation to the predictions of ordinary quantum mechanics. By making the clocks massive we can make this approximation as accurate as desired. For no finite mass, however, do we recover a Hilbert space formulation of the theory on surfaces of constant T. Formulated in terms of T and X there are no preferred surfaces, and the paths cross and recross a given T an arbitrarily large number of times.

My last example is general relativity and, in particular, the quantum mechanics of closed cosmologies. The histories, as will be described in more detail in the next section, are cosmological 4-geometries. There is no natural time parameter for a cosmological history. That is, there is no parameter constructed from the metric and matter fields which, for a general history, defines a foliating family of spacelike hypersurfaces such that the parameter takes a distinct value on each surface. Put differently, for a candidate parameter such as $\sqrt{h}$ or K, one can find histories that move forward and backward through a given value an arbitrarily large number of times. There is no preferred time. It seems unlikely, therefore, that one will recover a Hilbert space of states from sum over histories quantum mechanics by the analog of straightforward construction used for a particle earlier in this section. I would like to suggest, however, that, even in the absence of a Hilbert space formulation, one can formulate a predictive quantum mechanics of the single system which is our universe using the sum over histories formulation I have described. Correlations involving both conditions and observations on a spacelike surface can be investigated using the measure supplied by the sum over histories. Predictions of correlations verifiable for the single system can be made if the wave function is sufficiently peaked or sufficiently small. This formulation is but a slight modification of Hamiltonian quantum mechanics which coincides with that formulation when a Hamiltonian theory is available. In the following sections we shall discuss some first steps towards the implementation of this program.

3. LAWS FOR INITIAL CONDITIONS

3.1 The Sum Over Histories Formulation of Quantum Cosmology

To apply the sum over histories formulation of quantum mechanics to cosmology three things are needed: the histories, the action, and the measure. The histories are cosmological spacetimes with matter fields. For simplicity we shall take the spacetimes to be spatially closed, Lorentzian 4-geometries whose topology is of the form $\mathbf{R} \times M^3$ with M^3 a compact 3-manifold (typically the 3-sphere). There is no compelling reason for this restriction on topology, and indeed it is interesting to investigate other possibilities, [15] but this assumption will simplify our discussion without limiting the central ideas.

There is, today, no choice for the action of spacetime coupled to matter which yields a satisfactory quantum field theory judged by familiar local standards. Whatever the correct theory, we expect that its low energy limit will be Einstein gravity coupled to matter. The action for Einstein gravity on a spacetime region M is

$$\ell^2 S_E[g] = \int_M d^4x(-g)^{\frac{1}{2}} R - 2\int_{\partial M} d^3x(h)^{\frac{1}{2}} K. \tag{3.1}$$

Here, R is the scalar curvature, $\ell = (16\pi G)^{\frac{1}{2}}$ is the Planck length, ∂M is the boundary of M, h_{ij} is the metric induced on the boundary by $g_{\alpha\beta}$ and K is the trace of the extrinsic curvature of the boundary. The action for a free scalar field, Φ, with mass M is a simple representative of the many possible matter field actions,

$$\begin{aligned} S_\Phi[g,\Phi] = &-\frac{1}{2}\int d^4x(-g)^{\frac{1}{2}}\left[(\nabla\Phi)^2 + \xi R\Phi^2 + M^2\Phi^2\right] \\ &+ \xi\int_{\partial M} d^3x(h)^{\frac{1}{2}} K\Phi^2. \end{aligned} \tag{3.2}$$

Where concrete illustrations of the action are needed, we shall use (3.1) and (3.2).

Specification of the weights with which to carry out the sum over histories is just as important for quantum mechanics as the specification of the action. The sums over paths defining the quantum mechanics of a particle may be given a concrete meaning as the limit of sums over increasingly refined piecewise linear approximations to those paths. Weights can be assigned to each piecewise linear path defining concretely a "measure" on the space of paths. Sums over geometries may be given concrete meaning as the limit of sums over piecewise flat approximations to them using the methods of the Regge Calculus[16] and in this way a "measure" on geometries can be defined.[17]

Conditional probability amplitudes are formed from $exp(iS)$ by summing over geometries and field configurations. We have argued that representative predictions in cosmology involve observations made locally, "at one moment of time," with conditions specified, in part, at the same moment of time. Such predictions are extractable from the amplitude for observations on a spacelike surface, that is, from the wave function. The wave function for a spacelike surface is determined by a sum over histories restricted by conditions "in the past" of the spacelike surface. Specifically we write

$$\Psi_C[h_{ij}(\mathbf{x}), \Phi(\mathbf{x})] = \int_C \delta g \delta \Phi \exp\Big(iS[g, \Phi]\Big), \tag{3.3}$$

where the sum is over cosmological 4-geometries and field configurations which match the arguments of the wave function and satisfy the conditions C.

The wave function on a spacelike surface is not the only conditional probability amplitude which could be computed. It is, however, the one from which we expect to deduce most interesting cosmological predictions. We do this not by calculating probabilities, for the universe is a single system. Rather, as described in Section 2, we search the configuration space for regions where the wave function is sharply peaked. These correlations are the predictions of quantum cosmology. Given a set of conditions C, it is a largely open, but important question, what predictions one can expect. As we shall describe in Section 4, this problem is greatly simplified if there is a region of configuration space in which the wave function can be approximated semiclassically.

A law for initial conditions in quantum cosmology is a law for the conditions C. That is, a law for initial conditions is a specification of the class of geometries and field configuration which are summed over in (3.3) to yield the wave function of the universe.

3.2 Constraints

We are not free to specify any wave function as a theory of initial conditions. It must be representable as a sum over histories of the form (3.3) reflecting the underlying gravitational dynamics. In particular, it must satisfy certain *constraints* which are consequences of this dynamics. We shall now briefly review these using the example of pure Einstein gravity.[7]

There are four constraints in general relativity. Three of them arise from the requirement that the wave function, $\Psi[h_{ij}]$, depend only on three geometry and not on the choice of coordinates used to describe that geometry. Ψ must thus be the same on two three metrics which are connected by a diffeomorphism. Infinitesimal diffeomorphisms are generated by a vector ξ^k according to

$$h_{ij} \to h_{ij} + D_{(i}\xi_{j)}. \tag{3.4}$$

Thus, for infinitesimal ξ^k

$$\Psi[h_{ij} + D_{(i}\xi_{j)}] = \Psi[h_{ij}], \tag{3.5}$$

or equivalently

$$\int_{M^3} d^3x D_{(i}\xi_{j)} \frac{\delta\Psi}{\delta h_{ij}(\mathbf{x})} = 0. \tag{3.6}$$

Integrating by parts on the compact manifold M^3, and recalling that ξ^k is arbitrary, one arrives at the three constraints

$$D_i\Big(\frac{\delta\Psi}{\delta h_{ij}(\mathbf{x})}\Big) = 0. \tag{3.7}$$

These are called the "momentum" constraints.

The fourth constraint of general relativity arises because general relativity is an example of a parametrized theory in which time occurs as one of the dynamical variables. To illustrate the idea we begin with a simple model. [7]

Consider a non-relativistic particle whose dynamics is described by the action

$$S[X(T)] = \int dT\ \ell(dX/dT, X). \tag{3.8}$$

Express both X and T as functions of a parameter τ and thereby introduce the time T as a dynamical variable in the action

$$S[X(\tau), T(\tau)] = \int d\tau \dot{T} \ell(\dot{X}/\dot{T}, X), \tag{3.9}$$

where a dot denotes a τ-derivative. This action is invariant under reparametrizations of the label time

$$\tau = f(\tau'),\ X'(\tau') = X(f(\tau')),\ T'(\tau') = T(f(\tau')). \tag{3.10}$$

If we calculate the Hamiltonian associated with the Lagrangian in (3.9), we find first that

$$H = \dot{T}(p_T + h), \tag{3.11}$$

where p_T is the momentum conjugate to T and h is the Hamiltonian associated with the Lagrangian ℓ. Second, we find that, identically,

$$H = 0. \tag{3.12}$$

The vanishing of the Hamiltonian is a characteristic feature of theories which are invariant under reparametrizations of the time.

In the quantum mechanics of this model, we construct the wave function $\psi_C(X,T)$ for a particular moment of time as a sum of $exp(iS)$ over an appropriate class of paths, $X(T)$. We can carry out this sum in parametrized form – integrating over histories $X(\tau), T(\tau)$ and using the action (3.9). However, histories which differ only by a reparametrization of τ[eq. (3.10)] correspond to the same path. To count these only once in the sum over histories we can "fix" the parametrization by requiring a particular relation between τ and T

$$\tau = F(T), \tag{3.13}$$

for arbitrary increasing $F(T)$ and write the sum over histories as

$$\psi_C(X,T) = \int_C \delta X \delta T \left|\frac{dF}{dT}\right| \delta(\tau - F(T)) \exp\Big(iS[X,T]\Big). \tag{3.14}$$

The functional δ - function is the analog of the "gauge fixing δ - function" for gauge theories and $|dF/dT|$ is the analog of the "Faddeev-Popov determinant."

The familiar sum over histories for the quantum mechanics of a non-relativistic particle is recovered from (3.14) by doing the integral over T (most easily by choosing $F = T$). From this, and therefore from (3.14), the Schrödinger equation follows. Writing it in the form

$$\left(-i\frac{\partial}{\partial T} + h\right)\psi_C(X,T) = 0, \tag{3.15}$$

we see that it is the operator form of the constraint $H = 0$. Thus the classical constraints arising from invariance under reparametrization of the time are enforced as operator relations in quantum mechanics. One sees that the vanishing of the Hamiltonian for a parametrized theory does not mean the absence of dynamics, it *is* the dynamical relation.

General relativity is invariant under the group of diffeomorphisms in four dimensions. There are correspondingly four constraints. They can be written in "3+1 form" by choosing a family of spacelike surfaces and using as basic variables their intrinsic metric, h_{ij}, and extrinsic curvature, K_{ij}.

Three of the four constraints express the invariance under diffeomorphisms in the 3-surface. These are the constraints we have already discussed. The fourth constraint expresses the invariance of the theory under choice of the choice of spacelike surfaces, that is, under reparametrizations of the time. As in the simple model, the constraint is that the total Hamiltonian (density) vanish. For this reason it is called the *Hamiltonian constraint.* Classically its form is

$$H = \ell^2 G_{ijk\ell}\pi^{ij}\pi^{k\ell} + \ell^{-2}h^{\frac{1}{2}}\left(-{}^3R + 2\Lambda\right) = 0, \tag{3.16}$$

where ${}^3R$ is the scalar curvature of the 3-surface, π^{ij} are the momenta conjugate to h_{ij}

$$\ell^2\pi_{ij} = h^{\frac{1}{2}}(K_{ij} - h_{ij}K) \tag{3.17}$$

and $G_{ijk\ell}$ is the "supermetric"

$$G_{ijk\ell} = \frac{1}{2}h^{-\frac{1}{2}}\left(h_{ik}h_{j\ell} + h_{i\ell}h_{jk} - h_{ij}h_{k\ell}\right). \tag{3.18}$$

Quantum mechanically eq (3.17) becomes an operator constraint on the wave function called the Wheeler-DeWitt equation. It takes the form

$$\left[-\ell^2\nabla_x^2 + \ell^{-2}h^{\frac{1}{2}}\left(-{}^3R + 2\Lambda\right)\right]\Psi[h_{ij}] = 0, \tag{3.19}$$

where

$$\nabla_x^2 = G_{ijk\ell}\,\frac{\delta^2}{\delta h_{ij}(\mathbf{x})\delta h_{k\ell}(\mathbf{x})} + \begin{pmatrix}\text{linear derivative terms}\\ \text{depending on factor ordering}\end{pmatrix}. \tag{3.20}$$

The Wheeler-DeWitt equation follows formally from the sum over histories for quantum cosmology in much the same way that the Schrödinger equation follows from the sum over parametrized paths (3.14) (Problem 2). It may be thought of as a functional differential equation which expresses the dynamics of quantum cosmology in much the same way that the Schrödinger equation expresses the dynamics of particle quantum mechanics.

3.3 A Proposal for a Wave Function of the Universe

There have been a number of proposals for a quantum state of the universe[1–5]. Perhaps the most developed of these is the proposal of Stephen Hawking and his coworkers that the wave function of the universe is determined by a sum over compact Euclidean 4 - geometries. Detailed expositions of this idea can be found elsewhere[18,6]. Here we shall just state the proposal so that there is at least one concrete idea with which to illustrate the subsequent discussion.

Euclidean sums over histories as well as Lorentzian ones may be used to construct solutions to constraints. Consider, for example, the sum over particle paths

$$\psi_0(X_0) = \int_{C_0} \delta X \exp\Big(-I[X(T)]\Big), \tag{3.21}$$

where I is the Euclidean action for a non-relativistic particle in a potential $V(X)$

$$I[X(T)] = \int dT \left[\frac{1}{2}M\left(\frac{dX}{dT}\right)^2 + V(X)\right], \tag{3.22}$$

and the sum is over all paths which start at X_0 at Euclidean time $T = 0$ and proceed to a configuration of minimum action at large negative times. The wave function $\psi_0(X_0)$ so defined satisfies the constraint (3.15). It is, in fact, the ground state wave function.

Euclidean sums over 4 - geometries give solutions to the operator constraints of gravitational theories. Consider a cosmological manifold of the form $\mathbf{R}^+ \times M^3$, where $\mathbf{R}^+$ is half the real line. The manifold thus has an M^3 boundary. A sum over Euclidean 4-geometries and field configurations of the form

$$\Psi\Big[h_{ij}(\mathbf{x}), \Phi(\mathbf{x})\Big] = \int_C \delta g \delta\Phi \exp\Big(-I[g, \Phi]\Big), \tag{3.23}$$

where I is the Euclidean action for Einstein gravity coupled to matter, will formally satisfy the constraints (3.7) and (3.9) provided the metric and matter field induced on the boundary by each contributor to the sum match those in the argument of the wave function.[28] (Problem 2). A particular wave function is singled out by summing over *compact 4 - geometries with no other boundary* and over *matter field configurations which are regular on these geometries*. The proposal of Hawking and his coworkers is that this *is* the wave function of our universe.

The Euclidean action for Einstein gravity

$$\ell^2 I[g] = -\int_M d^4x g^{\frac{1}{2}} R - 2\int_{\partial M} d^3x h^{\frac{1}{2}} K \tag{3.24}$$

is not positive definite. Thus, for general relativity and other theories with this property the contour of integration in (3.23) cannot be over purely real metrics – the integral would diverge. The contour of integration must be taken in complex directions.[19] From the Hamiltonian perspective one is free to make this distortion as long as the correct sum over the true physical degrees of freedom is preserved. This seems to be possible for linearized gravity and for the sum over histories defining the ground state of isolated systems.[19] We shall presume that the analogous contour exists for closed cosmologies although this has yet to be demonstrated (Problem 3). The complex nature of the contour in the proposal (3.23) is not an inessential technicality. Were the contour purely real, the wave function would be positive and never oscillate. With a complex contour we expect oscillation in some regions of configuration space, and, as we shall see in Section 4, only in such regions does the wave function predict the correlations of classical physics.

4. THE LIMIT OF CLASSICAL GEOMETRY AND QUANTUM FIELD THEORY IN CURVED SPACETIME

In the context of quantum mechanics the predictions of classical physics are predictions of special kinds of correlations between special classes of observables.

For example, if we measure the position and momentum of a particle at one time with accuracies consistent with the uncertainty principle and then again at a later time the laws of classical physics predict a definite correlation between these two measurements.

In Section 2 we saw how a wave function predicts correlations among the observables of an individual system. It is a very special situation when the predicted correlations of some observables are classical, but also a very important situation. There are three reasons: First, fundamentally we interpret the world in classical terms. Second, certainly in cosmology our crude observations are of classical observables. Third, as we shall describe below, it is possible to give a simple criterion – the validity of the semiclassical approximation – for when a wave function predicts classical correlations. This feature greatly simplifies extracting predictions in quantum cosmology.

The discussion of this section is an attempt at one synthesis of ideas which have had a long history in general relativity. Some notable contributions have been those of Salecker and Wigner,[11] DeWitt,[12] Wheeler,[25] Peres,[13] Page and Wooters,[14] Banks,[21] Hawking and Halliwell,[22] D'Eath and Halliwell,[23] and Brout, Horwitz and Weil.[24]

4.1 The Semiclassical Approximation to Non-Relativistic Particle Quantum Mechanics

Let us recall the semiclassical approximation to the wave function of a state of definite energy in non-relativistic particle quantum mechanics. Assuming that the Hamiltonian has been normalized so that the energy is zero, we want to solve

$$H\psi(X) = 0, \tag{4.1}$$

where

$$H = -\frac{\hbar^2}{2M}\frac{d^2}{dX^2} + V(X). \tag{4.2}$$

To obtain the semiclassical approximation we write

$$\psi(X) = \exp\left[iS(X)/\hbar\right], \tag{4.3}$$

and expand $S(X)$ in powers of $\hbar$

$$S = S_0 + \hbar S_1 + \cdots \tag{4.4}$$

Writing out the Schrödinger equation to the lowest order, $\hbar^0$, one finds

$$\frac{1}{2M}\left(\frac{dS_0}{dX}\right)^2 + V(X) = 0, \tag{4.5}$$

so that S_0 obeys the classical Hamilton Jacobi equation and, indeed, is given by $\pm\int^x \sqrt{-2mV(X)}dX$. In regions where $V(X) < 0$ (the "classically allowed" regions for $E = 0$) there is a real solution for S_0 and the wave function oscillates. In the classically forbidden regions S_0 must be complex, $S_0 = iI_0$, where I_0 solves the "Euclidean" Hamiltonian Jacobi equation. The wave function in these regions is a sum of real exponentials.

The next order is also easy to compute. The order $\hbar$ part of the Schrödinger equation is

$$i\frac{d^2S_0}{dX^2} - 2\frac{dS_0}{dX}\frac{dS_1}{dX} = 0, \tag{4.6}$$

which is easily solved for S_1. The result, for example in the classically allowed region is

$$\psi = \left(\frac{dS_0}{dX}\right)^{-\frac{1}{2}} \exp(iS_0/\hbar). \tag{4.7}$$

In the classically forbidden region the order $\hbar^0$ approximation is modified by a prefactor in the same way. Approximate solutions satisfying given boundary conditions are built by taking linear combinations of (4.7) and its complex conjugate and of the two possible exponential behaviors and matching them across the boundaries between the classically allowed and classically forbidden regions.

In a classically allowed region the interpretation of the semiclassical approximation is straightforward. Suppose measurements of the particle's position and momentum are made with accuracies consistent with the uncertainty principle. For position measurements which yield the value X_0 the wave function (4.7) is sharply peaked about the momentum

$$p(X_0) = \left(\frac{dS_0}{dX}\right)_{X=X_0}. \tag{4.8}$$

This is because near X_0

$$S_0(X) \approx S_0(X_0) + \left(\frac{dS_0}{dX}\right)_{X=X_0}(X - X_0) + \cdots \tag{4.9}$$

with the higher terms being negligible because S_0 is slowly varying. In this approximation, (4.7) is a wave function of definite momentum (4.8). Thus, a semiclassical approximation of the form (4.7) predicts this classical correlation between position and momentum. In particular, if successive measurements are made which do not disburb either position or momentum by a large amount, values must be found which are consistent with the classical equations of motion

$$M\frac{dX}{d\tau} = p(X) = \left[-2MV(X)\right]^{\frac{1}{2}}. \tag{4.10}$$

In this way classical physics is recovered.

One should stress that classical physics is recovered only in the sense of certain correlations and that the nature of these correlations depends on the form of the semiclassical approximation. For example, the wave function (4.7) does not predict much about the position of the particle but only the correlation (4.8) between position and momentum, and that implied by (4.10) between present position and future position. Of course, there are wave packet states in which both position and momentum would be predicted, but these do not have definite energy. A semiclassical wave function of the form

$$\psi(X) = \left(\frac{dS_0}{dX}\right)^{-\frac{1}{2}} \cos\left[S_0(X)/\hbar\right], \tag{4.11}$$

would not even predict a correlation of position with momentum but only with its absolute value. That is, it would predict that three successive measurements of position would be correlated according to the equation of motion (4.10) with the sign of $p(X)$ determined from the first two.

4.2 The Born-Oppenheimer Approximation for Real Clocks

In non-relativistic quantum mechanics time is an external parameter labeling different measurements. It is not itself an observable. (See, for example, the preceding discussion of correlations in the semiclassical approximation.) We have access to this time only through the observation of correlations between the positions of clock indicators and the variables of the system. It should, therefore, be sufficient for prediction to formulate quantum mechanics entirely in terms of these variables. In the case of the quantum mechanics of a particle we would write

$$\psi = \psi(T, X), \tag{4.12}$$

where X is the particle's position and T the position of a clock indicator. We could then study ψ for the correlations between T and X or between T and other observations.

The wave function in such a formulation of quantum mechanics will satisfy a constraint reflecting invariance under the choice of parameter used to label the histories of T and X. Indeed, the parametrized time non-relativistic quantum mechanics of Section 3.3 is a model for this kind of theory. The constraint (3.15) of that theory was

$$H\Psi = \left(-i\frac{\partial}{\partial T} + h\right)\Psi = 0. \tag{4.13}$$

If we read this as the requirement that the total Hamiltonian vanish, the variable T may be thought of as the position of a kind of ideal clock whose Hamiltonian is

$$h_C = -i\frac{\partial}{\partial T} = P_T. \tag{4.14}$$

The Hamiltonian (4.14) is a rather poor model of a real clock. Among other unrealistic features, its spectrum is unbounded below. The Hamiltonians of more realistic clocks we expect to be quadratic in their momenta. A particle moving freely in one dimension with narrow dispersions in position and momentum is a simple example. The position of the particle is a measure of time. In such a theory, as we shall show below, we recover the classical notion of time only in the approximation in which the dynamics of the clock can be treated semiclassically.

The constraints of general relativity are also quadratic in the momenta. In the case of closed cosmologies it does not even seem possible to approximate the notion of an ideal clock.[25] Here too we shall recover a notion of time in the quantum theory only in the approximation in which spacetime is treated semiclassically. We shall discuss spacetime in the next subsection. Here, we begin with a simple model of a real clock due to Banks[21,24] which illustrates the central features of these ideas.

Consider a system consisting of a particle described by a position X and a clock with an indicator variable T. We consider a constraint of the form

$$H\Psi = (h_C + h)\Psi = 0, \tag{4.15}$$

where h is the Hamiltonian of the particle and h_C the Hamiltonian of the clock. The dynamics of the clock we take to be specified by the action

$$S[T(\tau)] = M\int d\tau\left(\frac{1}{2}\dot{T}^2 + V_C(T)\right), \tag{4.16}$$

so that h_C will be quadratic in the momenta. The quantity M is the mass of the clock although it also controls the coupling to the potential V_C.

In the limit $M \to \infty$ the clock motion can be treated classically. Physically, this is because, for given energy, as the mass becomes large the quantum fluctuations become small. Mathematically, it is because the classical limit in a sum over histories occurs as $\hbar \to 0$ in $exp(iS/\hbar)$ leading to destructive interference in the sum for all but the classical trajectory. The limit $M \to \infty$ is the same limit but for the clock part of the action alone. The approximation of large M is therefore not strictly the semiclassical approximation. Rather it is the analog of the Born-Oppenheimer approximation in molecular physics in which the motion of the massive nuclei are considered classically while the electronic cloud is treated quantum mechanically.

In the large M limit we look for a solution of the constraint equation (4.15) of the form

$$\Psi = e^{iS(T)}\chi(T, X) \tag{4.17}$$

where

$$S = MS_0 + S_1 + M^{-1}S_2 + \cdots \tag{4.18a}$$

$$\chi = \chi_0 + M\chi_1 + \cdots \tag{4.18b}$$

Since

$$h_C = -\frac{1}{2M}\frac{d^2}{dT^2} + MV_C(T), \tag{4.19}$$

we have, on writing out the constraint (4.15),

$$\left\{-\frac{1}{2M}\left[i\frac{d^2S}{dT^2} - \left(\frac{dS}{dT}\right)^2\right] + MV_C\right\}\chi - \frac{1}{2M}\frac{\partial^2\chi}{\partial T^2} - \frac{i}{M}\frac{dS}{dT}\frac{\partial\chi}{\partial T} + h\chi = 0. \tag{4.20}$$

We now insert the expansions (4.18) in (4.20) and systematically expand in powers of M. For the leading order we find

$$\frac{1}{2}\left(\frac{dS_0}{dT}\right)^2 + V_C(T) = 0. \tag{4.21a}$$

This is the clock's classical Hamilton-Jacobi equation. Defining $p_T = M(dS_0/dT)$ it can be written in the familiar form

$$\frac{1}{2M}p_T^2 + MV_C(T) = 0. \tag{4.21b}$$

In the next order, M^0, one finds

$$-\frac{1}{2}\left(i\frac{dS_0}{dT^2} - 2\frac{dS_0}{dT}\frac{dS_1}{dT}\right)\chi_0 - i\frac{dS_0}{dT}\frac{\partial\chi_0}{\partial T} + h\chi_0 = 0. \tag{4.22}$$

This is one equation for two unknowns. To fix the remaining freedom we rewrite (4.22) by *defining* a classical time from the solution to the classical Hamilton-Jacobi equation (4.21). We write

$$M\frac{dT_0}{d\tau} = p_T = M\frac{dS_0}{dT}. \tag{4.23}$$

Integration of this relation defines $T_0(\tau)$ up to an initial condition and hence τ as a function of T. Equation (4.20) can now be rewritten to read

$$-\frac{1}{2}\left(i\frac{d^2S_0}{dT^2} - 2\frac{dS_1}{d\tau}\right)\chi_0 - i\frac{\partial\chi_0}{\partial\tau} + h\chi_0 = 0, \tag{4.24}$$

where $\chi_0 = \chi_0(T_0(\tau), X) \equiv \chi_0(\tau, X)$. An inner product can be defined by integrating over X at constant τ

$$(\chi, \phi) = \int dX\, \chi^*(\tau, X)\phi(\tau, X) \tag{4.25}$$

and in it we can take the expectation value of (4.24). The result is

$$-\frac{1}{2}\left(i\frac{d^2S_0}{dT^2}-2\frac{dS_1}{d\tau}\right)(\chi_0,\chi_0)-i\left(\chi_0,\frac{\partial\chi_0}{\partial\tau}\right)+(\chi_0,h\chi_0)=0. \tag{4.26}$$

The imaginary part of this equation is, assuming that h is Hermitian,

$$\frac{1}{2}\left(-\frac{dS_0}{dT^2}+2\frac{dImS_1}{d\tau}\right)(\chi_0,\chi_0)-\frac{1}{2}\frac{d}{d\tau}(\chi_0,\chi_0)=0. \tag{4.27}$$

The real part is

$$\frac{d(ReS_1)}{d\tau}(\chi_0,\chi_0)-\frac{i}{2}\left[\left(\chi_0,\frac{\partial\chi_0}{\partial\tau}\right)-\left(\frac{\partial\chi_0}{\partial\tau},\chi_0\right)\right]+(\chi_0,h\chi_0)=0. \tag{4.28}$$

We recover a sensible quantum mechanics if we impose the condition that the inner product is conserved

$$\frac{d}{d\tau}(\chi_0,\chi_0)=0 \tag{4.29}$$

so that from (4.27)

$$-\frac{dS_0}{dT^2}+2\frac{dS_0}{dT}\frac{dImS_1}{dT}=0. \tag{4.30}$$

This is the usual next -after-leading-order equation for the semiclassical approximation to the T motion. Equation (4.26) becomes, assuming χ_0 normalized,

$$\frac{d(ReS_1)}{d\tau}+(\chi_0,h\chi_0)=0. \tag{4.31}$$

When (4.30) and (4.31) are substituted back into the original equation (4.22) we find

$$i\frac{\partial\chi_0}{\partial\tau}=\Big[h-(\chi_0,h\chi_0)\Big]\chi_0. \tag{4.32}$$

This is the Schrödinger equation for the particle moving in the "background" time τ to the extent $(\chi_0,h\chi_0)$ is constant or negligible.

The combined equation for ReS accurate to first order in M is from (4.31) and (4.21)

$$\frac{1}{2}\left(\frac{dReS}{dT}\right)^2+MV_C(T)+(\chi_0,h\chi_0)=0. \tag{4.33}$$

This is the classical Hamilton-Jacobi equation but with a small quantum correction to the energy. It is the semiclassical *back reaction* equation.

4.3 The Approximation of Quantum Field Theory in Curved Spacetime.

The structure of the Hamiltonian constraint for general relativity – the Wheeler-DeWitt equation – is similar both in structure and origin to the constraint of the simple model just discussed. Including the energy of a scalar matter field, the Wheeler-DeWitt equation reads

$$\left[-\frac{1}{2}\ell^2\nabla_x^2+\frac{1}{2}\ell^{-2}h^{\frac{1}{2}}(2\Lambda-{}^3R)+h^{\frac{1}{2}}T_{nn}\left(\Phi,-i\frac{\delta}{\delta\Phi}\right)\right]\Psi[h_{ij},\Phi]=0. \tag{4.34}$$

Here $T_{nn}(\Phi,\Pi)$ is the stress-energy of the matter field expressed in terms of the field's value and momentum and projected onto the normals of the spacelike hypersurface.

It is the Hamiltonian density for the scalar field. The inverse squared Planck length enters the constraint in exactly the same way as the mass of the clock did in the model problem [cf. (4.15),(4.19)]. We may therefore consider the limit when $\ell \to 0$ and expect to treat geometry semiclassically.[21–24] This is the limit when relevant length scales are large compared to the Planck length and when relevant energies are small compared to the Planck mass.

The model of Section 4.2 reveals the central features of the $\ell \to 0$ limit so clearly that we shall just sketch the parallel steps here. We write

$$\Psi[h_{ij}, \Phi] = e^{iS[h_{ij}]}\chi[h_{ij}, \Phi] \tag{4.35}$$

and systematically expand S, χ and the Wheeler-DeWitt equation in powers of ℓ. In the leading order we recover the Hamilton-Jacobi equation of general relativity for S_0. From its solution we can introduce the momenta

$$\ell^2\pi^{ij}(\mathbf{x}) = \frac{\delta S_0[h_{ij}(\mathbf{x})]}{\delta h_{ij}(\mathbf{x})}. \tag{4.36}$$

The π_{ij} are the tangent vectors to a set of integral curves in superspace which are solutions to the classical Einstein equation. For example, if we work in a gauge in which 4-metrics have the form

$$ds^2 = -d\tau^2 + h_{ij}(\tau, \mathbf{x})dx^i dx^j \tag{4.37}$$

then classically [cf. (3.17)]

$$\pi_{ij} = \frac{1}{2\ell^2 h^{\frac{1}{2}}}\frac{d}{d\tau}(hh_{ij}). \tag{4.38}$$

Integrating

$$\frac{dh_{ij}}{d\tau} = G_{ijk\ell}\frac{\delta S_0}{\delta h_{k\ell}}, \tag{4.39}$$

we recover a time dependent 4-geometry in the gauge (4.37) which satisfies the Einstein equation.

The values of χ_0 along an integral curve in superspace define χ_0 as a function of τ.

$$\chi_0 = \chi_0[h_{ij}(\tau, \mathbf{x}), \Phi(\mathbf{x})] = \chi_0[\tau, \Phi(\mathbf{x})]. \tag{4.40}$$

Then by exactly the same steps as led from (4.17) to (4.32) one finds starting from (4.35) that one ends at

$$i\frac{\partial \chi_0}{\partial \tau} = \left[h^{\frac{1}{2}}T_{nn}\left(\Phi, -i\frac{\delta}{\delta\Phi}\right) - \left(\chi_0, h^{\frac{1}{2}}T_{nn}\chi_0\right)\right]\chi_0. \tag{4.41}$$

This is a Schrödinger equation defining a quantum field theory in the curved background spacetime specified by a solution to (4.39). T_{nn} is the Hamiltonian density for the matter field which depends on the background metric in the usual way. The inner product with which this equation is derived is the standard inner product in the field representation on the spacelike surface with normal $\partial/\partial\tau$ in the background (4.37), i.e.

$$(\chi, \chi') = \int \delta\Phi \chi^*[\tau, \Phi(\mathbf{x})]\chi'[\tau, \Phi(\mathbf{x})]. \tag{4.42}$$

Accurate to order ℓ^2, the operator constraints of general relativity imply the classical constraint equations corrected by the quantum expectation value of the stress-energy of the matter field. The derivation is parallel to that of (4.33). In the Hamilton-Jacobi form in which they naturally emerge from (4.32) and (4.33) one has

$$\frac{1}{2}\left[-\ell^4 G_{ijk\ell}\frac{\delta S}{\delta h_{ij}}\frac{\delta S}{\delta h_{k\ell}} + h^{\frac{1}{2}}({}^3R - 2\Lambda)\right] = \frac{\ell^2}{2}\, h^{\frac{1}{2}}\Big(\chi_0, T_{nn}\chi_0\Big), \tag{4.43a}$$

$$D_j\left(\frac{\delta S}{\delta h_{ij}}\right) = \frac{\ell^2}{2}\, h^{\frac{1}{2}}\Big(\chi_0, T^i_n\chi_0\Big), \tag{4.43b}$$

or equivalently using eqs. (4.36) - (4.39)

$$\left(R^\alpha_n - \frac{1}{2}\delta^\alpha_n R\right) = \frac{\ell^2}{2}\Big(\chi_0, T^\alpha_n\chi_0\Big). \tag{4.44}$$

where $R_{\alpha\beta}$ and $T_{\alpha\beta}$ are the Ricci curvature and stress energy expressed in an orthonormal basis one member of which is n^α. These are the four constraint equations of the classical theory including the quantum corrections to the stress energy of the matter field.

The correspondence in form between (4.43) and (4.44) shows that any solution of the former will satisfy four of the quantum corrected Einstein equations. However, these Hamilton-Jacobi equations determine a solution of the *full* set of Einstein equations through (4.36) - (4.39). The reason is covariance. There was no intrinsic definition of the spacelike surface with normal n^α. The four constraint equations (4.44) must be satisfied on *any* spacelike surface in the geometry. This requirement is equivalent to the full set of Einstein equations.[26] Thus,

$$R_{\alpha\beta} - \frac{1}{2}g_{\alpha\beta}R = \frac{\ell^2}{2}(\chi_0, T_{\alpha\beta}\chi_0). \tag{4.45}$$

These are the quantum corrected ("backreaction") Einstein equations for the background geometry. (Problem 4).

The problem of extracting the predicted correlations from the wave function of the universe is in general a difficult one. As the results of this section argue, however, if the wave is well approximated by the form (4.35), with χ and S given by the first few terms in an expansion in powers of the Planck length, then the correlations of a classical 4-geometry containing quantum matter fields are predicted. The 4-geometry and quantum fields are defined precisely through equations (4.36 - 4.39) and (4.40 - 4.41). In the context of the discussion of Section 2 we envision this means the following: Imagine filling space with a system of rods, clocks and field meters which could define what is meant by measurements of distance and time to classical accuracies and measurements of field. The clocks, rods and meters are to be described by matter and gravitational fields so that it is possible to identify the regions in superspace consistent with various possible sets of values they might read. From the analog with non-relativistic quantum mechanics, we expect that, where the semiclassical approximation is valid, the wave function will be sharply peaked about a region consistent with values which define spacetimes satisfying the Einstein equations (4.45) and quantum fields satisfying (4.41).

In quantum cosmology, an important test of any theory of initial conditions is that there be a region of configuration space in which the wave function is well approximated semiclassically for we observe the present universe to behave classically. However,the wave function in the semiclassical approximation does not predict a unique classical history. Rather, it predicts a family of them. For example, in particle quantum mechanics a semiclassical wave function of the form $exp[iS(X)/\hbar]$ corresponds to classical trajectories with momentum $p = dS/dX$. Information about present position must be used to single out a unique classical history. A semiclassical wave function of the form $\cos[S(X)/\hbar]$ requires even more information for it corresponds to classical trajectories with $p = \pm dS/dX$. How much present information must be used to gain definite, classical predictions from the wave function of the universe is one of the subject's most important questions.

4.4 The Semiclassical Vacuum

The derivation of quantum field theory in curved spacetime presented in the preceding subsection clarifies a number of issues usually regarded as internal to that subject. For example, the derivation sheds light on the meaning of the metric in the semiclassical field equation (4.45). It emerges there not as the expectation value of a field operator in some *general* quantum state. Rather, $g_{\alpha\beta}$ is the metric that would be determined from *classical measurements of limited accuracy* in a regime of configuration space in which the semiclassical approximation to a *particular* wave function is valid.

As a second example, the derivation of quantum field theory in curved spacetime connects the choice of "vacuum" for the matter with a theory of initial conditions i.e. with a prescription for the wave function of the universe. If the semiclassical approximation is valid, $\chi_0[\tau, \Phi(\mathbf{x})]$ is determined by Ψ through (4.35), (4.40) and this functional defines a quantum state of the matter field in the classical background spacetime in the Hilbert space defined by (4.42). We shall call it the quantum state of the matter fields.

The determination of the quantum state of the matter fields by a prescription for the wave function of the universe may be instructively illustrated in a simple minisuperspace model. For the prescription we consider that of Section 3.3. For the minisuperspace model we restrict attention to homogeneous and isotropic geometries containing a single *conformally invariant* scalar field. This model is easy to analyse even it it is not very realistic.

The metric of a homogeneous isotropic geometry can be put in the form

$$ds^2 = \sigma^2\left[-d\tau^2 + a^2(\tau)d\Omega_3^2\right] \tag{4.46a}$$

$$= \sigma^2 a^2(\eta)\left[-d\eta^2 + d\Omega_3^2\right], \tag{4.46b}$$

where $\sigma^2 = \ell^2/24\pi^2$ is a convenient normalizing constant. The geometry of a three surface of constant τ is then characterized solely by its radius a_0. The wave function is a *function* of a_0 and a *functional* of the matter field on this surface.

$$\Psi = \Psi\left[a_0, \Phi_0(\mathbf{x})\right]. \tag{4.47}$$

Equivalently, we could expand any field configuration in harmonics

$$\Phi(\eta,\mathbf{x}) = \sum_n \Phi^{(n)}(\eta) Y_{(n)}(\mathbf{x}), \tag{4.48}$$

where the $Y_{(n)}(\mathbf{x})$ are standard harmonics on the 3-sphere. Then

$$\Psi = \Psi\left[a_0, \Phi_0^{(1)}, \Phi_0^{(2)}, \Phi_0^{(3)}, \cdots\right]. \tag{4.49}$$

The action is the sum of the Euclidean gravitational action (3.24) and the Euclidean action for the scalar field corresponding to (3.2) with $\xi = 1/6$. The Euclidean gravitational action restricted to the minisuperspace geometries is

$$I_E = \frac{1}{2}\int d\eta\Big[-(a')^2 - a^2 + H^2a^4\Big], \tag{4.50}$$

where a prime denotes an η-derivative and $H^2 = \Lambda/3$. The matter action is considerably simplified by a dimensional rescaling

$$\Phi(x) = \varphi(x)/(2\pi^2\sigma^2)^{\frac{1}{2}} \tag{4.51}$$

and by a conformal rescaling

$$\varphi^{(n)}(\eta) = \chi^{(n)}(\eta)/a(\eta). \tag{4.52}$$

The physics of the field, being conformally invariant, is essentially, the same in all conformally related spacetimes. The geometry is conformal to an Einstein static universe by a conformal factor $a(\eta)$. In that geometry, because of its timelike killing field, the analysis of the scalar field is considerably simplified. This is immediately apparent in the form of the matter action takes in terms of the variables $\chi^{(n)}$

$$I_M = \frac{1}{2}\int d\eta\Big[\Big(\chi^{(n)\prime}\Big)^2 + \omega_n^2\Big(\chi^{(n)}\Big)^2\Big], \tag{4.53}$$

where $\omega_n^2 = \gamma_n + 1$ and γ_n are the eigenvalues of the Laplacian on the 3-sphere.

In the minisuperspace model the wave function is given as

$$\Psi\Big[a_0, \chi_0^{(n)}\Big] = \int \delta a \prod_n \delta\chi^{(n)} e^{-\left(I_E + I_M\right)}. \tag{4.54}$$

The integral is over all $a(\eta)$ which correspond to compact geometries with boundary three sphere radius a_0, and over all matter mode configurations $\chi^{(n)}(\eta)$ which match $\chi_0^{(n)}$ on the boundary and are elsewhere regular. (Figure 3). A compact geometry will have one radius (the "south pole") at which a vanishes linearly in the polar angle r. Since $d\eta = dr/a(\eta)$, the coordinate η becomes logarithmically infinite at this point. We may use the last remaining gauge freedom to choose the boundary to be at $\eta = 0$. The relevant coördinate range for η is then $(-\infty, 0)$. We thus integrate over $a(\eta), \chi(\eta)$ which vanish at $\eta = -\infty$ and assume the prescribed values on the boundary.

In the simplicity of conformal invariance, the action separates into a sum of metric part and field part so that the two integrals can be done separately. The integral over the field part is trivial. One finds

$$\Psi\Big[a_0, \chi_0^{(n)}\Big] = \psi(a_0)\prod_n \exp\Big[-\frac{1}{2}\omega_n\Big(\chi_0^{(n)}\Big)^2\Big], \tag{4.55}$$

where

$$\psi(a_0) = \int \delta a e^{-I_E[a]}. \tag{4.56}$$

Let us now approximate the integral for $\psi(a_0)$ by the method of steepest descents. For this we must find the extrema of I_E through which the contour of integration

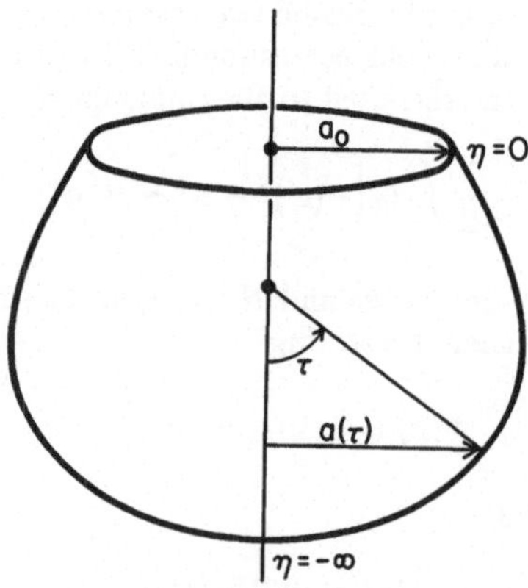

Fig. 3. A two dimensional representation of a homogeneous and isotropic 4-geometry contributing to the sum for the state of minimum excitation $\Psi(a_0, \Phi_0)$. Shown embedded in a flat 3-dimensional space is a 2-dimensional slice of such a geometry whose intrinsic geometry is

$$d\sigma^2 = d\tau^2 + a^2(\tau)d\varphi^2$$

τ is thus a "polar angle" and a the "radius from the axis." The geometry is compact and has only one boundary at which the radius is a_0, the argument of Ψ. The field configurations $\Phi(\tau, \mathbf{x})$ which contribute to the sum are those which are regular on this surface and which match the argument of the wave function Ψ on the boundary.

can be distorted. We begin with values of a_0 less than H^{-1}. The possible extrema of I_E are just the solutions of

$$a'' - a + 2H^2a^3 = 0. \tag{4.57}$$

The equation has an "energy integral" whose value may be found from the regular vanishing of the a at $\eta = -\infty$. Expressing this integral in terms of τ gives

$$\left(\frac{\dot{a}}{a}\right)^2 = \frac{1}{a^2} - H^2. \tag{4.58}$$

This is the Euclidean Einstein equation for a metric with the symmetries of the model as it must be. The solution is illustrated in Figures 4 and 5 and is just the 4-sphere of radius $1/H$. For $a_0 < 1/H$ there are thus two possible extrema which are compact 4-geometries with a 3-sphere boundary of radius a_0. One for which the boundary bounds less than a hemisphere of the 4-sphere and another for which it bounds more. The action for the 4-sphere is negative and therefore one might think that the extremum encompassing more 4-sphere should dominate. One must remember,

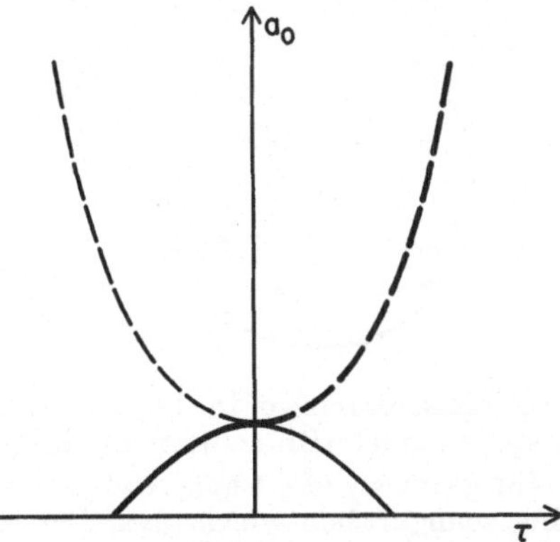

Fig. 4. The extremizing scale factor for the homogeneous, isotropic minisuperspace model with conformally invariant scalar field. The solid line is the solution of (4.58) for real Euclidean extrema of the action. The complete range of a from zero to maximum and back again describes the geometry of the 4-sphere (Figure 5). The dashed curve is the solution of (4.60) for complex Euclidean (Lorentzian) extrema. It describes the geometry of de Sitter space. For each value of a_0 there are thus two possible extremizing solutions. Choosing the trajectory to start on the left at $a_0 = 0$, the Euclidean prescription for the ground state singles out the heavy curve shown. This gives the semiclassical approximation to the wave function Ψ.

however, that because of the conformal rotation the contour of a integration is in the imaginary direction in the immediate vicinity of the extremum. Extrema of analytic functions are saddle points so that a maximum in a real direction is a minimum in an imaginary direction. The stationary configuration which contributes to the steepest descent evaluation of (4.56) is the one which is a maximum of the action in real directions and a least action configuration in imaginary directions. The extremum corresponding to the smaller part of the 4-sphere, therefore, provides the steepest descent approximation to the wave function. In fact, the contour cannot be distorted to pass through the other extremum. We thus have for $a_0 < 1/H$

$$\psi(a_0) \approx N\left[-1 + a_0^2 - H^2 a_0^4\right]^{-1/4} \exp\left[-\frac{1}{3H^2}\left(1 - H^2 a_0^2\right)^{3/2}\right], \tag{4.59}$$

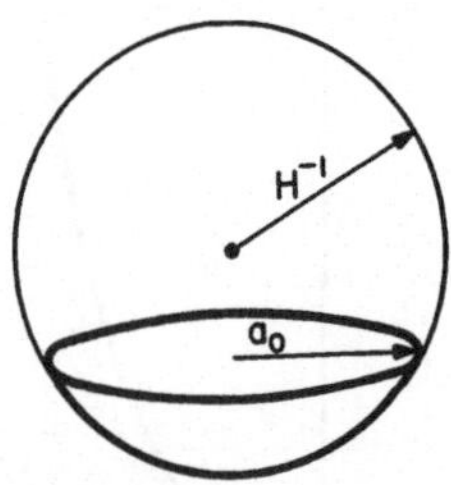

Fig. 5. The real Euclidean extrema of the homogeneous, isotropic minisuperspace model with conformally invariant scalar field have the geometry of a 4-sphere of radius H^{-1}. The extremizing configuration which gives the semiclassical approximation to ψ at $a_0 < 1/H$ is a part of the 4-sphere with a single 3-sphere boundary of radius a_0. There are two possibilities corresponding to more than a hemisphere or less. The Euclidean functional integral prescription for ψ identifies the smaller part of the 4-sphere as the contributing extremum. For $a_0 > H^{-1}$ there are no real extrema. The orientation of the 3-sphere in the 4-sphere is arbitrary. The semiclassical vacuum of the matter field is thus de Sitter invariant.

where N is an arbitrary normalizing factor.

If a_0 is increased to a value larger than $1/H$ there are no longer any real extrema because a 3-sphere of radius $a_0 > 1/H$ cannot fit into a 4-sphere of radius $1/H$. There are, however, complex extrema. These can be obtained by changing $\tau \to \pm\, it$ in Equation (4.58) so they solve

$$\left(\frac{\dot{a}}{a}\right)^2 = H^2 - \frac{1}{a^2}. \tag{4.60}$$

An extremum is then a solution of the Lorentzian Einstein equations with positive cosmological constant. This solution is called de Sitter space . These complex extrema must contribute in complex conjugate pairs so that the wave function is real. By a standard WKB matching analysis we can establish the form of the wave function for $a_0 > H^{-1}$

$$\psi(a_0) \approx 2N\left[H^2 a_0^4 - a_0^2 + 1\right]^{1/4} \cos\left[\frac{\left(H^2 a_0^2 - 1\right)^{3/2}}{3H^2} - \frac{\pi}{4}\right]. \tag{4.61}$$

This form could be derived by carefully following the extremum configuration as a_0 is increased along the heavy curve shown in Fig. 4.

In the region $a_0 > H^{-1}$ the steepest descents approximation to the wave function is a real linear combination of solutions of the semiclassical form (4.33). In this region we find the geometric correlations of classical de Sitter space. The state of the matter field can be read of (4.35) (4.40) and (4.56). It is, up to normalization,

$$\chi_0[\tau, \varphi_0^{(n)}] = \prod_n \exp\left[-\frac{1}{2}\omega_{(n)}\left(a(\tau)\varphi_0^{(n)}\right)^2\right]. \tag{4.62}$$

In the minisuperspace model, the prescription for the wave function of the universe of Section 3.3 predicts this particular quantum state of matter in the limit of classical geometry and quantum field theory in curved spacetime.

The state specified in the field representation by (4.62) is familiarly known as the Euclidean de Sitter invariant vacuum state. To see this we note that in terms of the variables $\chi_0^{(n)}$ the wave function (4.62) is that of a field in its ground state in the Einstein static universe. In Fock space the state is therefore annihilated by the modes which are positive frequency in the conformal time η. In terms of φ these are modes proportional to

$$[a(\eta)]^{-1}\exp(-i\omega_{(n)}\eta)Y_{(n)}(\mathrm{x}). \tag{4.63}$$

This is the conventional definition of the Euclidean, de Sitter invariant vacuum.[26]

As argued by D'Eath and Halliwell[23] the de Sitter invariance of the quantum state of the matter is an inevitable consequence of the symmetry of the Euclidean sum over histories prescription. In the defining region $a_0 < H^{-1}$ the extremizing configuration which supplies the steepest descents approximation is the smallest part of a 4-sphere bounded by a 3-sphere of radius a_0. However, the wave function does not depend on the orientation of the 3-sphere in the 4-sphere. There is thus an $O(5)$ invariance which in the Lorentzian region corresponds to the de Sitter group.

ACKNOWLEDGEMENTS

Preparation of these lectures was supported in part by the National Science Foundation under Grant PHY 85-06686.

PROBLEMS

1. Practice expressing cosmological and local observations in terms of correlations in the wave function of the universe. Are there observations which cannot be so expressed?
2. Provide a careful derivation of the Wheeler-DeWitt equation from the sum over histories which shows the connection between the factor ordering and the measure. Hint: See Ref. 28.
3. Is there a complex contour in the space of 4-geometries and matter field configurations along which the sum over histories for quantum cosmology is convergent?
4. Discuss the regularization of quantum field theory in curved spacetime in the context of the semiclassical limit of quantum cosmology described in Section 4. Hint: See Ref. 23.

NOTATION

For the most part we follow the conventions of Ref. 29 with respect to signature, curvature and indices. In particular:

Signature: (-,+,+,+) for Lorentzian spacetimes. (+,+,+,+) for Euclidean spacetimes.

Indices: Greek indices range over spacetime from 0 to 3. Latin indices range over space from 1 to 3.

Units: We use units in which $\hbar = c = 1$. The Planck length is $\ell = (16\pi G)^{\frac{1}{2}} = 1.15 \times 10^{-32}$ cm.

Covariant Derivatives: ∇_α denotes a spacetime covariant derivative and D_i a spatial one.

Traces and Determinants: Traces of second rank tensors $K_{\alpha\beta}$ are written as $K = K^\alpha_\alpha$ except when the tensor is the metric in which case g is the determinant of $g_{\alpha\beta}$ and h the determinant of h_{ij}.

Extrinsic Curvatures: If n_α is the unit normal to a spacelike hypersurface in either a Euclidean or Lorentzian spacetime, we define its extrinsic curvature to be

$$K_{ij} = \nabla_i n_j .$$

Intrinsic Curvatures: Intrinsic curvatures are defined so that the scalar curvature of a sphere is positive.

Metric on the unit n-sphere: This is denoted by $d\Omega_n^2$ and in standard polar angles is

$$d\Omega_2^2 = d\theta^2 + \sin^2\theta d\varphi^2 \qquad n = 2$$

$$d\Omega_3^2 = d\chi^2 + \sin^2\chi d\Omega_2^2 \qquad n = 3$$

REFERENCES

1. Discussions of the general problems encountered in the search for initial conditions may be found in R. Penrose, in *General Relativity: an Einstein Centenary Survey*, ed. by S. W. Hawking and W. Israel (Cambridge University Press, Cambridge, 1979), J. B. Hartle in *Inner Space/Outerspace: the Interface between Cosmology and Particle Physics*, ed. by E. W. Kolb, et.al. (University of Chicago Press, Chicago, 1986) and J. Barrow and F. Tipler, *The Anthropic Principle*, (Clarendon Press, Oxford, 1986). A sample of specific proposals for laws for initial conditions may be found in the article by Penrose cited above and in Refs 2-5.
2. S. W. Hawking, in *Astrophysical Cosmology: Proceedings of the Study Week on Cosmology and Fundamental Physics* ed. by H. A. Brüch, G. V. Coyne and M. S. Longair (Pontificiae Academiae Scientiarum Scripta Varia, Vatican City, 1982), *Nucl. Phys.* **B239**: 257, (1984) and J. B. Hartle and S. W. Hawking, *Phys. Rev.* **D28**: 2960, (1983).
3. A. Vilenkin, *Phys. Lett.* **B117**: 25, (1983), *Phys. Rev.* **D27**: 2848, (1983), *Phys. Rev.* **D30**: 509, (1984), *Phys. Rev.* **D32**: 2511, (1985), TUTP preprint 85-7.

4. J. V. Narlikar and T. Padmanabhan, *Physics Reports* **100**: 151, (1983), T. Padmanabhan "Quantum Cosmology - The Story So Far" (unpublished lecture notes).

5. W. Fischler, B. Ratra, L. Susskind, *Nucl. Phys. B* **259**: 730, (1985).

6. The author's lectures at Cargese were somewhat more extensive than presented here. Some of this material can be found in the author's lectures in *High Energy Physics 1985: Proceedings of the Yale Theoretical Advanced Study Institute* ed. by M. J. Bowick and F. Gürsey (World Scientific, Singapore 1985).

7. For a clear review of the basic ideas in canonical quantum gravity in greater detail than can be presented here see K. Kuchar, in *Relativity Astrophysics and Cosmology* ed. by W. Israel (D. Reidel, Dordrecht, 1973) and in *Quantum Gravity 2* ed. by C. Isham, R. Penrose and D. Sciama (Clarendon Press, Oxford, 1981).

8. See, e.g., H. Everett, *Rev. Mod. Phys.* **29**: 454, (1957), the many articles reprinted and cited in *The Many Worlds Interpretation of Quantum Mechanics*, ed. by B. DeWitt, and N. Graham, (Princeton University Press, Princeton, 1973), M. Gell-Mann (unpublished), and the lucid discussion in R. Geroch, *Noûs* **18**: 617, (1984).

9. D. Finkelstein, *Trans. N.Y. Acad. Sci.* **25**: 621, (1963); N. Graham, unpublished Ph.D. dissertation, University of North Carolina 1968 and in *The Many Worlds Interpretation of Quantum Mechanics*, ed. by B. DeWitt, and N. Graham (Princeton University Press, Princeton, 1973); and J. B. Hartle *Am. J. Phys.* **36**: 704, (1968).

10. For a very clear discussion of this equivalence and its consequences see C. M. Caves, *Phys. Rev.* **D33**: 1643, (1986) and to be published.

11. H. Salecker and E. P. Wigner, *Phys. Rev.* **109**: 571, (1958).

12. B. DeWitt, *Phys. Rev.* **160**: 1113, (1967).

13. A. Peres, *Am. J. Phys.* **48**: 552, (1980).

14. D. Page and W. Wooters, *Phys. Rev.* **D27**: 2885, (1983), W. Wooters, *Int. J. Th. Phys.* **23**: 701, (1984).

15. See, e.g., J. B. Hartle and S. W. Hawking, *Phys. Rev.* **D28**: 2960, (1983), S. W. Hawking (to be published), D. N. Page (to be published) J. B. Hartle, *Class. Quant. Grav.* **2**: 707, (1985), A. Anderson and B. DeWitt, *Found. Phys.* (to be published).

16. See, e.g., H. Hamber and R. Williams, *Phys. Lett.* **157B**: 368, (1985), *Nucl. Phys.* **B267**: 482, (1986), *ibid* **B269**: 712, (1986); H. Hamber in *Critical Phenomema, Random Systems and Gauge Theories: Les Houches 1984* ed. by R. Stora and K. Osterwalder (Elsiever Science Publishers, Amsterdam, 1986); J. B. Hartle, *J. Math. Phys.* **26**: 804, (1985).

17. H. Leutwyler, *Phys. Rev.* **134**: B1155, (1964), B. S. DeWitt, in *Magic Without Magic: John Archibald Wheeler*, ed. by J. Klauder, (Freeman, San Francisco, 1972), L. Faddeev, and V. Popov, *Usp. Fiz. Nauk.* **111**: 427, (1973) (*Sov. Phys. -Usp.* **16**: 777, (1974)), E. Fradkin and G. Vilkovisky, *Phys. Rev.* **D8**: 4241, (1973), M. Kaku, *Phys. Rev.* **D15**: 1019, (1977).

18. E.g., S. W. Hawking in *Relativity Groups and Topology II* ed. by B. DeWitt and R. Stora (Elsevier, Amsterdam, 1984).

19. G. Gibbons, S. W. Hawking and M. Perry, *Nucl. Phys.* **B138**: 141, (1978), J. B. Hartle and K. Schleich in the Festschrift for E. Fradkin (to be published), K. Schleich (to be published).

20. J.A. Wheeler in *Problemi dei fondamenti della fisica, Scoula internazionale di fisica "Enrico Fermi"* Corso 52 ed. by G. Toraldo di Francia (North Holland, Amsterdam, 1979).

21. T. Banks, *Nucl. Phys.* **B249**: 332, (1985).

22. S. W. Hawking and J. Halliwell, *Phys. Rev.* **D31**: 1777, (1985).

23. P. D'Eath and J. Halliwell, (to be published).

24. R. Brout, G. Horwitz and D. Weil (to be published).

25. J. B. Hartle (to be published).

26. S. Hojman, K. Kuchar, and C. Teitelboim, *Ann. Phys. (N.Y.)* **76**: 97, (1976).

27. See, e.g., N. D. Birrell and P. C. W. Davies *Quantum Fields in Curved Space* (Cambridge University Press, Cambridge, 1982).

28. A. Barvinsky and V. N. Ponomariov, *Phys. Lett.* **167B**: 289, (1986), A. Barvinsky, *Phys. Lett.* **175B**: 401, (1986).

29. C. Misner, K. Thorne, and J. A. Wheeler, *Gravitation* (W. H. Freeman, San Francisco, 1970).

THE QUASI-ISOTROPIC UNIVERSE

D. J. Raine
Astronomy Department
University of Leicester, England

1. INTRODUCTION

The standard picture of the Universe is of a system highly isotropic and spatially homogeneous on scales larger than the observed clustering scale of local luminous matter. Evidence for this is derived from the isotropy of the microwave background radiation and the abundance of helium. In this picture the Universe is represented by a Friedman-Robertson-Walker (FRW) model with small perturbations. However, while exact homogeneity and isotropy imply a FRW model, observations of approximate homogeneity and isotropy in a finite region do not, as we shall see, imply approximate FRW behaviour for all time even locally. We therefore find that the universe need not be of FRW type in order to satisfy the observational constraints. This has important consequences for proposed explanations of approximate homogeneity and isotropy. Such explanations have been focussed on (i) the choice of initial conditions; (ii) dynamical dissipation of anisotropy and imhomogeneity and (iii) the anthropic principle. Under (i) I shall discuss the role of gravitational entropy and I shall mention Mach's Principle. (Quantum initial conditions are considered by Hartle in this volume.) I shall also comment on the anthropic explanation. Investigations of the behaviour of certain cosmological models of non-FRW type suggest that these approaches are unlikely to be successful. Under (ii) it is well-known that physical dissipation of arbitrary amounts of anisotropy or inhomogeneity leads to conflict with the entropy per baryon in the microwave background (unless the baryon production can be suitably controlled). Inflationary models are discussed in the context of (ii) by Kolb and by Barrow in this volume. I shall comment on their conclusion that arbitrary initial conditions cannot give rise to the observed universe.

2. THE MICROWAVE BACKGROUND

Observations of large angular scale variations in the brightness temperature of the microwave background radiation show a dipole component of about 300 km s^{-1} in a direction some 45^{0} away from the Virgo cluster. The motion of the Earth in the Local Group is known to be 300 km s^{-1} in roughly the opposite direction. At least part of the difference must be due to the motion of the Local Group in the Local Supercluster: estimates of this range between 300 and 400 km s^{-1} in a direction approximately 30^0 from the Virgo cluster at the centre of the Local Supercluster[1,2]. So this alone appears not to account for the whole of the measured dipole. An additional 200-300 km s^{-1} is required in roughly the direction of our nearest neighbouring supercluster, the Hydra-Centaurus supercluster[3], at ~ 30 Mpc. In fact, Burstein et al. (as reported at this meeting by Lynden-Bell) have shown that the *whole system* of galaxies within 60 Mpc (for H_0 = 100 km s^{-1} Mpc^{-1}) is moving relative to the microwave background with a velocity of around 700 km s^{-1} in roughly the same direction, and this is consistent with the streaming motion derived from infrared photometry of galaxies at a mean distance of 50 Mpc.[4]

The measured microwave dipole should disagree with the observed peculiar motion at least to the extent of the contribution from fluctuations at last-scattering due to the Sachs-Wolfe effect[5,6,7]. According to this, fluctuations in the gravitational potential, $\Delta\Phi$, arising from density inhomogeneities, induce temperature fluctuations of order $\Delta T/T \sim \Delta\Phi$. A contribution to $\Delta T/T$ on large angular scales can arise either from the tail of the correlation function, $\xi(r) = \langle\rho(r+s)\rho(s)\rangle/\langle\rho\rangle^2$, reflecting weak non-random clustering on large scales[8], or from *random* clumping of correlated structures, such as clusters or superclusters[6,10]. For random clumping the contribution to the observed dipole is of order 100 $(J_3/10^3\ \mathrm{Mpc}^3)^{1/2}\ \Omega_0^{-0.2}$ km s^{-1} for Ω_0, the present value of the density parameter, of order unity, and $J_3 = \int\xi(r)r^2dr$. Another source of an intrinsic variation in the microwave temperature would be anisotropic expansion in spatially homogeneous models or source-free shear in inhomogeneous ones. At present the uncertainty in our peculiar motion means that this intrinsic component is constrained only by the total observed microwave dipole.

A more stringent limit to an intrinsic anisotropy in the background can be obtained from the quadrupole variation, $Q \lesssim 3 \times 10^{-5}$, corresponding roughly to $\Delta T/T = \langle T(\theta)T(0)\rangle^{1/2}/T \lesssim 10^{-4}$ for $\theta \sim 90^0$. This can be used to limit anisotropic expansion (shear) and rotation in both spatially homogeneous and inhomogeneous models[11].

3. HELIUM ABUNDANCE

The production of helium in the early universe is sensitive to the expansion timescale. The existence of shearing motions acts like an energy density to speed expansion and reduce helium production. Thus the helium abundance can be used as a test of isotropy. In this way, Barrow[12] and Olson[13] predict $(\Delta T/T)_{quad} \lesssim 10^{-8}$ in cosmologies close to the open Robertson-Walker model. Barrow[9,14] shows the requirement is much less stringent in more general models (those with anisotropic spatial curvature). In some of these the helium limit is less restrictive than the microwave limit.

4. ANISOTROPIC SPATIALLY HOMOGENEOUS COSMOLOGIES

For detailed reviews of spatially homogeneous cosmologies see, e.g. MacCallum[15], or Ellis[16]. Briefly, the essential points are these. The universe is assumed to contain a fluid with velocity u^μ (μ = 0,1,2,3); the mean expansion of the fluid is given by the Hubble parameter

$$H \equiv \frac{1}{3}\Theta \equiv \frac{1}{3}u^\mu{}_{;\mu}. \qquad (4.1)$$

In the case that the fluid flow is normal to the surfaces of homogeneity the shear is

$$\sigma_{ij} = \frac{1}{2}(u_{i;j} + u_{j;i}) - \frac{1}{3}\theta\delta_{ij} \tag{4.2}$$

in hypersurface orthogonal coordinates (i,j ... = 1,2,3). The symmetry of the spatial hypersurfaces is determined by the Killing vectors, $\xi_A{}^{\mu}$, the type of space being characterised by the number (and form) of the independent parameters in the structure constants of the Lie algebra of Killing vectors. Those models in which the Lie algebra of Killing vectors is a subalgebra of the algebra in a FRW model can be continuously deformed to a FRW model as the shear goes to zero.

This distinction is important: in the group types which contain FRW models the spatial curvature is isotropic, i.e. if $R_{ij}{}^*$ is the Ricci tensor constructed from the 3-space metric then

$$R_{ij}{}^* = \frac{1}{2}\delta_{ij} R^* . \tag{4.3}$$

For fluid flows with u^{μ} orthogonal to the homogeneous space sections the shear evolves according to

$$\dot{\sigma}_{ij} + 3\theta\sigma_{ij} + R_{ij}{}^* - \frac{1}{3}R^*\delta_{ij} = 0 ; \tag{4.4}$$

so, for isotropic curvature, $\sigma = (\frac{1}{2}\sigma_{ij}\sigma^{ij})^{\frac{1}{2}} \propto \ell^{-3}$ if $\theta \equiv 3H = 3\dot{\ell}/\ell$. Here the dot denotes differentiation with respect to the cosmic time, t, defined such that the surfaces of homogeneity are t = constant. For Ω = 1 we get $\sigma/H \propto (1 + z)^{3/2}$ at a redshift z. Thus only a decaying mode of shear is present. In other models, anisotropic spatial curvature gives rise to a growing mode of shear (relative to expansion)[11] with $\sigma/H \propto (1 + z)^{-1}$. So no matter how small σ may be at any time such models are not close to Robertson-Walker; we say they are not of FRW type.

The existence of the growing mode led Collins and Hawking[17] to rule out those models as realistic cosmologies. But, as Barrow has pointed out, such models may contain an extended period of low anisotropy compatible with present observations.

We therefore have the following possibilities:

(i) the Universe contains only a decaying mode with a sufficiently short timescale t_d to be compatible with helium abundances. From this point of view the present age $t_o \gg t_d$ and the universe is isotropic because it is old.

(ii) the Universe contains both a decaying mode and a growing mode with timescale $t_g \gg t_o$, sufficiently long to be compatible with the microwave background. In this case the universe is isotropic because it is middle-aged.

(iii) a third possibility is the quiescent cosmology of Barrow[18]. In this model we have only the growing mode, with $t_o \ll t_g$, so the universe is isotropic because it is young.

Not all models can be compatible with the observed limits on helium abundance and the microwave anisotropy, but we <u>cannot</u> expect to show from observation alone that the Universe must be of FRW type.

5. THE WAINWRIGHT AND ANDERSON SOLUTION

To illustrate these remarks we consider here and in the next section two particular examples of universes of non FRW type. The first is a solution for a perfect fluid equation of state first found by Collins[19] and developed by Wainwright and Anderson[20]. The metric is

$$ds^2 = -dt^2 + T^{4/(3\gamma)}\,[A^{2q_1}\,dx^2 + A^{2q_2}\,e^{2r[s+(3\gamma-2)]x}\,dy^2 + A^{2q_3}\,e^{2r[s-(3\gamma-2)]x}\,dz^2] \tag{5.1}$$

where

$$A(t) = [1 + \gamma^2\,\alpha_c\,T^{2-4/(3\gamma)} + \alpha_s\,T^{1-2/\gamma}\,]^{1/(2-\gamma)}\ , \qquad \frac{dT(t)}{dt} = A^{1-\gamma} \tag{5.2}$$

and the constants q_i, s and r are given by

$$q_1 = \tfrac{1}{2}\ , \quad q_2 = (2-\gamma+s)/4\ , \quad q_3 = (2-\gamma-s)/4$$
$$s^2 = (3\gamma+2)(2-\gamma)\ , \quad r^2 = \alpha_c(3\gamma+2)/[36(2-\gamma)]\ . \tag{5.3}$$

Our units here are $c = 1$, $8\pi G = 1$. The energy density and pressure are given by

$$\mu = \frac{4}{3\gamma^2 T^2 A^\gamma}\ , \quad p = (\gamma-1)\mu\ . \tag{5.4}$$

The parameter α_s is related to the singularity structure: it controls the decaying mode of shear; the parameter α_c is related to the spatial curvature and controls the growing mode. The ratio of shear, $\sigma = (\tfrac{1}{2}\sigma_{ij}\sigma^{ij})^{\frac{1}{2}}$, to expansion is

$$0 < \frac{\sigma}{H} = \frac{1}{2\sqrt{3}}\left[-\frac{2\gamma^2(3\gamma-2)}{3(2-\gamma)}\,\alpha_c\,T^{2-4/(3\gamma)} + \alpha_s\,T^{1-2/\gamma}\right] \times\left[1 + \frac{4\gamma^2}{3(2-\gamma)}\,\alpha_c\,T^{2-4/(3\gamma)} + \frac{1}{2}\,\alpha_s\,T^{1-2/\gamma}\right]^{-1} < \sqrt{3}\ . \tag{5.5}$$

For suitably chosen values of α_c and α_s the model has a period of quasi-isotropic expansion for which $\sigma/H < \epsilon$ for any ϵ. The spatial curvature $\tilde{R}^* = (\tfrac{1}{2}R^*{}_{ij}R^{*ij})^{\frac{1}{2}}$ is also restricted by $\tilde{R}^*/H^2 < \epsilon$, while the matter density satisfies $\mu/H \sim 1$ (so $\Omega \sim 1$). For $\gamma = 1$, appropriate to the present epoch at t_0, the quasi-isotropic phase lasts until a time $\sim (\alpha_c{}^{3/5}\,\alpha_c{}^{2/5}/\epsilon)^{-3/2}\,t_0$, at which time distortions in the local Hubble flow of order ϵ would become apparent. The microwave anisotropy, however, depends on the cumulative effect of the shear and may therefore be detectable on a much shorter timescale.

To investigate this we have numerically integrated the null geodesics for this metric to obtain the radiation temperature as a function of angle[21]. The redshift in a direction $k^\mu(0)$ is[16]

$$1 + z = \frac{(k^\mu u_\mu)_{obs}}{(k^\mu u_\mu)_{\ell.s.}} , \quad (5.6)$$

where the numerator on the right is evaluated at the present and the denominator at last scattering. The homogeneity implies that last scattering occurs at T = constant (implied by t = constant). We took last scattering to occur at a redshift around 700 (depending on direction), the precise value being unimportant. The microwave temperature in a direction (θ,ϕ) is given by

$$T_r(\theta,\phi) = \frac{T_{r,o}}{1 + z(\theta,\phi)} . \quad (5.7)$$

For the numerical results we fixed α_s by the helium abundance constraint. Note that at early times (equivalent to $T \to 0$ for $\alpha_s \geqslant 0$) for $\gamma = 4/3$, we have

$$A(t) \propto T^{-3/4} . \quad (5.8)$$

Therefore

$$T \propto t^{4/3} \quad (5.9)$$

and $T_r \propto \mu^{1/4} \propto t^{-1/3}$. Thus, in this case the anisotropic curvature does not balance the shear to give the FRW behaviour, $T_r \propto t^{-1/2}$, so there is a significant constraint from the helium abundance, namely that the shear must decay sufficiently early[14]. If this epoch is described by parameters α_s', α_c', $\gamma = 4/3$, then provided the growing mode can be neglected ($\alpha_c' T << 1$), we have

$$aT_r^4 = \tfrac{1}{3} T^{-2} (1 + \alpha_s' T^{-1/2})^{-2} . \quad (5.10)$$

Then $\sigma/\theta \lesssim 0.5$, as required by the helium abundance[18], provided $\alpha_s' \lesssim a^{-1/4} T_r^{-1} \lesssim 10^{-8}$. The radiation dominated model ($\gamma = 4/3$) must join smoothly on to the dust model ($\gamma = 0$), for which $T \propto t$ in the quasi-isotropic era. If we impose this condition at $t \sim t_0/30$, the time at which the radiation and matter densities are equal, hence $T \sim T_0/30$, in this model, we find $\alpha_s \sim 3\alpha_s'$, and hence $\alpha_s \lesssim 3\ 10^{-8}$. For the numerical computations we took $\alpha_s = 10^{-8}$; in fact, we found the distortion of the microwave background to be essentially independent of α_s provided $\alpha_s \lesssim 10^{-6}$.

Fitting an expansion in spherical harmonics of degree $\leqslant 2$ to the computed temperature,

$$\Delta T(\theta,\phi) = \sum_{\ell=1,2} a_{\ell m} Y_{\ell m}(\theta,\phi), \quad (5.11)$$

in the case $\alpha_s \leqslant 10^{-6}$ we found, for $\alpha_c \gtrsim 10^{-9.5}$, a dipole contribution which could be fitted numerically by

$$P = (\Sigma\, a_{1m}{}^2/4\pi)^{1/2} \sim 2.6 \times 10^{-4} (\alpha_c/10^{-8})^{1/2} \quad (5.12)$$

and a quadrupole

$$Q = (\sum_m a_{2m}{}^2/4\pi)^{1/2} \sim 3 \times 10^{-5}(\alpha_c/10^{-8})^{1/2} \quad (5.13)$$

at the present time.

6. AN INHOMOGENEOUS MODEL

We can think of the shear in the above model as "source-free" in the sense that we have $\mu/\theta^2 \to 0$ as $t \to 0^+$ and $t \to \infty$, but $\sigma/\theta \to \infty$ in both of these limits: hence the shear does not depend on the presence of matter. This leads us to ask what limits the uniformity of the microwave background imposes on the existence of <u>in</u>homogeneous source-free shear.

To investigate the restriction on inhomogeneous shear we consider a so-called Bondi or Bondi-Tolman metric,

$$ds^2 = dt^2 - X(r,t)dr^2 - Y^2(r,t)[d\theta^2 + \sin^2\theta \, d\phi] \, , \tag{6.1}$$

which represents a spherically symmetric expanding universe[22]. The field equations for dust (P = 0) can be integrated in terms of a parameter η to give[23]:

$$Y = \lambda(r)\beta^{-1}(r)\sinh^2\eta/2 \tag{6.2a}$$

$$t - t_*(r) = \tfrac{1}{2}\lambda(r)\beta^{-3/2}(r)(\sinh\eta - \eta) \tag{6.2b}$$

$$X = (1 + \beta(r))^{-1/2} \, \partial Y/\partial r. \tag{6.2c}$$

We choose $\beta > 0$ to correspond to the open FRW models which are obtained if $t_*'(r) = 0$ and $(\lambda\beta^{-3/2})' = 0$ (where the prime denotes $\partial/\partial r$). The second condition corresponds to the absence of density perturbations, so we retain this here. But we drop the first condition to allow the presence of source-free shear[24]. The shear is then:

$$\frac{\sigma}{H} = -\sqrt{3}\, c\beta^{\frac{1}{2}}\, t_*'[\cosh^2\eta/2 + \tfrac{1}{2}]\{3(\lambda/\beta)'\sinh^3\eta/2\cosh\eta/2 - c\beta^{\frac{1}{2}} t_*'[2\cosh^2\eta/2 - \tfrac{1}{2}]\}^{-1} \tag{6.3}$$

with $\beta = r(1 - r^{2/4})^{-1}$. Note that $\sigma/H \to 0$ as $t \to \infty$ $(\eta \to \infty)$ so only the decaying mode is present.

For the numerical integration of the null geodesics[25] we have to choose a functional form for $t_*(r)$. Let us take a Gaussian profile

$$t_*'(r) = \Sigma \, e^{-(r/r_0)^2} \tag{6.4}$$

where r_0 determines the coordinate radius of the region and Σ the maximum amplitude:

$$(\sigma/H)_{max} \sim 6 \times 10^{-5} \, \Sigma \, r_0 \tag{6.5}$$

at the present time. The redshift z_L at the centre $(r = 0)$ is used to characterise the location of the region. We also need the microwave temperature on the last scattering surface since this now depends on r. We assume the perturbation to be adiabatic, so $T(z) \propto \rho^{1/3}(z)$. The last scattering surface is given by unit optical depth to Thomson scattering:

$$1 = \sigma_T f \int_0^{s_*} n k^0 ds \tag{6.6}$$

where fn is the electron density and $k^0 = dt/ds$ with s an affine parameter along null rays.

The numerical results give limits on the range of Σ, r_0 and z_L compatible with a given observed fluctuation $(\Delta T)_{max}$. For illustration we took $\Omega_0 = 0.2$, $f = 0.1$, but the conclusions are not sensitive to this choice. If $(\Delta T)_{max} \leqslant 0.5$ mK we can approximate the result analytically, for $z_L > 1$, as

$$\ell \lesssim 2\ 10^3\ z_L^{-2/3}\ \Sigma^{-1/6} \tag{6.7}$$

where $\ell = 4500\ r_0$ Mpc gives the proper size of the shearing region to a sufficient approximation. Note that for fixed ℓ, z_L must decrease as Σ increases. This is because a more distant region influences the background radiation further in the past when the shear was larger. The most effective way to hide the shear is therefore in a region of a size comparable to our distance from its centre. We then find $(\sigma/H)_{max} \lesssim 10^{-2}$.

This is rather a weak constraint compared with that from helium abundance for which $(\sigma/H)_0 \lesssim 10^{-12}$ in spatially homogeneous open FRW type models[12,13]. In fact, since the Bondi model contains only the decaying mode, we obtain similar restrictions on localised regions of shear from the homogeneity of the helium abundance. Requiring $(\sigma/H)_0 \lesssim 10^{-12}$ for z ~ 1 restricts the shear to $(\sigma/H)_{max} \lesssim 10^{-2}$ in domains of a few hundred Mpc in regions of space a few thousand Mpc in radius.

7. DISCUSSION OF RESULTS

7.1 *The Anthropic Principle*

We return to the Wainwright and Anderson solution. To satisfy the dipole (and quadrupole) constraint we have $\alpha_c \lesssim 10^{-8}$. As well as the present age, t_0, the model therefore contains two timescales $t_d = \alpha_c t_0$ and $t_g = \alpha_c^{-2/3}\ t_0$, which must differ by at least thirteen orders of magnitude in the observed universe. Can we attribute this to an accident which might be accounted for by the anthropic principle? The difficulty of so doing arises from the fact that we do not appear to need an exceedingly high degree of isotropy for our existence, so ϵ need not be very small, in which case the quasi-isotropic expansion lasts for a time >> t_0 probably well after there is any life around to need explanation. The duration of the quasi-isotropic phase is $\Delta t_0 \sim (\alpha/\epsilon)^{-3/2}\ t_0$ (from above), where $\alpha = \alpha_c^{3/5}\ \alpha_s^{2/5}$. Suppose, as generously as possible, for the sake of argument, we require $\epsilon = (\sigma/H)_0 = 3\ .\ 10^{-7}$ now on anthropic grounds (!). For $\alpha_c = \alpha_s = 10^{-8}$ our numerical computation shows the anisotropy would show up in the background radiation, with present sensitivities, in 5 t_0 years and the quasi-isotropic phase would also last around 5 t_0 years. At this time we should discover the existence of the growing mode (and abandon theories that forbid its existence), and might try to argue that the anthropic principle had prevented it from appearing sooner. But this depends on detector technology. Suppose we improve this by a factor x in the near future (i.e. in less than a Hubble time). If $(\Delta T/T)_0 \lesssim 10^{-4}/x$, i.e. if we still do not find the growing mode, then $\alpha_c \lesssim 10^{-8}/x^2$ and $\alpha \sim 10^{-8}/x^{6/5}$. Thus $\Delta t_0 \sim x^{6/5} t_0$. For $x \gtrsim 10$ it would be difficult to justify such a universe on anthropic grounds: it would require not only that the quasi-isotropic phase be accidentally sufficiently extended, compared with an evolutionary timescale, to accommodate us, but also, it would seen, that the universe be designed with our engineering potential in mind.

Qualitatively similar remarks apply to the existence of inhomogeneous shear, where an anthropic argument would not be able to restrict the regions of the Universe in which we might not have been able to evolve to be substantially smaller than that in which we have.

Penrose[26] gives an alternative probabalistic argument against the anthropic principle.

7.2 *Initial Conditions*

The history of the use of initial conditions to restrict possible cosmological solutions goes back at least to Einstein's attempt to implement 'Mach's Principle' as the requirement of closed spatial sections. Nowadays two facts about Mach's Principle are well known: first that no-one can agree what it says, and second that everyone agrees it is satisfied in the observed Universe. The essence of the idea is to restrict the excitation of the free gravitational degrees of freedom[27]. This differs in spirit from the approach through the 'wave function of the Universe' only in as much as only the free gravitational modes are considered. We have seen in both homogeneous and inhomogeneous models how this is required for compatibility with observation. But Mach's Principle could not rule out mass distribution more inhomogeneous than the one we have, so cannot provide a complete set of initial conditions.

7.3 *Gravitational Entropy*

The matter entropy of the Universe within the horizon is of order 10^{88}k, almost all of which resides in the 10^{8}k per baryon of the microwave background radiation. As Penrose[28] has pointed out, this apparently large number is in fact small compared to the entropy obtained if gravity acts to clump matter into black holes. If all the matter within the horizon is put into a single black hole one obtains a maximum entropy of around 10^{123}k. Why then is it, as Penrose puts it, that only 10^{-34} of the available chaos is used initially? A possible clue comes from the observations (i) that if the initial low entropy is not to be associated with matter (because the matter is in thermal equilibrium) it must be associated with a special geometry; (ii) that the high entropy clumping of matter is associated with a non-zero Weyl tensor (because an absence of clumping implies isotropy, which in turn implies no gravitational principal null directions); and (iii) that the special nature of the FRW singularity is connected with the vanishing of the Weyl tensor. One would expect a generic (= high entropy?) singularity to have a large Weyl tensor, at least relative to the Ricci tensor. This suggests a possible connection between entropy and the Weyl tensor such that the vanishing at the initial singularity of the Weyl tensor, or the ratio of the Weyl tensor to some power of the Ricci tensor, is an expression of the gravitational arrow of time.

In the Wainwright and Anderson solution we find $C^2 \equiv (C_{\lambda\mu\nu\rho}C^{\lambda\mu\nu\rho})$ diverges as $T \rightarrow 0$. Thus the vanishing of the Weyl tensor is not a necessary condition for compatibility with observation. If $\alpha_s \neq 0$ we find also that $P = (C_{\lambda\mu\nu\rho}C^{\lambda\mu\nu\rho})^{1/2}/(R_{\mu\nu}R^{\mu\nu})^{1/2} \rightarrow$ constant $\neq 0$ as $T \rightarrow 0$; thus the observations are also compatible with a weaker violation of the vanishing 'entropy' requirement. If $\alpha_s = 0$, so there is no decaying mode, then the Weyl tensor still diverges, $C^2/H^4 \rightarrow$ constant $\neq 0$ as $T \rightarrow 0$, but $P \rightarrow 0$ at the initial singularity. Goode and Wainwright[29] have shown this to be a necessary consequence of their definition of an *isotropic singularity*. The initial unimportance of the Weyl tensor, expressed through the condition $P \to 0$, is therefore related, at least, to the quiescent cosmology explanation of isotropy ($t < t_g$, (iii) above). This is interesting, because the low gravitational entropy constraint, $P \to 0$, would force $\alpha_s = 0$, but it cannot be the whole story, because the isotropic singularity is compatible with the presence of a growing mode of arbitrary timescale.

Bonner[30] has considered the Bondi-Tolman solution in the case $t_*'(r) = 0$, $(\lambda\beta^{3/2})' \neq 0$, corresponding to density perturbations of a FRW model in the absence of source-free shear. In the recollapsing case, $\beta < 0$, he finds that

$$P = \frac{4}{3}\left[1 - \frac{3\lambda Y'}{Y\lambda'}\right] \tag{7.1}$$

is an increasing function of the cosmic time, t, ranging from zero at the big bang to infinity at the final crunch. The low 'entropy' constraint, $P \to 0$ at the initial singularity, forces an isotropic singularity, $t_*'(r) = 0$, in accordance with the theorem of Goode and Wainwright, and hence imposes the absence of the decaying mode.

In fact, Bonnor shows that $Pu^\lambda = P\delta_0{}^\lambda$ does not have a non-negative covariant divergence, so cannot be an entropy current. He suggests instead that $\mu P^2 \equiv (C^2/(R_{\mu\nu}R^{\mu\nu})^{3/2})^2$, which does satisfy $(\mu P^2\delta_0{}^\lambda)_{;\lambda} = 0$ and is also increasing, is a better measure of entropy. In the Wainwright and Anderson solution we find, for example, if $\alpha_s \neq 0$, $\gamma = 4/3$, then $\mu P^2 \to \infty$ as $T \to 0$ as $\mu P^2 \to 0$ as $T \to \infty$. i.e. precisely the opposite of the hoped-for behaviour (P itself evolves non-monatonically between two finite values). One might argue that this is grounds for ruling out the presence of the decaying mode. Then the $\gamma = 4/3$ solution has $\mu P^2 \to 0$ as $T \to 0$ or ∞, i.e. the growing mode does not show up in increasing 'entropy' (although it is reflected in the behaviour of P itself which increases from zero to a finite value). The $\gamma = 1$ model behaves rather better as one might expect, since Bonnor chose the function μP^2 for a dust model, but the sign of time cannot be allowed to depend on the matter content. It seems not unfair to conclude that, so far, applications of the entropy constraint seem to rule out a decaying mode of shear, which, as we have seen, is compatible with but not justified by observation, and to permit the growing mode, which is certainly ruled out by observation.

The problem that emerges from this discussion in accounting for the observed isotropy seems to be that while the observational data cannot rule out small departures from exact FRW, constraints on initial conditions either imply exact FRW (e.g. $C^2 = 0$), or do not distinguish qualitatively between small and less small departures (is a natural amount of gravitational entropy 0 or 10^{88}k?). But this widening of our view to include non-FRW type models for the observed universe suggests a possible alternative approach to the evolutionary explanation. Namely, that we might look towards a theory that evolves a large class of initial conditions to quasi-isotropy, but evolves others *away from* such states, perhaps even not to classical universes at all.

REFERENCES

1. Hart, L. & Davies, R.D., *Nature*, 297, 191 (1982).
2. Aaronson, M., Mould, J., Schechter; P.L. & Tully, R.B., *Astrophys. J.* 258, 64 (1982).

3. Aaronson et al., *Astrophys. J.* 302, 536 (1986).
4. Collins, C., Joseph, R. & Robertson, N., *Nature* 320, 506 (1986).
5. Sachs, R.K. & Wolfe, A.M., *Astrophys. J.* 147, 73 (1967).
6. Peebles, P.J.E., "The Large Scale Structure of the Universe" Princeton Univ. Press (1980).
7. Anile, A.M. & Motta, S., *Mon.Not.R.astr.Soc.* 184, 319 (1978).
8. Silk, J. & Wilson, M.L., *Astrophys. J.* 244, L37 (1981).
9. Barrow, J.D. & Senoda, D.H., *Phys. Rep.* 139, 1 (1986).
10. Peebles, P.J.E., *Astrophys. J.* 243, L119 (1981).
11. Barrow, J.D., Juszkiewicz, R. & Senoda, D.H., *Nature* 305, 397 (1983).
12. Barrow, J.D., *Mon.Not.R.astr.Soc.* 175, 359 (1976).
13. Olson, D.W., *Astrophys. J.* 219, 777 (1978).
14. Barrow, J.D., *Mon.Not.R.astr.Soc.* 211, 221 (1984).
15. McCallum, M.A.H., in "Cargese lectures in Physics, Vol. 6", ed. E. Schatzman (Gordon & Breach), (1973).
16. Ellis, G.F.R., in "General Relativity and Cosmology: Proc. Int. Sch. Theor. Phys. 'Enrico Fermi' Cse 47." Ed. R.K. Sachs, Academic Press (1971).
17. Collins, C.B. & Hawking, S.W., *Mon.Not.R.astr.Soc.* 162, 307 (1973).
18. Barrow, J.D., *Nature* 272, 211 (1978).
19. Collins, C.B., *Comm. Math. Phys.* 23, 137 (1971).
20. Wainwright, J. & Anderson, P.J., *Gen. Rel. Grav.* 16, 609 (1984).
21. Raine, D.J. & Thomas, E.G., *Astrophys. Lett.* (1986).
22. Bondi, H., *Mon.Not.R.astr.Soc.* 107, 410 (1947).
23. Raine, D.J. & Thomas, E.G., *Mon.Not.R.astr.Soc.* 195, 649 (1981).
24. Eardley, P., Liang, E.& Sachs, R.K., *J. Math. Phys.* 13, 99 (1972).
25. Raine, D.J. & Thomas, E.G., *Astrophys. Lett.* 23, 37 (1982).
26. Penrose, R., in "Quantum Gravity 2, A Second Oxford Symposium" ed. by C.J. Isham, R. Penrose & D.W. Sciama, Oxford U.P., 244 (1981).
27. Raine, D.J., *Rep. Prog. in Phys.* 11, 1151 (1981).
28. Penrose, R., in "General Relativity, An Einstein Centenary Survey", ed. by S.W. Hawking & W. Israel, Cambridge U.P. (1979).
29. Goode, S.W. & Wainwright, J., *Class. Quantum Grav.* 2, 99 1985.
30. Bonnor, W.B., *Physics Lett. A*, 112, 26 (1985).

SEMICLASSICAL QUANTUM GRAVITY IN TWO AND FOUR DIMENSIONS

Norma Sánchez

Groupe d'Astrophysique Relativiste
CNRS – Observatoire de Paris
92195 Meudon Principal Cedex, France

INTRODUCTION

While a full quantum theory of gravity is still non-existent, continuous effort over the last years has shown some of the properties which an eventually complete theory will have to possess. At present, what is called "semiclassical quantum gravity" refers to different approaches and approximations:

i) Q.F.T. in curved space-time, in which matter fields are quantised on classical gravitational backgrounds, one of the first important examples being the Hawking radiation by black holes; this is also of conceptual and practical interest in early Cosmology and Inflation.

ii) Semiclassical Einstein equations, in which quantised matter fields react back (through the expectation value of the energy-momentum tensor) on the geometry (the so-called "back-reaction problem"); important problems being the resolution of the late time evolution of black holes due to the reaction of Hawking radiation, and the reaction of particle production in the early-time evolution of the Universe.

iii) Semiclassical approximation to the path integral of gravity and matter fields, developed in the context of euclidean gravity with instanton and partition function methods (Gibbons and Hawking), recently combined with the Wheeler-DeWitt equation of canonical quantisation and applied to Cosmology for the problem of initial conditions and the ground state (Hartle-Hawking wave function).

(One of the well known examples in the previous approaches is that of the thermal properties of black holes: first suspected at the very classical (and purely formal) level, properly found at the level of Q.F.T. in curved spacetime, then recovered at the tree-level from the path integral of gravity and matter fields.)

Here we report our recent work in connection with the above mentioned problems:

1) We investigate the vacuum fluctuations $\langle\Psi^2\rangle$ and the energy momentum tensor $\langle T_{\mu\nu}\rangle$ of massless fields of spins 0, $\frac{1}{2}$ and 1 near static distorted black holes. We describe these four dimensional quantities near the even horizon entirely in terms of the two dimensional spatial geometric invariants of the horizon surface for both the

Hartle-Hawking vacuum (thermal) and the Boulware ("non-thermal") vacua. (This work is in collaboration with V.P. Frolov from the Lebedev Institute of Moscow.)[1]

2) We investigate the role of discrete symmetries (in particular, the antipodal identification map) and the modification of the spacetime topology for Q.F.T. in curved spacetime. We analyze the implications for *Antipodal Identified Black Holes,* and for deSitter space at the level of Green functions, Fock space and thermal properties. (Work done in collaboration with B.F. Whiting from The University of North Carolina at Chapel Hill and with A. Folacci at Meudon.) [2,3]

3) We show that *Semiclassical Quantum Gravity in two dimensions* is exactly solvable. The general solution of the two dimensional semiclassical Einstein equations is presented and analyzed in terms of analytic mappings. The connection with the Liouville and string theory is analyzed.[4]

1. QUANTUM EFFECTS NEAR DISTORTED BLACK HOLES

Q.F.T. in curved space allows one to determine non-linear effects of the coupling between matter fields and geometry through the (renormalised) expectation value of the stress-energy tensor $[\langle T_{\mu\nu}\rangle_{ren}]$ of quantum fields. In flat spacetime, the vacuum state is uniquely defined. One has a clear idea of what the zero energy state is and therefore can give a well defined procedure for rendering quantum operators finite. In a general curved space, this is not the case. It is not clear how to indicate which state might be selected as "the vacuum" (or even whether there is any such state). Different choices of boundary conditions for the propagators (equivalently, for the definition of the positive frequency modes) lead to different possible vacuum states (for a recent discussion, see reference 5). The quantity $\langle\varphi^2\rangle_{ren}$ (for a scalar field φ) near black holes is also of interest as describing the quantum fluctuations of this field. Quantum fluctuations of the spacetime itself are neglected for $M \gg m_{pl}$ (M being the black hole mass and $m_{pl} = \sqrt{\hbar c/G}$). At the one-loop approximation, the contributions of different fields can be considered separately and summed additively. The contributions of massive fields are essentially local and can be studied within an expansion in $\varepsilon = m_{pl}^4/m^2M^2$ for $\lambda = \hbar/mc \ll 2GM/c^2$).[6] The contributions of massless fields are essentially non-local and much more complicated. A rather simple (and extremely good) approach for calculating $\langle T_{\mu\nu}\rangle_{ren}$ in static spacetimes satisfying $R_{\mu\nu} = \Lambda g_{\mu\nu}$ (Λ = constant) has been proposed by Page[7] and independently by Brown.[8] These approaches are based on the possibility of obtaining $\langle T_{\mu\nu}\rangle_{ren}$ in the spacetime of interest (ds^2) from that calculated in an auxilliary, conformally related spacetime ($d\bar{s}^2$) where trace anomalies vanish. [Here $d\bar{s}^2 = (-\xi_{\mu\nu}\xi^\mu)ds^2$, $\xi^\mu = (\partial/\partial t)^\mu$ being the Killing vector; t is the Killing time.] The conformal factor $\Omega^2 \equiv -\xi_\mu\xi^\mu$ is defined up to the transformation $\bar{\Omega} = \Omega e^{at}$ (a = constant). This freedom is associated with the choice of the vacuum state. We deal here with the Hartle-Hawking ($|\,\rangle_H$) and the Boulware ($|\,\rangle_B$) vacua corresponding to a thermal and to an empty state at large radii respectively. ($|\,\rangle_B$ is pathological at the horizon in the sense that $\langle T_{\mu\nu}\rangle_B^{ren}$ diverges there.) [The choice $a = 2\pi\kappa_0$ ($\kappa_0 = 1/(4M)$ being the surface gravity of the black hole) corresponds to $|\,\rangle_H$; $a = 0$ is associated with $|\,\rangle_B$.] We investigate the contribution of massless fields of spins 0, 1/2, and 1 to the vacuum polarisation near the event horizon of static Ricci-flat spacetimes. Within the Page-Brown approximation we calculate $\langle\varphi^2\rangle_{ren}$ and $\langle T_{\mu\nu}\rangle_{ren}$ near static distorted black holes, for both the $|\,\rangle_H$ and $|\,\rangle_B$ vacua. [Such a field arises when there are massive bodies outside of the black hole. Their gravitational field changes the metric near the event horizon and distorts the black hole.] Using the Israel description of static spacetimes,[9] we express these *four dimensional* quantities in an invariant geometric way. We obtain the result that $\langle\varphi^2\rangle_H^{ren}$ and $\langle T_{\mu\nu}\rangle_H^{ren}$ near the horizon depend only on the two dimensional geometry of the horizon surface. We find:[1]

$$\langle\varphi^2\rangle_H = \frac{1}{48\pi^2}K_0 \tag{1.1}$$

$$\langle T_0^0\rangle_B = (7\alpha + 12\beta)K_0^2 - \alpha^{(2)}\triangle K_0 \tag{1.2}$$

and

$$\langle\varphi^2\rangle_B = -\frac{\kappa_0^2}{48\pi^2 X} + \frac{K_0}{48\pi^2} + \mathcal{O}(X),, \quad (X \equiv -\xi_\mu \xi^\mu) \tag{1.3}$$

$$\langle T_{\mu\nu}\rangle_B = 2\frac{\kappa_0^2}{X^2}(\beta - \alpha/3)\ \mathrm{diag}(-3,1,1,1)_\mu^\nu \tag{1.4}$$

In a number of cases, these formulas coincide identically with the exact values of $\langle\varphi^2\rangle_H$ and $\langle T_{\mu\nu}\rangle_H$. In particular, it happens for $\langle\varphi^2\rangle_H$ and $\langle T_{\mu\nu}\rangle_H$ of the electromagnetic field at the horizon of the Schwarzschild black hole and for $\langle\varphi^2\rangle_H$ at the pole of the event horizon of the axially symmetric distorted black hole. The reason for this as well as the reason for the remarkable accuracy of Page's approximation in the case of the scalar field until now remains unknown (see reference 4 and references therein).

2. Q.F.T. AND THE ANTIPODAL IDENTIFICATION OF BLACK HOLES AND OF DESITTER SPACE

The antipodal points (U, V, θ, φ) and $(-U, -V, \pi - \theta, \pi + \varphi)$ of the Schwarzschild-Kruskal manifold, usually interpreted as *two different* events (in two different worlds) are considered here as *physically identified* (to give *one single* world). This has fundamental consequences for the Q.F.T. formulated on this manifold. The antipodal symmetric fields have (globally) zero norm. The usual particle-antiparticle Fock space definition breaks down. The antipodal symmetric Green functions have the *same periodicity* $\beta = 8\pi M$ in imaginary (Schwarzschild) time as the usual (non-symmetric) ones. (Identification with "conical singularity" would give a period $\beta/2$.) In any case, no usual thermal interpretation is possible for $T = \beta^{-1}$ (nor for $2/\beta$ or any other value) in the the theory. We also investigate the consequences of this identification for Q.F.T. as formulated in de Sitter space and its implications for inflation. Antipodally symmetric and antisymmetric fields and Green functions are described. We calculate for these fields the expectation values of the square of the field operator and stress-tensors in the family of de Sitter O(1,4) invariant vacua and study limiting cases of interest. In the inflationary regime the antipodal identification gives for $\langle\varphi^2\rangle$ a value *which differs by a factor of 2 from the ordinary Bunch-Davies value.* The modifications introduced to $\langle T_{\mu\nu}\rangle$ vanish in the confomally invariant case. The antipodally identified theory also allows a better understanding of the massless and minimally coupled ordinary theory (without identification). We found new vacuum states which are O(4), O(1,3) (and E(3)) invariant and calculate the stress energy tensors for them.

It is known that a given local spacetime geometry with metric g satisfying the Einstein equations admits in principle different possible associated global topologies; however most of our present understanding of Quantum Field Theory in a given non-trivial manifold refers to a particular choice of *spacetime* topology. (Important though a modification of the *spatial* topology is, fundemental changes to the known features of Q.F.T. in non-trivial manifolds would not be expected.) Spaces $\bar{\mathcal{M}}_g$, locally identical to $\mathcal{M}_g$ but with different (large scale) spacetime topology can be obtained by identifying points in $\mathcal{M}_g$ equivalent under a discrete isometry without fixed points. The field theories as formulated in the spaces $\mathcal{M}_g$ and $\bar{\mathcal{M}}_g$ are essentially different. In de Sitter space the simplest such identification is to identify the antipodal points (the antipode $\bar{P}$ of a point P is defined as having its light cone without intersecting that of P.) This is an old idea first proposed by Schrödinger and called the "elliptic interpretation." For black holes, it has been shown recently by Gibbons,[11] and by Whiting and the author,[2] that the antipodal identification destroys the thermal features and the usual Fock space construction.

The usual Fock space construction breaks down in the identified space because, even at the classical level, a complete set of positive (negative) frequency basis functions cannot be constructed: eigenmodes have zero norm! (on global sections). It means that quantum creation and anihilation cannot be defined and no vacuum Fock state exists. [In the usual theory (without identification) there are different well defined possibilities (and thus an ambiguity) in choosing a positive frequency basis, but such an unity normalised basis always exists.]

The usual thermal features in this context are destroyed. The spacetime identification considered here does not manifest itself as a mere change in the value of temperature (a "naive" consideration could lead to a modification of it by a factor of two) but the usual notion of temperature in this context (i.e. the inverse imaginary time periodicity of the Green functions) does not apply. The zero norm fields here do not allow us to describe any thermal properties in the normal way.

In Schwarzschild spacetime, the antipodal transformation

$$J : P(X) \rightarrow \bar{P}(\bar{X}), \tag{2.1}$$

without fixed points takes the form:

$$J : (U, V, \Omega) \rightarrow (-U, -V, \bar{\Omega}) \tag{2.2}$$

Here

$$\begin{aligned} U &= X - T, \qquad & V &= X + T \\ \Omega &= (\theta, \varphi), \qquad & \bar{\Omega} &= (\pi - \theta, \varphi + \pi), \end{aligned} \tag{2.3}$$

are global (Kruskal) type coordinates.

Because of the "double universe" nature of the Kruskal Schwarzschild geometry, (usually referred to as an "eternal black hole") two Schwarzschild coordinate patches are needed to cover all of spacetime. There are two (past and future) $r = 0$ singularities (corresponding to $T = \sqrt{1 + X^2}$ and $T = \sqrt{1 - X^2}$) and two exterior asymptotically flat regions corresponding to $X > |T|$ and $X < |T|$. For $M = 0$, each of these Kruskal regions becomes all of Minkowski spacetime. (The dynamic spacetime of a black hole formed by gravitational collapse only has one future horizon and one exterior region.)

In the usual (conventional) interpretation,[12,13] antipodal points $P(X)$ and $\bar{P}(\bar{X})$ of Kruskal space are *physically distinct* events which are causally disconnected. Every point mass (would) split the universe in two: the real world and its mirror (inaccessible) copy. A Kruskal space is a wormhole, connecting two distant regions of ordinary space.

In the elliptic interpretation, antipodal points $P(X)$ and $\bar{P}(\bar{X})$ are physically identified. They are considered as different representations in Kruskal space of *one and the same* Schwarzschild (and ultimately Minkowski) event (r, t, θ, φ). Thus, there is only a single one world with only one singularity and only one exterior region and no wormhole needed. The price of the more "economical" picture is that $\mathcal{M}$ is not time-orientable. There is a breakdown of the global distinction between past and future in the interior region to the event horizon ($r < r_H$). However, no problem arises for $r > r_H$.

We consider fields on the identified spacetime which are symmetric under the action of the antipodal operator J (equation 2.1). Thus we define

$$\Psi_{JS} = \tfrac{1}{2}[\Psi(X) + \Psi(JX)]. \tag{2.4}$$

We will build up these symmetric fields from fields with arguments specified in the right hand wedge. These building blocks will be either positive or negative frequency components with an inner product given by the usual Klein-Gordan one

$$\begin{aligned} &\langle \Psi, \Phi \rangle = -i \int \Psi^* \overleftrightarrow{j^\mu} \Phi \, d\Sigma_\mu \\ &(\overleftrightarrow{j^\mu} = \sqrt{g}\, g^{\mu\nu} \overrightarrow{\partial_\nu} - \overleftarrow{\partial_\nu} g^{\mu\nu} \sqrt{g}) \end{aligned} \tag{2.5}$$

taken over the right wedge. Each of these fields (whether positive or negative frequency on the right wedge) can be extended to be a positive or negative frequency field on global spacelike surfaces. But in the inner product defined on the original (full) Hilbert space, our symmetric fields defined by equation (2.4) have zero norm in global spacelike surfaces. This is so because on a global spacelike surface, time orientation has been reversed in the left hand wedge, relative to the local orientation in the right hand wedge.

Our identified manifolds have non-trivial (multiply connected) spacetime topology. Instead of choosing symmetric fields on these manifolds we could equally well have choosen antisymmetric fields, that is fields which change sign under the antipodal map J on the global manifold. This would give a situation somewhat analogous to the twisted fields considered by several authors some years ago.[14,15] However, previously, it was only the *spatial* topology which was non-trivial and the problem of zero norm states never arose.

This property of the symmetric field theory can be understood in terms of its projections on separate halves of the global (e.g. Kruskal) manifold. Although our symmetric fields have zero norm on global spacelike sections, they have positive (negative) norm on the half space to the future (past) of the past horizon. Particles and anti-particles can be well defined on each half space but with conjugate roles. Thus, from a global point of view, all distinction between particles and antiparticles is removed by the identification.

We define the symmetric Green function as

$$G_{JS}(X,X') = \langle \Psi_{JS}(X)\Psi_{JS}(X')\rangle \tag{2.6}$$

and express it in terms of the Green function for the ordinary field operators

$$G(X,X') = \langle \Psi(X)\Psi(X')\rangle. \tag{2.7}$$

Here $\langle\ \rangle$ stands for expectation value in the ordinary Fock states of the ordinary field, i.e. $|0\rangle$. Thus,

$$G_{JS}(X,X') = \tfrac{1}{4}[G(X,X') + G(X,JX') + G(JX,X') + G(JX,JX')] \tag{2.8}$$

It is necessary to indicate rather carefully what are the properties of the symmetric Green function G_{JS}. To do this, we use Schwarzschild-type coordinates u, v covering the right hand wedge, i.e.

$$\begin{aligned} U &= e^{\alpha\mu}, \qquad u = x - t \\ V &= e^{\alpha\nu}, \qquad v = x + t \end{aligned}$$

It is clear that the Green function G is unchanged if $t \to t + i\beta$ where $\beta = 2\pi/\alpha$, and G_{JS} similiarly has this property, i.e.

$$G(t,t' + i\beta;\Omega) = G(t,t';\Omega) \tag{2.9}$$

$$G_{JS}(t,t' + i\beta;\Omega) = G_{JS}(t,t';\Omega). \tag{2.10}$$

(For the Schwarzschild black hole, $\beta = 8\pi M$.) In addition, we have

$$G_{JS}(t,t' + \tfrac{i\beta}{2};\bar{\Omega}) = G_{JS}(t,t';\Omega). \tag{2.12}$$

but

$$G_{JS}(t,t' + \tfrac{i\beta}{2};\Omega) \neq G_{JS}(t,t';\Omega). \tag{2.12}$$

If G satisfies equation (2.9), then G_{JS} satisfies equation (2.10). If we had not mapped to antipodal points on the sphere under J, i.e. if instead of J, we had taken

$$J_0 \ :\ (U,V,\Omega) = (-U,-V,\Omega) \tag{2.13}$$

we would have had a conical singularity (at $U = V = 0$) in our spacetime and the corresponding symmetric Green function $[G_{JOS}]$ would be unchanged under $t \to t + i\beta/2$, i.e.

$$G_{JOS}(t, t' + \tfrac{i\beta}{2}; \Omega) = G_{JOS}(t, t'; \Omega). \tag{2.14}$$

Usually, in (euclidean) Q.F.T. one can regard G as satisfying equation (2.9) as a thermal Green function at temperature $1/\beta$. However, even though our symmetric Green function G_{JS} is periodic with period β (and not $\beta/2$), this does *not* mean that we can interpret it as a thermal Green function at temperature β^{-1} (or at any other temperature). Our zero norm fields do not allow us to construct any thermal properties in the normal way. [And the same results apply to G_{JOS} satisfying equation (2.14).]

The main result in this section which we would like to emphasize is that the discrete antipodal symmetry of the classical Schwarzschild-Kruskal manifold cannot be implemented without problems at the level of quantum Fock space. One cannot construct (local) Fock states symmetric under the antipodal transformations by a quantum operator acting on the usual (non-symmetric) states. In particular, there is no antipodal symmetric vacuum state. We can however, impliment this symmetry on the field operators in the space configuration and work on the usual (non-symmetric) Fock vacuum state (it is not necessary for the vacuum of a quantum theory to have a symmetry of the classical manifold). In that case, an antipodal symmetric Green function (G_{JS}) can be defined (eq. 2.6) and expressed in terms of the Green functions G of the ordinary theory.

For the antipodal identification map J (eq. 2.3) without fixed points, if the ordinary Green function G has period β, then G_{JS} has the same periodicity. For the antipodal map J_O (eq. 2.13) with conical singularity, G_{JOS} has period $\beta/2$. In any case, the important result here is that the zero norm states of the theory do not allow us to interpret, as in the usual way, $T = 1/\beta$ (nor $T_O = 2/\beta$) as a temperature of the theory.

From the results presented here, it is clear to us what is the precise context in which the results of references (16,17) arise. The *actual* framework of references (16,17) is that of Q.F.T. in curved (and non quantum) spacetime. Identification of classical spacetime has been (implicitly or explicitly) adopted, precisely, the identification map J_O (the presence of conical singularity is not a relevant criticism here). The bras in the right hand wedge of spacetime were implicitly identified to kets in the left hand wedge. A factor 2 for the quantity $T_O = 2/\beta$ (twice the usual $T = 1/\beta$) was found. However, usual Fock space, non-zero norm states and an operator projection relating the usual states to the identified (symmetric) ones were *assumed to exist (and used) for the theory.* The quantity $T_O = 2/\beta$ was interpreted as a temperature. The factor 2 for T_O found in references (16,17) is correct (*within* the above precise context) *but* its assumed derivation and interpretation there, are not correct. While such conclusions might still hold in the full theory of quantised Gravity, a framework for determining this does not exist in references (16,17). (Nor here!)

In de Sitter space (considered as a hyperboloid embedded in a five-dimensional flat space), J is simply an inversion in R^5:

$$J \,:\, x^a \to -x^a \qquad (a = 1, \ldots, 5).$$

In the different known coordinate systems on the hyperboloids, J takes the form

$$\begin{aligned} J(t, \chi, \theta, \phi) &\to (-t, \pi - \chi, \pi - \theta, \pi + \phi) \\ J_E(\tau, r, \theta, \phi) &\to (\tau + \beta/2, r, \pi - \theta, \pi + \phi) \\ J(u, v, \theta, \phi) &\to (-u - v, \pi - \theta, \pi + \phi). \end{aligned}$$

J_E stands for the Euclidean version ($t = i\tau$) of J in "static coordinates," where $\beta = 2\pi\sqrt{3/\Lambda}$.

We calculate $\langle\Psi^2_{\substack{JS\\JA}}\rangle$ and the energy momentum tensor $\langle T_{\mu\nu\ \substack{JS\\JA}}\rangle$. We obtain[3]

$$\langle\Psi^2_{\substack{JS\\JA}}\rangle = \langle\Psi^2\rangle \pm \langle\bar{\Psi}^2\rangle$$

$$\langle T_{\mu\nu\ \substack{JS\\JA}}\rangle = \langle T_{\mu\nu}\rangle \pm \langle\bar{T}_{\mu\nu}\rangle$$

(JA here stands for the antisymmetric values) where

$$\langle\bar{\Psi}^2\rangle = \frac{1}{16\pi\cos\pi\nu}[m^2 + (\xi - \tfrac{1}{6})R]$$

$$\langle\bar{T}_{\mu\nu}\rangle = -\frac{m^2}{64\pi\cos\pi\nu}[m^2 + (\xi - \tfrac{1}{6})R]g_{\mu\nu}.$$

$\langle\Psi^2\rangle$ and $\langle T_{\mu\nu}\rangle$ stand for the usual (so called Bunch-Davies or euclidean[18]) values of the ordinary theory. $\langle\bar{\Psi}^2\rangle$ and $\langle\bar{T}_{\mu\nu}\rangle$ stand for the new terms introduced by the elliptic theory. Here $\nu = (\frac{9}{4} - \frac{M^2}{H^2})^{1/2}$, $M^2 = m^2 + \xi R$, m^2 is the mass of the fields, ξ is the coupling and $R = 12H^2$, $(H = \sqrt{\Lambda/3})$. We study limiting cases of interest

$M^2 \ll H^2$ (inflationary)
$M^2 \gg H^2$ (massive)
$M^2 = H^2$ (conformal invariant)

In the inflationary regime, we find

$$\langle\Psi^2_{JS}\rangle = 2\langle\Psi^2\rangle, \qquad \left(\langle\Psi^2\rangle = \frac{R^2}{384\pi^2 M^2}\right)$$

$$\langle T_{\mu\nu\ JS}\rangle = \frac{61}{151}\langle T_{\mu\nu}\rangle, \qquad \left(\langle T_{\mu\nu}\rangle = \frac{-61R^2}{138240\pi}g_{\mu\nu}\right).$$

The J-symmetric theory allows good inflation. On the contrary, for the J-antisymmetric theory for $M^2 \ll H^2$ we find

$$\langle\Psi^2_{JA}\rangle = 0$$

$$\langle T_{\mu\nu\ JA}\rangle = -\frac{29}{61}\langle T_{\mu\nu}\rangle$$

which does not satisfy the fundamental hypothesis of inflation.[19] However, the J-antisymmetric theory allows a better understanding of the massless and minimally coupled ($m^2 = 0$, $\xi^2 = 0$) ordinary theory (without identification). $G_{JA}(X, X')$ is not infrared divergent. We find new vacuum states which are $O(4)$ and $O(1,3)$ invariant [together with $E(3)$[20] these are the maximal subgroups of $O(1,4)$]. We obtain

$$\lim_{\substack{m^2\to 0\\ \xi\to 0}} \langle\Psi^2\rangle_{O(1,4)} \neq \langle\Psi^2\rangle_{E(3)} \neq \langle\Psi^2\rangle_{O(4)} \neq \langle\Psi^2\rangle_{O(1,3)}$$

and

$$\lim_{\substack{m^2\to 0\\ \xi\to 0}} \langle T_{\mu\nu}\rangle_{O(1,4)} \neq \langle T_{\mu\nu}\rangle_{E(3)} = \langle T_{\mu\nu}\rangle_{O(4)} = \langle T_{\mu\nu}\rangle_{O(1,3)}.$$

Similar results hold for the JS and JA-theories. This is a manifestation of the fact that the $O(1,4)$ invariant Green function is infrared divergent in the massless and $\xi = 0$ limit and there is no Fock state which is $O(1,4)$ invariant in that case. Therefore, the correct values of $\langle T_{\mu\nu}\rangle$ for the massless minimally coupled field are those corresponding the $E(3)$, $O(4)$, and $O(1,3)$ vacua and not to the limiting case of the $O(1,4)$ vacuum which does not exist in such limit.

More results and detailed derivations are given in ref. 3. The J-symmetric field theory in the "elliptic" de Sitter space applies equally well to other scenarios of inflation as the Starobinsky model[21] and the models of quantum creation of the universe as proposed by Vilenkin,[22] Linde,[19] and Hartle-Hawking.[23] This will be discussed elsewhere.[24]

3. THE BACK-REACTION PROBLEM IN TWO DIMENSIONS: LIOUVILLE AND SCHROEDINGER EQUATIONS

We present here a complete solution to the semi-classical Einstein equations in two dimensions. It is given by a constant curvature metric parametrised by solutions ("wave functions") of a zero energy Schrodinger equation.[4] We analyze *global, thermal and topological properties of the universe as a function of its quantum matter content including the graviton contribution.* Recently, two dimensional gravity has raised interest, mainly in connection with Polyakov's work on strings.[25] It has been proposed that Liouville theory which in geometric form reads $R + \Lambda = 0$, ($\Lambda =$ const.) could govern the dynamics of two dimensional gravity.[24,31] In ref. 4 we consider semi-classical Einstein equations in two dimensions and derived Liouville theory as one of the dynamical equations of the gravitational field: the other equation involved appears to be Schroedinger's one. (The quantum trace anomoly $\langle T^{\mu}_{\ \mu}\rangle \neq 0$ allows for a non trivial dynamics in two dimensions. Semi-classical in the context means that matter fields ϕ including the graviton are quantised to one-loop level and coupled to (c-number) gravity through the equations

$$R_{\mu\nu} - \tfrac{1}{2}Rg_{\mu\nu} + \Lambda g_{\mu\nu} = 8\pi G\langle T_{\mu\nu}(\phi, g_{\mu\nu})\rangle.$$

These equations for $g_{\mu\nu}$ are highly complicated and need to be treated within some type of self consistent framework. $\langle T_{\mu\nu}\rangle$ is a non-local geometrical object: it depends on the geometry and on the quantum state $|\ \rangle$, fixed by the boundary conditions of matter fields. In two dimensions we have

$$\Lambda g_{\mu\nu} = 8\pi G\langle T_{\mu\nu}\rangle. \tag{3.1}$$

These equations reduce to

$$4\partial_u\partial_v \ln C + \tilde{\Lambda}C = 0 \tag{3.2}$$

and

$$\begin{aligned} d_u^2\tilde{X}_u(u) - 12\pi\gamma^{-1}\tilde{U}(u)\tilde{X}_u(u) &= 0 \\ d_v^2\tilde{X}_v(v) - 12\pi\gamma^{-1}\tilde{V}(v)\tilde{X}_v(v) &= 0. \end{aligned} \tag{3.3}$$

Here $\tilde{\Lambda} = 6\Lambda/\gamma$ where γ is the trace anomaly. Equation (3.2) is the Liouville equation. $R + \tilde{\Lambda} = 0$, for the curvature scalar $R = 4C^{-1}\partial_u\partial_v \ln C$; C is the conformal factor of the metric $dS^2 = C(u,v)du\,dv$ ($u = x - t$, $v = x + t$). For $C = e^{\beta\psi}$ ($\beta =$ const.) equation (3.2) reads

$$4\partial_u\partial_v\psi + \tilde{\Lambda}\beta^{-1}e^{\beta\psi} = 0.$$

This equation corresponds to the uv component of equation (3.1). As is well known, the general solution is

$$C = \frac{f'(u)\ g'(v)}{[1 + (\tilde{\Lambda}/8)f(u)g(u)]^2}.$$

Here f, g are not totally arbitrary functions but determined by equations (3.3).

Equations (3.3) are (zero-energy) Schroedinger type equations for the wave functions

$$\tilde{X}_u = (\sqrt{f'})^{-1}, \qquad \tilde{X}_v = (\sqrt{g'})^{-1}.$$

By giving the "potentials" $\tilde{U}(u)$ and $\tilde{V}(v)$ we determine $\tilde{X}_u$, $\tilde{X}_v$. $\tilde{U}$ and $\tilde{V}$ describe the choice of the boundary conditions on the matter fields. In particular, $\tilde{U} = 0$, $\tilde{V} = 0$ determine $f(g)$ as $\tilde{X} =$ const., $f = (\alpha u + \beta)/(\sigma u + \delta)$, ($\alpha$, β, σ, $\delta =$ constants) and a "minimal" vacuum state at zero temperature. A constant potential $\tilde{U}(u) = \tilde{U}_0$ such that $\tilde{U}_0/\gamma > 0$ gives $\tilde{X} = Ae^{-\tilde{\kappa}u}$, $f = (2\tilde{\kappa}A^2)^{-1}e^{2\tilde{\kappa}u}$, ($A = (\sqrt{2\tilde{\kappa}})^{-1}$) and a thermal state.[4,26]

The zero-energy transmission coefficient $\tilde{\kappa} = \sqrt{12\pi\gamma^{-1}\tilde{U}_0}$ is related to the temperature $T = \pi^{-1}\tilde{\kappa}$. On the contrary, if $(\tilde{U}_0/\gamma) < 0$, there is no transmission coefficient ($\tilde{\kappa}$ becomes imaginary) and no event horizon is formed. The geometry does not carry an intrinsic temperature in this case. More generally, each positive discontinuity in the "effective" potential $\tilde{U}_0/\gamma$ gives rise to an event horizon in the spacetime, the transmission coefficient $\tilde{\kappa} = |\frac{X'}{X}|_{\text{horizon}}$ playing the role of the "surface gravity" $\kappa = 2\tilde{\kappa}$ of the horizon.

The cosmological constant $\tilde{\Lambda} = 6\Lambda/\gamma$ in the Liouville equation (3.2) is modified with respect to the classical one by the trace anomaly factor γ. The character of the solution depends on the sign of Λ/γ. Vector fields in two dimensions do not contribute to γ. The scalar contribution is positive and that of gravitons is negative. Therefore:

I) If sign$\Lambda \neq$ signγ, i.e. $\Lambda > 0$ and $\gamma < 0$ (graviton dominated universe) or $\Lambda < 0$ and $\gamma > 0$ (matter dominated universe), the geometry has $R > 0$ with one event horizon.

II) If sign$\Lambda =$ signγ, the geometry has $R < 0$ without horizon.

This means that for a given sign of Λ, the presence or absence of event horizons depends on the number of matter fields. The universe could change from an Anti-de-Sitter to a de-Sitter phase (or vice-versa). The graviton contribution is crucial here to arise these possibilities. This contrasts with the standard classical situation (in four dimensions) in which R and the presence or not of event horizon only depends on Λ. If N (the number of matter fields)$\to \infty$ then the Hawking temperature $T \to 0$ and the semiclassical geometry is flat even if $\Lambda \neq 0$. If $\gamma = 0$ the dynamics is not determined by the semiclassical Einstein equations. In ref. 27 the Liouville equation has beeen derived in the semiclassical context but the graviton contribution so crucial to this problem has been overlooked. The total value of γ as calculated in refs. 28 and 29 (denoted $\gamma_{(GKT)}$ and $\gamma_{(CD)}$ following the notation of ref. 30) is

$$\gamma_{(GKT)} = (N_0 - 1 + N_{1/2} - \tfrac{15}{2}N_{3/2}), \quad \gamma_{(CD)} = (N_0 - 1 - N_{1/2} + N_{3/2}) \tag{3.4}$$

Here, the graviton interacts with N_s massless fields of spins s, $s \leq 3/2$. [The graviton contribution to γ was also obtained equal to -1 in ref. 30.] In the context of quantised strings,[25] the trace anomaly coefficient for a theory with N matter fields coupled to two dimensional gravity was obtained equal to[25]

$$\begin{aligned} \gamma_{(P)} &= N - 26 \quad \text{for bosons,} \\ \gamma_{(P)} &= N - 10 \quad \text{for fermions (with supersymmetric coupling).} \end{aligned} \tag{3.5}$$

We denote it $\gamma_{(P)}$ because of ref. 25. These values were calculated at the one loop level in the conformal gauge $g_{\mu\nu} = e^{\phi}\eta_{\mu\nu}$. The "critical dimension" 26 (10) in equation (3.5) is only for the ghost part (Faddeev-Popov determinant) of the graviton contribution. It does not take into account the quantisation of the conformal factor (the Liouville field ϕ) that remains fixed. This should explain the difference between the values 1 in equation (3.4) and 26 (10) in equation (3.5). The value of γ that should be considered in the Liouville equation of two dimensional gravity is that given by equation (3.4) and not that of equation (3.5). The connection with the quantisation of the Liouville theory in this context deserves future investigation. It would be interesting to connect the results found here with those obtained from a semiclassical limit of the Hawking "wave function" approach[30] and the Jackiw model.[31,32]

More details about this work are given elsewhere.[4]

REFERENCES

1. V.P. Frolov, N. Sánchez, Phys. Rev. **D33**, 1604 (1986).
2. N. Sánchez, B.F. Whiting, "Quantum Field Theory and the Antipodal Identification of Black Holes," Nucl. Phys. B (to appear).
3. A. Folacci, N. Sánchez, "Quantum Field Theory and the Elliptic Interpretation of deSitter Space," Meudon preprint (July, 1986) .
4. N. Sánchez, Nucl. Phys. **B266**, 487 (1986).
5. N. Sánchez, B.F. Whiting, Quantum Fields, *Curvilinear Coordinates and Curved Spacetimes*, in: "Quantum Concepts of Space and Time." C. Isham and R. Penrose eds., Oxford University Press, Oxford (1986).
6. V.P. Frolov, A.I. Zel'nikov, Phys. Rev. **D29**, 1057 (1984).
7. D. Page, Phys. Rev. **D25** 1499 (1982).
8. M.R. Brown, A.S. Ottewill, Phys. Rev. **D31**, (1985).
9. W. Israel, Phys. Rev. **164**, 1776 (1967).
10. E. Schrödinger, "Expanding Universes," Cambridge Univ. Press, Cambridge (1957).
11. G.W. Gibbons, "The Elliptic Interpretation of Black Holes and Quantum Mechanics," Cargèse Lectures (1985), Plenum Pub. Co. (to appear).
12. M.D. Kruskal, Phys. Rev. **119**, 1743 (1960).
13. C.W. Misner, K.S. Thorne, J.A. Wheeler, "Gravitation," p. 840, Freeman, San Francisco (1973).
14. C.J. Isham, Proc. Roy. Soc. Lond. **A362**, 383 (1978).
15. R. Banach, J.S. Dowker, J. Phys. **A12**, 2527 and 2545 (1979).
16. G. 't Hooft, J. Geom. Phys. **1**, 45 (1984).
17. G. 't Hooft, Nucl. Phys. **B256**, 727 (1985).
18. T.S. Bunch, P.C.W. Davies, Proc. Roy. Soc. Lond. **A360**, 117 (1978).
19. A.D. Linde, Rep. Progr. Phys. **47** (1984).
20. B. Allen, Phys. Rev. **D32**, 3136 (1985).
21. A.A. Starobinsky, Phys. Lett. **B91**, 99 (1980).
22. A. Vilenkin, Phys. Rev. **D30**, 509 (1984).
23. J. Hartle, S.W. Hawking, Phys. Rev. **D28**, 2960 (1983).
24. A. Folacci, N. Sánchez, in preparation.
25. A.M. Polyakov, Phys. Lett. **B103**, 207 and 211 (1981) .
26. N. Sánchez, Phys. Rev. **D24**, 2100 (1981).
27. R. Balbinot, R. Floreanini, Phys. Lett. **B151**, 401 (1985).
28. R. Gastmans, R. Kallosch, C. Truffin, Nucl. Phys. **B133**, 417 (1978).

29. S.M. Christensen, M.J. Duff, Phys. Lett. **B79**, 213 (1978).

30. S.W. Hawking, Lectures on Quantum Cosmology in: "Les Houches Session XL Relativity, Groups, and Topology II," B. de Witt and R. Stora, eds., pp. 333-379, North Holland (1984).

31. R. Jackiw, Nucl. Phys. **B252** 343 (1985) and references therein.

32. M. Henneaux, Phys. Rev. Lett **54**, 959 (1985).

TOWARDS A THEORY FOR THE QUANTUM MECHANICS OF GRAVITATIONAL COLLAPSE

G. 't Hooft

Institute for Theoretical Physics
Princetonplein 5, P.O. Box 80.006
3508 TA Utrecht, The Netherlands

ABSTRACT

The difficulty in reconciling the principles of general relativity with those of quantum mechanics shows particularly clearly when black holes are compared with elementary particles. In a viable theory these notions should be placed on an equal footing. In this lecture we show that indeed the mathematical structure of modern string theories for elementary particles suggests an interpretation in terms of dynamical properties of the black hole horizon.

1. INTRODUCTION

In its standard form, the theory of general relativity contains equations of motion that correspond to an extremum principle. There is a lagrangean,

$$\mathcal{L} = \sqrt{-g}\, R \,, \tag{1.1}$$

whose integral can be required to be stationary under infinitesimal variations:

$$\delta \int \mathcal{L}\, d^4x = 0 \,. \tag{1.2}$$

As is well-known, such a system can, in principle, be "quantized" by replacing Poisson brackets by commutators. At first sight, the resulting "quantum theory of gravity" seems to be quite reasonable: the gravitational force is transmitted by a graviton, a particle with spin 2, which, due to the correct sign of the gravitational constant G, has positive energy and production probabilities.

Nevertheless, there are problems. The statement that quantum gravity is not renormalizable is equivalent to saying that at very short distances the interactions run out of control. When particle physicists talk about "renormalizable theories of gravity", what they really mean is that in a perturbative expansion with respect to G the individual terms do not get infinite contributions from the small-distance limit; but stil, at distance scales

$$\ell = (\frac{G\hbar}{c^3})^{\frac{1}{2}} = 1.6.10^{-33} \text{ cm} , \qquad (1.3)$$

or smaller, the interactions are so strong that their cumulative effects defy any description. Certainly, the perturbation expansion itself starts to diverge badly.

A key ingredient in quantum field theory, as much as in general relativity, is locality: of any pair of points in space-time, regardless how close together, one can tell which is earlier and which is later. Precisely this locality must be given up in a true theory of quantum gravity: two points can be separated only by using particles at an energy scale inversely proportional to their distance:

$$mc^2 = E \gtrsim \hbar c/\ell , \qquad (1.4)$$

but if we try to describe the gravitational field of such particles we get a Schwarzschild horizon at a distance

$$\ell = 2Gm/c^2 = (2G\hbar/c^3\ell) \qquad (1.5)$$

so that indeed the distance (1.3) (the Planck scale) is never surpassed.

Paradoxically, the most impressive results in understanding physics at the Planck length scale were obtained by ignoring the problem. String theory originated in a different branch of particle physics and its proponents claim that it should produce the resolution of our problems of locality, among all others. In the next section we give a very brief outline of string theory. Much more extensive reviews can be found elsewhere [1] . What we really need is the "classical" string equation (2.21).

Next we turn to an apparently quite different subject: black holes. It is also quite well-known that black holes are expected to emit radiation with a thermal spectrum. In sect. 3 we argue that this radiation is actually due to a very delicate boundary condition in the Schwarzschild coordinate system and is perhaps not as well understood as some authors suggest . In any case it is not so unreasonable to suspect that the tiniest black holes merge into the spectrum of elementary particles, which are also unstable, but do not behave thermally.

If one includes gravitational self-interaction between in- and outgoing particles a quite different picture emerges and in the last sections we show that black holes indeed look very much like strings... .

2. STRING THEORY IN A NUT SHELL

Only a few notions from string theory will be needed in the sequel. We here summarize some of its basic features; more extensive and "complete" accounts can be found in the enthusiastic literature [1,2] .

If in a scattering experiment a virtual particle is exchanged, its effect can be seen in the form of a pole in the amplitude. Consider for instance the collision of two particles, 1 and 2, yielding after the collision particles 3 and 4. Let p^1, p^2, p^3,p^4 be their 4-momenta. Define

$$k_\mu = p_\mu^1 + p_\mu^2 = p_\mu^3 + p_\mu^4 ;$$

$$q_\mu = p_\mu^3 - p_\mu^1 = p_\mu^2 - p_\mu^4 ; \tag{2.1}$$

$$s = - k_\mu^2 ; \; t = - q_\mu^2 . \tag{2.2}$$

Here, s is the $(\text{energy})^2$ in the rest frame; q_μ is the exchanged momentum. t can be seen to be directly related to the scattering angle θ in the center of mass frame:

$$t = A(m^i,s) + B(m^i,s) \cos\theta . \tag{2.3}$$

A Lorentz-invariant amplitude could look like

$$G(s,t) = \sum_i \frac{\lambda_i^s}{k^2+m_i^2-i\varepsilon} + \sum_j \frac{\lambda_j^t}{q^2+M_j^2-i\varepsilon} . \tag{2.4}$$

So there are poles of the form

$$\frac{1}{m_i^2-s} \quad \text{and} \quad \frac{1}{M_j^2-t} . \tag{2.5}$$

Here, m_i and M_j are masses, and we note that the poles are located at those values of the external momenta for which the particles could be actually produced.

Notice the symmetry $s \leftrightarrow t$ (crossing symmetry). We ignore for simplicity the possibility of producing particles in the u-channel:

$$- u = (p_\mu^4-p_\mu^1)^2 , \tag{2.6}$$

which could be suppressed because of unfavorable quantum numbers in that channel.

If the residues of the poles (2.5) are independent of the other parameter t or s, respectively, then the exchanged particles are scalars. Otherwise we get

$$\frac{\mathrm{Pol}_i(t)}{m_i^2-s} \quad \text{and} \quad \frac{\mathrm{Pol}_j(s)}{M_j^2-t} \tag{2.7}$$

where the degrees of the polynomials correspond to the spins of the exchanged particles.

A model amplitude having an infinite series of poles was suggested by G. Veneziano :

$$G(s,t) = \frac{\Gamma(-s)\Gamma(-t)}{\Gamma(-s-t)} = B(-s,-t) \tag{2.8}$$

where $\Gamma(x)$ is Euler's gamma function. One easily checks that the pole structure is as in (2.7), where

$$m_i^2 = M_i^2 = n = 0,1,2,\dots , \tag{2.9}$$

and the degrees of the polynomials, hence the spins, increase with increasing n. In fact, because the polynomials are not pure Legendre polynomials we have superpositions of various spins. The model found its first applications in strong interaction theory where indeed bound states have a mass spectrum resembling (2.9).

The physical interpretation of Veneziano's formula (2.8) is quite interesting. Let us assume that the external particles are massless:

$$(p^i)^2 = 0 \qquad \text{for each } i \ . \tag{2.10}$$

Then (2.8) can be rewritten as

$$\begin{aligned} B(-s,-t) &= \int_0^1 x^{-s-1}(1-x)^{-t-1}dx \\ &= \int_0^1 \frac{dx}{x(1-x)} \exp[2(p^1.p^2)\log x - 2(p^1.p^3)\log(1-x)] \ . \end{aligned} \tag{2.11}$$

The logarithms can be regarded as Green functions in a two-dimensional space, which can either be taken to be the entire complex plane, or a compact, simply connected subspace, such as the unit circle. Take the latter case and transform it conformally into the upper half plane (Fig. 1).

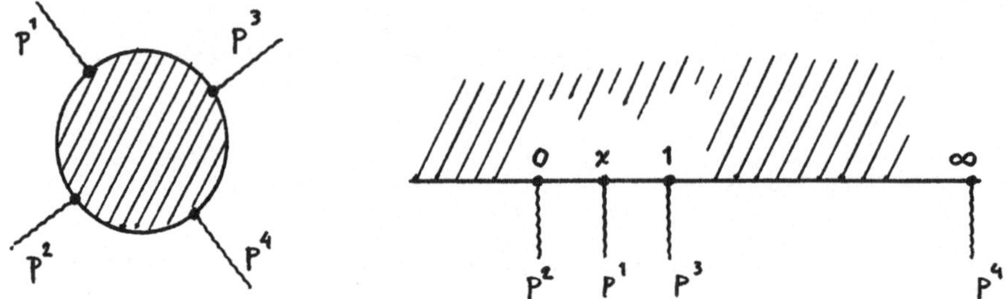

Fig. 1. Electronic simulation with conducting sheet, reproducing the exponent in eq. (2.11).

The exponent in (2.11) can be seen as the dissipated power in an electronic circuit where four different kinds of currents p_μ ($\mu = 1,\ldots,4$) are sent into or out from a uniformly conducting sheet, at the points, 0, x, 1 and ∞. The voltage due to the currents 2 and 3 will be minus the real part of

$$f_\mu(z) = 2p_\mu^2 \log z - 2p_\mu^2 \log (1-z) \ , \tag{2.12}$$

which satisfies

$$\partial_z^2 f_\mu(z) = 2\pi\rho_\mu(z) \ ;$$

$$\rho_\mu(z) = p_\mu^2\ \delta(z) - p_\mu^3\ \delta(z-1) \ . \tag{2.13}$$

(If the source were not at, but inside the border, the 2π would have to be replaced by 4π.)
We now notice that the amplitude (2.8) can be rewritten as

$$G(s,t) = \int_0^1 \frac{dx}{x(1-x)} \int D\eta_\mu(z) \exp \int d^2z \left(\frac{1}{4\pi} (\vec{\partial}_z \eta_\mu(z))^2 + \rho_\mu(z)\eta_\mu(z)\right) . \tag{2.14}$$

The integral over η_μ is a functional integral over its imaginary values. We will not go into the problem of defining the measure of this functional integral (requiring this measure to be truly invariant under conformal transformations leads us into the complexities of superstring theory). Notice that, using the restriction (2.10), we can replace (2.13) by

$$\rho_\mu(z) = \sum_i \pm \rho_\mu^i \, \delta(z-x^i) , \tag{2.15}$$

with $x^i = (x,0,1,\infty)$, and the sign depending on whether the particle was in- or outgoing.

Identifying $e^{\int \rho_\mu \eta_\mu(z) dz}$ in (2.14) with a particle wave function e^{ipx}, we see that η_μ must be considered to be i times a coordinate x_μ. Apparently, we have a functional integral over all configurations

$$- i\eta^\mu(\sigma+i\tau) = x^\mu(\sigma,\tau) . \tag{2.16}$$

The corresponding classical equation is

$$\vec{\partial}_z^2 x^\mu = - 2\pi i \, \rho_\mu(\sigma,\tau) . \tag{2.17}$$

The quantity (2.16) describes a string-like object moving around in 4-space. The reason why Veneziano's model works well in theories such as QCD is that indeed in QCD one has quarks held together by field configurations in the form of vortices, and we may view the strings as idealizations of the vortices.

Eq. (2.14) can be seen to follow from a string action:

$$S = \int d\sigma \, d\tau \sqrt{\left(\frac{\partial x^\mu}{\partial\sigma} \cdot \frac{\partial x^\mu}{\partial\tau}\right)^2 - \left(\frac{\partial x^\mu}{\partial\sigma}\right)^2 \left(\frac{\partial x^\mu}{\partial\tau}\right)^2} , \tag{2.18}$$

if the parametrization z is chosen sufficiently carefully. The energy of a string is

$$\frac{1}{2\pi} \int \frac{d\sigma}{\sqrt{1-v_\perp^2}} , \tag{2.19}$$

where σ is a coordinate along the string and $v_\perp$ the velocity perpendicular to the string. In QCD the mass unit for the spectrum (2.9) is roughly 1 GeV. The string tension T, as an energy per unit of length, then turns out to be

$$T \approx \frac{1}{2\pi} (\text{GeV})^2 = 14 \text{ tons} . \tag{2.20}$$

Putting the dimensions right, (2.17) becomes

$$\partial^2 x^\mu = - i\rho_\mu / T . \tag{2.21}$$

In unified superstring theories one takes T to be enormous:

$$T = \mathcal{O}(M_{PL}^2) \ . \tag{2.22}$$

Often, the string will be without end points (closed string theory). In this case, (2.21) still applies, but ρ_μ then becomes a distribution defined in the interior of the complex z plane.
A consequence of (2.22) is that the Schwarzschild radius R of a superstring must be

$$R > 2GTL = \mathcal{O}(L) \tag{2.23}$$

where G is the gravitational constant and L the total length of a string. Apparently, in superstring theories, gravitational collapse should not be ignored! Paradoxically, strings are usually treated as if they are immune to gravitational collapse. Could this be because superstrings "are" black holes?

3. THE BLACK HOLE

We write

$$M = Gm \ , \tag{3.1}$$

and the black hole metric as

$$ds^2 = -\left(1 - \frac{2M}{r}\right)dt^2 + \left(1 - \frac{2M}{r}\right)^{-1}dr^2 + r^2 d\Omega^2 \ . \tag{3.2}$$

The Kruskal coordinates x and y, defined by

$$\begin{aligned} xy &= \left(1 - \frac{r}{2M}\right)e^{r/2M} \ , \\ x/y &= -\ e^{t/2M} \end{aligned} \tag{3.3}$$

can replace r and t, after which we can extend analytically [5] to the entire space $(x,y \mid xy < 1)$.
That this metric may arise naturally when a large amount of matter is accumulated is being discussed extensively at this School. The physical interpretation of the various regions in x-y space has also been discussed at various places [5,6].

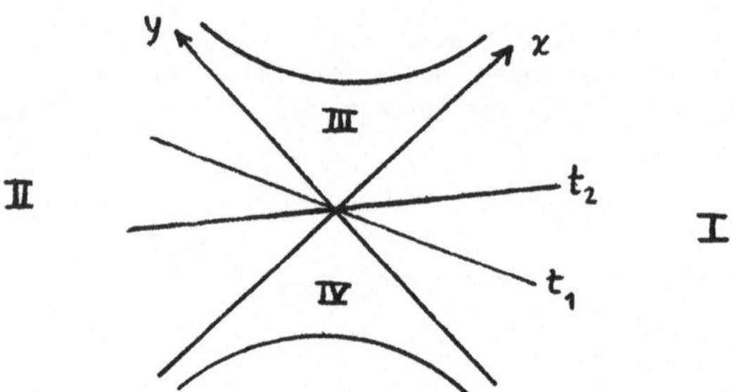

Fig. 2. Black hole in Kruskal coordinates.

Of particular interest is the region close to the origin. Writing

$$\zeta^2 = \left(\frac{r}{2M} - 1\right)e^{r/2M} \,, \qquad \tau = t/4M \,, \tag{3.4}$$

we see that

$$\tfrac{1}{2}(x+y) = \zeta \sinh \tau \,, \qquad \tfrac{1}{2}(x-y) = \zeta \cosh \tau \,. \tag{3.5}$$

At the origin the metric is regular:

$$ds^2 \to -\frac{16M^2}{e}\, dxdy + 4M^2 d\Omega^2 \,. \tag{3.6}$$

Since we shall mainly concentrate on the origin of xy space we could just as well replace (3.6) by a flat Minkowski metric. Consider a flat Minkowski space with coordinates x_{tr}, z, t, and

$$ds^2 = dx_{tr}^2 + dz^2 - dt^2 \tag{3.7}$$

(where x_{tr} is a two component vector replacing the angles θ and φ, and where t is not to be confused with the Schwarzschild time t in (3.2)). The Rindler coordinates ζ and τ are now defined by [7]

$$z = \zeta \cosh \tau \,, \qquad t = \zeta \sinh \tau \,. \tag{3.8}$$

An observer for whom τ acts as a time coordinate feels a strong gravitational field, singular at $\zeta = 0$. Since his time-translations correspond to Lorentz transformations in the original Minkowski space, the observer in this new world, called Rindler space [7], experiences laws of nature that are constant in time.

Let the Minkowski Hamiltonian be the integral over a Hamilton density:

$$H_M = \int \mathcal{H}(\vec{x})\, d^3\vec{x} \,. \tag{3.9}$$

Of course, (3.9) is the generator of time-translations. The generator of Lorentz transformations in Minkowski space is the Rindler Hamiltonian:

$$H_R = \int d^3\vec{x}\, \left(\mathcal{H}(\vec{x})z - P_3(\vec{x})t\right) \,, \tag{3.10}$$

where $P_3(\vec{x})$ is the momentum density. We can split H_R into two parts (taking $t = 0$):

$$H_1 = \int_{z>0} \mathcal{H}(\vec{x})z\, d^3\vec{x} \,; \tag{3.11}$$

$$H_2 = \int_{z<0} \mathcal{H}(\vec{x})\, |z|\, d^3\vec{x} \,; \tag{3.12}$$

$$H_R = H_1 - H_2 \ . \tag{3.13}$$

In most field theories one finds easily:

$$[H_1,H_2] = 0 \ . \tag{3.14}$$

The physical interpretation of (3.14) is that no signals can be transmitted between the regions I ($z > |t|$) and II ($z < -|t|$).

Suppose we had a scalar field φ in Minkowski space satisfying the Euler-Lagrange equations generated by the Lagrangean

$$\mathcal{L} = -\tfrac{1}{2}(\partial_z\varphi)^2 + \tfrac{1}{2}(\partial_t\varphi)^2 - \tfrac{1}{2}(\partial_{tr}\varphi)^2 - \frac{m^2}{2}\varphi^2 \tag{3.15}$$

If we write

$$\zeta = e^{\sigma} \qquad (-\infty < \sigma < \infty) \tag{3.16}$$

then

$$\mathcal{L}\, dz\, dx_{tr}\, dt = \ell\, d\sigma\, dx_{tr}\, d\tau \ , \tag{3.17}$$

with

$$\ell = -\tfrac{1}{2}(\partial_\sigma\varphi)^2 + \tfrac{1}{2}(\partial_\tau\varphi)^2 + e^{2\sigma}\left(-\tfrac{1}{2}(\partial_{tr}\varphi)^2 - \frac{m^2}{2}\varphi^2\right) \ . \tag{3.18}$$

Close to the horizon ($\sigma \to -\infty$), this Lagrangean generates plane waves:

$$(\partial_\tau^2 - \partial_\sigma^2)\varphi \to 0 \ . \tag{3.19}$$

The left-going waves will never reach the horizon itself, whereas the right-going ones are infinitely old. Therefore at first sight no boundary condition at $\sigma = -\infty$ (or $\zeta = 0$) seems to be necessary. Nevertheless, we expect "Hawking radiation". This is a thermal flux of particles ariving from $\sigma = -\infty$. This follows if one makes one extra assumption: total energy in the original Minkowski space must be finite. Since this inevitably involves linear combinations of different eigenstates of H_1 and H_2, one ends up with a quantummechanically mixed state in terms of H_1 alone, hence the thermal nature of the radiation .
Any "eigenstate" $|E_1,E_2\rangle$ with

$$\begin{aligned} H_1|E_1,E_2\rangle &= E_1|E_1,E_2\rangle \ ; \\ H_2|E_1,E_2\rangle &= E_2|E_1,E_2\rangle \ , \end{aligned} \tag{3.20}$$

being different from the Minkowski vacuum yet Lorentz-invariant, must have infinite energy in Minkowski space.

The infinite spectrum of equation (3.19) in the space (3.16), not bounded for small σ values, causes a nasty divergence at the horizon. Also, in Rindler space, but also for finite size black holes, no correlation whatsoever is expected between ingoing and outgoing particles. In a free theory all these states are orthogonal to each other, and nothing can affect the truly stochastic nature of the thermal radiation.

As stated in the introduction, this situation seems to be unacceptable if we were to consider black holes as just some sort of elementary particles. What one expects there is a scattering matrix S:

$$|\phi_{out}\rangle = S|\phi_{in}\rangle \ . \tag{3.21}$$

The reason why (3.21) may perhaps not be incompatible with Rindler space dynamics is that the assumption of having only free fields in Minkowski space is clearly wrong. At sufficiently far negative values of the coordinate σ, sooner or later, gravitational forces (which are extremely nonlinear), will dominate. If we would build such a combination of states (3.20) that the total energy in Minkowski space would become large, then gravitational interactions will cause severe curvature of space-time. Could the particles that went in, not be knocked out again by this curved space-time? Perhaps values of ζ smaller than the Planck length are forbidden. Are particles bounced back somehow?

4. THE SHIFTING HORIZON

Shifts in the parameter τ of transformation (3.8) correspond to Lorentz transformations in Minkowski space. Even the lightest particles become exceedingly energetic after a certain amount of "time" τ. It is therefore particularly important to study the gravitational effect of such energetic particles on space-time [8,9].

The metric of a very light particle at rest can be approximated by

$$g_{\mu\nu} \approx \delta_{\mu\nu}\left(1 + \frac{2m_o}{r}\right) + \frac{4m_o}{r}\, u_\mu u_\nu \ , \tag{4.1}$$

with

$$r^2 = x^2 + (x.u)^2 \ , \tag{4.2}$$

$$u^2 = -1 \ . \tag{4.3}$$

u_μ is the 4-velocity of the particle. Since (4.1) - (4.3) have been written in a Lorentz-invariant way, they remain true also when the particle is boosted to tremendous energies. We let $m_o \to 0$, $u_\mu \to \infty$, with

$$P_\mu = m_o u_\mu \approx (0,0,p,ip) \tag{4.4}$$

fixed. If $(x.u) \neq 0$ we have

$$r \to |x.u| = |x.p|/m_o \ , \tag{4.5}$$

so in most of space-time, r is large and indeed $|m_o/r| \ll 1$.
Now consider the new coordinates $y^\mu_\pm$ with

$$y^\mu_\pm = x^\mu \pm 2m_o u^\mu \log r \ . \tag{4.6}$$

At

$$(x.u) > 0 : \quad ds^2 \to dy_+^2 \ , \tag{4.7}$$

And at

$$(x.u) < 0 : \quad ds^2 \to dy_-^2 \ . \tag{4.8}$$

So space-time is flat at points x_μ with $(x.p) \neq 0$. At $(x.p) = 0$ we must make the transition from the y_+ to the y_- coordinates: At $(x.p) = 0$:

$$y_+^\mu = y_-^\mu + 2p^\mu \log y_{tr}^2 . \tag{4.9}$$

The second derivative of this shift with respect to y_{tr} produces real curvature across the seam. Notice that the shift, Δy^μ, obeys the two-dimensional Laplace equation

$$\partial_{tr}^2(\Delta y^\mu) = 8\pi \, p^\mu \, \delta^2(y_{tr}) . \tag{4.10}$$

Thus, spacetime can still be represented by flat coordinates y^μ, provided that we draw the geodesics that cross the plane $(y.p) = 0$ in a special way (Fig. 3)

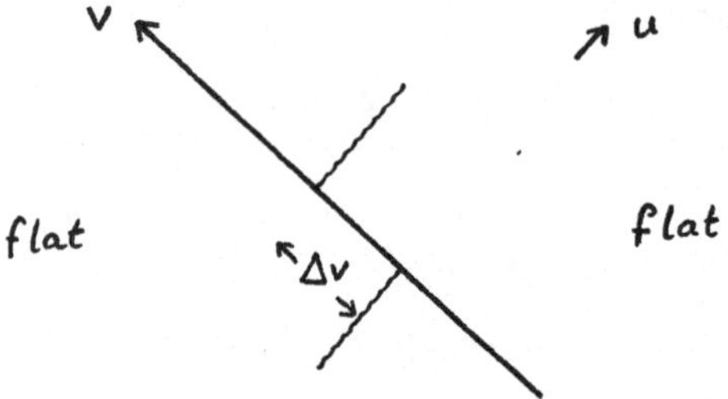

Fig. 3. A fast particle moving in the plane $u = 0$ causes a drag Δv in all geodesics that cross the plane $u = 0$, whose value depends on the transverse distance y_{tr} from the particle according to eq. (4.10).

In a finite-size black hole a particle approaching the horizon causes a similar drag. The laplacian ∂_{tr}^2 is then replaced by an angular laplacian plus a constant [9], see Fig. 4.

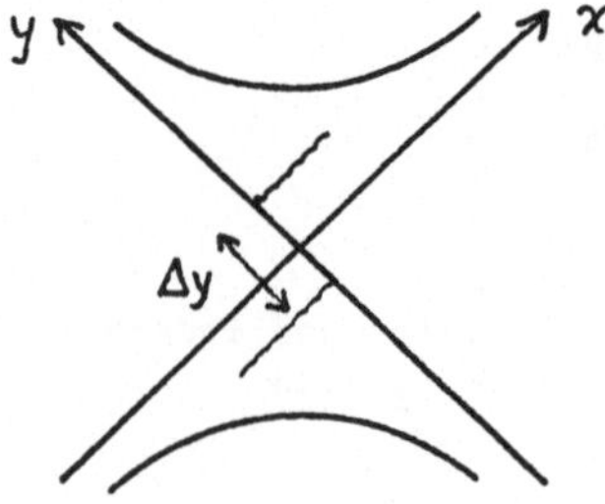

Fig. 4. A particle that moved into a black hole a long time ago causes a shift Δy in the trajectories of outgoing particles.

If we have very fast particles both on the u- and on the v-axis, complications arise. Clearly the two sets of particles disturb each other's motion, and the surrounding space-time will no longer be flat. As a first approximation however one may assume that the shifts due to the u-particles and the v-particles can be added linearly.

5. A LINK WITH STRING THEORY

Consider now a black hole that has been formed in the distant past. A freely falling observer detects very few particles. It seems to be very well possible to consider the analytic extension of his space-time to the entire Kruskal frame ($xy < 1$). Of course we realize that this analytic extension went beyond the line $t = -\infty$, which is before the black hole was actually made. In reality a large amount of collapsing matter had accumulated on that line and the analytic extension the mathematical model one gets was invalid. Nevertheless the mathematical model one gets is important. For example, if an outgoing particle is actually detected one enters into an element of Hilbert space which contains the past singularity of Kruskal space, a feature further discussed in ref. [10].

The future horizon is defined to be the boundary of that region of space-time from which escape out of the black hole is still possible. The past horizon is the boundary of the region of Kruskal space that can be reached from outside the black hole starting at $t = -\infty$. The analytic extension discussed above is realistic exactly up to the past horizon. The space-like horizon is defined to be the intersection of future and past horizon.

Suppose we take that part of the world that is accessible to us and add to it some ingoing particles with momenta p^-_{in}, and some outgoing particles with momenta p^+_{out}. How do these particles affect the shape of the space-like horizon?

As long as p^-_{in} and p^+_{out} are not too large we may assume their effects to be additive (as will turn out later, the products of the interesting values for p^-_{in} and p^+_{out} will be negligibly small). From the previous section, eq. (4.10), we deduce for the coordinates $x^{\pm}(\theta.\varphi)$ of the space-like horizon:

$$\partial^2_{tr} x^- = 8\pi G p^-_{in}\ \delta^2(\Omega-\Omega_{in}) \ ; \tag{5.1}$$

$$\partial^2_{tr} x^+ = -\ 8\pi G p^+_{out}\ \delta^2(\Omega-\Omega_{out}) \ , \tag{5.2}$$

where $\Omega_{in}, \Omega_{out}$ are the angles (θ,φ) at which the particles crossed the horizon. ∂^2_{tr} is the laplacian with respect to θ and φ. The constant term is neglected.
Writing the 4-vector

$$p^{\mu}_{ex} = \left(p^-_{in}, p^+_{out}\right) \qquad ((5.3)$$

and replacing θ and φ by any set σ,τ via conformal transformations, we see that

$$\partial^2_{tr} x^{\mu} = 8\pi G \rho^{\mu}_{ex} \tag{5.4}$$

where ρ^{μ}_{ex} is the distribution of external momenta on the θ,φ (or $\sigma\tau$) plane.

The analogy with the string theory in section 2 is striking, apart from a factor $-i$ that at first sight seems to be very troublesome, but may find a quite natural explanation: a factor −1 may be needed if, instead of the coordinates $x^\mu(\sigma,\tau)$ we consider the image of an ideal surface as seen through the "gravitataional lens" caused by the in- and outgoing particles, and the factor i may be seen as resulting from transformations of the sort

$$x^{\pm} \rightarrow \pm i\, x^{\pm} , \tag{5.5}$$

which are Lorentz transformations with complex arguments. Also one must bear in mind that string theory is usually defined in a $\sigma\tau$ space with metric (+,−), in order to obtain timelike surfaces.

A complete formalism that explains string theory from black hole dynamics is still lacking. The above however suggests that these concepts are closely related. If indeed the factor $\pm i$ can be accounted for we expect the string constant T to be

$$T = 1/8\pi G , \tag{5.6}$$

simply by comparing (2.21) with (5.4).

6. CONCLUSION

Even without understanding all technical details of the suggested link between string theory and black hole dynamics, we can consider its topological aspects. A conventional particle being a point in 3-space traverses a trajectory in 4-space. A virtual particle, giving rise to an instantaneous interaction between two points (such as a photon producing the Coulomb force), is a one-dimensional subspace of 3-space, but instantaneous in time (Fig. 5).

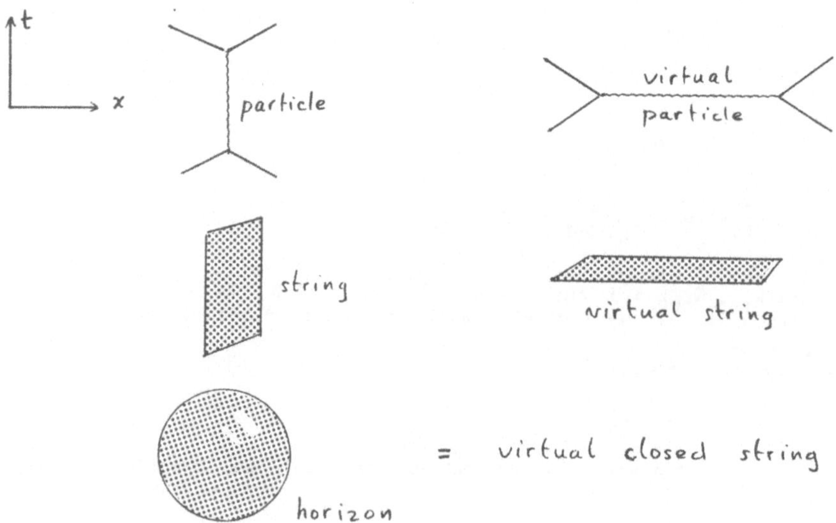

Fig. 5. A virtual string is sheet-like in 3-space and instantaneous in time. The space-like horizon can be a virtual closed string.

A virtual string can therefore be a sheet in 3-space. This is why the space-like horizon, being an S_2 sphere, has the same topology as a virtual closed string.

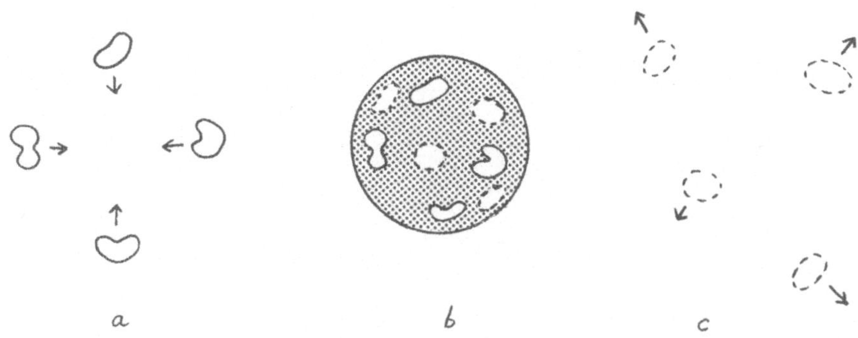

Fig. 6. Imploding particles (a), arrive at the horizon (b) and produce Hawking radiation (c).

In Fig. 6 we show how a black hole can perhaps be seen as a virtual closed string. Particles in the form of little closed stringscome together, exchange a virtual string that leaves holes; these holes connect to the outgoing Hawking particles, againclosed strings. Our aim was to bring black hole physics and string theory together. What we see is that in some sense the functional integral (2.14) for a string may actually describe the oscillations of a black hole horizon.

REFERENCES

1. J. Scherk, Rev. Mod. Physics **47**:123 (1975).
   J.H. Schwarz, Phys. Rep. **89**:223 (1982).
2. M.B. Green and J.H. Schwarz, Nucl. Phys. **B181**:502 (1981).
   Phys. Lett. **149B**:117 (1984).
   D.J. Gross et al., Phys. Rev. Lett. **54**:46 (1985).
3. G. Veneziano, Nuovo Cim. 57A:190 (1968).
4. Y. Nambu, Proc. Int. Conf. on Symm. and Quark Models, Wayne State Univ. (1969).
5. S.W. Hawking and G.F.R. Ellis, "The large scale structure of space-time", Cambridge Univ. Press (1973).
6. S. Chandrasekhar, "The Mathematical Theory of Black Holes", Clarendon Press, Oxford 1983.
   C.W. Misner, K.S. Thorne, J.A. Wheeler, "Gravitation", Freeman, San Francisco, 1973.
   R.M. Wald, "General Relativity", Univ. of Chicago Press, 1984.
7. W. Rindler, Am. J. Phys. **34**:1174 (1966).
8. P.C. Aichelburg and R.U. Sexl, Gen. Rel. Grav. **2**:303 (1971).
9. T. Dray and G. 't Hooft, Nucl. Phys. **B253**:173 (1985).
10. G. 't Hooft, "Gravitational Collapse and Quantum Mechanics", lectures given at the 5th Adriatic Meeting on Particle Physics, Dubrovnik, June 1986.
    See Also: G. 't Hooft, "Strings From Gravity", Second Nobel Symposium on Elementary Particle Physics, Marstrand (Göteborg), June 1986.

INDEX